Historical & Cultural Astronomy

More information about this series at http://www.springer.com/series/15156

Only portrait of Simon Marius from *Mundus Iovialis*, Nürnberg 1614,)( )( )(2^{v}; Credit: Jay and Naomi Pasachoff Collection

Hans Gaab • Pierre Leich
Editors

Simon Marius and His Research

Springer

Editors
Hans Gaab
Simon Marius Society
Fürth, Germany

Pierre Leich
Simon Marius Society
Nürnberg, Germany

ISSN 2509-310X ISSN 2509-3118 (electronic)
Historical & Cultural Astronomy
ISBN 978-3-319-92620-9 ISBN 978-3-319-92621-6 (eBook)
https://doi.org/10.1007/978-3-319-92621-6

Original German Edition published by AVA, Leipzig, (2016)

Cover illustration: Only portrait of Simon Marius from *Mundus Iovialis*, Nürnberg 1614,)()()(2^{v}. Credit: Jay and Naomi Pasachoff Collection

This Springer imprint is published by the registered company Springer Nature Switzerland AG.
The registered company address is: Gewerbestrasse 11, 6330 Cham, Switzerland

Foreword

A little more than 400 years ago, the German astronomer Simon Marius (1573–1625) was one of the first to use the newly invented telescope for the study of planets, stars and the universe. Born in Gunzenhausen and working mainly in Ansbach, he discovered the four major moons of Jupiter independently of Galileo. Marius was the first to propose, in print, naming them Io, Europa, Ganymede and Callisto, as originally suggested to him by Johannes Kepler. It is not (yet) so well known that Simon Marius also worked on many other interesting astronomical topics; e.g. he investigated the Andromeda nebula and described it as an extended source of pale light with its intensity decreasing outwards from the centre.

This edition is an expanded English version of the original German volume *Simon Marius und seine Forschung* from 2016, which in turn was based on a conference about Simon Marius's life and work jointly presented by the Nuremberg Astronomical Society and the Study Group for History of Astronomy of the German Astronomical Society in 2014. It contains more than a dozen very interesting and highly elucidating chapters on Marius and his research. The articles reflect the wealth and breadth of Simon Marius's investigations and explorations; e.g. he studied and examined sunspots and comets, he worked on calendar making and astrology and he also considered the "world at large".

Simon Marius accepted that Jupiter with its moons was something like a planetary system on its own. In this sense, he overcame the purely geocentric worldview. However, he still believed that Jupiter with its moons orbited the Earth, as would the other planets and the sun. So Simon Marius wouldn't go the full distance to the heliocentric system, but rather got stuck halfway, similar to Tycho Brahe; hence he is considered a "Tychonic" (and the readers of this volume will learn and understand why!).

This book adds significant new research to the life and work of Simon Marius. To pick just one article, Jay Pasachoff's account of the parallel and independent discoveries of Jupiter's moons by Galileo and by Marius reads like a thriller and is very relevant for every researcher up to the present. Everyone interested in the history of astronomy in general and in the transition between the geocentric and the heliocentric worldviews in particular will benefit a lot by reading this book. This English version will offer these interesting new insights to a worldwide audience and hence broaden the potential readership significantly. It is my pleasure to thank the authors for their insightful contributions as well as the two editors Hans Gaab and Pierre Leich—who were in charge of the German version as well—for their very careful and meticulous work. May this book find many interested readers!

Heidelberg University, Heidelberg
Germany

Astronomische Gesellschaft
(German Astronomical Society)
February 2018

Joachim Wambsganß

Preface

Although the history of astronomy of the early seventeenth century has been well researched, the margravial court mathematician Simon Marius from Ansbach in Southern Germany has till now attracted little attention. This was above all the result of the—as we now know—unjustified accusation of plagiarism by Galileo Galilei. However, a critical appraisal of his early observations of comets, the moons of Jupiter, the phases of Venus and sunspots has been somewhat lacking.

For this reason, the Nürnberger Astronomische Gesellschaft (Nuremberg Astronomical Society) initiated the Simon-Marius-Anniversary 2014 on the 400th anniversary of his opus magnum *Mundus Iovialis*. 66 cooperation partners participated in 60 events on which more than 250 reports appeared in the media.

The central project was the Marius Portal (www.simon-marius.net), which, with a multilingual-menu navigation, documents everything by and about Marius. The webpage was launched in the Staatsarchiv Nürnberg (States Archive Nuremberg) on 18th February and in the meantime forms the most extensive and most important presentation of Marius.

The mathematician, physician, astronomer and calendar maker experienced in 2014 a delayed gratification through the naming of asteroid (7984) Marius by the International Astronomical Union.

The concluding climax to the year was the conference "Simon Marius and his Times" in cooperation with the Arbeitskreis Astronomiegeschichte in der Astronomischen Gesellschaft (the Study Group for the History of Astronomy in the Astronomical Society)—the professional Association for German Astronomy and Astrophysics—as well as the Dr. Karl Remeis Observatory. It presented the actual state of research on Marius in the Nicolaus-Copernicus-Planetarium Nuremberg.

Simon Marius—as became very clear during the conference—belonged to the small group of astronomers, who in the year following the introduction of the telescope in the Netherlands undertook observations and were aware that the new discoveries effected the question as to which was the correct cosmology. Also, although Marius rejected the Ptolemaic system, he was not prepared to accept heliocentrism preferring the Tychonic system. It is, however, exactly this intermediate position, which makes his argumentation especially interesting for comprehending the Copernican Revolution and an engagement with Marius profitable.

This volume resulted from the conference and was published in German with the title, *Simon Marius und seine Forschung* by the publisher AVA Akademische Verlagsanstalt, Leipzig in 2016. The volume was taken up in the leading German series *Acta Historica Astronomia* by the series editors Wolfgang R. Dick and Jürgen Hamel.

The book is now presented in a complete English translation, whereby several contributions were originally written in English. We are very pleased that authors could be persuaded further to contribute chapters, and through a completed translation of *Mundus Iovialis*, both the most important text and the actual secondary literature are gathered together in a single volume.

Contact with Butler Burton was brought about through the good offices of Jay Pasachoff, so that Springer is publishing the volume in their series *Historical and Cultural Astronomy*. We thank Ramon Khanna the editor for astrophysics at the Springer office in Heidelberg and Maury Solomon the relevant editor at Springer in New York for their support.

- This volume begins with the English translation of *Mundus Iovialis*. This was largely carried out by Arthur Octavius Prickard in the early twentieth century and has now been completed by Albert van Helden, for this edition, who has translated those sections left out by Prickard, such as the second appendix.
- *Hans Gaab* delivers a long overdue biography as the second contribution, which illuminates Simon Marius's environment and disposes of a series of legends.
- *Wolfgang R. Dick* extends these observations with details about Hans Philip Fuchs von Bimbach, Simon Marius's patron, who was responsible for his early acquisition of a telescope.
- *Dieter Kempkens* outlines the obligation of a court astronomer through the example of Marius's predecessor Georg Caesius.

Following Marius's personal environment, the next section deals with his astronomical observations.

- *Jay M. Pasachoff* tackles the issue of the simultaneous discovery of the moons of Jupiter by Galileo and Marius, whereby the latter first presented his discovery in detail in his *Mundus Iovialis* in 1614.
- The comets of 1596 and 1618 and the comet research of the period form the focus of *Jürgen Hamel's* investigations, which reveal a development from "Alumnus" in Heilsbronn in 1596 to experienced scholar in Ansbach in 1618.
- *Dagmar L. Neuhäuser* and *Ralph Neuhäuser* examine Marius's sunspot observations and are able to correct the current research in several points.

The next section addresses Marius's abilities in the construction of astronomical theories.

- *Christopher M. Graney* clarifies with the question "Was Marius too good an astronomer?" why it was the perception of star size that led him to adopt the Tychonic system.
- *Pierre Leich* establishes the connection between the telescopic observations at the beginning of the seventeenth century and the dispute about which cosmology to adopt and documented on this basis the scientific legitimacy of Marius's argumentation.

The following essays investigate a field that was mandatory for a court mathematician of the period.

- *Klaus Matthäus* presents Marius as the editor of yearly calendars and shows how Marius distances himself from the prevalent calendars issued in his times and claims to utilize the latest calculation method.
- *Richard L. Kremer* examines the mathematical and astronomical basis for Marius's calendar calculations, uncovers his sources and describes how Marius shaped his relationships to contemporary astronomers.
- *Thony Christie* undertakes a preliminary examination of Marius's largest astrological work, his *Tabulae Directionum Novae*, in which he accuses Regiomontanus of having misunderstood Ptolemy.

The final section collects contributions on reception and didactics.

- *Joachim Schlör* explains the motivation for a German translation, with the help of which students learning Latin can develop an understanding of the dramatic new orientation of the early modern comprehension of the world.
- *Albert van Helden* sketches the debate about Marius at the start of the twentieth century, which led to the first English translation of *Mundus Iovialis*.
- *Pierre Leich* follows up questions of priority, reception and rehabilitation of Simon Marius and traces the dispute from the charge of plagiarism up to the Marius Portal as virtual "Collected Works".
- *Albert van Helden* and *Huib J. Zuidervaart* contribute a cautionary note about the rehabilitation of Marius and on the involvement of the Hollansche Maatschappij der Wettenschappen.
- *Norman Anja Schmidt* and *Pierre Leich* present in detail the Marius Portal as representation of Simon Marius in the digital age and go into its origins, contents, technique, reception and planned future developments.
- The naming of the asteroid (7984) Marius gave *Thomas Müller* occasion to examine more closely those members of the asteroid belt between Mars and Jupiter.
- *Rudolf Pausenberger* explains that the observations of Galileo and Marius can be simulated in a planetary model 1:50 billion, which suggests the orbiting satellite conclusion drawn by the two observers.
- Finally, *Olga Sinzev* reports on the travelling exhibition "The Sun, the Moon and Marius" which resulted from the confrontation between the Russian Youth Art School, "Obraz", and Marius's findings.

The bibliographies are at the end of each chapter. In addition, a detailed bibliography of Marius's writings with links is included at the end of the book. *Prog. yyyy* refers to *Prognosticon astrologicum for the year yyyy*. As far as the authors didn't write their contributions in English themselves, Christian Gottschall, Nicola Neumann and Edith Wagner helped by providing raw translations. Finally, Thony Christie did his best to turn all the translations into acceptable English free of glitches and grammatical howlers. His expertise in history of science was of invaluable help. Chris Graney, Rich Kremer and Jay M. Pasachoff as well as Naomi Pasachoff have kindly read through several of the texts and their comments also led to improvements in English. We thank the publisher's copyeditors for the final revision.

We would like to thank the rights holders of the works and photographs used as illustrations for permission to do so. As well as the support of the archives, libraries, publishers and universities, it is not least the sponsors and supporters, without whom the production of this book would not have been possible. Above all our thanks go to the HERMANN GUTMANN STIFTUNG, the Stiftung NÜRNBERGER Versicherungsgruppe and the Vereinigten Sparkassen Gunzenhausen. To this group we count the town of Ansbach, the Kost-Pocher'sche Stiftung and the N-Ergie. We would also like to thank the district of Weißenburg-Gunzenhausen, the towns of Gunzenhausen and Nuremberg as well as Norman Anja Schmidt NOSCC.

We hope that this collected volume, which is also Volume 2 of the *Edition Simon Marius* of the Simon Marius Society, presents not only the results of his researches but will also serve as a motivation for an intensive engagement with those researches and will as well bring an astronomer, who has been undervalued in the history of science, into the awareness of both the experts and an interested public.

Fürth, Germany Hans Gaab
Nürnberg, Germany Pierre Leich

Contents

1 **The World of Jupiter, English Translation of *Mundus Iovialis*** 1
Arthur Octavius Prickard and Albert Van Helden

2 **Concerning the Biography of Simon Marius (1573–1624)** 55
Hans Gaab

3 **Hans Philip Fuchs von Bimbach (ca. 1567–1626), Patron of Simon Marius** 139
Wolfgang R. Dick

4 **Georg Caesius as Official Court Astronomer of the Margrave Georg Friedrich of Brandenburg-Ansbach** 179
Dieter Kempkens

5 **Simon Marius's *Mundus Iovialis* and the Discovery of the Moons of Jupiter** 191
Jay M. Pasachoff

6 **Simon Marius's Reports on the Comets of 1596 and 1618, in the Context of the Comet Research of His Times** 205
Jürgen Hamel

7 **Sunspot Observations by Simon Marius** 223
Dagmar L. Neuhäuser and Ralph Neuhäuser

8 **An Astronomer Too Excellent: Simon Marius, the Telescope, and the Problem of the Stars During the Copernican Revolution** 239
Christopher M. Graney

9 **In the Turmoil of the Early Seventeenth-Century Cosmology Debate: Simon Marius as a Supporter of the Tychonic System** 249
Pierre Leich

10 **Simon Marius as a Calendar Writer** 273
Klaus Matthäus

11 Simon Marius as a Tychonic Calendar Maker 321
Richard L. Kremer

12 Simon Marius: *Tabulae Directionum Novae*—A First Approach . . . 363
Thony Christie

13 Translating the *Mundus Iovialis* into German 371
Joachim Schlör

14 Prickard's English Translation of *Mundus Iovialis*, Completed 387
Albert Van Helden

15 Priority, Reception, and Rehabilitation of Simon Marius: From the Accusation of Plagiarism to the Marius-Portal as His Virtual Collected Works 389
Pierre Leich

16 A Word of Caution About the "Rehabilitation" of Simon Marius 413
Albert Van Helden and Huib Zuidervaart

17 The Marius Portal: Simon Marius as Digital Human in the Twenty-First Century 417
Norman Anja Schmidt and Pierre Leich

18 The Franconian Asteroid 7984 Marius 437
Thomas Müller

19 Planet Model 1: 50 Billion 449
Rudolf Pausenberger

20 "Sun, Moon, and Marius" 455
Olga Sinzev

Simon Marius: Works 461

About the Authors 465

Author Index 475

Subject Index 479

Chapter 1
The World of Jupiter, English Translation of *Mundus Iovialis*

Arthur Octavius Prickard and Albert Van Helden

The World of Jupiter

Discovered in the year 1609 by means of a Belgian spy-glass:
The Theory of the four Jovian planets with tables, founded mainly on personal observations, from which their position relatively to Jupiter, for any given time, may be calculated with the utmost readiness and ease,
by SIMON MAYR of Guntzenhausen, Astronomer in Franconia to the Margraves of Brandenburg, and student of the higher medicine,
By permission and privilege of his Sacred Majesty the Emperor,

---

Arthur Octavius Prickard (1843–1939) was a fellow of New College Oxford.

Source: *Mundus Iovialis Anno M.DC.IX. Detectus Ope Perspicilli Belgici, Hoc est, Quatuor Jovialium Planetarum, Cum Theoria, Tum Tabulæ, Propriis Observationibus Maxime Fundatæ, Ex Quibus situs illorum ad Iovem, ad quodvis tempus datum promptissimè & facilimè supputari potest. Inventore & Authore Simone Mario Guntzenhusano, Marchionum Brandenburgensium in Franconiâ Mathematico, puriorisque Medicinæ Studioso*. Nürnberg: Johann Lauer 1614

English translation from *Mundus Iovialis Anno M.DC.IX.* by Arthur Octavius Prickard (The 'Mundus Jovialis' of Simon Marius, *The Observatory. A review of astronomy* 39 (1916), pp. 367–381, 403–412, 443–452, 498–503)

English translation from *Mundus Iovialis Anno M.DC.IX.* by Albert Van Helden

German translation from *Mundus Iovialis Anno M.DC.IX.* by Joachim Schlör: *Mundus Iovialis – Die Welt des Jupiter. Die Entdeckung der Jupitermonde durch den fränkischen Hofmathematiker und Astronomen Simon Marius im Jahr 1609 – lateinisch und deutsch* (= *Fränkische Geschichte*, vol. 4). Gunzenhausen: Johann Schrenk 1988

German translation from *Mundus Iovialis Anno M.DC.IX.*, Appendix of the second edition (in: Hans Gaab, Pierre Leich, Marius' Replik auf Scheiner – Der Anhang zum Mundus Iovialis von Simon Marius, *Globulus – Beiträge der Natur- und kulturwissenschaftlichen Gesellschaft e.V.*, 18 (2014), pp. 11–14)

A. Van Helden (✉)
University of Utrecht, Utrecht, Netherlands
e-mail: a.vanhelden@uu.nl

H. Gaab, P. Leich (eds.), *Simon Marius and His Research*, Historical & Cultural Astronomy, https://doi.org/10.1007/978-3-319-92621-6_1

At the charges and from the type of Johann Laur, citizen and printer of Nuremberg in the year 1614.

To the most illustrious Princes and Lords, the Lord Christian and the Lord Joachim Ernest, brothers, Margraves of Brandenburg, Borrusia, Stettin, Pomerania, the Cassubii, the Vandals, Dukes of Crosna and Jägerndorf in Silesia, Burggraves of Nuremberg and Princes of Rugia, etc., my most clement Lords.

Most high and exalted Princes, my most clement Lords: It is now 63 years since, under the name and authority of the most illustrious Prince Albert, Margrave of Brandenburg, Duke of Borrusia, of most honoured memory, the very excellent and illustrious Astronomer Erasmus Reinhold published his tables of the second movables entitled 'Prutenic,' hitherto in sole or principal use throughout the whole of Europe, because no age has ever seen similar tables of greater correctness, and more nearly corresponding to the heavens. For from them the numerous volumes of Ephemerides have been deduced; they are the foundation upon which rests the reform of the Calendar; they have been employed by all who, since the date of their publication, have been in the practice of computing forecasts of the heavenly bodies for the year and their natal combinations. Thus it is that the most renowned and immortal name of that great Prince passed forth with these tables into all provinces of the whole of Europe where the studies of the liberal arts have flourished. But the good Reinhold was not satisfied to spread abroad over the whole world the memory of his Sovereign Prince by a mere dedication; he also so contrived that the tables have taken their own name from his, choosing that of 'Prutenic,' in order that every time the tables were mentioned the memory of the most honourable Prince should be renewed.

What was the cause of such gratitude on the part of Reinhold? It may be inferred, not only from. the dedication of the 'Tables Prutenic,' but also from his other published writings. There were two main causes: firstly, the love of that most high Prince for Mathematical Studies and for those who were pursuing them; and, secondly, the great benefits and the liberality with which he abundantly rewarded the professors of this art, especially Reinhold himself. There is no doubt that this most bountiful Prince had at his own Court many on whom he bestowed gifts and honours, as is usual in the Courts of Princes. But, for anything which they did to help it, the memory of His Highness might have perished long ago; whereas through the liberality which he bestowed upon Reinhold not only has his great renown penetrated through entire Europe, but the very name of that most illustrious Prince has been rendered immortal.

To what end am I recounting all this? Most illustrious Princes and most clement Lords, I gratefully acknowledge the great benefits which have been bestowed upon me by your Highnesses, with the utmost clemency and abundance; I have mentioned them already in my dedication of the yearly forecast, and to some extent in the chapter of this work which deals with the names to be given to these Jovian planets. But, to put it in a word, they are such and so great that I can never sufficiently repay them. Moved therefore by Reinhold's praiseworthy example, since it is impossible to make any other return to your Highnesses for so great benefits, inasmuch as I have no silver or gold, and poverty is as it were an inseparable accident of all astronomers, yet, that these great benefits may never be given over to oblivion, I have desired, by thus dedicating and most humbly offering to your Highnesses the JOVIAN WORLD, to ascribe their memory to Heaven, that so, with these Jovian stars, the benefits given by your Highnesses may, till the end of the world, be passed down to the full knowledge of all posterity, which shall have any interest in the Stars of Heaven. So shall it be seen that the outlay borne by your Highnesses, and also my own watchings and toils, have been bestowed to the best advantage. For since my first discovery of this Jovian world, made with a Belgian spy-glass, I have spent, as God has ordained, more than four whole years; and have endured incredible labours in watching, in observing, and in calculating, until I have, as I think, ascertained all the apparent diversities of their movements, and then have accounted for what I had ascertained by a suitable theory, and out of that theory have constructed tables, from which, by an easy process, the position of these stars with respect to Jupiter may be calculated and made plain for any given time. It is true that Reinhold took seven entire years to frame his 'Tabulae Prutenicae,' yet he was assisted by the observations of 2000 years, more or less, also by the 'Alphonsine Tables,' and again by the discoveries and observations of Copernicus. All these advantages were lacking to me. Moreover, the planets for which Reinhold framed canons have been known from the creation of the world. Mine down to the year 1609 were absolutely unknown, and unobserved. I am not saying this from any wish to disparage the labours and authority of that excellent man Reinhold, but rather that my own labours may be placed beside those of Reinhold in a clearer light, and so may themselves attain some further measure of authority.

Receive, therefore, most Illustrious Princes, most clement Lords, receive, I pray you, with a favourable and clement mind, this Jovian world discovered by me your most obedient Astronomer, and to the best of my power worked out and illustrated.

For my part, I do not seek from your Highnesses any compensation for my labours, or advancement in any office, since whatever can proceed from me in this department is all yours, all produced and provided at your charges. Also, I am entirely content with that favour which it has hitherto been my happy lot to receive from your Highnesses. Court life, and dignity, to which some, to their own very great injury too eagerly aspire, I reckon, as nothing worth; a life of privacy and philosophic studies are rather my delight, and so I perform my proper duties.

This one boon I pray from your Highnesses with a mind more than humble, that you [continue] the grace and favour with which you have thus far most clemently treated me; and do not now desert me, the one Heilsbron student out of so large a number who has been incited, doubtless by Heaven, to these sublime studies of Astronomy, in advanced age, with a family, and in a condition of great infirmity of brain, brought on chiefly by this study, which demands the whole of a man, and in which I have been self—taught, and have never had the assistance of any living man as tutor. This trouble has been greatly aggravated by a fall from a height in Italy, which might have killed me.

In return, I faithfully promise, for the whole of my life, complete obedience, absolute integrity, and all possible diligence in such things as shall be demanded of me.

And now, I pray the eternal God and Father of our Lord and Saviour Jesus Christ, with genuine sighs and a faithful heart, that he may be pleased long to preserve your Highnesses in happiness of every kind, and to load you abundantly with the good things of soul, body, and estate.

Lastly, next to God, I commend myself to the affection of your most clement Highnesses.

Dated at Ansbach, in my Astronomical Observatory, on the day of Concord, 18 Feb. 1614.

Your Highnesses,
Most obedient humble Astronomer,
SIMON MAYR.

Preface

To the Candid Reader

It had been my intention, Candid Reader, to deal with you at some length in this preface, and to give a lengthened statement of all the objects which I have observed up to the present time through the Belgian instrument commonly called a spy-glass, in the Sun, the Moon, the other stars, and in the heavens generally, as you may see in various passages of this little book. But, as bad health and interruptions caused by other business have kept me back, and also the Frankfurt fair was close at hand, and the book was already going through the press, I have been unable to keep my promise, and find myself unwillingly compelled to reserve for another time the publication of my observations. In what now follows I will briefly explain when and how I came to make acquaintance with this instrument, and to use it.

In the year 1608, when the Frankfurt autumn fair was going on, it happened that there was at the same place the most noble, gallant, and energetic John Philip Fuchs, of Bimbach in Mohr, Lord and Knight, and a dauntless General, Privy Councillor of my own most illustrious Princes, not only a patron and lover but also an eminent student of all Mathematics and other kindred sciences. Various things went on there, and among others it chanced that a certain merchant met the nobleman mentioned above, whose acquaintance he had formerly made, and told him that there was then present in Frankfurt at the fair a Belgian, who had invented an instrument by means of which the most distant objects might be seen as though quite near. Hearing this, he begged the merchant to bring the Belgian to him, which the merchant at last consented to do. Our nobleman had a long discussion with the Belgian first inventor, and felt doubts as to the reality of the new invention. At last the Belgian produced the instrument, which be had brought with him, and one glass of which was cracked, and told him to make trial of the truth of his statement. So he took the instrument into his hand, and saw that objects on which it was pointed were magnified several times. Satisfied of the reality of the instrument, he asked the man for what sum he would produce one like it. The Belgian demanded a large price, and when he understood that he could not get what he first asked, they parted without coming to terms. When he returned to Ansbach, the Nobleman sent for me, and told me that an instrument had been devised by which very remote objects were seen as though quite near near. I heard the news with the utmost surprise. He frequently talked the matter over with me after supper, and at last came to the conclusion that such an instrument must necessarily be composed of glasses, of which one was concave, the other convex. He took up a piece of chalk and with his own hand drew a sketch on the table to show what sort of glasses he meant. We afterwards took glasses out of common spectacles, a concave and a convex, and arranged them one behind the other at a convenient distance, and to a certain extent ascertained the truth of the matter. But as the convexity of the magnifying-glass was too great, he made a correct mould in plaster of the convex glass, and sent it to Nuremberg to the makers of ordinary spectacles that they might prepare glasses like it; but it was no good, as they had no suitable tools, and he was unwilling to reveal to them the true principle of the process. No expense was spared, and several months elapsed. If we had been acquainted with the method of polishing glasses, we should have produced excellent spy-glasses immediately after our return from Frankfurt. In the meantime, glasses of the same kind were becoming common in Belgium, and a fairly good one was sent, with which we were highly pleased. This was in the summer of 1609. From this time I began to look into the heavens and the stars with this instrument, whenever I was at the house of the nobleman so often mentioned, at night time; sometimes he used to allow me to carry it home, and in particular about the end of November, when I was observing the stars according to my custom in my own observatory. Then for the first time I looked at Jupiter, who was in opposition to the Sun, and made out some tiny stars, sometimes following, sometimes preceding Jupiter in a straight line with him. First, I thought that they were of the number of those fixed stars which cannot be seen without this instrument in other parts, such as those which I was finding in the Milky Way, the Pleiades, the Hyades, Orion, and elsewhere. However, as Jupiter was then retrograding, and still I saw these stars accompanying him throughout December, I was at first much astonished; but by degrees arrived at the following view, namely, that these stars moved round Jupiter, just as the five solar planets, Mercury, Venus, Mars, Jupiter, and Saturn revolve round the Sun. I therefore began to record my observations. The first was taken on December 29, when three stars of this description were visible in a straight line from Jupiter towards the west. At this time, as I frankly confess, I thought that there were only three such stars accompanying Jupiter, since I several times saw three of them near Jupiter. Meanwhile, two glasses extremely well polished, a convex and a concave, were being sent from Venice by that most distinguished and accomplished man, the Lord John Baptist Leuccius, who had returned from Belgium to Venice after peace was made, and who had already been thoroughly acquainted with the instrument. These glasses were fitted into a leaden tube, and made over to me by the most noble and active nobleman whom I have mentioned already, in order that I might try what they would show among the constellations and stars near Jupiter. Accordingly, from this time until January 12, I gave my diligent attention to these Jovian stars, and somehow ascertained that there were four such bodies, which themselves revolved about Jupiter. At length, about the end of February or beginning of March, I felt entirely confirmed in my view as to the definite number of these stars. From January 13 till February 8, I was at Hal in Swabia, and left the instrument at home, fearing injury to it on the journey. After my return, I resumed my accustomed observations. In order that I might observe the Jovian stars with greater closeness and diligence, the illustrious nobleman whom I have frequently mentioned, out of his special affection towards these astronomical studies, placed the instrument entirely at my disposal. From that time accordingly to the present, I have made continuous observations with this instrument and with others afterwards constructed. This is the exact truth. I should never be allowed thus in a public document to say what was not true about so great a man who is alive and is here, a man of the highest celebrity, not only for his ancient and noble lineage, but also and chiefly for his great deeds, his heroic exploits, and his consummate skill in war throughout France, Hungary, Belgium, and Germany. All that has been observed by me in this department, now worked out and given to the public, I owe entirely to this most excellent and noble man, my own worshipful patron. In recounting all this, I am not to be understood as wishing to lessen Galileo's reputation, or to snatch from him the discovery of these Jovian stars among his countrymen in Italy—far from it. My object rather is, that it may be understood that these stars were not shown to me by any mortal in any way, but were discovered and observed by me, by my own investigation, in Germany, almost at the very time, or slightly before it, at which Galileo first saw them in Italy. The credit, therefore, of the first discovery of these stars in Italy is deservedly assigned to Galileo and remains his. Whether any one among my German countrymen has discovered and seen them before me, I have not yet been able to make out. My experience is altogether to the contrary, for there have been those who have unblushingly accused both Galileo and myself of being mistaken. However, I have no doubt that these persons are already penitent and ashamed of their own mistake, and of their hasty judgment upon the labours of others. So if this little book of mine shall reach Florence and come into Galileo's hands, I pray that he will receive it in the same spirit in which it is written by me. I am so far from wishing in any way to detract from his authority and his discoveries, that I rather thank him greatly for publishing his 'Nuncius Sidereus,' for in it he has done much to confirm my view. His own observations have been especially useful to me, because they were made at the particular time when I was at Hal, and my observations were discontinued. Although they do not appear to me exact in all respects, yet as to the Eastern and Western direction, and the relative position of these stars to one another, they were of great assistance to me. Galileo's own method for getting these distances from Jupiter has not succeeded with me, but I have kept to my own, which I had employed before I was acquainted with the 'Sidereus Nuncius,' and which I will explain in another place, when I publish my more important observations.

It had been my intention, according to my former proposal, to deal now with the spots on the Sun, setting out all my observations upon them from August 3, 1611 to the present time. However, I do not wish—and, indeed, am unable—to make any definite statement about them at present, not only from the causes originally pointed out, but for the further reason that I find the greatest authorities in disagreement, and am unable to satisfy myself. I therefore pass these matters by, and will take up here four other points not yet mentioned by me in the dedications of my yearly forecasts.

Among them the first is that with the spy-glass, from 15 December 1612 I discovered and observed a fixed star with a certain wonderful shape that I cannot find in the entire heavens. It is near the third and northernmost [star] in the belt of Andromeda. Without the instrument the same is seen as some sort of little cloud; and with the instrument no distinct stars are seen as in the nebular star in Cancer and other nebular stars, but rather only white rays, which the closer to the center the brighter they come out; in the center there is a dull and pale light; and its diameter is about a quarter of a degree. About the same brilliance appears when a bright candle is observed through a clear lantern from a long distance. It appears not unlike that comet observed by Tycho Brahe in 1586. In September of last year, when the most learned man, Mr. Lucas Brunnius, the mathematician of the most illustrious Elector of Saxony, was with me [visited me], among other mathematical conversations, thanks to clear weather I also showed him that same monstrous star, which he saw with the greatest admiration. But whether it is a new star or not I cannot assert for certain, let others consider and judge it. I am amazed by the fact that the very sharp-sighted Mr. Tycho, who measured the longitudes and latitudes of the more southern stars in the belt of Andromeda with his instruments, left this nebulosity untouched although it was near these stars.

The second is, what all physicists and astronomers disagree about among themselves, that is, what is the cause or manner of the scintillation of stars. Almost al our predecessors thought that scintillation only happened with stars, but very little with the planets. Experience and observation with the Dutch instrument proves that this is false, for all stars in the sky except the Moon scintillate sometimes more, sometimes less, even the Sun itself. Among the planets, Saturn scintillates the least, then Jupiter, third Mars followed by Venus. But Mercury scintillates most strongly, which can be clearly seen with and without the spy-glass. Next, about the Sun I would argue that here again there is no lack among the learned who loudly rebuked and accused me of foolish and gross errors: *faciant sane quod illis libuerit.* I will nevertheless communicate to the candid reader what I have seen with my eyes and what I have diligently observed. For him who has access to a good spy-glass and who wishes to explore the truth of things, let him remove the concave glass and apply the part of the instrument left empty by the glass to the eye and direct the spyglass to a star or planet whose scintillation he wants to inspect, and he will then see with admiration what I am talking about, provided that the sky is very clear and the air very quiet. For it can happen that the fixed stars and planets appear pierced by many holes. This is caused by the material of the convex glass, which makes the masses of the bodies of the fixed stars and planets very large, and the scintillation appears like lightning or a bubbling of the material of the stars. Often definite and different colors will appear, in some more and in some less. And in the stars hitherto believed to be of the nature of Mars the color red surpasses that in others, namely Mars, Aldebaran, and other similar stars.

In Sirius, however, all colors, green, gold, blood-red and blue, follow each other in order with equal strength and fullness, so that they bring forth great admiration joined with the greatest pleasure. Mr. Kepler writes in his *Optics* that he has seen these colors with the naked eye, and the same was confirmed by that Most Illustrious man, Mr. Johannes Matthias Wacker von Wackenfels, the Imperial Court Councilor of his Sacred Catholic Majesty, in Regensburg after dinner [a meal], when we talked about this. I do not say here what my opinion is about what causes scintillation, but reveal faithfully what I saw; I leave it to be discussed and explained by other, more subtle, minds. But I maintain that the nature and quality of the fixed stars can be explored and determined more easily and certainly by this method than has happened until now.

The third is that not so long ago, that is, after my return from Regensburg I acquired an instrument through which not only the planets, but also all the principal fixed stars are seen exquisitely round, but especially great and small dog-star, the brighter stars in Orion, Ursa Major, etc. which I had never been able to see before. I am indeed amazed that with his very excellent instruments Galileo has not seen the same. For he writes in his *Sidereus Nuncius* that the peripheries of the fixed stars are never seen circularly terminated, which was then held to be the greatest argument that it confirmed the Copernican system of the world, that is that because of the immense distance of the fixed stars from the Earth the round shape of the the fixed stars can never be seen from the Earth. But now, because it is established with certainty that the fixed stars, too, are seen round from the Earth with this spy-glass, this argument certainly fails and the contrary is confirmed, that is certainly the sphere of the fixed star is not at all removed from the Earth by such an incredible distance as Copernicus's speculation has it. Rather, the separation of the sphere of the fixed stars from the Earth is such that with this instrument the masses of their bodies can nevertheless be seen distinctly circular, and this confirms the Tychonic spherical arrangement of the heavens and mine, as in the second part of this little book will be confirmed in the fifth phenomenon. But these things must be discussed and explained elsewhere. But that the fixed stars shine with their own light I will easily concede to Galileo, because they have a much greater brilliance and brightness than the planets.

The fourth is a certain peculiar observation of the Sun, besides the spots, about which on several occasion there has been a discussion in letters between me and David Fabricius, the theologian in East Frisia, a most excellent astronomer, and a singular friend of mine. For many, in churches and other dark places where through a hole or broken glass orb [*orbem vitreum fractum*] a ray from the sun fell on an opposite wall a sufficiently far removed from the opening, I have seen a strongly trembling motion of the solar ray so that it did not proceed evenly but was carried in a vibrating, waving and jumping fashion. And since the above mentioned Mr. Fabricius always contradicted me, asserting that this motion of the rays does not arise from the Sun itself but from the intervening air, I attended to this matter more diligently and also used the spy-glass, which I affixed firmly and immovably to the opening in the wall so

that no other rays could enter the darkened room except through the spyglass. I received the beam on an upright board sufficiently far opposite and covered with a white sheet of paper which I also kept fixed. With this arrangement I diligently observed the beam and the sunspots and discerned three distinct motions in it: one, on the surface of the beam, as it were some flashing change in the brightness of the Sun which generally appears in the fixed stars, especially in the Big Dog, as has been seen before. This motion I believe to be the scintillation of the Sun, and I am persuaded that if someone were to see the Sun from Saturn, then without doubt he would see the Sun scintillate very strongly. For the light and size of the same Sun is not as great as it appears to us on Earth, since its diameter would appear as 3 minutes at least, and moreover this makes the angle of its lightning and bubbling motion much larger near Saturn that near us. I have seen the same motion not infrequently without the instrument with the naked eye, and indeed better than with the instrument, when the Sun was declining. I observed through an obliquely rolled up piece of black paper, whose narrow end was held near the eye and the wider end towards the Sun. With this method I saw the surface of the Sun moves just like gold liquefied in the greatest heat, in which a similar motion of the liquid and also a lightning on the surface of the gold, where, however, the surface always remains the same and does not bubble as much as other liquid substances or water.

The second motion is observed at the outer edge of the beam, and I think that it can be called the proper undulation, and this in my judgment arises from the motion of the air outside the aperture. A similar motion is observed above field in the summer, when the heat is greatest. The same is also observed with the the spy-glass this very snowy and cold winter in fields and woods during very clear and very cold weather.

I wonder very much about the third motion which is observed to be unequal. When one diligently attends to it one may see that the beam is gradually carried forward this progress is not even but very varied. For sometimes the beam appears almost to stand still in its motion, which is elsewhere called diurnal, but at other times it moves forward in a sort of jump made in one moment. Sunspots are subject to the same unequal motions, and after much experimentation I have found out that this leaping motion of the solar beam involves less than the 200th part of the beam. To be sure, let it be the 100th part. Terefore this motion does not belong to the Sun, nor to the Earth, nor, finally, to the air. I think it cannot come from the air because this motion is clearly different from that other one that I have called something of an undulation. Copernicans will say that this inequality of motion is due to the diurnal motion of the Earth, but I reject this on the basis of this probable reason. If the diameter of the Sun is 8876 German miles—see part 3, below—and two lines are drawn from the center of the Earth touching the Sun, then these two lines intercept 7 German miles of the

Earth's surface or 3595 geometrical paces,[1] which arc is the diameter of the the rays entering the camera obscura, and its 200th part is 182 geometrical paces or 960 feet, by which a tower in any moment moves from East to West. This jumping, unequal motion of the surface of the Earth would without doubt be felt on on the highest mountains if the speculation of Copernicus were true. Moreover, if this motion were a property of the Earth it would also be discerned in a ray of the Moon, although it would be difficult. But this does not occur, and therefore the motion is not a property of the Sun. However, this is not not therefore put forward by me as if I wanted it to be a paradox, but rather that others pay close attention to this matter, since no one that I know, who has observed the rays of the the Sun or the Sun itself has done up to now.

These are the things, candid reader, with which I wanted at this time and on this occasion to impress upon you, asking at the same time that you honestly interpret al this that I have published with good heart and in good faith and accept it as the first foundation of the World of Jupiter, on which may be built ever more correctly. Fare well, and happily enjoy these my [nightly] vigils and labors[2]

First Part

A General View of the Dimensions of the Jovian World

In undertaking to write a history of the Jovian world, I have thought that it will not be amiss to subdivide my whole treatment into three parts. The first will deal with a general view of this Jovian world—its dimensions, the magnitude of the four bodies which it includes, and the velocity of their motion round Jupiter, will be determined with probable accuracy. In the second the different velocities of their several movements will be explained. In the third all these phenomena will be explained by a suitable theory; all which will be followed by the framing and application of tables, and this is the principal scope of all my little book. I will begin then with a general view of this Jovian world, unknown to any mortal man since the first construction of the machine of the Universe. By the most diligent observation within my power, and that given daily, I have ascertained that Jupiter contains in his diameter about 35 sixtieths of the diameter of the Earth. For I have many times and day after day, seen that at his mean distance from the Earth his diameter in the heavens does not subtend more than an angle of 1 minute. With this as a foundation, I shall endeavour to investigate the whole dimensions and the outermost circumference of the Jovian world, and I shall proceed in the following manner.

[1] A geometrical pace equals 5 feet, and 1000 of these paces equals 1 (Roman) mile. The German mile was a little less than 4.7 modern (English) miles.

[2] English translation from *Mundus Iovialis Anno M.DC.IX.* by Albert Van Helden.

If we allow for one degree of the equator on the surface of the Earth, 15 common German miles, then there will be 1718 $^{2}/_{11}$ such miles in the diameter of the Earth. For the sake of simplicity, reject the fraction, so that the diameter is equal to 1718 miles (Tycho assumed 1720),[3] and let the proportion (of circumference to diameter) be 22 to 7. Hence the diameter of Jupiter is found by Detrus's Golden Rule in the same German miles. Thus: 60, i.e. the whole diameter of the Earth, gives 1718, how many will 35 give? This works out to 1002. Not to be too precise, take it at 1000 German miles. Thus we have a method for investigating the circumference of the Jovian world, as will appear in the sequel.

By my own observations, and those of Galileo, it has been ascertained that the fourth Jovian wanderer—that is, the one which reaches the greatest elongation from Jupiter—passes out to about 13 minutes in either direction from him, when he is at his mean distance from the Earth. I will take 14 minutes for the present, so as to be quite liberal and not to confine too closely the Jovian stage. Now, since Jupiter, at this distance from the Earth, covers 1 minute, with his visible diameter, and 1000 German miles correspond to 1 minute, it follows that the semidiameter of the whole system of the Jovian world will then cover 14,000 German miles, and the whole diameter 18,000. Now, applying the same proportion inversely (that is, as 7–22), the whole outer circumference of the Jovian world is ascertained to be 88,000 German miles. No contemptible dimensions for this Jovian world, unknown and unobserved by human eye, so far as history can tell us, since the Universe was framed!

II

Of the Dimensions of the Spheres of the Four Jovian Planets

Now that we are quite clear as to the extreme outer circumference of the Jovian world, I will pass to the other bodies contained in it, taking them in order, and, beginning with the Fourth, "the Saturn of Jupiter," as I shall henceforth call it, will investigate the circumference of its orbit, and the velocity of its movement.

Of the Fourth

Observation shows, as I have already stated, that the fourth Jovian Rambler, that is, the Saturn of Jupiter, passes to 13 minutes in either direction from Jupiter, when he is at his mean distance from the Earth, and returns thence to him. Accordingly, the semidiameter of its orbit will be 13,000 German miles, and the whole diameter 26,000. Then, by the proportion given, the whole circumference of its sphere will be 81,714–2/7 German miles. Now it is established, by my own latest correction, that

[3] Source.

this Saturn of Jupiter, that is, the Fourth Jovian wanderer, traverses this circumference in a period of 16 days, 18 hours, 9 minutes, 15 seconds nearly. Hence, by a calculation, it will travel about 206 German miles in 1 hour, a speed truly incredible, yet undeniable. If, then, these small bodies have so rapid a motion in the heavens with regard to other bodies, what are we to conclude—or rather, what are we to doubt—as to the other larger ones?

Of the Third

The third Jovian planet, or the Jupiter of Jupiter, as my own observations testify, which are not contradicted by those published by Galileo, passes out to 8 minutes of arc in either direction from Jupiter, when he is at his mean distance from the Earth. Thus the semidiameter of his orbit will be 8000 German miles, the whole diameter 16,000, and the extreme outer circumference 50,286. Now it is proved that this Third Jovian member traverses this circumference in 7 days, 3 hours, 56 minutes, 34 seconds. Thus about 292 common German miles will be the rate for 1 hour. This Third has, therefore, a greater velocity than the Fourth—of course, because it is nearer to Jupiter.

Of the Second

The Second Jovian Rambler, or the Venus of Jupiter, as my observations show, passes to 5 minutes in either direction from Jupiter, when he is moving at his mean distance from the Earth. Thus the semidiameter of his orbit is 5000 German miles, and the whole diameter 10,000, and, by the proportion 7–22, we get the whole circumference or circuit 31,429 German miles. The distance is completed by this Jovian Wanderer in a period of 3 days, 13 hours, 18 minutes, a rate of about 369 or 370 German miles in one hour—I do not want to be too precise in this enquiry.

Of the First

The first Jovian planet, that is, the Mercury of Jupiter, moves 3 minutes away from Jupiter in either direction, at his above-mentioned distance from the Earth. Thus the semidiameter of the actual orbit will be 3000 German miles, the whole diameter 6000, and the whole circumference 18,857. It traverses this space in 1 day, 18 hours, 28 minutes, 30 seconds, and will, therefore, pass over about 440 German miles in an hour.

Such, then, are the dimensions of the Jovian world, of its extreme circumference, and of the orbits of the four wandering bodies, the speed of which has at the same time been shown in popular measurements—that is, in German miles. From all this it appears that their speed increases with their proximity to Jupiter, exactly what is seen to happen to the planets already familiar to us, according to their proximity to the Sun. For the Jovian Mercury is more rapid than the Venus, and so is the Venus than the Jupiter, the Jupiter than the Saturn. Whether, however, this increase or decrease of speed depends on the revolution of the real Jupiter or not, as Kepler, the Imperial Astronomer, has argued with some probability about the Sun and his planets, Mercury, Venus, Mars, Jupiter, and Saturn, is so far unascertained by me, and unobserved. As I cannot assert it with certainty, so I cannot absolutely deny it. I therefore suspend my judgment on this point. But, to speak the honest truth, I wholly disapprove of this method of reckoning speed or its opposite. For what have heavenly bodies to do with our measurements—furlongs, miles, and the like—which we apply to the surface of the Earth? One method is in place when I consider the mass of any body as a whole, another when I consider a single particle of it. I will give a ludicrous example. Imagine a bull; let him move in a straight line, and complete in a minute a distance such that its 30th part is the length of the bull. Now let a wasp be sitting somewhere on the bull—say, on his forehead—such that a 100 wasps, if placed in a row, would be equal to the length of the bull. Now, if I were to choose to infer the wonderful swiftness of the bull from the fact that in less than a minute of time he has passed over the whole length of 3000 wasps, I should be laughed at by every one, and quite rightly. Whereas, if I say that, within a minute, the bull completes a distance of 30 times the length of his own body, no one will find any marvel in the speed of the bull. Just so about heavenly bodies; their speed is to be reckoned by considering the whole mass, not a single point or central part of it, yet the latter has hitherto been the practice of all astronomers. Having thus expressed my disapproval and rejection of the earlier method of measurement, I will now pass to that other method of which I have already made a passing mention in the dedication of my forecast for the year 1613. I shall inquire first into the extent of the orbit of each of these Jovian Wanderers, and afterwards into their speed in terms of their own diameters. Hence it will clearly appear that I am not attributing to these little bodies any incredible rate of speed, but am rather expounding a rational method of justifying that stupendous speed of the heavenly bodies on which followers of Aristarchus and Copernicus base an objection to those who maintain that the Earth is at rest, according to the testimony of all Holy Scripture, and in particular of the first chapter of Genesis. Let no one think me so insane as to wish to deal with the orbits of these secondary planets, while the magnitude of the planets already familiar to us is still undetermined. If there is anyone who has such a conception of me, I would have him know that I am here merely approximating to the truth, and that I hold it better to approach truth in a rough fashion than to despair of truth itself by seeking to go to its roots. I know also that a thorough and minute measurement of these little bodies is wholly out of the question; but that, meanwhile, it is not absurd to form a conjecture of their dimensions relatively to other heavenly bodies as to whose magnitude we are more certainly agreed. And it is in this sense that what is now to follow about the dimensions of these Jovian Wanderers is to be understood.

I have ascertained by frequent, diligent, and daily observation, that Jupiter, at his mean distance from the Earth, subtends with his diameter an angle of about 1 minute, as I have already several times remarked. I have also found by observation that three of the Jovians—that is to say, the fourth, the second, and the first—are as nearly as possible equal as to their apparent size; also, so far as can be made out by conjecture (for this subject is not open to precise observation at all) each is about equal to a 12th part of the diameter of Jupiter, while the 3rd, which is noticeably larger and more brilliant than the rest, covers an 8th of the same diameter. From this the orbit of these heavenly bodies is investigated in the following manner:

Jupiter, as has already been pointed out, has a diameter equal to 1000 German miles, the 12th part of which is 83, and this is the length of the diameter of the three Jovian planets—the fourth the second, and the first—because these are presumed to be all equal. Then, as 7–22, so is 83–261, and that is the whole circumference of the mass of the fourth, the second, and the first.

The eighth part of 1000 German miles (that is, of the whole diameter of Jupiter) is 125, which is the diameter of the third Jovian Wanderer; and the same calculation as before gives the circumference of this planet 393 German miles.

Observe that when, here and in other parts of this treatise, I speak of the circumference of bodies, I do not mean the whole bodily mass of a planet, but the full circuit which passes round the surface of the planet, having the same centre as the centre of the planet. For when we are measuring a spherical body by the common method, we first ascertain the diameter on a certain scale of measurement, and afterwards the circumference on the same scale. What I have so far said is to be taken as a rough statement as to the dimensions, or rather the circumference, of the four Jovian planets, which is what I had proposed. I now approach the investigation of the speed or slowness of these bodies according to my own method.

Of the Fourth

From what has been said above, it is clear that the hourly motion of this fourth Jovian planet is 200 German miles. Also its diameter contains 83 German miles. Dividing this hourly motion by this figure, we get as the quotient 2 1/2 times its own diameter, and that is its advance in 1 hour. Thus there is no speed to be considered, but rather a slowness of movement; it is as though a wheel were to move in a whole hour over two of its own diameters, with a half added.

Of the Third

The hourly movement of this third Jovian stroller has already been found to be 292 German miles. Dividing this by the 125 German miles contained in its own diameter, we get about 21/3 diameters as the advance in 1 hour, and it is a little faster than the fourth.

Of the Second

The hourly movement of this Jovian planet has been found to be 370 German miles, which, divided by 83, gives a movement in 1 hour of 4 of its own diameters, with nearly a half added.

Of the First

Dividing, in the same manner, 440, the hourly movement of this planet, by 83, we get five of its own diameters, with nearly a third added, and that is its advance in 1 hour.

I have now given, in the fewest possible words, sufficiently for the general view, all that I had to say under this head, of this Jovian world, its bodies and their circumferences—I have now to pass to the explanation of the difference between each individual movement. I will add, however, some remarks about these planets, dealing with their names.

Of the Names to be Assigned to These Four Jovian Planets

In the dedication of my Forecast for the year 1613, as also in previous ones and in tables calculated by me, I distinguished these four attendants of Jupiter only by numbers, or rather by the order in which they are placed with reference to Jupiter: calling "First" the one which makes the narrowest circuit about Jupiter, and only reaches a distance of 3 minutes in either direction; " Second" that which reaches 5 minutes from Jupiter, in its own greatest elongation; "Third" that which travels as far as 8 minutes away from Jupiter in either direction; "Fourth" that which has for the appointed limit of its excursion a distance of 13 minutes, or of 14 (of which more in its proper place).

Galileo, in his 'Nuncius Sidereus,' calls them "Medicean Stars," chiefly for this reason, that he was himself born and educated at Florence, under the dynasty of the great Dukes of Tuscany, who during many years past have sprung from the illustrious Medicean family.

If I name these same Jovian wanderers "Brandenburg Stars," who is to find fault with me, seeing that I have far more just causes for doing so? For not only was I born under the dynasty of this most illustrious and exalted family, but also, from my fourteenth year to the present time, I have been most liberally supported at the charges of those Illustrious Princes, Margraves of Brandenburg, George Frederic, of honoured memory, and, after his lamented death, of the brothers the Lords Christian and Joachim Ernest; I have been trained to familiarity with liberal arts and languages, have been maintained in Italy more than three years for medical study, and am to this day supported, with my family, by the most Illustrious Prince Albert Margrave of Brandenburg, Duke of Prussia, as a mark of that singular love for the Mathematics, for which they are indebted, as though of hereditary right, to him, from whom too the

'Tabulae Prutenicae' take their name. Herein, for myself and my descendants, if any there shall be, I gratefully acknowledge the extreme liberality of all these Princes sprung from this most illustrious line, and I bequeath my gratitude to my posterity as I ought to do. And by this name I make them, so far as in me lies, what they are most worthy to be immortal. So then let these stars, "Medicean stars" to Galileo, their first observer in Italy, be for me, who first saw and observed them in Germany (as is plain to the reader from my Preface) "Brandenburg Stars"; and this in memory, as I have said, of all those benefits bestowed most liberally upon me by this illustrious and in part electoral family.

Kepler, in a certain letter to myself, calls them "Jovian Wanderers"; David Fabricius, also in a letter to myself, names them "Jovial," others "Circumjovian" or "Circumplanetary," according to the humour of each. If, however, there are any who contend for separate names to be given to each, I hope that they will be satisfied by the course I have taken, in calling the one which makes the longest outward passages the Saturn of Jupiter. For as the real Saturn, that of the Sun, passes to the greatest distances from the Sun, compared with all the rest, and so moves in his own revolutions, so does this body from Jupiter.

The Second, which, in the majesty and quantity of his apparent light, surpasses the other satellites, and which from the outset I have called "Third," shall be the Jupiter of Jupiter; the second the Venus of Jupiter; the first the Mercury of Jupiter. The reason why I altogether exclude Mars is this. The real Jupiter is held to be the most fortunate among all the planets, as to his influence on sublunary bodies. Mars, on the contrary, is traditionally held by all astrologers to be an unlucky planet, and can in nowise, or at any rate only with the greatest difficulty, be associated with Jupiter. To Jupiter are ascribed the following qualities: Justice, Piety, Equity, Integrity, Gentleness, Temperance, Seriousness, and similar virtues; to Mars all the opposites. Moreover, a diligent scrutiny of these Jovians shows no trace of the ruddiness of Man, who is therefore deservedly excluded from this happy Jovian fellowship. Then as to Saturn, though he, too, is acknowledged by astrologers to be an unlucky planet, yet he agrees much better with Jupiter as to certain virtues, as Gravity, Patience, Authority, Majesty, and the like. The colour, too, of this Fourth is not very dissimilar from that of the Solar Saturn. Moreover, there are times when Jupiter, if in a bad situation, is thought by astrologers to signify quarrels and hypocrisy; but this may be supposed to come about from combination with this Jovian Saturn. Perhaps, however, some will be found, who are dissatisfied with the names so far enumerated, but call on astronomers for a separate name for each of these Jovian stars. I think that they may be satisfied by the following proposal, which, however, I desire to make without any superstition and by license of the Theologians. Jupiter is much blamed by the poets on account of his irregular loves. Three maidens are specially mentioned as having been clandestinely courted by Jupiter with success. Io, daughter of the River Inachus, Callisto of Lycaon, Europa of Agenor. Then there was Ganymede, the handsome son of King Tros, whom Jupiter, having taken the form of an eagle, transported to heaven on his back, as poets fabulously tell, and notably Ovid (*Metam.* x. 6). I think, therefore, that I shall not have done amiss if the First is called by me Io, the Second Europa, the Third, on account of its majesty of light, Ganymede, the Fourth Callisto. These names are included in the following distich:

Io,[4] Europa, Ganymede, Callisto—all of Jove
Preferred on Earth, around his orb in Jovian radiance move.

This fancy, and the particular names given, were suggested to me by Kepler, Imperial Astronomer, when we met at Ratisbon fair in October 1613. So if, as a jest, and in memory of our friendship then begun, I hail him as joint father of these four stars, again I shall not be doing wrong. As, however, these names have been freely invented by me, so I would have every one free either to reject or to accept them.

So much as to the First Part of the Treatise; I now pass to the Second.

Second Part

Particular Investigation of the World of Jupiter

Thus far I have set forth, shortly and succinctly, rather than at any length, the points usually, and rightly, considered by an Astronomer. It is now time that I should turn to the particular differences in movement between each of the four Jovian planets, placing before the eye of the reader, and proving by geometrical methods, the results and observations hitherto obtained as to their motion. These are the following seven phenomena:

I

The first phenomenon, or appearance, in these planets, is that they are not fixed in one place, or at one constant distance from Jupiter, but move round Jupiter, being sometimes to the East of him, sometimes to the West.

II

Every one of these four Jovian bodies observes a special limit of maximum elongation from him on either side. This follows from the fact of observation, that I have never seen two or more together near the maximum distance of the Fourth. The amount of the elongation of each partly comes out of what I have already said; I shall also deal with this matter at greater length presently.

[4]In the original it inadvertently says "In" instead of "Io".

III

They attain their greatest velocity when near Jupiter; at the limits of maximum distance they are slow and, as it were, stationary.

IV

I have found their periods of return in their revolution round Jupiter to be unequal—quicker for a nearer planet, slower for one more remote.

V

After making very many observations and ascertaining as nearly as was possible the periods of the revolution of each, I have noticed another phenomenon. The equality of their motion is relative mainly to Jupiter; and next to Jupiter, not to the Earth, but to the Sun.

VI

These secondary Jovian planets move in a line parallel to the Ecliptic, as regards their total revolution; but, in the course of it, they are deflected from this parallel, at one time towards the North, at another to the South, by an appreciable difference—especially when two are seen in conjunction, one approaching Jupiter, the other receding from him.

VII

These Jovian Wanderers are not always seen as of equal magnitude but at one time larger, at another smaller.

These seven phenomena have been, up to the present time, discovered by me in the motion of these Jovian stars. I shall speak, in what follows, of each by itself, and that shortly and succinctly, because Nature has denied me eloquence. At the same time I shall endeavour to "save" and demonstrate them by what I think is a convenient hypothesis, thought out by myself.

Of the First

I think that there is no occasion for me to repeat here what I have already said in the Preface. Here I only enforce this point: that this phenomenon, as it was the first, so it has been the easiest of observation. From day to day—I may say, from hour to hour—the bearings of these bodies to Jupiter were found to change during my earliest observations, taken in the autumn of 1609, but more particularly about the end of that year and the beginning of the following. For although I at first supposed that they belonged to those fixed stars which are in other parts of the heavens invisible without such an instrument, yet because Jupiter was then retrograding, this sudden and also varying change of bearings towards him surprised me exceedingly, until I began to have doubts whether they could really be fixed stars. However, when Jupiter had passed backward through several degrees, and still I saw these stars keeping company with him, I was greatly amazed that it should be so, and began to keep notes of what I observed. My first observation so made was on the 29 December, 1609. On that day about 5 o'clock in the evening I saw three bodies to the West of Jupiter, almost in a straight line with him. After this I made continuous observations till the present time.

From this I have come to be certain that these stars have Jupiter for their acknowledged centre, and are borne around him, exactly as Mercury, Venus, Mars, Jupiter, and Saturn move round the Sun as their centre.

Of the Second

This phenomenon was more difficult to observe than the first. Very numerous observations were necessary, for the following reason. First, it was necessary to be clear as to the number of these stars, and on this I scarcely reached certainty before March of 1610. Then, when I had ascertained that there were at least four such bodies moving around Jupiter, I found the greatest assistance from the maximum elongation of the Fourth, and I gave my closest attention to these stars when they were at their greatest distance outwards. The Third, as being conspicuously larger than the others, was also the easiest to observe, and was distinguished at its own maximum distance without much trouble or attention. Within 6 months I had made myself certain as to the maximum elongation of these two. The other two caused me a great deal of trouble and labour. I was compelled to wait for a time when all four were seen at once, and then to observe them continuously for several hours, sometimes, if a clear sky allowed it, during the whole time that Jupiter remained above the horizon. By this method I ascertained that the Fourth passes out to a distance of 13 minutes from Jupiter, in either direction, is almost stationary there, and thence ret urns towards Jupiter—the Third to a distance of 8 minutes, the Second of five, the First of three.

My own Tables of distances are calculated for these maximum elongations.

It is to be observed, however, that these maximum outward passages are correct when Jupiter is at quadrature to the Sun, and at his mean elongation from the Earth. About his times of opposition to the Sun these distances are manifestly increased. This is especially the case with the Fourth, which, as I have ascertained, not only approaches, but even slightly exceeds, a distance of 14 minutes. When the Sun was approaching Jupiter, or when Jupiter was outside the solar rays, so that these stars could be seen and observed, I found these distances to be perceptibly diminished and contracted. However, my instrument has not so far allowed me to measure this augmentation or diminution; thus I am uncertain whether the observations show as much of either as the variation in Jupiter's elongation from the Earth requires. I have, therefore, preferred to settle nothing for the present about this, intending to leave it for future and more elaborate observations. Accordingly, the distances given in my tables are to be taken as averages, which await an accurate determination of this further difference, or, to speak more correctly, of this defect and excess. The candid thinker, and the admirer of this new aspect of the heavens, must be content to have a theory and tables from which, as I hope, he may, by an easy process, know which of these stars are in the East, which in the West, and at what approximate distance from Jupiter. I have actually begun this further and more delicate speculation on the excess and defect in the present year 1613. I have taken the mean elongation of the Fourth from Jupiter to be 12 minutes 30 seconds. Then I have taken the actual distance of Jupiter from the Sun at 15 degrees, and calculated in units of which there are 11 between the Earth and the Sun, 60 between Jupiter and the Sun. Besides this, I have investigated, for the distances so found, the excess over the mean elongation from Jupiter and the defect from it, which result when Jupiter is equidistant from Sun and Earth. But the calculation would have been rendered too intricate. I have therefore preferred to reserve this task for another time; meanwhile, I will make more observations, to arrive at a fuller certainty as to this phenomenon also.

Of the Third

This phenomenon also can be ascertained and observed with very great facility, especially so far as relates to the position of the Fourth. For it has sometimes been discovered by me at the same maximum distance from Jupiter during nearly three complete days, and no appreciable difference could be detected. Near Jupiter, however, there was a sudden variation of distance special to the Third, as being the one which is more easily recognised than the others, from the majesty and abundance of its light. This body would be seen to the west of Jupiter at such an hour of one day, and at the same hour of the following was to the east of him, and *vice versa.* The very high velocity of its motion is especially observable when two moons are in conjunction near Jupiter, the one approaching, the other receding from him. For when thus placed, they are clearly seen to separate and take their different ways, within the course of 1 or 2 hours.

This phenomenon has been of great assistance to me in thinking out a theory to suit the facts, which is that the motion of the bodies around Jupiter is circular. For a circular motion has this property, that bodies attain their greatest velocity near a line passing through the centre, but become very slow, and are almost stationary, on the tangent—as will hereafter be shown in my explanation of the theory, and as is elsewhere proved in ordinary treatises upon the sphere.

Of the Fourth

"The task, the toil is here!" Unless I had put the second and third phenomena on a certain basis, I should never have succeeded in determining the times of the periodical return.

For by no method could I have arrived at a knowledge of the periods of revolution, unless I had somehow ascertained the limit of maximum elongation from Jupiter. Accordingly, my first inquiry into the periodic movement was directed to the Fourth Jovian Wanderer, as being the one which reaches a greater elongation than the others. By numerous observations, therefore, I ascertained the time of the half revolution—that is, the time between the greatest eastern and the greatest western elongation, and this only in days.

For it could not be accurate, on account of the slowness of the movement, in such a position relative to Jupiter. However, it was necessary to begin with those positions which were simpler and more readily open to observation. By doubling the period after the half, the result was the whole period of recovery of motion of the Fourth Jovian—that is, of the Saturn of Jupiter, or Callisto—and this I found within some seven or eight months to be about 17 days. During the course of this enquiry, lo! the Third also betrays himself, partly by the intrinsic majesty of his light, partly by the fact that he sometimes appeared simultaneously with the Fourth, each almost stationary, the latter at a distance of 13 minutes, the former of 8. As this happened several times, I began to arrive at the investigation of the periodic recovery of the Third also. This I discovered after a whole year, that is, about the end of 1610 to be this: he completes a revolution in his own orbit round Jupiter in 7 days. While thus occupied, I gradually arrived at a more accurate figure for the return of the Fourth to its starting point, so that in March 1611 I came to believe that its periodic time embraces more than 16 days 18 hours. I then supposed that the period of the Third was 7 days, 3 hours, 53 minutes. These times, though not exact, were nevertheless of great assistance in investigating and clearing up that of the revolution of the Second Jovian, which I had meantime ascertained to pass not more than 5 minutes of arc away from Jupiter on either side; this was in the main done when all four were seen at once, the Fourth being with the Third, each at its greatest elongation. Having thus completed this hitherto incredible task, to make the conclusion of my story short, I reached a full acquaintance with the periodic times of all the Planets of Jupiter, owing, I hope, to the bounty of God the good fortune of my success; I know that up to the present time they continue to satisfy my more painstaking observations. The results for following years will be attested by the observations which are to follow. I do not now guarantee absolute certitude. I have laid foundations of this whole matter, not unserviceable to a diligent observer, if any such there shall be hereafter. It will be very easy for him to add whatever is in defect, to cut back what is in excess.

For it is necessary to keep as many observations as possible, and these at sufficiently long intervals, especially observations giving the same position of Jupiter relatively to Sun and Earth. The reason of this will be shown in the sequel.

It was not, however, sufficient for the investigation of the periodic time to use only the maximum elongation; I used also observations taken nearer Jupiter, when the movement of these secondary planets is more rapid. I am unwilling to speak of the great labours which I underwent; I think that only one who has hazarded a similar enquiry will understand them. So I think it needless to say more about this phenomenon.

Of the Fifth

Having ascertained the periodic time, and the limit at either end of maximum elongation from Jupiter, as I have pointed out in the previous pages, I now cast up and tabulated, first, the mean rates of movement around Jupiter and, secondly, the distances from him in either direction, and began, in the words of the adage, to triumph in advance of the victory, as the candid reader will clearly see from what follows.

I settled the epochs of the mean motions at the beginning of the year 1610, at the midnight preceding the first day of January, according to the Julian Calendar, because, as will be understood, only one observation of these stars bad been noted by me in the preceding year, that of the 29 December. In the meantime, there had also appeared the 'Sidereus Nuncius' of Galileo, which first came into my hands in the June of the same year. I began to draw up a calculation from my own recently prepared tables, and to compare it with my own and Galileo's observations. In the course of this more accurate comparison, I find that in some places, and these at a sufficiently long distance from one another, the calculation sufficiently agrees with the observations, while at others it deviates from them by an appreciable distance. This disturbed me greatly, and almost reduced me to despair and hopelessness of ever finding a suitable hypothesis. For at that time I still thought that these Jovian bodies had an even motion relatively to Earth. At last I examined observations made about the time of Jupiter's opposition to the Sun and adjusted the epochs to them. For a doubt was gradually arising in my mind as to the behaviour of these stars. I therefore took into my counsels observations made about the quadratures of Jupiter and of the Sun, and soon found a manifest difference—in fact, the excess of the observed figure over that of my calculation in the one case was equal to the deficiency in the other. So I took heart again, and began to reflect as to the cause; and discovered, without much effort, the reason of this phenomenon. Afterwards I sought from Copernicus the proportion of the Great Orbit to that of Jupiter, which I found to be as 11–60 or thereabouts. I then entirely neglected the first inequality arising from eccentricity, as being, on my own view, insignificant for the present purpose. I supposed, too, that the eccentricity of the Sun was here a vanishing quantity or incapable of observation. I therefore took the proportion which I have mentioned, and calculated a table of equations. The method which I employed I shall show later on. The discovery was actually suggested to me by my own view of the system of the Universe, in general identical with that of Tycho, at which I arrived in the winter of 1595–6, when I first read Copernicus. At that time I was still at school at Sacrifontan (Heilbronn[5]), and Tycho was not known to me even by name, much less his hypothesis, which I only saw in the following year, in outline, at the house of the reverend and learned Francis Raphael, pastor of the Church at Onold (Ansbach), now with God, the sketch having been sent to him by a student of Wittenberg. I have many witnesses of the fact of my discovery. First, the very learned man whom I have already named. Then all the Assessors to the illustrious Consistory at that time, to whom I presented my hypothesis, with an explanation, after Easter 1596, and under whose advice a special lodging was assigned to me in the monastery mentioned above by the Illustrious Prince George Frederick, Margrave of Brandenburg, of most honoured memory, for my greater convenience in pursuing this study. Further, I call as witnesses my own beloved teachers, whose other courses of lectures did not allow them to follow it out themselves, but who helped me greatly with books, such as the following: Wenceslas Gunkfelder, George Hirchbauer, and John Neser. As they have given me full permission, I have felt it a duty and a pleasure to add their names in honour and gratitude; they have been of the utmost service to me, not only in this matter, but in very many others.

I say nothing now of my dear brother James, of beloved memory, who was also thoroughly satisfied as to my studies in Astronomy.

Among the others, not the last place is due to the learned and deeply read Augustin Lanius, now living in retirement at Halle in Saxony, then an organist at Heilbronn,[6] whom proximity and a friendship of long standing brought into almost continuous observation of my doings. I bring in these names, not from any motive of ambition, but because of the dull and sometimes wicked cavillings of certain persons, of one in particular, whom, though I had intended otherwise, I pass over as unworthy of refutation, lest any words of mine should make his name known to other honourable men. I now return to my subject,

Of the Sixth

This phenomenon also brought itself plainly before the eye, especially when there was a conjunction of two Jovian bodies, one approaching Jupiter, the other receding from him. For when two of them come into conjunction near Jupiter, both moving in the same sense (that is, towards or away from Jupiter), they are so near one another that they may be said to touch, and the two may be taken to be one very bright body.

This phenomenon was at first established from a conjunction of the Fourth and Third, when the Third was at its greatest elongation and had absolutely no latitude, as will be shown later on by examples and observations when I come to the explanation of my theory. I came very slowly to recognise it, because not only is a conjunction of the Third and Fourth, with the Third at its extreme distance, a somewhat rare occurrence, but also because such an observation is often hindered by a cloudy

[5]Correct would be Heilsbronn, not Heilbronn.

[6]Heilsbronn.

sky. It is true that when nearer Jupiter this conjunction admits of more exact observation; yet for me, with my instrument, it was more difficult, for reasons already pointed out in the Preface.

When once satisfied, however, as to this phenomenon—namely, that these Jovian bodies do not always move in a direct line drawn through Jupiter parallel to the Ecliptic, but with a perceptible deflection, sometimes to the North, at others to the South—I began to enquire into it with greater diligence; and at length ascertained that the Jovian bodies at their greatest elongation are always found on the parallel line just mentioned, but outside those limits always decline from it, to the South in the upper part of their orbit, to the North in the lower, and that this inclination is greatest near Jupiter. How great the maximum inclination of each is I was unable to measure, because it is a matter of seconds only, which I do not profess to observe. I did, however, notice this: that none of these planets reached so great an angle from the parallel line mentioned hat I saw it pass above or below Jupiter at the time of conjunction with him. For the latitude of the Fourth is greater than that of the Third, and that of the Third greater than that of either of the others. However, by a probable conjecture from this conjunction of the Fourth and the Third, I put the greatest latitude of the Fourth at15 seconds, of the Third at 12, of the Second and First at 10. On this basis I have drawn up a table of latitudes for these Jovian stars, from which, taken with the simple motion of the planet and the addition of 90°, the latitude of any one can easily be deduced, as will be clearly explained later on, when I speak of the use of the tables.

Of the Seventh

This phenomenon has given much trouble, not only to me, but also, as appears from the 'Sidereus Nuncius,' to Galileo. I also confess that in the early stage of my observations, and especially in the first year, 1610, I several times failed to notice the Fourth, or even to see it, at the time of its greatest elongation, the reason being; that it was so small that it was somewhat difficult to make it out. That the reason alleged by Galileo, with some probability, why these Jovian stars appear greater and smaller at different times, does not satisfy the phenomenon, I will now explain. His inference is that a lunar body moves in an orbit which is vaporous and denser than the rest of the atmosphere, just as a similar orbit surrounds our Earth. He takes this last as proved, so that it is strictly analogous that a similar vaporous orbit should have its place around Jupiter. This being interposed, the moons would appear smaller at their apogee, larger at their perigee, because of the absence or attenuation of such an orbit. That such a view is here inadmissible I prove in the following manner. If the reasoning were sound, then this visible shrinkage of the Jovian moons would always and only occur at the time of apogee—that is, of their greatest elongation from the planet. Except in that position, they would always appear as of the same magnitude. Neither is, in fact, true. Observations show that the same thing happens not only in that position, but even at the maximum distance from Jupiter, especially in the case of the Fourth. If, therefore, this shrinkage were caused by the supposed vaporous orbit, it would necessarily follow that such an orbit extended beyond the extreme distance of the Fourth. And if, at that distance, it were able by its own density to reduce the light of the Fourth, almost to invisibility, it would certainly follow that its proportionate density near Jupiter would never allow the Fourth to be seen when near Jupiter at its apogee, contrary to my own observations, which show that the Fourth has very frequently been seen and observed by me near Jupiter, though with varying quantity of light. We must, therefore, put away from Jupiter this "vaporous orbit," and look out for a very different cause. Moreover, I also deny—what Galileo treats as proved and granted—about a vaporous orbit appearing around our Moon. For I have never, since I began to use this instrument, detected any variation other than that which arises from the quality of the atmosphere surrounding the Earth; nor have I seen any shifting spots on the Moon, such as are visible on the Sun.

As to the statement that no gaps or inequalities are seen on the extreme circumference of the Moon, this is not true in all cases, though commonly it appears to be so. I have seen not unfrequently, when sky or atmosphere was very calm and pure, certain breaks and gaps on the upper or northern, as well as on the southern, side of the waxing Moon, very narrow, no doubt, so narrow that they could only be observed by very close attention. Also on the western side of the Moon's circumference, when she is a little more than half full, a gap is distinctly seen about the breadth of a finger across. Kepler also saw similar gaps in the circumference of the Moon during a solar eclipse, or on the disk of the Sun, in May 1612, as is stated by him in a letter to me. The true cause why the circle (or part of a circle when the Moon is not full) generall y appears perfect, without those breaks or excrescences, I take to be that which Galileo states on his page 21; his other cause I entirely disapprove, as has been already said. Nor do I find any difficulty in the phenomenon of the Solar Eclipse of 1567, on which Kepler has much to say in his 'Optics.' Meanwhile, I am not denying that sometimes exhalations are expelled from the Earth, either by her own proper-motion or by one originated by the stars, which are borne to the highest part of the atmosphere, and remain there a long time, until the thinness of that highest atmosphere bordering upon the ether causes a recoil, and they recover their density, and return to Earth, and are the cause of copious rains. This most often happens after a long and continuous drought.

Thus I hold the true and genuine cause of increase and decrease in the apparent magnitude of these stars to be this: That they receive their light from the Sun in the same way as our Moon and the other planets, including Jupiter himself, that the half opposite the Sun is always in light, that turned away from him in darkness, and that the body of Jupiter throws a shadow. I also conclude that the four Brandenburg Stars imitate the Moon exactly, and receive their light in two ways, both from the Sun and from their neighbour Jupiter; moreover, that they differ one from another in subtlety and excellence of material, and that in the high polish of its surface and in the excellence of its material the Third far excels the others, inasmuch as it most vigorously returns the solar rays which it receives, especially when it is moving in the lower part of its orbit, near the extreme limits. The Fourth I take to consist of a more obscure material and to have a less polished surface; hence it has not so much force in returning the solar rays. This is like what we see in Venus; when she is in her crescent phase, she yet returns the solar rays with great vigour, because of the excellence of her material and the high polish of her surface. With Saturn it is far otherwise. The cause of the fact that our Jovian stars appear larger and smaller at different times, lies in their varied position relatively to the Sun, Jupiter, and the Earth. For it is probable that the same thing happens between these Jovian planets and Jupiter as between the Earth and the Moon. It is a discovery of Maestlin that the Moon receives light the Earth on her dark side, as is clear from Kepler's 'Optics.' And so these Jovian wanderers are illuminated in two ways, both by the Sun and by Jupiter. But the power of Jupiter to throw out his borrowed light to his satellites is very feeble—firstly, because Jupiter is smaller than the Earth; but also, and chiefly, because Jupiter is much farther from the Sun, his distance being six times, or five times, greater, and therefore the light of the Sun is less effectually imparted to Jupiter himself and also to his planets, and again reflected from them. For these reasons I think that the amount of the visible light of these stars should be referred to their varying position towards Jupiter, the Sun, and also the Earth, especially when they are at or near their maximum elongation from Jupiter, and this is best observed in the Fourth. For these four stars are, as it were, so many moons, and to an observer on Jupiter have the same appearance as the moon from our Earth; only with this difference, that at every single revolution, or period of their full moon, an eclipse takes place; of which more presently. That when near Jupiter something similar happens to them, so that they not only appear smaller, but, as seems probable, are actually obscured or eclipsed, is clear from the following consideration. Jupiter is not a transparent body, any more than Venus or Mercury; therefore throws a shadow on the side turned away from the Sun. How far such a shadow extends, and whether all four pass into it and are eclipsed once in each revolution, I will now show in as few words as may be.

According to the observations and discoveries of that great astronomer Tycho Brahe, the diameter of the Sun covers approximately $5^1/_6$ Earth diameters.[7] Taking, therefore, the diameter of the Earth at 1718 German miles,[8] as I have assumed from the beginning, the diameter of the Sun will be 8876 of the same miles. Now I have stated from the first that the diameter of Jupiter covers 1000 German miles; I have also assumed that the distance from Jupiter to the Sun is to that from Earth to Sun as 60 to 11. Following this ratio, I will very shortly investigate the length of the axis of the shadow thrown on the side of Jupiter away from the Sun, that is, the length of the line GE on the figure appended (Fig. 1.1).

Let AF be the semidiameter of the. Sun, CG the semidiameter of Jupiter, FG the distance of the Sun from Jupiter. Subtracting CG, 500 German miles, from AF, 4438, we have the remainder AH 3938. I proceed: As is AH (3938) to HC or FG

[7]In his *Astronomiae Reformatae Progymnasmata* (1602), Tycho made the solar diameter $5^{14}/_{75}$ times as large as the Earth's, but for all the sizes and distances indicated that these measures were approximations. See *Tychonis Brahe Dani Opera Omnia*, II: pp. 422–426, 431.

[8]Assuming 15 German miles to 1° at the equator, or 1 German miles is about 7 km. The diameter of the Earth, according to Marius, was about 12,000 km.

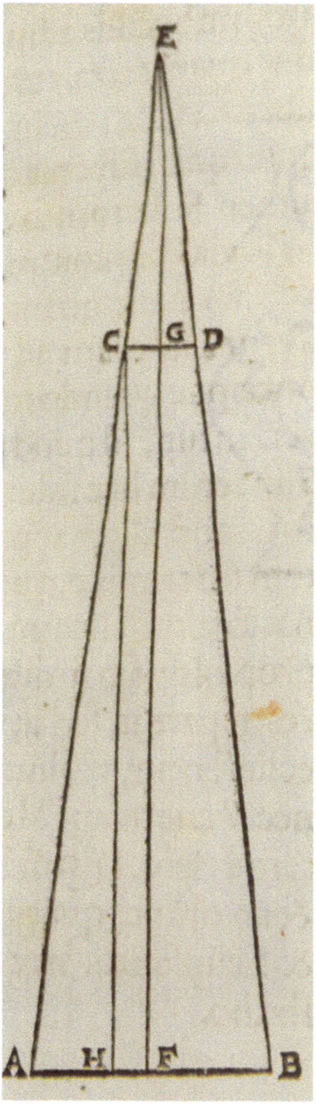

Fig. 1.1 The length of the shadow in *Mundus Iovialis*, Nuremberg 1614, sig. D2^{v}; Municipal Archive Gunzenhausen

(60), so is CG (500) to GE. This works out to 7 $^{67}/_{100}$, or for simplicity say 7 $^{7}/_{10}$, for GE, when FG is 60. To state this length also in German miles, we proceed thus: Eleven parts (that is, the distance of the Sun from the Earth) give 1150 semidiameters of the Earth,[9] required the number of such semidiameters which will be given by the 7 $^{7}/_{10}$ parts found above. When the sum is worked out we get 805, which multiplied

[9]Nicholas Copernicus and Tycho Brahe both used approximately this figure. See Van Helden, *Measuring the Universe*, pp. 46–50.

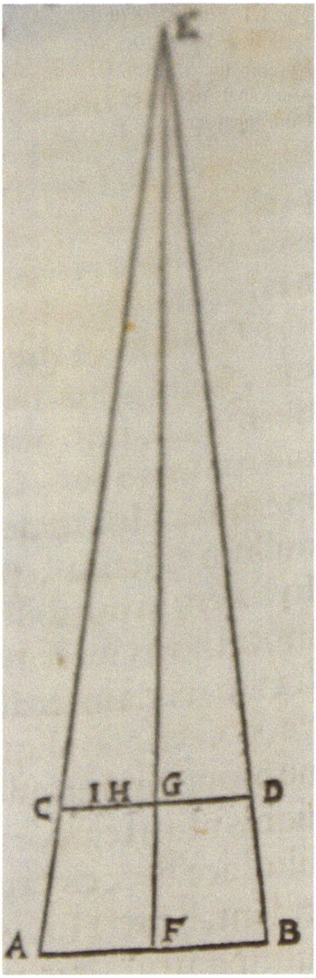

Fig. 1.2 Entry in shadow of the fourth moon *Mundus Iovialis*, Nuremberg 1614, sig. D3r; Municipal Archive Gunzenhausen

by the 859 German miles in a semidiameter of the Earth, give 691,495 German miles, the length of the line GE. Now I have already assumed, as the result of observations, 13 minutes (that is, 13,000 German miles) for the greatest elongation of the Fourth from Jupiter (FG on the second figure). This subtracted from the whole FE leaves GE 678,495 German miles. We have now to find the diameter CD of the shadow of Jupiter where the Fourth is at his greatest elongation from the Sun, which is done as follows: As is EF 691,49 5 to FA, the semidiameter of Jupiter; 500, so is EG, 678,495 to the semidiameter of the shadow CG, which comes to 491 German miles (Fig. 1.2).

We have now at last to see whether the Fourth, at its own greatest distance from the Sun (that is, at the beginning of its uniform motion), moves within the shadow of Jupiter or whether it passes beyond it at the side. About the other three there is no doubt, because of their proximity to Jupiter and the slightness of their distance in latitude. I have a little above, in dealing with the sixth phenomenon, reckoned the extreme distance in latitude of the Fourth at 15 seconds. Taking the diameter of Jupiter as 1000 parts, GH will be 250. Further, in our general consideration of the Jovian system, we found the diameter of the Fourth to be 83 German miles, and its semidiameter HI 41, which added to the extreme distance in latitude, GH 250, gives 291 for GI, and that is the distance of the Fourth, at the extremity of its circumference, from the axis of the shadow. Since, then, GI is much less than the semidiameter of the shadow GC, it follows that the whole of the Fourth moves within the shadow of Jupiter, and is eclipsed in the passage. Therefore all the four Jovian bodies are within the shadow of Jupiter at the beginning of their movement, and are eclipsed.

With regard to the first figure, it is to be observed that, for greater facility of working, I have used it as though the line GO drawn from the centre of Jupiter to the tangent ACE, parallel to the line FA, were the same as the line drawn from that centre to the point of contact. For the difference here is almost inappreciable, and causes no difficulty in the attainment of my purpose. For if the line ACE were drawn through the point of contact, the shadow would be lengthened, and GO, the semidiameter of the shadow at the place of transit, would be made rather longer. To meet the cavils of ill-disposed persons, I have thought it well to add this here.

When, therefore, the Fourth is moving near the shadow of Jupiter, and receives the Sun's rays with greater difficulty, it appears smaller than at other times—indeed, it is totally eclipsed—a fact which Galileo saw with his own most perfect instrument, as is attested by Kepler's letters to my self. It has, however, happened to me not infrequently to see no Jovian satellite near Jupiter, but, after a few hours, to see one at a noticeable distance from him, which did not correspond to the movement corresponding to the interval of time, being much greater. Conversely, I have sometimes seen a planet at a noticeable distance from Jupiter; after the lapse of some hours it has disappeared, although according to its own rate of motion it ought to have been still in sight. However, I did not notice the times of this observation, when it was made: It must necessarily happen near the quadrature of Jupiter and the Sun, on the western side near their first quadrature, on the eastern side near their last. For a year from this time on, I paid careful attention to this matter, especially in the case of the Fourth; in the case of the others it is impossible, with my instrument, to observe such an eclipse happening. However, I have not, up to the present time, had the good fortune to get a similar observation; I shall endeavour to do so in future, that I may have confirmation of this fact also.

Whether a mutual eclipsing of these bodies, or at least some interception of the Sun's light, is possible, I am uncertain; it seems to me likely that it is. At any rate, I have an observation made at 10 P.M. on the 7/17 February of the present year 1613, when all four were visible, three to the East, and one, the First, to the West. All were very clear except the Fourth, which was extremely near the Second, in the direction of Jupiter, more to the south, and very meagre, so much so that it was scarcely visible. The Fourth was in the upper part of its orbit, and moving away, the Second was in the lower part, and approaching near them was the Third, also approaching; moreover, the shadow of Jupiter was on its western side, so that it could not be the cause of the meagreness of the light. It is probable, therefore, that these two bodies, the Third and more especially the Second, prevented the rays of the Sun from being able to reach the Fourth and flow on in full force.

So much on the seventh and last phenomenon; and thus conclude the second part of this treatise. There remains the explanation of Theory, which shall be the Third Part, wherein the diversities of motion already mentioned are explained and proved.

Third Part

The first two parts of this treatise having now been set out to the best of my ability, there remains the third, which is to deal with the statement of Theory and its adjustment with observations, and is chiefly occupied with calculation. This I shall shortly and succinctly develop in what follows.

According to my imagination, my theory about these four bodies is this. I think that they are borne with an equable and simple motion around Jupiter as a centre, and that Jupiter with his satellites takes for his acknowledged centre not the Earth but the Sun. I assume that the Sun himself moves in a sort of concentric orbit round the Earth, not that he is actually borne in a circle concentric with the Earth's orbit, but that the eccentricity of his orbit vanishes and becomes almost inappreciable in the apparent movement of the Brandenburg Stars. Further, I place the beginning of the simple motion of these bodies in the middle of their apogee, when they are at their greatest distance from Jupiter, so that they pass on thence towards the East, and so complete their periods.

To make this more readily intelligible, observe the next figure. A is Jupiter, around it as centre are drawn the orbits of the four Jovian planets, B is the Sun, BA the distance of Jupiter from the Sun, which I take as 60 when the distance of the Earth from the Sun is 11. In D these planets are at their apogee. Hence they pass with simple and equable motion eastwards to E, at which place they are stationary—that is to say, at their greatest elongation from Jupiter. From this point they return towards Jupiter until they reach F, where they are in conjunction with Jupiter; thence they pass outwards to G, where they are again stationary, apparently, so far as has been observed, because they are then moving on the line of contact. From this goal they at last return to their own initial point at D, and so complete the time of periodic restitution (Fig. 1.3).

By this figure phenomena 1, 2, and 3 are accounted for and proved. The first is explained as follows. Because these bodies are borne in a circular orbit round Jupiter, they are therefore not seen fixed at one point, but sometimes in one position relatively to Jupiter, sometimes in another. The second phenomenon was given above, that any one of these four Jovian bodies has its own special and recognised terminal distance from Jupiter. Thus the Fourth has a distance of 13 minutes, the Third of 8, the Second of 5, the First of 3; the reason of this is clear from the diagram annexed. For because experience shows that these bodies revolve round Jupiter in a circular orbit, it is manifest that, after leaving D, they can only recede from Jupiter

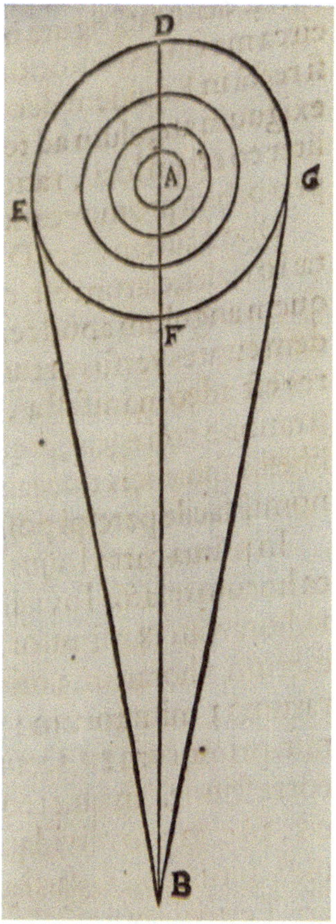

Fig. 1.3 Orbits and orbital times of the moons in *Mundus Iovialis*, Nuremberg 1614, sig. E1[r]; Municipal Archive Gunzenhausen

towards the East until they have reached E by simple movement. The same happens at the point G. What is said here about the orbit of the Fourth is to be understood also about the orbits of the others. In dealing with the Third Phenomenon we have said that at their extreme limits of distance (that is, at E and G) these Jovian bodies are slowest and almost stationary, and fastest near Jupiter. The reason is clear from the figure here drawn. For when they are near their points of greatest distance E and G they are moving almost in a straight line, at E downwards, at G upwards, making no angle towards Earth or Sun, or a very small one, although even there they are advancing evenly, because of their own simple movement upon their own orbit.

That these Jovian planets reach their utmost velocity near D and F is clear, because at those points they suddenly make by their own movement an angle with Earth or Sun, when near D towards the East, when near F towards the West. This is so evident that it needs no fuller statement or proof, more especially as the same thing is set forth in the common treatises of Spherical Astronomy, and may easily be perceived even by beginners.

In the first part of this treatise the periodic times of these four Jovian bodies have been pointed out. The Fourth has 16 days 18 hours 9 minutes 15 seconds nearly, the Third 7 days 3 hours 56 minutes 34 seconds, the Second 3 days 13 hours 18 minutes, the First 1 day 18 hours 28 minutes 30 seconds. These figures are according to my latest correction. If the complete circle of 360° is divided into these periodic times, the simple movement of each in one day comes out thus:

Of the Fourth, 21° 29′ 3″; of the Third, 1 sign 20° 14′ 57″; of the First, 6 signs 23° 25′. Upon this basis the tables of the simple movements of the Brandenburg Stars have been computed by me; out of which the simple movement of each for any given time after the year 1608 may be obtained by an easy process, as will presently be shown.

Note.—Here we might explain at greater length all that happens in the comparison of the greatest elongations of these satellites of Jupiter. For the maximum elongation of the Third is as nearly as possible a mean proportional between that of the Fourth and that of the Second. So the maximum elongation of the Second is nearly a mean proportional between the maximum distance of the Third and that of the First. But, as I have said, I am unwilling to deal with these points at greater length here. I will reserve them for another time, since the first foundations at least of the Jovian system have now been laid in a manner above contempt. I have only wished to call attention.

Now that we have thus determined the equable and simple movement of these bodies, my next business is to state further the other points essential to the investigation of the apparent motion. The first is to ascertain the several distances of these bodies from Jupiter, to the East and West of him, which correspond to their equable motion in their own orbits. This is done thus (Fig. 1.4).

In this figure let A be the Earth, or the Sun, which is the same thing, because these distances are computed for the mean distance of the Earth from Jupiter, which is the distance of Jupiter.

From the Sun. Let B be the centre of Jupiter, D the initial point of the simple motion of the Fourth (the same method applies to the others), and, in the present example, let DE be 45°, BE the semidiameter of the orbit of the Fourth, 13′. We have to find from this the line EC, the distance of the Fourth from Jupiter to the East, corresponding to the simple motion: For greater ease in working, I shall employ the most simple method, which is this:

In the right-angled triangle ECB we have given, besides the right angle ECB, the angle DBE, the simple motion of the satellite, and the side BE 13′: thus the side EC will be known. For as is the whole sine 100,000 to the side BE 13′, so is the sine of the angle EBC 70,711 to the side EC 9′ 12″ which is the distance of the Fourth from Jupiter to the east, corresponding to the equable motion over 45°, as is found also in the table of distances. On this method all the distances of all four satellites have been calculated by me, and placed in the tables, for every five degrees.

I am aware that these distances should have been computed in another way; this method, however, is sufficient for my present purpose. If anyone is not satisfied, let him try the other and usual process; he will find the difference al most inappreciable, because of the very great distance of Jupiter from the Earth, and the very small

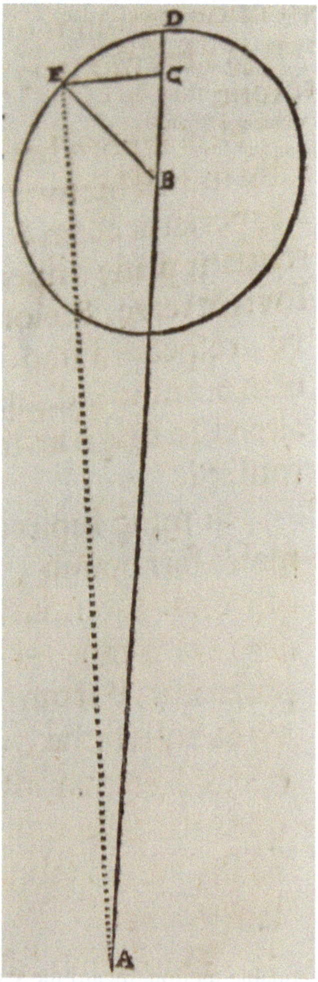

Fig. 1.4 Displacement with reference to the orbital position in *Mundus Iovialis*, Nuremberg 1614, sig. E2[r]; Municipal Archive Gunzenhausen

elongations of these satellites from Jupiter relatively to the distance of Jupiter from the Earth. Any small difference which there may be, is, as I have already pointed out, quite undeserving of the great labour which would have been involved in the other and more artificial process.

Now if Jupiter had the Earth for the centre of its orbit according to the Ptolemaic view, the calculation would now be completed, and the method of determining the distance of the Jovian satellites from Jupiter in either direction; for that distance could be found from the tables just constructed, through the simple movement of each. However, my own observations made near the quadrature of Jupiter and the Sun testify that there is yet another inequality here, that Jupiter has not the Earth but the Sun as his centre, and that these satellites with Jupiter himself, in their own

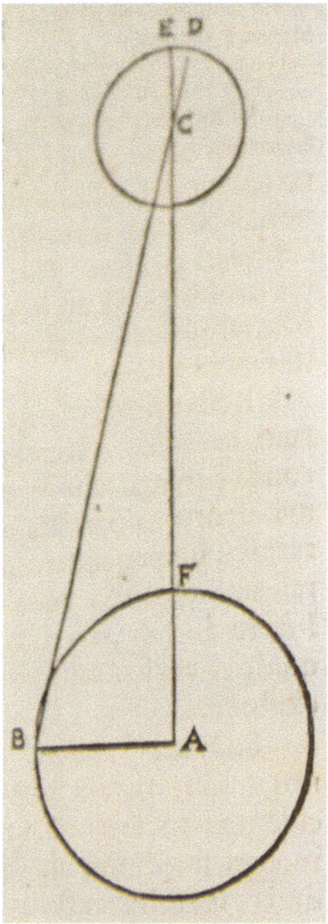

Fig. 1.5 Compensation due to Sun as reference point in *Mundus Iovialis*, Nuremberg 1614, sig. E3r; Municipal Archive Gunzenhausen

equable motion, regard the same centre. Accordingly another table had to be drawn up, which I have called a "table of equation."

This equation is thus ascertained. On the diagram here given, let A be the Earth, B the Sun, C the centre of Jupiter, and let the line BO, the distance of Jupiter from the Sun, always be 60 when BA, the distance of the Earth from the Sun, is 11. We must now find the angle ACB, which is the equation required. In the triangle ABC we have three data (that is to say, AB 11, BO 60, and the angle BAC, which is the true distance of Jupiter from the Sun—that is, the arc FB). Let this, in the present instance, be a complete quadrant of 90° (Fig. 1.5).

I proceed: as the side BC 60 to the sine of the angle BAF 90° (100,000), so is the side BA 11 to the sine of the required angle BCA. Working this out, we have for the sine of this angle 18,333, to which corresponds an arc of 10° 34′, namely DE, and this must be subtracted. For BCD is the line of Mean Apogee, from which the true motion is reckoned. Then subtracting the arc DE from the equable motion, we have remaining the corrected length of the satellite from the point E, with which the true distance of the satellite from Jupiter is to be taken from its own table.

I know that the distance of the Earth from the Sun (that is, the semi-diameter of the Great Orbit of Copernicus) is not exactly 11, but nearly 30′ more, and so the angle BCA would come out to 11° 3′. But the fact is that in the whole of this treatise I have neglected complete accuracy, just as I have preferred, in both cases, that of Jupiter and that of the Sun, to keep 11 for the figure of the eccentricity, for the sake of greater ease in working; moreover, the 29′, which, as well as the difference between the angle in either case, are to be extracted from the proper table, introduce no appreciable error.

When we have at last reached a certain result for all the apparent phenomena in the motion of these bodies, it will be possible for these and other points to be minutely calculated, either by myself or by some other astronomer, so that no pharisaical critic of the labours of other men may have any ground for complaint. Thus, starting from the assumed hypothesis, we have explained, and reduced to tables, the movements in longitude of these four bodies; and let this suffice for the candid lover of astronomy.

Now, then, remains the composition, and method of using, of a table of latitude for these Jovian wanderers. It has been said in the Second Part (phenomenon VI) that these satellites, at the extreme point of elongation from Jupiter on either side of him, move in a line which passes through the centre of Jupiter, parallel to the Ecliptic or path of the Sun. But when they are stationary on a line drawn from the Sun through Jupiter, they are at their extreme latitude—in apogee towards the South, in perigee towards the North.

I have computed this table of latitude for every 15° of equable motion, beginning from the limit of greatest distance towards the West, in the following manner: As is an entire quadrant, or a whole sine, to the sine of maximum latitude, so is the sine of any given elongation from either extreme point of maximum distance from Jupiter to the sine of the latitude corresponding to the given distance or arc.

FOR EXAMPLE: I want to know the latitude of the Fourth at a distance of 45° from its western terminal point. I say: as the whole sine, 10,000,000, to the sine of 15″ which is 717, so is the sine of 45°, or 7,071,068, to the sine of the latitude required, 514, to which corresponds an arc of 11″, which is the latitude required, as may also be seen on the table of latitude.

I have now completed, by the help of God, all the essentials for calculating the latitude and longitude of these Jovian satellites. It remains that I should point out the method of using these tables, which is the first and foremost aim of this treatise, and which is no doubt expected with avidity by the studious calculator.

I will embody the whole method of the calculus in a few definite rules, in order that it may be made easier and may be the better grasped by the memory.

A Method of Computing from the Tables Following the Position of These Bodies Relatively to Jupiter, Both in Longitude and Also in Latitude

I

The time given, of every description, is to be taken as completed: this is done by subtracting one from the year, month, and day given, as is done in other cases also.

II

I reckon year and day from the midnight of the preceding First of January. This was the Roman method to which Reinhold also kept in his tables.

III

When the time has thus been adjusted to the tables, the epochs are taken with the years completed out of the table of epochs. To these are added the simple movements corresponding to the months, day, hours, and minutes given. Afterwards, let all be reduced to one total, just as Astronomers usually do. It would be puerile to repeat and explain all this in detail. Thus we get the equable movement of any satellite. Account must be taken of leap-year in the table of months.

IV

Let the distance of Jupiter from the Sun be found, by subtracting the position of Jupiter from that of the Sun.

V

With this distance, let the equation be found from the table of equations. It is to be observed, however, that if the figure for the distance of Jupiter from the Sun is not given exactly in the table, the proportional part must be taken. It is to be noted, also, by how many degrees a number differs from the next on the table mentioned.

VI

This equation is to be subtracted from the equable movement of the satellite in cases where the distance of Jupiter from the Sun is less than six signs. In the other half it is to be added; and so we get the co-equated movement of the satellite.

VII

With this movement co-equated, let the distance be extracted from the table of distances; this will be towards the east when the movement is less than six signs, towards the west in the remaining six signs, or the other semicircle; and we have the required distance of the satellite from Jupiter. The proportional part must not be neglected in extracting the distance.

These tables, and the epochs of mean movements have been computed for the meridian of Ansbach, which is 34° 45′ from the prime meridian, or 2′ to the West from the meridian of Nuremberg.

For Finding Latitude

To the simple movement not co-equated add three signs. The result is the distance of the planet from the terminal point of extreme remoteness to the west; with which let the required latitude be found from the table of latitude—this will be southern when the given distance is less than six signs, northern when it it greater.

Example

In the year 1613, the 1 April, at 8 P.M., all four Jovian satellites were at their maximum distance from Jupiter. The Fourth and the Second were at their extreme eastern, the Third and First at their extreme western distance.

The time when made complete and adjusted to the tables is year 1612, month March, day 0, hours 20. To this correspond the following equable movements:

	Sign	°	′
First	8	29	21
Second	3	2	31
Third	9	6	3
Fourth	2	13	18

Jupiter was in Virgo 18°, the Sun in Aries 22°. Thus the distance between them is 7 signs 4°. I disregard minutes for the present. To this corresponds the equation 5°

52′ plus (as is also indicated by the letter A in front of the table). Therefore the co-equated movements are as follows:

	Sign	°	′
First	9	5	13
Second	3	8	23
Third	9	11	55
Fourth	2	19	10

To these co-equated movements correspond the following distances, according to the table of distances, after making the correction for each:

	′	″	
First	2	59	West
Second	4	56	East
Third	7	49	West
Fourth	12	44	East

That is, all four are as nearly as possible at their maximum elongation.

Another Example

In the same year 1613, day 14 February, hour 7 P.M., the position of these bodies relatively to Jupiter was as follows: The Third, or brilliant satellite, was about 7 minutes to the east. The Second was near it, about 5 minutes off in the direction of Jupiter; a small one came more to the north, the Fourth. Towards the west the First was about 3 minutes off Jupiter.

The time when adjusted to the tables is year 1612, month January, day 13, hours 19. The simple movements are:

	Sign	°	′
First	8	23	42
Second	3	18	57
Third	4	2	29
Fourth	5	14	49

The distance between Jupiter and the Sun was 5 signs, 12°. The equation 3° 15′ minus; Therefore the co-equated movements are:

	Sign	°	′
First	8	20	27
Second	3	15	24
Third	3	29	14
Fourth	5	11	34

To which correspond the following distances:

	′	″	
First	2	57	West
Second	4	48	East
Third	7	0	East
Fourth	4	8	East

For Latitude

In the preceding example the Fourth was in the North, whilst the Second was almost at its extreme distance to the East, and therefore they have no latitude. I add, therefore, 3 signs to the simple movement of the Fourth; the result is 8 signs 14° 44′ of elongation from the extreme western point, to which corresponds a latitude of 14″ North.

Another Example for Latitude

In the same year 1613, day 20 January, at 6 o'clock of the morning, the Third was at its maximum western elongation; near it in the direction of Jupiter, at a distance of about 1 minute, was the Fourth, but to the South. At this time the mean movement of each was:

	Sign	°	′
Third	9	9	3
Fourth	11	5	24

The equation on account of the distance of Jupiter from the Sun was 7° 34′ minus. Thus the co-equated movements were:

	Sign	°	′
Third	9	1	29
Fourth	10	27	50

Therefore, the distance of the Third was 8′, of the Fourth 6′ 53″, both towards the West; the Second was towards the East, and the First, very near one another. Now, adding three signs to the mean movement of the Fourth, and discarding 12 signs, or a whole circle, we have for the elongation of the Fourth from its extreme western point 2 signs 5° 24′, to which corresponds a latitude of 13″ South.

In the next edition of this treatise, which shall, if necessary, be made more correct, I will place before the candid calculator numerous observations, especially some made near both quadratures, of Jupiter and of the Sun, and near the opposition of both bodies, by means of which it will be possible in future years for these tables of mine to be corrected and improved.

A Useful Appendix

Or

A Method of Ascertaining the Places of JUPITER and the Sun, and So the Distance Between Them, Without an Almanack

Seeing that the Belgian spy-glass is now in the hands of very many persons, so that those who are unable to have an Almanack can use my tables, I have thought it well to subjoin the following brief Appendix:

I

To Find the Place of the Sun

In my yearly Calendars the day on which the Sun enters any sign of the Zodiac is noted for the several months. If, therefore, you count from that day exclusive to the given day inclusive, and take one degree for a day, you will have the sign and degree of the Sun.

II

To Find the Place of Jupiter

First of all, you must know the sign of Jupiter, which will also be easily got from my Calendars, for times when Jupiter is in conjunction with the Moon, for then Jupiter and the Moon are in the same sign. But in order to get the degree of Jupiter in all circumstances, proceed thus: for the given day, or that next to it (for the position of Jupiter varies very little within 2 or 3 days), see whether the Moon bears any relation to Jupiter. If the hours of such relation be after noon, that is, when the hour-number is shown at the right of the symbol of Jupiter, then for every 2 hours take one degree, add these degrees to the degrees of the Moon marked for each day under the heading "Monds lauff," always keeping the sign of Jupiter. But if the hour-number is found to the left of the symbol of Jupiter, subtract that number from 12, and the remainder is the number of hours to be counted backwards from noon; for every 2 hours take one degree, as before, subtract the sum from the number of degrees set against noon of the given day; thus you have the degree in the sign in which Jupiter is.

Now subtract the sign and degree of Jupiter from the sign and degree of the Sun; the distance between Jupiter and the Sun is obtained.

First Example

The given day is the first of April. On the tenth day of the preceding month of March the entry of the Sun into Aries is marked. Counting from that day exclusive to the first of April inclusive, we have 22 days. Accordingly the Sun is on the given day in the 22nd degree of Aries.

Further, from the conjunction of Jupiter and the Moon, which takes place on the 24th of March, it is clear that Jupiter was then in Libra. But on the given day, the 1 April, at 5 P.M., was the trine[10] of Jupiter and the Moon; to these hours correspond two degrees and a half, which, added to the 15 degrees marked for noon of the same day, give18·9 degrees. Thus Jupiter was then at the 18th degree of Libra. Subtracting, now, the sign and degree of Jupiter from the sign and degree of the Sun, the remainder, 7 signs 4°, is the required distance between Jupiter and the Sun, as is also shown above from the Calendar.

Now follow the tables themselves, computed for the Meridian of Ansbach, because it is distant from Nuremberg 2 minutes of longitude[11] to the West (Figs. 1.6, 1.7, 1.8, 1.9, 1.10, 1.11, 1.12, 1.13, 1.14, 1.15, 1.16, and 1.17).

To the Candid Reader,

Here you have, candid reader, the improved tables of Jupiter's planets, that I promised I would make if I found in them something that needed correction. As you know, even the smallest error in the lengths of some periods are noticed in the beginning but with time appear and become manifest. This has also happened with these tables, and even now I do not promise their complete perfection because some things still appear to be missing in the theory of the Jovian planets that can only be detected by repeated observations over long periods by reason of the great distance between Jupiter and the Earth. Because of this the entire width of the Jovian world is compressed that particular differences easily escape detection. For that reason, the Ingolstadt Jesuit Scheiner, from whom I had expected otherwise, has done me a very great and undeserved injustice when in his recent *Disquisitiones Mathematicae* he unjustly railed against me. For on 4/14 July the most learned Petrus Saxo Holsatus, a devotee of the mathematical subjects, visited me on his way straight from Ingolstadt, from the said Scheiner, told me among other things that Scheiner was publishing a book (which had been printed the previous year) about some novelties in astronomy, and he gave me to expect shortly to receive a letter from Scheiner in which he would engage me amicably about similar astronomical matters. This prospect

[10]P has *trinus*.

[11]Here Prickard wrote latitude by mistake.

EPOCHÆ QUATUOR PLANETARUM JOVIALIUM IN ANNIS COMPLETIS.

	Primi			Secundi			Tertij			Quarti		
	sig.	gr.	m.	sig.	gr.	m.	sig.	gr.	m.	sig.	gr.	m.
1608	10	20	35	7	22	20	1	26	13	7	3	13
1609	1	17	40	4	3	11	1	8	40	4	15	0
1610	4	14	45	0	14	2	0	19	37	1	26	47
1611	7	11	50	8	24	53	0	0	34	11	8	34
1612	5	2	20	8	17	1	1	1	45	9	11	50
1613	7	29	25	4	27	52	0	12	42	6	23	37
1614	10	26	30	1	8	43	11	23	38	4	5	24
1615	1	23	35	9	19	34	11	4	35	1	17	11
1616	11	14	5	9	11	42	0	5	47	11	20	27
1617	2	11	10	5	22	33	11	16	44	9	2	14
1618	5	8	15	2	3	24	10	27	41	6	14	1
1619	8	5	20	10	14	15	10	8	38	3	25	48
1620	5	25	50	10	6	23	11	9	50	1	29	4
1621	8	22	55	6	17	14	10	20	47	11	10	51
1622	11	20	0	2	28	5	10	1	44	8	22	38
1623	2	17	5	11	8	56	9	12	41	6	4	25
1624	0	7	35	11	1	4	10	13	53	4	7	41
1625	3	4	40	7	11	55	9	24	50	1	19	28
1626	6	1	45	3	22	46	9	5	47	11	1	47
1627	8	28	50	0	3	37	8	16	44	8	13	2
1628	6	19	20	11	25	45	9	17	56	6	16	18
1629	9	16	25	8	6	36	8	28	53	3	28	5
1630	0	13	30	4	17	27	8	9	50	1	9	52

Fig. 1.6 Times of the upper conjuctions of the four Moons of Jupiter in full years in *Mundus Iovialis*, Nuremberg 1614, sig. F4^r; Municipal Archive Gunzenhausen. Identical in 2nd edition

certainly pleased me. And while I was waiting for this letter, behold, the treatise that Holsatus had mentioned was sent to me from Nuremberg. When I eagerly read it, instead of heavenly friendship, I found calumnies, disparagements, and many other dishonorable things falsely belched out against me, none of which I deserve from him, so that I was totally surprised. And what is most base is that he also brings up my religion in an astronomical matter and in an insulting

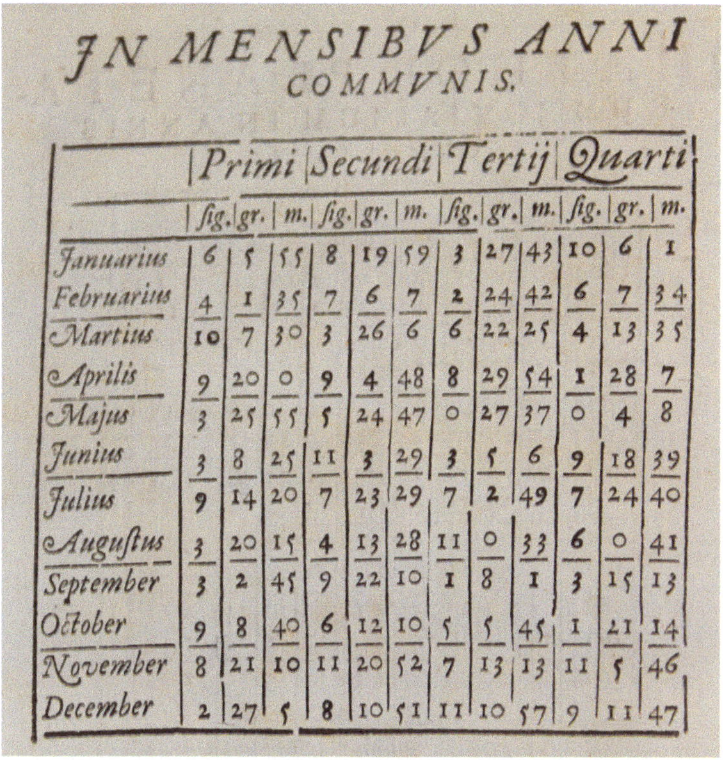

IN MENSIBVS ANNI COMMVNIS.

	Primi			Secundi			Tertij			Quarti		
	sig.	gr.	m.	sig.	gr.	m.	sig.	gr.	m.	sig.	gr.	m.
Januarius	6	5	55	8	19	59	3	27	43	10	6	1
Februarius	4	1	35	7	6	7	2	24	42	6	7	34
Martius	10	7	30	3	26	6	6	22	25	4	13	35
Aprilis	9	20	0	9	4	48	8	29	54	1	28	7
Majus	3	25	55	5	24	47	0	27	37	0	4	8
Junius	3	8	25	11	3	29	3	5	6	9	18	39
Julius	9	14	20	7	23	29	7	2	49	7	24	40
Augustus	3	20	15	4	13	28	11	0	33	6	0	41
September	3	2	45	9	22	10	1	8	1	3	15	13
October	9	8	40	6	12	10	5	5	45	1	21	14
November	8	21	10	11	20	52	7	13	13	11	5	46
December	2	27	5	8	10	51	11	10	57	9	11	47

Fig. 1.7 Times of the upper conjuctions of the four Moons of Jupiter in normal years in *Mundus Iovialis*, Nuremberg 1614, sig. F4v; Municipal Archive Gunzenhausen. Identical in 2nd edition

passage calls me a Calvinist, which I never was.[12] This is how contempt, jealousy, arrogance and religious hatred have led him astray. I will not answer such frivolous calumnies, lest I stir up this hornet's nest, for which I have not caused any occasion, even further. Let my *Mundus Iovialis*, which is not at all affected by these calumnies, be the answer. Indeed, its main points remain unchallenged, and they will never be undermined or destroyed by Scheiner if he does not produce more solid arguments. At any rate, I add and solemnly affirm that except for *Sidereus Nuncius* I had heard nothing about Galileo nor read anything by him, and I have not been able to obtain Apelles's book up to now. I don't know how this happened, for I diligently inquired about it in Nuremberg. The first discoverers of sunspots are the two Fabriciuses, father and son, but because they are considered heretics their names are suppressed.

[12]*Disquisitiones Mathematicae*, p. 78: the companions of Jupiter were first detected a few years ago by the outstanding, brilliant Italian mathematician Galileo (for in vain did some Calvinist nearly persuade us of the contrary, very unseasonably for the first time this year) [...].

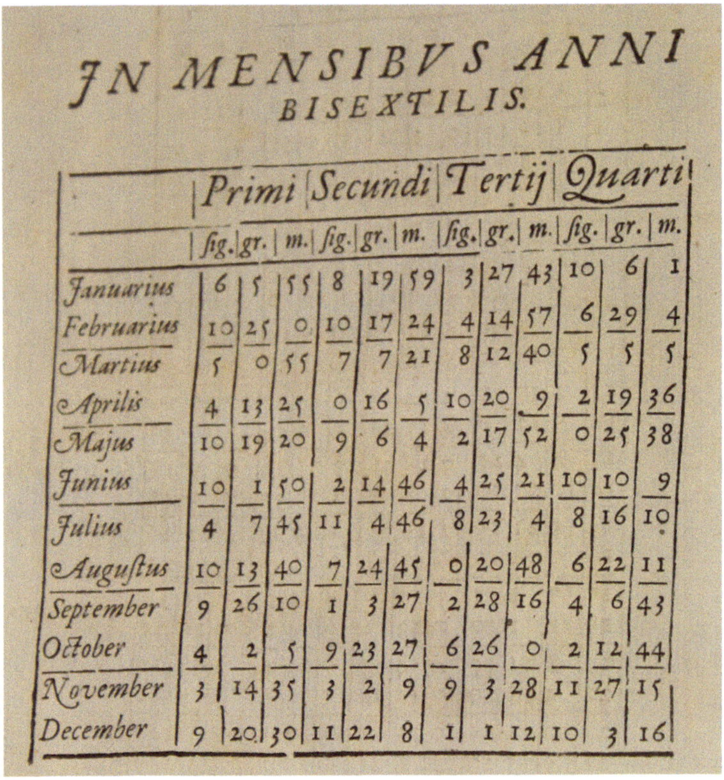

JN MENSIBVS ANNI BISEXTILIS.

	Primi			Secundi			Tertij			Quarti		
	sig.	gr.	m.	sig.	gr.	m.	sig.	gr.	m.	sig.	gr.	m.
Januarius	6	5	55	8	19	59	3	27	43	10	6	1
Februarius	10	25	0	10	17	24	4	14	57	6	29	4
Martius	5	0	55	7	7	21	8	12	40	5	5	5
Aprilis	4	13	25	0	16	5	10	20	9	2	19	36
Majus	10	19	20	9	6	4	2	17	52	0	25	38
Junius	10	1	50	2	14	46	4	25	21	10	10	9
Julius	4	7	45	11	4	46	8	23	4	8	16	10
Augustus	10	13	40	7	24	45	0	20	48	6	22	11
September	9	26	10	1	3	27	2	28	16	4	6	43
October	4	2	5	9	23	27	6	26	0	2	12	44
November	3	14	35	3	2	9	9	3	28	11	27	15
December	9	20	30	11	22	8	1	1	12	10	3	16

Fig. 1.8 Epochs in months of a year, which contain a intercalary day in *Mundus Iovialis*, Nuremberg 1614, sig. G1ʳ; Municipal Archive Gunzenhausen

And what Scheiner claims as new, among other things that the Sun rises and sets in an oval figure, he should know that this has been known to me since 1596, and that this does not happen in this way. For the middle of the upper limb is nearly spherical while the lower part is compressed, and this is because the middle of the [lower] limb is more affected by refraction. Further, he writes that the third star of Jupiter is 10 semi-diameters distant from Jupiter's center and the fourth 20, and for the following reason: when the third and fourth are at their maximum elongations on the same side [of Jupiter], then the third is exactly in the middle between the fourth and the center of Jupiter. This is so wrong that one should be ashamed [even to bother] to refute it, and that it is not a mental slip is shown by [Scheiner's] adjoined diagram. All observations, from 1609 to the present disagree with this.

What he says about the observation method for establishing the period of the one closest to Jupiter is childish: it is correct in theory but in no way practical. The method of observing the colors of stars I discovered in 1611, just as on 3/13 October of that year I found a way of observing sunspots on the sun itself

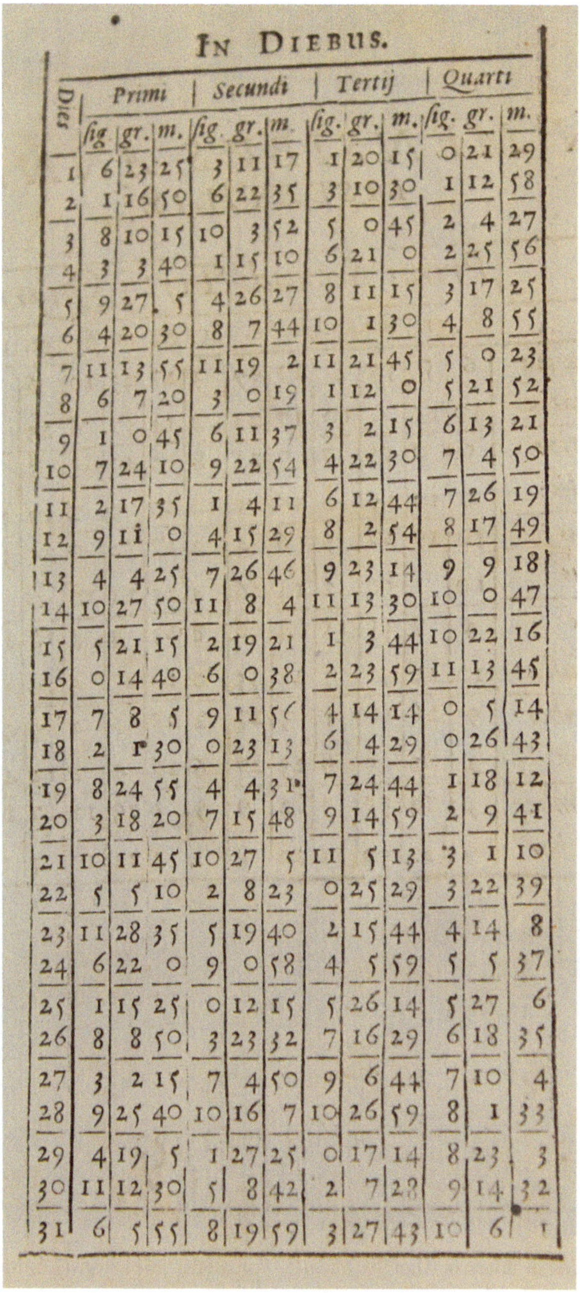

IN DIEBUS.

Dies	Primi			Secundi			Tertij			Quarti		
	sig.	*gr.*	*m.*	*sig.*	*gr.*	*m.*	*sig.*	*gr.*	*m.*	*sig.*	*gr.*	*m.*
1	6	23	25	3	11	17	1	20	15	0	21	29
2	1	16	50	6	22	35	3	10	30	1	12	58
3	8	10	15	10	3	52	5	0	45	2	4	27
4	3	3	40	1	15	10	6	21	0	2	25	56
5	9	27	5	4	26	27	8	11	15	3	17	25
6	4	20	30	8	7	44	10	1	30	4	8	55
7	11	13	55	11	19	2	11	21	45	5	0	23
8	6	7	20	3	0	19	1	12	0	5	21	52
9	1	0	45	6	11	37	3	2	15	6	13	21
10	7	24	10	9	22	54	4	22	30	7	4	50
11	2	17	35	1	4	11	6	12	44	7	26	19
12	9	11	0	4	15	29	8	2	54	8	17	49
13	4	4	25	7	26	46	9	23	14	9	9	18
14	10	27	50	11	8	4	11	13	30	10	0	47
15	5	21	15	2	19	21	1	3	44	10	22	16
16	0	14	40	6	0	38	2	23	59	11	13	45
17	7	8	5	9	11	56	4	14	14	0	5	14
18	2	1	30	0	23	13	6	4	29	0	26	43
19	8	24	55	4	4	31	7	24	44	1	18	12
20	3	18	20	7	15	48	9	14	59	2	9	41
21	10	11	45	10	27	5	11	5	13	3	1	10
22	5	5	10	2	8	23	0	25	29	3	22	39
23	11	28	35	5	19	40	2	15	44	4	14	8
24	6	22	0	9	0	58	4	5	59	5	5	37
25	1	15	25	0	12	15	5	26	14	5	27	6
26	8	8	50	3	23	32	7	16	29	6	18	35
27	3	2	15	7	4	50	9	6	44	7	10	4
28	9	25	40	10	16	7	10	26	59	8	1	33
29	4	19	5	1	27	25	0	17	14	8	23	3
30	11	12	30	5	8	42	2	7	28	9	14	32
31	6	5	55	8	19	59	3	27	43	10	6	1

Fig. 1.9 Epochs in days in *Mundus Iovialis*, Nuremberg 1614, sig. G1^{v}; Municipal Archive Gunzenhausen

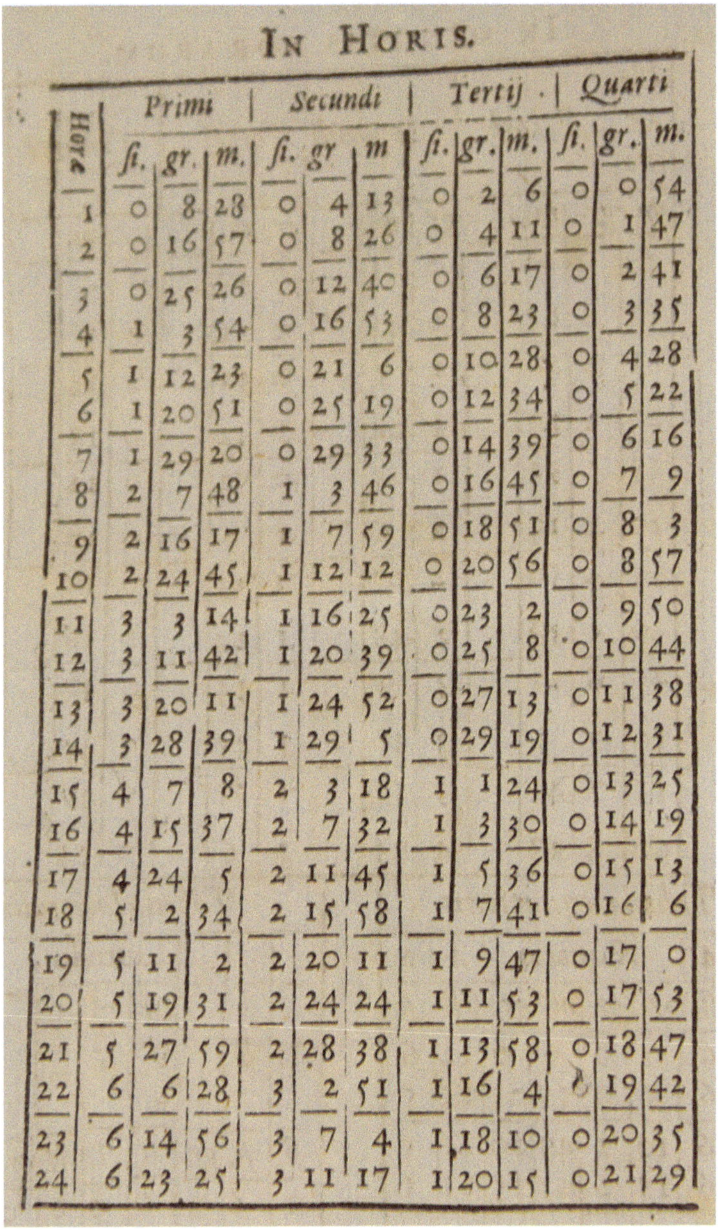

IN HORIS.

Horæ	Primi			Secundi			Tertij			Quarti		
	si.	gr.	m.	si.	gr	m	si.	gr.	m.	si.	gr.	m.
1	0	8	28	0	4	13	0	2	6	0	0	54
2	0	16	57	0	8	26	0	4	11	0	1	47
3	0	25	26	0	12	40	0	6	17	0	2	41
4	1	3	54	0	16	53	0	8	23	0	3	35
5	1	12	23	0	21	6	0	10	28	0	4	28
6	1	20	51	0	25	19	0	12	34	0	5	22
7	1	29	20	0	29	33	0	14	39	0	6	16
8	2	7	48	1	3	46	0	16	45	0	7	9
9	2	16	17	1	7	59	0	18	51	0	8	3
10	2	24	45	1	12	12	0	20	56	0	8	57
11	3	3	14	1	16	25	0	23	2	0	9	50
12	3	11	42	1	20	39	0	25	8	0	10	44
13	3	20	11	1	24	52	0	27	13	0	11	38
14	3	28	39	1	29	5	0	29	19	0	12	31
15	4	7	8	2	3	18	1	1	24	0	13	25
16	4	15	37	2	7	32	1	3	30	0	14	19
17	4	24	5	2	11	45	1	5	36	0	15	13
18	5	2	34	2	15	58	1	7	41	0	16	6
19	5	11	2	2	20	11	1	9	47	0	17	0
20	5	19	31	2	24	24	1	11	53	0	17	53
21	5	27	59	2	28	38	1	13	58	0	18	47
22	6	6	28	3	2	51	1	16	4	0	19	42
23	6	14	56	3	7	4	1	18	10	0	20	35
24	6	23	25	3	11	17	1	20	15	0	21	29

Fig. 1.10 Epochs in hours in *Mundus Iovialis*, Nuremberg 1614, sig. G2[r]; Municipal Archive Gunzenhausen

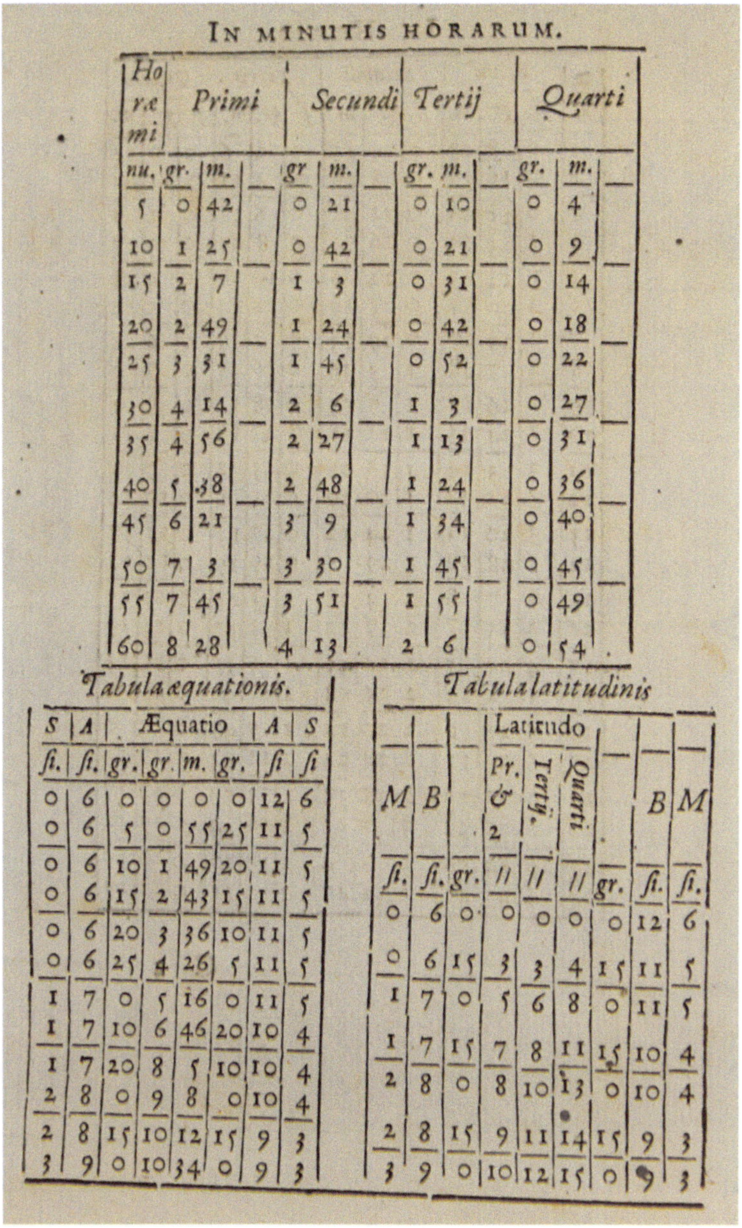

IN MINUTIS HORARUM.

Horae mi	*Primi*		*Secundi*		*Tertij*		*Quarti*	
nu.	*gr.*	*m.*	*gr*	*m.*	*gr.*	*m.*	*gr.*	*m.*
5	0	42	0	21	0	10	0	4
10	1	25	0	42	0	21	0	9
15	2	7	1	3	0	31	0	14
20	2	49	1	24	0	42	0	18
25	3	31	1	45	0	52	0	22
30	4	14	2	6	1	3	0	27
35	4	56	2	27	1	13	0	31
40	5	38	2	48	1	24	0	36
45	6	21	3	9	1	34	0	40
50	7	3	3	30	1	45	0	45
55	7	45	3	51	1	55	0	49
60	8	28	4	13	2	6	0	54

Tabula æquationis.

S	*A*	Æquatio				*A*	*S*
si.	*si.*	*gr.*	*gr.*	*m.*	*gr.*	*si*	*si*
0	6	0	0	0	0	12	6
0	6	5	0	55	25	11	5
0	6	10	1	49	20	11	5
0	6	15	2	43	15	11	5
0	6	20	3	36	10	11	5
0	6	25	4	26	5	11	5
1	7	0	5	16	0	11	5
1	7	10	6	46	20	10	4
1	7	20	8	5	10	10	4
2	8	0	9	8	0	10	4
2	8	15	10	12	15	9	3
3	9	0	10	34	0	9	3

Tabula latitudinis

M	*B*		Latitudo *Pr. & 2*	*Tertij.*	*Quarti*		*B*	*M*
si.	*si.*	*gr.*	//	//	//	*gr.*	*si.*	*si.*
0	6	0	0	0	0	0	12	6
0	6	15	3	3	4	15	11	5
1	7	0	5	6	8	0	11	5
1	7	15	7	8	11	15	10	4
2	8	0	8	10	13	0	10	4
2	8	15	9	11	14	15	9	3
3	9	0	10	12	15	0	9	3

Fig. 1.11 Epochs in time minutes, Table of the compensation and Table of the Latitude in *Mundus Iovialis*, Nuremberg 1614, sig. G2^{v}; Municipal Archive Gunzenhausen

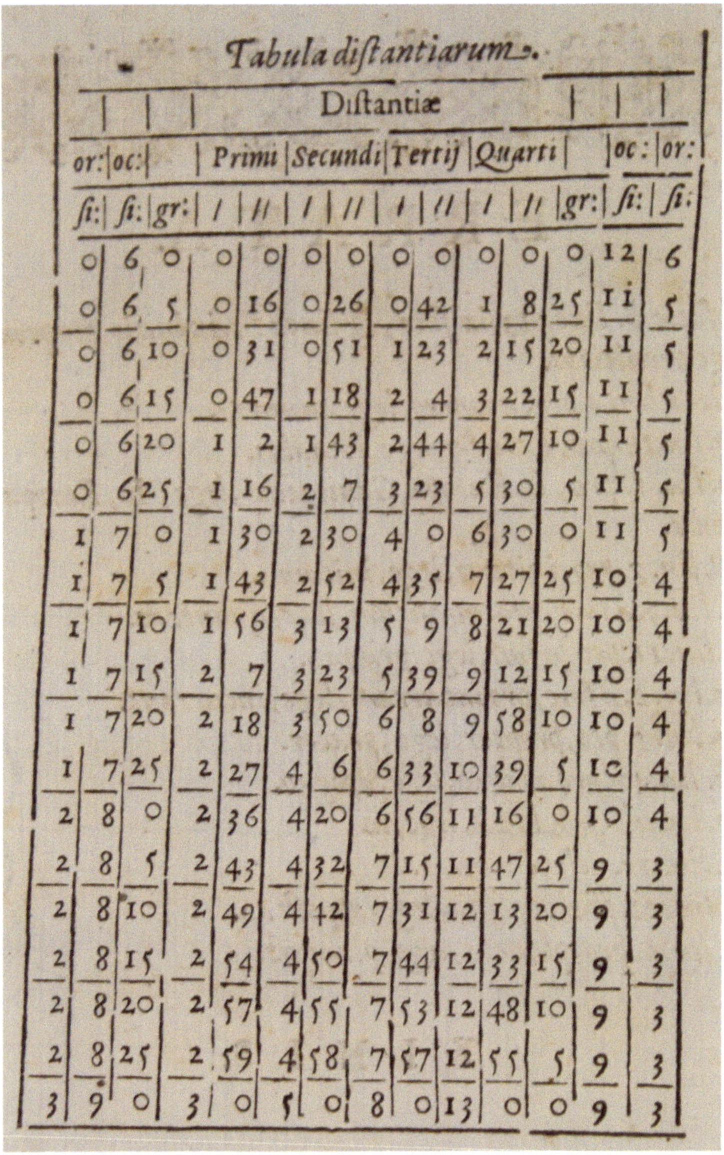

Tabula diſtantiarum.

			Diſtantiæ									
or:	oc:		Primi		Secundi		Tertij		Quarti		oc:	or:
ſi:	ſi:	gr:	'	''	'	''	'	''	'	''	gr:	ſi:	ſi:
0	6	0	0	0	0	0	0	0	0	0	0	12	6
0	6	5	0	16	0	26	0	42	1	8	25	11	5
0	6	10	0	31	0	51	1	23	2	15	20	11	5
0	6	15	0	47	1	18	2	4	3	22	15	11	5
0	6	20	1	2	1	43	2	44	4	27	10	11	5
0	6	25	1	16	2	7	3	23	5	30	5	11	5
1	7	0	1	30	2	30	4	0	6	30	0	11	5
1	7	5	1	43	2	52	4	35	7	27	25	10	4
1	7	10	1	56	3	13	5	9	8	21	20	10	4
1	7	15	2	7	3	23	5	39	9	12	15	10	4
1	7	20	2	18	3	50	6	8	9	58	10	10	4
1	7	25	2	27	4	6	6	33	10	39	5	10	4
2	8	0	2	36	4	20	6	56	11	16	0	10	4
2	8	5	2	43	4	32	7	15	11	47	25	9	3
2	8	10	2	49	4	42	7	31	12	13	20	9	3
2	8	15	2	54	4	50	7	44	12	33	15	9	3
2	8	20	2	57	4	55	7	53	12	48	10	9	3
2	8	25	2	59	4	58	7	57	12	55	5	9	3
3	9	0	3	0	5	0	8	0	13	0	0	9	3

Fig. 1.12 Table of the Distances in *Mundus Iovialis*, Nuremberg 1614, sig. G3[r]; Municipal Archive Gunzenhausen. Identical in 2nd edition

Epochæ & tabulæ mediorum motuum correctiores Planetarum Iovialium.

Annor.	*Primi.*			*Secundi.*			*Tertij.*			*Quarti.*		
Comp.	*sig*	*gr.*	*m.*	*sig.*	*gr.*	*m.*	*sig*	*gr.*	*m.*	*sig*	*gr.*	*m.*
1608	10	23	37	7	23	58	1	27	13	7	3	13
1609	1	18	16	4	4	0	1	7	9	4	17	4
1610	4	12	55	0	14	2	0	17	5	2	0	55
1611	7	7	34	8	24	4	11	27	1	11	14	46
1612	4	25	38	8	15	24	0	27	12	9	20	6
1613	7	20	17	4	25	26	0	7	8	7	3	57
1614	10	14	56	1	5	28	11	17	4	4	17	48
1615	1	9	35	9	15	30	10	27	0	2	1	39
1616	10	27	39	9	6	50	11	27	11	0	6	59
1617	1	22	18	5	16	52	11	7	7	9	20	50
1618	4	16	57	1	26	54	10	17	3	7	4	41
1619	7	11	36	10	6	56	9	26	59	4	18	32
1620	4	29	40	9	28	16	10	27	10	2	23	52

In Mensibus Anni Communis.

	Primi.			*Secundi.*			*Tertij.*			*Quarti.*		
	sig	*gr.*	*m*	*sig.*	*gr.*	*m.*	*sig*	*gr.*	*m.*	*sig*	*gr.*	*m.*
Ianuarius.	6	5	43	8	19	55	3	27	38	10	6	11
Februarius,	4	1	11	7	5	59	2	24	82	6	7	55
Martius.	10	6	54	3	25	54	6	22	11	4	14	6
Aprilis.	9	19	12	9	4	32	8	29	34	1	28	48
Majus.	3	24	55	5	24	27	0	27	12	0	4	59
Iunius.	3	7	13	11	3	5	3	4	36	9	19	41
Iulius.	9	12	55	7	23	0	7	2	14	7	25	53
Augustus.	3	18	38	4	12	56	10	29	52	6	2	4
September.	3	0	56	9	21	34	1	7	16	3	16	46
October.	9	6	38	6	11	29	5	4	54	1	22	58
November.	8	18	56	11	0	7	7	12	18	11	7	40
December.	2	24	39	8	10	2	11	9	56	9	13	51

G

Fig. 1.13 Epochs of the Moon from the second edition of the *Mundus Iovialis*, Nuremberg [2]1614, sig. G1[r]; Bibliothek des Evangelischen Predigerseminars Wittenberg: Diss 52/6

In Mensibus Anni Bisextilis.

	Primi.			Secundi.			Tertij.			Quarti.		
	sig	gr.	m.	sig.	gr.	m.	sig	gr.	m.	sig	gr.	m.
Ianuarius.	6	5	43	8	19	55	3	27	38	10	6	11
Februarius,	10	24	36	10	17	16	4	14	47	6	29	24
Martius.	5	0	19	7	7	11	8	12	25	5	5	35
Aprilis.	4	12	37	0	15	49	10	19	49	2	10	17
Majus.	10	18	19	9	5	44	2	17	27	0	26	29
Iunius.	10	0	37	2	14	22	4	24	51	10	11	11
Iulius.	4	6	20	11	4	18	8	22	29	8	17	22
Augustus.	10	12	2	7	24	13	0	20	7	6	23	34
September.	9	24	20	1	2	51	2	27	31	4	8	36
October.	4	0	3	9	22	46	6	25	9	2	14	27
November.	3	12	21	3	1	24	9	2	32	11	29	9
December.	9	18	4	11	21	20	1	0	11	10	5	20

In minutis Horarum.

Horæ.	Primi.		Secundi.		Tertij.		Quarti.	
m.	gr.	m.	gr.	m.	gr.	m.	gr.	m
5	0	42	0	21	0	10	0	4
10	1	25	0	42	0	21	0	9
15	2	7	1	3	0	31	0	14
20	2	49	1	24	0	42	0	18
25	3	31	1	45	0	52	0	22
30	4	14	2	6	1	3	0	27
35	4	56	2	27	1	13	0	31
40	5	38	2	48	1	24	0	36
45	6	21	3	9	1	34	0	40
50	7	3	3	30	1	45	0	45
55	7	45	3	51	1	55	0	49
60	8	28	4	13	2	6	0	54

Fig. 1.14 Epochs of the Moon in Leap Years from the second edition of the *Mundus Iovialis*, Nuremberg ²1614, sig. G1^{v}; Bibliothek des Evangelischen Predigerseminars Wittenberg: Diss 52/6

IN DIEBUS.

Dies	*Primi*			*Secundi*			*Tertij.*			*Quarti.*		
	si:	*gr.*	*m.*	*si:*	*gr.*	*m.*	*si:*	*gr.*	*m.*	*si.*	*gr*	*m.*
1	6	23	25	3	11	17	1	20	15	0	21	29
2	1	16	49	6	22	34	3	10	30	1	12	59
3	8	10	14	10	3	52	5	0	44	2	4	28
4	3	3	38	1	15	9	6	20	59	2	25	58
5	9	27	3	4	26	26	8	11	14	3	17	27
6	4	20	28	8	7	44	10	1	29	4	8	56
7	11	13	52	11	19	1	11	21	43	5	0	26
8	6	7	17	3	0	18	1	11	58	5	21	55
9	1	0	41	6	11	35	3	2	13	6	13	25
10	7	24	6	9	22	53	4	22	28	7	4	54
11	2	17	31	1	4	10	6	12	43	7	26	23
12	9	10	55	4	15	27	8	2	57	8	17	53
13	4	4	20	7	26	44	9	23	12	9	9	22
14	10	27	44	11	8	2	11	13	27	10	0	52
15	5	21	9	2	19	19	1	3	42	10	22	21
16	0	14	34	6	0	36	2	23	57	11	13	50
17	7	7	58	9	11	53	4	1[illegible]	11	0	5	20
18	2	1	23	0	2	11	6	4	27	0	26	49
19	8	24	47	4	4	28	7	24	41	1	18	19
20	3	18	12	7	15	45	9	14	56	2	9	48
21	10	11	37	10	27	3	11	5	10	3	1	17
22	5	5	1	2	8	20	0	25	25	3	22	47
23	11	28	26	5	19	37	2	15	40	4	14	16
24	6	21	50	9	0	54	4	5	55	5	5	46
25	1	15	15	0	12	12	5	26	10	5	27	15
26	8	8	40	3	23	29	7	16	24	6	18	44
27	3	2	4	7	4	46	9	6	39	7	10	14
28	9	25	29	10	16	3	10	26	54	8	1	43
29	4	18	53	1	27	21	0	17	9	8	23	13
30	11	12	18	5	8	38	2	7	24	9	14	42
31	6	5	43	8	19	55	3	27	38	10	6	11

G 2

Fig. 1.15 Epochs of the Moon in Leap Years in Days from the second edition of the *Mundus Iovialis*, Nuremberg [2]1614, sig. G2[r]; Bibliothek des Evangelischen Predigerseminars Wittenberg: Diss 52/6

In Horis.

Horæ	Primi			Secundi			Tertij			Quarti		
	si.	gr.	m.	si.	gr.	m.	si.	gr.	m.	si.	gr.	m.
1	0	8	28	0	4	13	0	2	6	0	0	54
2	0	16	57	0	8	26	0	4	11	0	1	47
3	0	25	26	0	12	40	0	6	17	0	2	41
4	1	3	54	0	16	53	0	8	23	0	3	35
5	1	12	23	0	21	6	0	10	28	0	4	28
6	1	20	51	0	25	19	0	12	34	0	5	22
7	1	29	20	0	29	33	0	14	39	0	6	16
8	2	7	48	1	3	46	0	16	45	0	7	9
9	2	16	17	1	7	59	0	18	51	0	8	3
10	2	24	45	1	12	12	0	20	56	0	8	57
11	2	3	14	1	16	25	0	23	2	0	9	50
12	2	11	42	1	20	39	0	25	8	0	10	44
13	3	20	11	1	24	52	0	27	13	0	11	38
14	3	28	39	1	29	5	0	29	19	0	12	31
15	3	7	8	2	3	18	1	1	24	0	13	25
16	3	15	37	2	7	32	1	3	30	0	14	19
17	4	24	5	2	11	45	1	5	36	0	15	13
18	4	2	34	2	15	58	1	7	41	0	16	6
19	4	11	2	2	20	11	1	9	47	0	17	0
20	5	19	31	2	24	24	1	11	53	0	17	53
21	5	27	59	2	28	38	1	13	58	0	18	47
22	6	6	28	3	2	51	1	16	4	0	19	42
23	6	14	56	3	7	4	1	18	10	0	20	35
24	6	23	25	3	11	17	1	20	15	0	21	29

Fig. 1.16 Epochs of the Moon in Leap Years in Hours from the second edition of the *Mundus Iovialis*, Nuremberg [2]1614, sig. G2[v]; Bibliothek des Evangelischen Predigerseminars Wittenberg: Diss 52/6

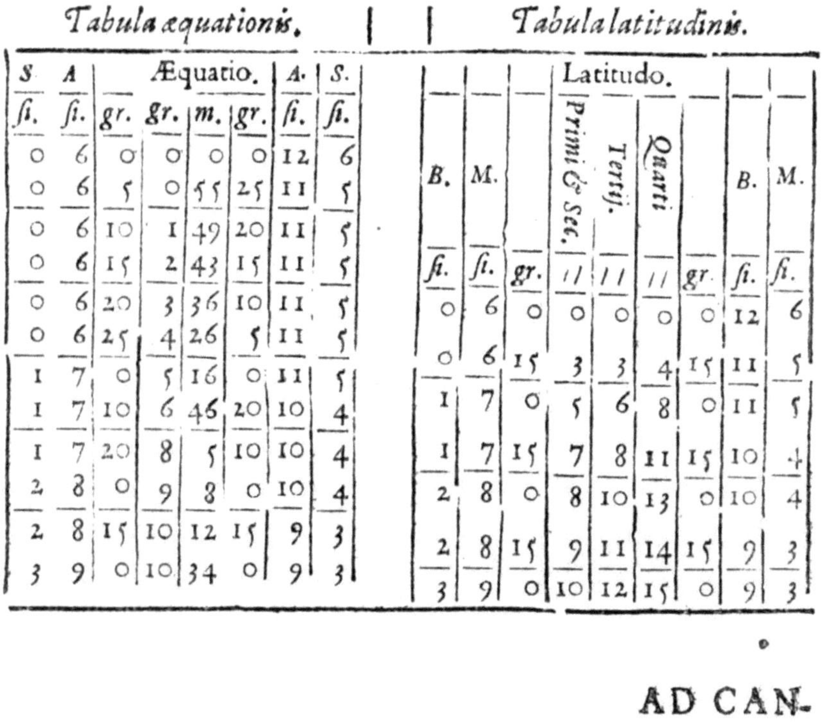

Tabula æquationis.

S	A	Æquatio.				A.	S.
ſi.	ſi.	gr.	gr.	m.	gr.	ſi.	ſi.
0	6	0	0	0	0	12	6
0	6	5	0	55	25	11	5
0	6	10	1	49	20	11	5
0	6	15	2	43	15	11	5
0	6	20	3	36	10	11	5
0	6	25	4	26	5	11	5
1	7	0	5	16	0	11	5
1	7	10	6	46	20	10	4
1	7	20	8	5	10	10	4
2	8	0	9	8	0	10	4
2	8	15	10	12	15	9	3
3	9	0	10	34	0	9	3

Tabula latitudinis.

			Latitudo.					
B.	M.		Primi & Sec.	Tertij.	Quarti		B.	M.
ſi.	ſi.	gr.	′′	′′	′′	gr.	ſi.	ſi.
0	6	0	0	0	0	0	12	6
0	6	15	3	3	4	15	11	5
1	7	0	5	6	8	0	11	5
1	7	15	7	8	11	15	10	4
2	8	0	8	10	13	0	10	4
2	8	15	9	11	14	15	9	3
3	9	0	10	12	15	0	9	3

AD CAN-

Fig. 1.17 Table of the Compensations and the Latitudes from the second edition of the *Mundus Iovialis*, Nuremberg ^{2}1614, sig. G3^{v}; Bibliothek des Evangelischen Predigerseminars Wittenberg: Diss 52/6. Similar in 1st edition., sig. G2^{v}

through a tube without any damage to the eyes, and further at that time that sunspots do not move along the Sun's ecliptic but make an angle with it, as shown in the figure made on 17/27 November 1611, which I showed to the said Holsteiner, who saw it with admiration and said that this secret had been entrusted to him by Scheiner. Where is the monstrous star in Andromeda? Where are the daily observations of Jupiter, Mars, and the heart of the Lion, and many other things that were first discovered and observed by me? If the same have been discovered by others, I praise them, I just do not want to be accused of having stolen them from others, which it completely alien to me. The more diligent astronomers now earnestly vie with each other to observe these new astronomical phenomena.

But you, candid lover of my labors will judge them sincerely since happily they are honest and true [*utpote ingenui germani feliciter* fuere].[13] Farewell.[14]

[13]Note the play on words in *germani*.

[14]English translation from Mundus Iovialis Anno M.DC.IX. by Albert Van Helden.

Chapter 2
Concerning the Biography of Simon Marius (1573–1624)

Hans Gaab

In this chapter, we will consider the local situations in which Simon Marius lived his life, focusing especially on his education. His published work is discussed in other chapters and therefore is only taken into account as far as it seems appropriate to illuminate his intellectual and social surroundings. His astronomical worldview is also discussed elsewhere and even his dispute with Galileo moves into the background for this chapter.

Youth in Gunzenhausen

Simon Marius was born on January 10th 1573, shortly before midnight[1]:

> On this day anno 1573 shortly after 12 o'clock in the night/I was born to endure much pain and suffering/in Gunzenhausen at the Altmühl/at the latitude 49′ 6″, longitude 35′ 0″.

[1]"Eben an diesem Tag Anno 1573. halbweg 12 Uhr nach Mittag in der Nacht/bin ich auf die Welt zu viel Creutz und Leyden geboren worden/zu Guntzenhausen an der Altmühl/dessen latitudo ist 49. grad 6. minuten, longitudo 35. grad 0 minuten" (*Prog. 1609*, sig. B2^r; cf. Klug 1906, p. 395 footnote 1; Zinner 1942, p. 55). According to Vocke 2 (1797/2001, p. 414), Marius's birthdate couldn't be verified, because the Gunzenhausen sixteenth-century baptismal registers were burned during the Thirty Years' War. However the records were found again later; Klug (1906, p. 396) and Clauß (1922, p. 18f.) quote them; an illustration can be found in Mühlhäußer 2012, p. 40. In old lexica entries, the birth year 1570 is mentioned. Marius's portrait, printed in the *Mundus Iovialis*, is captioned with "Anno M.DC.XIV AETATIS XLII." Georg Christoph Oertel was in 1775 the first to note that the year 1570 cannot be the birth year, so he changed it to 1572. But in the portrait's caption, it says 1614 in his 42nd year; Marius therefore was 41 years old at that time, corresponding with the given date.

H. Gaab (✉)
Simon Marius Society, Fürth, Germany

H. Gaab, P. Leich (eds.), *Simon Marius and His Research*, Historical & Cultural Astronomy, https://doi.org/10.1007/978-3-319-92621-6_2

The following day, he was baptized in the town church Mariä Virginis by dean Jodokus Braun (1522–1584)[2]:

Father Reichart Mayr, child Simon, godfather Simon Keiser, all from Guntzenhausen.

His godfather Simon Keiser belonged to a large family of butchers and innkeepers. Werner Mühlhäußer, town archivist of Gunzenhausen, has researched the Marius family history. According to him, the grandfather Michael Mayr was a cooper, who owned a house in the town's center around 1532 and had been a council member since 1532. He held important positions, for example, he served as mayor four times between 1541 and 1550.[3] His father Reichart (Reinhard, ca. 1529–1599) was a councilman ("Raths-Verwandter") and cooper (Hocker 1739, p. 43; Zinner 1942, p. 25). For a long time he was a member of the council and held several related positions. In 1585 he held the town's highest position as a mayor (Mühlhäußer 2012, p. 36f.).

On March 13th 1553, the father became a citizen of the town of Gunzenhausen. In the same year he married Veronica Fischer from the neighboring town Cronheim,[4] the widow of Sebastian Fischer from Gunzenhausen. On November 25th their daughter Barbara had already been baptized, shortly after her mother died. This daughter probably died very early, because in 1562 another daughter from a second marriage was also named Barbara (Clauß 1922, p. 18; Mühlhäußer 2012, p. 36).

In 1556 the father married his second wife, Elisabeth (Elisabetha, ca. 1534–1599), the daughter of an innkeeper from Sammenheim, south of Gunzenhausen. The father died at the age of 70 and was buried on December 12th 1599, the mother at the age of 65 the following day. Presumably they had succumbed to an epidemic (Mühlhäußer 2012, p. 36, 42).

Simon was the seventh and last child from this second marriage.[5] His three sisters Elisabetha, Barbara, and Margaretha all stayed in Gunzenhausen, where they married into prestigious bourgeois and councilors's families (Mühlhäußer 2012, p. 36f.).[6] To this day descendants of Barbara are living in the town.[7]

---

[2]"Vater Reichart Mayr, Kindt Simon, Gevatter Simon Keiser, alle zu Guntzenhausen" (Mühlhäußer 2012, p. 39f.; cf. Clauß 1922, p. 18f.; Klug 1906, p. 396). Braun studied in Leipzig and Wittenberg and was the first pastor in Gunzenhausen since 1567 (Simon 1957, p. 48 entry 293).

[3]Mühlhäußer 2012, p. 36. Dannheimer (1952 p. 95 entry 322) says his father was a principal in Gunzenhausen; Simon calls him a baker in the parish records of Ansbach (Simon 1957, p. 303 entry 1831).

[4]Cronheim is part of Gunzenhausen since 1971.

[5]According to Clauß 1922, p. 18, the baptismal register of Gunzenhausen lists the following siblings "Barbara, baptized 25. Nov. 1553, Elisabeth 17. April 1555, Michael 11. Juni 1560, Barbara 17. Januar 1562, Jakob 14. Juni 1565, Leonhard 13. Juni 1567, Margareta 5. Mai 1570." Mühlhäußer (2012, p. 36) corrected Elisabetha's birth year to 1557.

[6]Elisabetha married Leonhard Kistner from Gunzenhausen on February 13th 1582 (for Barbara see the following footnote); Margaretha married the locksmith Zeislein from Gunzenhausen on January 7th 1605. These statements can be found in the family tree records in possesion of Wolfgang Marius from Graz.

[7]Barbara Mayr married the baker Georg Kretzer on April 24th 1599. Their son Leonhard (1604–1669), who himself was a baker, married Margarete Stozt (1614–1675) on October 26th 1635. Their daughter Eva Barbara (1642–1681) married the master stove fitter Johann Kaspar

The oldest brother Michael (1560–1624) was "schoolmaster of Creglingen on the Tauber, two miles [northeast] of Rotenburg" (Zinner 1942, p. 45).[8] On May 8th 1581, he matriculated in Wittenberg.[9] In 1585 he was sent to teach in Creglingen, where he died after 39 years of service on June 2nd 1624.[10] On October 19th/20th 1596, Simon Marius resided "in Cräglingen on the Tauber," probably he was visiting his brother (*Prog. 1612*, sig. B5^r). In 1613 he thought about his "many children, and his body's weakness" (Zinner 1942, p. 44).[11] He recommended his brother's oldest son Johannes (1586–?) in a letter dated December 6, 1609, to Michael Mästlin (1550–1631) in Tübingen. Johannes didn't pay his debts in Tübingen, which accounts for most of the correspondence between Marius and Mästlin (Zinner 1942, pp. 40–45).[12]

The middle brother Jakob (1565–1607) matriculated in Wittenberg on November 2nd, 1586, where he, in contrast to his older and younger brothers, was exempted from paying a matriculation fee.[13] He became a pastor of the villages

Mußolt (1644–1717) on February 28th 1671. The son Georg Leonhard (1671–1727) married Maria Kohler (1679–1736) on February 3rd 1705. Their daughter Eva Maria (1710–1749) finally married master tailor Johann Christoph Elterlein (1702–1773) on April 25th 1729. They're the ancestors of the Elterlein, who live in Gunzenhausen to this day. I owe this information to Wilhelm Elterlein.

[8]Marius in a letter to Mästlin June 15th 1614. According to the family tree records in the possesion of Wolfgang Marius from Graz, Michael Marius married a Barbara Dürr in Gunzenhausen on June 16th 1685. But according to the records in the municipal archive of Creglingen, he mas married to a midwife called Helene (approx. 1575–1633).

[9]"Micaelus Meyer Guntzenhusanus" (Foerstemann, Hartwig, Naetebus 1894, p. 298a). He matriculated together with Johann Sebastian Brendel from Gunzenhausen (reads the entry in the register, but he hailed from Pfofeld east of Gunzenhausen). He became principal in Gunzenhausen in 1588 and as of 1611 pastor in Alerheim east of Nördlingen, where he died 1634 (Simon 1957, p. 51 entry 314).

[10]In that time the town was infested with the plague, with 400 people succumbing to it, among them "two chaplains and two school master." Michael Marius received a special payment "for mercy." Details given by Claudia Heuwinkel from the municipal archive of Creglingen (e-mail from February 13th 2015).

[11]Marius in a letter to Mästlin from August 1st 1613. According to Claudia Heuwinkel from the municipal archive of Creglingen, Michael Marius suffered from a disease in 1601 "and lost half of his voice because of it," rendering him barely able to continue to carry out his profession. Fritz Mägerlein (1989, p. 449) verified the following 13 children in the parish registers of Creglingen: (1) Johann July 24th 1586 (baptism); (2) Dorothea August 25th 1587; (3) Jakob July 25th 1588; (4) Dorothea August 29th 1589, she died unmarried on September 14th 1637; (5) Christoph April 26, 1591; (6) Helene February 14th 1594; (7) the twins Michael and Georg January 9th 1596, Michael died on January 31st 1596, Georg on May 15th 1600; (8) Veit Dietrich July 9th 1598; (9) Barbara February 6th 1601; (10) Georg December 13th 1602, he died after long and serious illness on February 24th/25th 1609; (11) Ulrich December 7th 1604; (12) Konrad January 23rd 1608.

[12]The son can't be found in the register of Tübingen.

[13]"Nomina gratis inscriptorum [...] Jacobus Marius Francus Guntzenhusanus" (Foerstemann, Hartwig, Naetebus 1894, p. 346a; cf. Clauß 1922, p. 18 f.; Zinner 1942, p. 25). He matriculated together with "Johannes Joachimus Cardus Guntzenhusanus," about whom nothing more is known (Foerstemann, Hartwig, Naetebus 1894, p. 343b).

Gräfensteinberg and Pfofeld, east of Gunzenhausen (Mühlhäußer 2012, p. 37; Simon 1957, p. 303 entry 1831).[14] Jakob appears to have been very close to Marius, because he wrote about him in 1614[15]:

> I say nothing now of my dear brother Jakob, of beloved memory, who was even well informed about my astronomical work.

Jakob recorded the weather for Simon during his stay in Italy.[16] He died in 1607 from the plague, as did his son and two daughters (Mühlhäußer 2012, p. 46; Simon 1957, p. 303 entry 1831).

The youngest brother Leonhard (1567–1613) resided at the "Fürstenschule" of Heilsbronn 1587/1588, where he "relocated at his own cost."[17] On May 26th, 1590, he matriculated in Wittenberg[18]; 4 years later he started his new job as principal in Solnhofen. In 1601 he became chaplain in Feuchtwangen, 1608 pastor in Reubach south of Rothenburg, where he died 5 years later. He had seven children.[19] In contrast to their youngest brother, the older sons all studied in Wittenberg.

At first his father sent Simon to the "townschool" (Vocke 2 (1797), p. 414). In his *Prognosticon for 1618*, he commemorated (Prog. 1618, sig. A4^{r}; vgl. Zinner 1942, p. 66)[20]:

> [...] the dignified and well-taught Mister Vogtherr, vicar of Meynheim at the Altmühl, [who] was my preceptor during my youth in Gunzenhausen, whom I want to remember in honor as a good occultist and aficionado of astronomy

This Georg Vogtherr (1556–1623) was descended from a dynasty of pastors, who were heavily engaged in ophthalmology; he himself was a well-known ophthalmologist. He studied in Wittenberg and was employed as schoolmaster in

[14]Jakob married Regina Brendel (?–1604) on October 16th 1592 in Gunzenhausen, daughter of the pastor Sebastian Brendel (1528–1600) from Pfofeld, whose successor was Jakob Marius (Simon 1957, p. 51 entry 315).

[15]"Taceo nunc Charissimum Fratrem meum Jacobum, pie memoriae, cui etiam optime de meo studio Astronomico constabat" (Marius 1614, sig. C3^{v}).

[16]"In July 1605 after the 14th/24th great heat followed by heavy thunder and hail everywhere, as my blessed brother noted." ("Anno 1605 im Julio nach dem 14./24. grosse Hitz mit folgentem hefftigen Donner an etlichen orten Hagel, wie mein Bruger Jacob seliger verzeichnet," *Prog. 1628*, sig. B3^{r}; cf. Klug 1906, p. 398 footnote 2; also cf. *Prog. 1612*, sig. B4^{v}, B5^{r}; 1615, sig. A4^{r}; 1626, sig. A4^{v}, B1^{v}, C2^{r}; 1627, sig. B2^{r}, C3^{r}).

[17]Hocker 1739, p. 43. Dannheimer (1959, p. 173) lists him at the Fürstenschule for the years 1587/88.

[18]"Leonhardus Maior Guntzenhusanus" (Foerstemann, Hartwig, Naetebus 1894, p. 374b).

[19]Mühlhäußer 2012, p. 37; Simon 1957, p. 306f. entry 1852. He married Ursula Krafft from Zimmer north of Solnhofen on January 9th 1598 (Landeskirchliches Archiv, Nuremberg: Pfarrerbuch Gunzenhausen LAELKB 315_17, sig. 30).

[20]Without giving a name, he wrote in the *Prognosticon for 1602* (sig. A3^{r}): "[...] ex astrorum motibus (to which my nature has a certain affinity since youth)."

Fig. 2.1 Portrait of Georg Vogtherr (1556–1623). From Vogtherr 1908

Pfarrer Georg Vogtherr in Meinheim.

Gunzenhausen, before he was assigned to be the pastor in Meinheim near Altmühl, a few kilometers south of Gunzenhausen (Simon 1957, p. 525 entry 3129; Vogtherr 1908, pp. 52–55) (Fig. 2.1). Marius used his weather observations in his calenders from 1617 on.[21]

[21]In his *Prognosticon* of 1625, Marius reports about a planetary constellation on December 20th 1595, which "brought early fog according to my and Mister Vogthers register [. . .] But Mr. Vogther alone, says, that he saw on the day before at Meinheim on the Altmül between 12 midday and 1 o'clock three beautiful suns while in Heilsbron it was cloudy, although it is situated only 4 miles north to Meinheim" ("Nach meiner vnd Herrn Vogthers verzeichnuß [gab es] früe Nebel [. . .] Allein Herr Vogther setzet/daß er den 19 zuvor zu Meinheim an der Altmül/von 12. vhr zu Mittag biß umb 1. vhr drey Sonnen gar schön gesehen hab zu Heilßpronn war es trüb/vnnd ist doch nur 4. Meil versus boream davon" (*Prog. 1625*, sig. A4^{v})). Similarly the Prognosticon of the following year says that Marius himself noted for the February 17th 1596 in Heilsbronn a nice and warm day. "Mr. Vogther has recorded, that this day was very bright, although we were situated only 4 miles apart" ("H. Vogtherus hat verzeichnet/es sey solchen tag zimblich hell bey ihme gewesen/da wir doch nur 4. Meilwegs voneinander gewesen" (*Prog. 1626*, sig. B2^{v})). Cf. auch *Prog. 1618*, sig. A4^{r}, printed in Zinner 1942, p. 66; 1626, sig. A4^{v}, B2^{r}, B3^{v}, D3^{r}; 1629, sig. A3^{v}, B4^{r}; 1627 sig. B1^{v}, B2^{r}, C4^{v}. The weather observations by Vogtherrs were later transmitted by his son Andreas (1584–1650), pastor in Eyb near Ansbach, to the lecturer for mathematics in Altdorf, Abdias Trew (1597–1669), who favorably regarded astrological meteorology as did Marius. Even the astrologer Andreas Goldmayer (1602–1665) from Gunzenhausen is supposed to have used these observations (Cf. Matthäus 1969, col. 1098 footnote 804; Trew 1643, sig. P3^{v}).

Admission to the "Fürstenschule" of Heilsbronn

Legend has it that Margrave Georg Friedrich[22] (1539–1603), who had a hunting lodge in Gunzenhausen (J. Meyer 1892, p. 54), heard the then 13-year-old Simon singing, which fascinated him so much that he admitted the young Marius to the "Fürstenschule" of Heilsbronn and "employed him at his princely chapel shortly thereafter" (Vocke 2 (1797), p. 415; cf. Oertel 1775, p. VIII). He didn't come back to Heilsbronn (Fig. 2.2) until 1589, where he "was received in place of his brother Leonhard [...]" (Hocker 1739, p. 43).

This story, frequently appearing in older reference works, has been challenged in recent times. According to Mühlhäußer, it cannot be verified and "should probably be expelled to the realm of legends" (Mühlhäußer 2012, p. 39). Although Marius left several personal anecdotes in his writings, he never talked about his singing skills. Not until his *Prognosticon* for 1622 did he write[23]:

> I'm a medicus, mathematicus, musicus and in my heart a foolish theologus.

He signed the dedication of his comet treatise on September 11th 1596; it was addressed to the margrave[24]:

> [. . .] for I have richly received for ten years food and drink/and other necessities/from E.F. Duchl. [your princely highness]/and have been taught with diligence in the good and useful arts/in your well appointed Fürstenschule of Heilsbronn

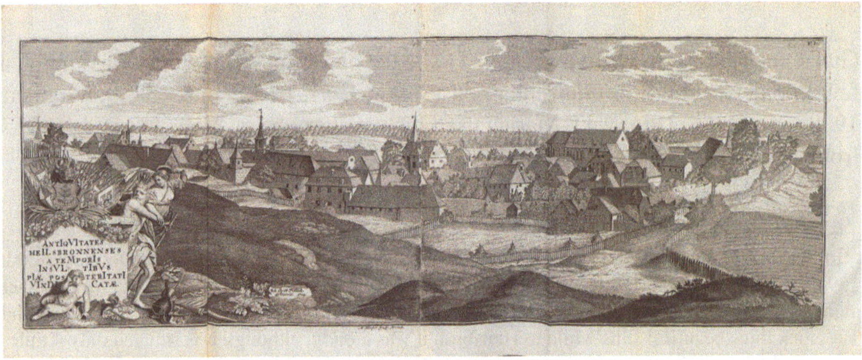

Fig. 2.2 Overall view of Heilsbronn from Hocker's antique treasure 1731. Courtesy of ETH Zürich: Rar 658 q

---

[22]For the Margrave Georg Friedrich, see also Schuhmann 1980, p. 101–124.

[23]"Ich bin ein Medicus, Mathematicus, Musicus vnd in meinem Hertzen ein einfältiger Theologus," *Prog. 1622*, sig. A3^r; cf. Matthäus 1969, col. 1098. Elsewhere Marius defends the philosophers, about whom is often said: "He is a philosophus, meaning/he is a fantast/or stockfish" ("Er ist ein Philosophus, das ist/er ist ein Fantast/oder Stockfisch," *Prog. 1608*, sig. A3^r).

[24]"[. . .] weil ich nun zehen Jar lang mit Speiß vnd Tranck/vnd andern notwendigen sachen/von E. F. Durchl. bin reichlich erhalten/vnnd inn guten und nützlichen Künsten/in derer wolbestellten Fürstenschul Hailsbronn/mit fleiß vnterricht worden" (Marius 1596, Widmung, sig. A2^v).

No word about having resided in Ansbach for 3 years. In *Mundus Iovialis* he confirms that[25]:

> [. . .] from my fourteenth year to the present time, I have been most liberally supported at the charges of those Illustrious Princes, Margraves of Brandenburg, George Frederic, of honoured memory, and, after his lamented death, of the brothers the Lords Christian and Joachim Ernest [. . .]

Marius turned 13 years old, starting the 14th year of his life, on January 10th 1586, which means he would have come to Heilsbronn in 1586. According to Dannheimer Marius resided there from 1586 until 1599 (Dannheimer 1959, p. 173 entry 764). He evaluated the supper registers; the year mentioned arise from the first and the last entries he could find about Marius. Marius can be first found on Trinitatis 1586, meaning the Sunday after Pentecost, which fell on May 22nd according to the Julian calendar. In the following year, one can find his brother Leonhard.[26] Additional entries from the years 1587 and 1588 couldn't be found for either of them; the books for 1589 to 1593 are missing (Dannheimer 1959, p. 155). Therefore a stay in Ansbach is possible for Marius, but a definite proof has not been found.

It was only 85 years after his death that we can find a reference connecting Marius with music. According to the *Allgemeines historisches Lexicon* (1709) by Johann Franz Buddeus (1667–1729), Simon's father urged his son to learn about the sciences[27]:

> . . . because he understood music very well. And by this he was favored by the margrave of that time/Georg Friedrich/so that he admitted him to the newly established Fürstenschule in Heilsbron in 1582. He was not allowed to stay there for long but was sent to the court chapel in Ansbach/where he stayed for four years.

Here the dates got mixed up. The "Fürstenschule" was ceremoniously opened on the margrave's birthday in 1582; Marius only arrived "4 full years" later at Heilsbronn. The references given by Buddeus only deal with the discovery of the Jupiter moons and include no biographical information.

Buddeus is the source for a short entry in the *Compendiöses Gelehrten-Lexikon* (1715) by Johann Burckhardt Mencke (1674–1732) which does not include any

[25]"[. . .] ab anno 14. aetatis meae usque in praesens tempus, sumptibus Illustrissimorum Principum, Marchionum Brandenburgensium, GEORG FRIDERICI, laudatissimae memoriae, & hoc piè defuncto, DOMINI CHRISTIANI & JOACHIMI ERNESTI fratrum, &c. liberalissimè sum enutritus" (Marius 1614, sig. B1^{r}). English Marius 1614/1916/2019, of the names to be assigned to these four Jovian planets.

[26]Landeskirchliches Archiv Nürnberg: 1. Kirchenbuch Heilsbronn 1499–1706, sig. 147^{v}, 152^{v}.

[27]"[. . .] da er dann absonderlich die music sehr wohl begriffe. Und eben dadurch brachte er sich des damaligen Marggrafens/George Friedrichs/gnade zu wege/daß ihm dieser an 1582 in die neue aufgerichtete Fürsten=Schule zu Heilsbronn aufnehmen liesse. Doch durffte er nicht lange daselbst bleiben/sondern wurde nach Anspach in die Fürstliche hof-capelle gethan/darinn er sich 4 gantzer jahr gebrauchen lassen" (Buddeus 3/4 (1709), p. 460). The entry in Iselin (1729, 1747, p. 392) is identical to Buddeus'.

dates regarding Marius's time in Heilsbronn.[28] The paragraph in the *Allgemeine Gelehrten-Lexikon* (1751) by Christian Gottlieb Jöcher (1694–1758) is identical to Mencke's. The entry in the 19th volume of the *Universal-Lexicon* (1739) by Heinrich Zedler (1706–1751) is almost identical in wording (Jöcher 3 (1751), col. 188f.; Zedler 19 (1739), col. 1588). Johann Friedrich Weidler (1691–1755) mentions in his *Historia astronomiae* of 1741 the discovery of the moons of Jupiter, but his biographical information is only a Latin translation of Buddeus's entry.[29] In 1739 professor Johann Ludwig Hocker[30] (1670–1746) at least adjusted a few dates in his Supplement to the *Heilsbronn antique treasure* (*Supplementa zu dem haylßbronnischen Antiquitäten-Schatz*)[31]:

> [Simon Marius] was admitted in 1586 into the local high school, but was soon comandeered and sent to the court chapel, but after three years in 1589 came back [to Heilsbronn] in exchange for his brother (who had moved on his own cost).

As mentioned, Dannheimer could prove that the brother Leonhard was in Heilsbronn in 1587 (Dannheimer 1959, p. 173 entry 761) and matriculated in Wittenberg in May 1590. As far as it goes Hocker's reference is trustworthy. However on the same page, he writes about Marius having studied in Königsberg, which is untenable. Characteristically, Hocker says virtually nothing about the margrave having heard Marius' singing.

The first detailed biography about Marius was published in 1775, 150 years after his death, by Georg Christoph Oertel[32] (1715–1790), principal of the high school of Neustadt an der Aisch. Oertel's work was discussed in the *Erlangische gelehrte Anmerkungen* (*Erlangen Erudite Notes*)[33]:

> The editor, because he endeavored to obtain reliable local sources, received from the grandson of the former meritorius Heilsbron professor Joh. Fridr. Krebs, namely the through his thorough knowledge of natural history renowned Herr Magister Esper, a still existing Latin manuscript and the Latin vita read at Marius's funeral oration.

---

[28]"Through music he received the margrave's mercy, who also gave him money, so he could learn about astronomy with the famous Tycho Brahe" (Mencke 1715, col. 1275).

[29]"Adolescens musicam amauit, in eaque sic excelluit, ut Georgii Friderici, Marchionis Anspacensis, gratiam mereretur, eiusque iussa a. 1582 inter alumnos Heilbrunnenses [= Heilsbrunnenses] reciperetur" (Weidler 1741, p. 430).

[30]For Hocker see Chr. Meyer 1993; Vocke 2 (1797/2001), pp. 95–97.

[31]"Dieser [Simon Marius] ist An. 1586. in hiesiges Gymnasium aufgenohmen, aber so balden wider abgefordert, und zur Fürstl. Capell gezogen, nach dreyen Jahren aber An. 1589. an seines Bruders Leonhardi Stell (der auf eigene Kosten hinaus gezogen) wider hier recipirt worden" (Hocker 1739, p. 43).

[32]For Oertel see Schlichtegroll 1791, pp. 374–378.

[33]"Es hat der Herr Verf. da er sich um sichere hieher gehörige Quellen bewarb, von dem Enkel des ehemaligen verdienten Professors zu Heilsbron Joh. Fridr. Krebs, nehmlich dem durch seine gründliche Känntnis in der NaturHistorie [...] berühmten Herrn Magister Esper, ein noch vorhandenes lateinisches Manuskript, und den ebenfalls geschriebenen und bey dem LeichBegängniß des Marius selbst abgelesenen Lebenslauf desselben erhalten." *Erlangische gelehrte Anmerkungen und Nachrichten* 30/XVI. Stück (18.04.1775, p. 122.)

Starting in 1675 Johann Friedrich Krebs[34] (1651–1721) had been principal of the high school in Heilsbronn, where he remained until his end of life. He had a funeral sermon for Marius in his possession inherited by his grandson, Johann Friedrich Esper (1732–1781),[35] which is no longer traceable. Esper was a pastor in Uttenreuth near Erlangen from 1763 until 1778, before he transferred to Wunsiedel. Today he is known as a speleologist. With the funeral sermon, Oertel had access to at least one excellent source. The fact that he questioned the then known birth date of 1570 for Marius shows his critical approach. The story about the margrave liking Marius's voice can be found here for the first time, quoted from a short summary in the *Erlangische Anmerkungen*[36]:

> He had a very pleasant voice and when the ruling Margrave Georg Friedrich heard him singing, he cared for him and sent him to the famous school in Heilsbronn [. . .]. Because of his beautiful voice he was taken from there to the court chapel, but after four years was sent back to Heilsbronn [...]

If Marius in 1589 actually did replace his brother Leonhard in Heilsbronn, then he would have spent a maximum of 3 years in Ansbach, not 4. Or did his brother stay until 1590, contrary to what had been stated by Hocker, which would explain why he only matriculated in Wittenberg in that year? And if the margrave was really fascinated by his voice, why did he send him to Heilsbronn to begin with and not to his court chapel?

Oertel is the first to tell the story that Marius had impressed the margrave with his singing talent. At the same time, he had access to the funeral sermon, from which he could have gathered this information. The problem is, however, that he also used older lexica entries.[37] He does not include any sources for single passages. Therefore

[34]For Krebs see Fikenscher 5 (1803), pp. 139–149; Vocke 1 (1796/2001), pp. 185–190. He had broad interests, including astronomy, as we know from his correspondence with the astronomer and observatory founder Georg Christoph Eimmart from Nuremberg (1638–1705) (Staatsbibliothek St. Petersburg: Fond No. 998: Estate of Eimmart. Vol. 1, sig. 365–383).

[35]For Esper see *NDB* 4 (1959), p. 655f. (author: Florian Heller).

[36]"Er hatte eine sehr angenehme Stimme, und als ihn der damals regierende Marggraf Georg Friedrich von ungefehr singen hörte, nahm er denselben in Gnaden, so daß er ihn in die damals berühmte Schule nach Heilsbronn schickte [. . .]. Seiner schönen Stimme wegen wurde er von da weg, und in die HofKapelle genommen, doch nach vier Jahren wieder nach Heilsbronn geschickt [. . .]" (*Erlangische gelehrte Anmerkungen und Nachrichten* 30/XVI. Stück (18.04.1775), p. 122f.). In Oertel (1775, p. VIIIf.), the section reads: "Voce praeditus fuit admodum canora et suaui. Haec adeo mouebat Serenissimum eo tempore principem, Dominum GEORGIUM FRIDERICVM, Marchionem Brandenburgicum [. . .] cum cantantem eum forte audiuisset, ut inter alumnos scholae illustris ad Salutis Fontes, die suo natali 5. Aprilis, A. 1582. erectae adoptari puerum clementissime iuberet. Datus igitur est Heilsbronnensi Gymnasio, ex quo praeter tam multos excellentes atque idoneos uiros etiam BOECLER, FRISCHMANNI, TAVBMANNI, alli, olim prodierunt, A. 1586, d. 24. Junii, quo ipso mense annum aetatis suae XVI. s. potius XIV. forte exegerat [. . .] At inter symphoniacos Principis suauis ille desideratur cantor, Heilsbronnam missus: reuocatur igitur in aulam, et quamquam non sponte sua, quatuor annos musicorum choro uocem suam commodat."

[37]He cites p. 4: "WEIDLERVS [. . .] BVDDEVM, ISELINVM, MENCKENIVM, JOECHERVM, WALCHIVM." With Walchium he refers to the *Philosophische Lexicon*, published by Johann Georg Walch (1693–1775) 1726, reissued 1775. However, the discovery of the moons of Jupiter is only mentioned very briefly (cf. Walch 1726, col. 1530). For the other authors, see the footnotes above.

it is possible that the funeral sermon contained information about Marius having a beautiful voice and liking to sing, along with the older lexica entries, leading to the establishment of this famous story. It is noted that according to Oertel Marius was admitted to the "Fürstenschule" on June 24th, 1586 (Oertel 1775, p. VII). But actually he can already be found in the supper register on May 22nd.

Since Marius himself never mentioned a stay in the Ansbach court chapel, the question mark behind this story remains. One thing seems certain, however; he resided at the "Fürstenschule" with short interruptions[38] from 1589 (at the latest) until 1601.[39]

Marius's fondness for the muses emerges in an *Anagrammatismus* by Johann Hartmann[40] (1577–1634), "Pastore Rinderfeldensi," which was printed by Marius in his calendar of 1614 under the horoscope on the back of the front cover. Here Hartmann rearranged the letters of the name *Simon Marius* to *Amor in Musis*[41]:

> Simon Marius.
> Amor in Musis.
>
> Est Amor in Musis; sunt Musae in amore vicißim:
> Nihil est amore suavius reciproco.
> If Amor is the Muses, the Muses are also in love.
> There is nothing more beautiful than mutual love.

Rinderfeld is a small village south of Creglingen. Through his brother Michael, Marius could have come into contact with Hartmann. But before 1608 the latter was a pastor in Rügland, a few kilometers north of Ansbach, so he probably met Marius during his initial period in Ansbach. Admittedly, the cited muses do not necessarily refer to the arts; they could also allude to Marius' scholarship.

Marius at the "Fürstenschule" of Heilsbronn

The "Fürstenschule" (Figs. 2.3 and 2.4) was established in 1581 by Margrave Georg Friedrich through a deed of foundation. It was supposed to train "religious and moral church, school and civil servants and through their efficacy improve the religious and moral life of the people" (Muck 1880, p. 13f.). The deed of foundation wished[42]:

---

[38]As cited above, he, for example, stayed in Creglingen on October 19th/20th 1596 (*Prog. 1612*, sig. B5^r).

[39]Considerations such as his voice changing in 1589, that is, his voice breaking at puberty, require no further attention (Christianson 2000, p. 320).

[40]For Johann Hartmann see Flood 1 (2006), pp. 774–776. He studied in Jena and was pastor in Rügland from 1601 to 1608 and pastor in Rinderfeld from 1608 to 1617, subsequently in Rothenburg o.d. Tauber.

[41]My attention was drawn to this poem by Klaus Matthäus, Joachim Schlör helped with the translation.

[42]"[...] daß fürnehmlich armer, sowohl auch der Kirchen- und Schul-Diener und um die Herrschafft wohl verdienter Leut Kinder, die entweder arm und unvermöglich, oder von Gott mit vielen Kindern begabt seyn, in diese Unsere Schule angenommen werden sollen." The foundation letter is printed in Junger 1971, pp. 44–49. The nutritional regulations can be found in Muck 3 (1880), p. 25.

Fig. 2.3 The "Fürstenschule" in Heilsbronn. Image taken by the author

[. . .] that above all the children of the poor and of church and school servants and those of rightful people, who are either poor and without resources or gifted with many children by God, should be admitted to the school.

The students should have been between 12 and 16 years old and had to pass four classes, whereby the duration of a class was 2 to 3 years. On average the school education took 10 years. After that period exams would be held, "that those who qualify for church or scholar service by the means of their knowledge and age, can receive their duties, and the others can move on to university according to their propensity" (Fuchs 1837, p. 44).

In the four classes, the students essentially had to learn Latin and Greek to accompany their theological instruction. Additionally in the first class, music was taught, in the third class for 1 hour per week studied "Quaestiones Sphaericas, Calendarium Ecclesiasticum or Computum, and also Arithmeticam," i.e., arithmetic, an introduction in spherical geometry as well as instructions in computus, calculating the moveable church feast days, especially the date of Easter.

The school education was free; in exchange, students were obligated "to be available for church or school service at the prince's pleasure, and [. . .] not to

Fig. 2.4 Inscription on the "Fürstenschule." Image taken by the author

assume other responsibilities without permission" (Junger 1971, p. 55).[43] It is therefore not only empty rhetoric when Marius writes in his dedication of the *Mundus Iovialis* to the margrave, "since whatever can proceed from me in this department is all yours, all produced and provided at your cost" (Marius 1614/1916/2019, Dedication). In 1631 the school was closed because of the Thirty Years' War but was solemnly reopened on January 30th 1655 until it was finally relocated to Ansbach or Bayreuth (Junger 1971, p. 62, 65, 121, 153, 202f.).[44]

[43]"In the Fürstenschule in Heilsbronn excellent Latin poetry, mathematics and music were practiced. Therefore it was said: Omnis Heilsbronnensis aut Musicus, aut Poeta, aut Magus (i.e. Mathematicus). The music was practiced because of the court chapel in Ansbach" (E. F. Ch. Oertel 1831, p. 27).

[44]A report about the closing of the Fürstenschule from a contemporary perspective can be found in Ch. Meyer 1893, pp. 510–512. Also here on p. 504, a list with all of the monastery's and high school's employees of the 1720s can be found.

Karl Heinrich von Lang[45] (1764–1835) gives account of the monastic lifestyle of the students (Lang 1811, p. 346):

> At 6 o'clock in the morning class started, at 7 o'clock choir and after that was breakfast. From 8–10 classes, at 10 o'clock midday meal with plenty and almost excessive food on wooden plates, namely midday and evening 3 dishes each and ½ a pound of meat per head calculated. At 12 o'clock the instruction in the classes began again until 2 o'clock, when there was once again choir and after this was supper without libation. At 3 o'clock were classes again, at 4 o'clock personal repetition; at 5 o'clock the evening meal was held. Wednesday and Saturday afternoon were free time. The pupils had to serve themselves, make their beds, clean their rooms and polish their shoes. They were only allowed to speak in Latin, the language of instruction. Not only the lectures were started with spiritual hymns but also at table religious lectures were held and it was expected that every student should pray and read the Bible in his cell for one hour in the morning, i.e., from 5 to 6 o'clock and another hour in the evening.

Marius seems to have had a good relationship to his "valued teachers." In particular he mentions "Markus Wenzeslaus Gunkfelder [!], Markus Georg Hirschbauer and Johannes Neser" (cf. the following listing). In *Mundus Iovialis* he addressed them[46]:

> I have felt it a duty and a pleasure to add their names in honour and gratitude, because they are already deceased and they have been of the utmost service to me, not only in this matter, but in very many others. (Marius 1614/1916/2019, Second Part, Particular Investigation of the World of Jupiter, Of the Fifth)

Marius's Teachers at the "Fürstenschule" of Heilsbronn

Principal

From 1584 until his death in 1588, *Johann Hertel* was principal of the "Fürstenschule." Before that he had held the same position at the Ansbach high school, starting in 1575. One of his praised merits was the translation of the *Epistles of John* into Hebrew (E. F. Ch. Oertel 1831, p. 34f.; Schreibmüller 1928, p. 35f.; Vocke 2 (1797), p. 367).

His successor was *Johann Codomann* (1548–1616), born 1548 in Schauenstein west of Hof, who had studied in Wittenberg. In 1577 he became principal in Kulmbach. In 1582 he transferred as deputy head to Heilsbronn. In 1602 he returned as Konsistorialrat and superintendent to Kulmbach (E. F. Ch. Oertel 1831, p. 22; Vocke 2 (1797), p. 41).

In 1577 Codomann had taken in the poor but gifted student Friedrich Taubmann (1565–1613) as a boarder at the Kulmbach high school, with whom he left for Heilsbronn in 1582. Taubmann became professor for poetry

(continued)

[45]For Lang see *NDB* 13 (1982), p. 542 f. (author: Bernhard Sicken).

[46]"[. . .] quorum nomina, quia satis jam concessere, honoris & gratitudinis ergo asscribere debui & volui, quia de me non saltem in hac parte, sed etiam in alijs quam plurimis, optimè meriti sunt" (Marius 1614, sig. C3^{v}).

in Wittenberg and devoted his long life to a revival of humanist studies. In his poems he praised his teachers, Hertel and Codomann. Dannheimer could prove Taubmann's presence in Heilsbronn from 1582 to 1588, but it is also said that he resided there until 1592. Marius never mentions him so presumably did not get to know him (*ADB* 37 (1894), pp. 433–440; Dannheimer 1959, p. 165 entry 300; Sommer 1842, p. 15).

Usually, the principal taught the highest of the four classes.

Deputy Head

Johann Codomann was the deputy head from 1582 until he became principal in 1588 (see above).

His successor as deputy was *Wenzeslaus Gurkfelder* (1562–1608), who already worked as the third teacher in Heilsbronn since 1582. In 1601 he was called to be consistorial secretary in Ansbach, where he died in 1608 (E. F. Ch. Oertel 1831, p. 35f.; Hocker: Antiquitäten-Schatz 1731, p. 194f.; Vocke 1 (1796), p. 130f.).

Marius observed the lunar eclipse on August 6/16, 1617 "in the presence of the honorable and highly regarded Friderici Gurkfelderi from Ansbach, Princely Secretary,"[47] perhaps a son of Wenzeslaus Gurkfelder. Friderici became personal secretary of the margrave in 1617 (cf. Vocke 1 (1796), p. 138f.; Zinner 1942, p. 67).

The deputy usually taught the third and fourth grades.

Teacher of the Second Grade

Balthasar Bernhold (1564–1648) was one of the first "Fürstenschüler." After his studies in Wittenberg, he became teacher of the second grade, despite his bad reputation. Shortly after his recruitment, an investigation was started against him, because of his "secretly being married," leading to his transfer to the Ansbach school. As of 1602 he worked as a pastor in Ansbach and enjoyed a good reputation. He delivered a short contribution to Marius's wedding tract (Groß 1 (1727), pp. 30–32; Muck 3 (1880), p. 27, 35; Schlund 1987, p. 30 enrty 4; Simon 1957, p. 33 entry 203; Vocke 1 (1796), p. 183).[48]

His classes were probably taken over by Markus Georg Hirschbauer (1555–1607), who has been at the Latin school of Ansbach since 1583/ 1584. In January 1605 he became the pastor of Blaufelden, south of Rothenburg, where he died 2 years later.[49]

(continued)

[47]"[...] in beysein dess Ehrenvesten vnnd wolgeachten Herrn Friderici Gurckfelderi, allhier zu Anspach, Fürstlichen Secretraij" (*Prog. 1620*, sig. C6^{v}).

[48]Bernhold married Margarethe Oettinger (?–1617) from Heilsbronn on November 10th 1589, probably eliminating the problems with his secret marriage (Groß 1 (1727), p. 31).

[49]Fuchs 1837, p. 47, gives the year 1583, Simon 1957, p. 197 entry 1202 the year 1584.

Teacher of the Cantor's Grade

The lowest grade was the cantor's, managed by the composer *Johann Neser* (before 1560–1602) from the foundation of the "Fürstenschule" until his death in 1602. He is said to have collected the "Hymnos/so zu Anfang und Schluß der Lectionum zu singen sind angeordnet worden/gesammelt" (*ADB* 23 (1886), p. 441f. (author: Robert Eitner); Fuchs 1837, p. 47f.; Stübner 1690, p. 54). His son of the same name[50] (1593–1659) published them in 1620 as *Hymni sacres*.[51]

Abbot of the Heilsbronn Monastery

The Abbot of Heilsbronn was responsible for the theological instruction of the higher grades. As of 1579 this position was held by *Conrad Limmer* (1522–1592) from Neustadt on the Oder. He had studied in Leipzig, Wittenberg, and Jena. In 1570 he became monastery preacher in Ansbach; in 1579 he came to Heilsbronn where he probably retired in the autumn of 1589 as a Melanchtonian (Hocker 1731, pp. 149–151; Simon 1957, p. 285 entry 1732).

Limmer was succeeded by *Adam Francisci* (1540–1593) in 1590. Francisci came from the Silesian Jägerndorf,[52] then in possession of the Margrave of Ansbach. During a visit, the margrave became aware of the poor but talented son of a wainwright and enabled him to study in Wittenberg. As such this could have been a possible parallel to Marius's life. Later, Francisci taught in Wittenberg and supervised students of the Latin school in Ansbach. In 1572 he was called to Ansbach as Georg Karl's (1512–1576) assistant.[53] Five years later he became Karl's successor as general superintendent. In the meantime, Karg was discharged because of heresy but was put back in charge after a revocation in 1570. As a "strict representative" of orthodox Lutheranism, Francisci was placed at his side. Because of his weak health, Francisci was transferred, as the successor of Limmer, to Heilsbronn, where he lived a more peaceful life as the abbot until he died 3 years later (Hocker 1731, pp. 151–153; Muck 3 (1880), pp. 7–9; Simon 1957, p. 128 entry 771; Vocke 2 (1797), p. 44f.; Vogtherr 1927, p. 36). In 1592 he published his *Margarita Theologica*,[54] which was reissued several times. This theological

(continued)

[50]For Neser see Simon 1957, p. 342 entry 2066.

[51]Neser, Johann: *Hymni Sacri In Usum Ludi Illustris Ad Fontes Salutares: Melodiis & Numeris Musicis compositi & collecti; Hymni sacri in usum ludi illustris ad fontes salutares*. Nuremberg: Matthäus Pfeilschmidt 1612 [SUB Göttingen: 8 CANT GEB 205]; reissue Nuremberg: Johann Friedrich Sartorius 1620 [BSB München: Liturg. 1372 d].

[52]Today Jägerndorf is the city Krnov in the region Okres Bruntál, Czech Republic.

[53]For Karg see Simon 1957, p. 230 entry 1402; Vocke 2 (1797/2001), p. 332f.

[54]Francisci, Adam: *Margarita Theologica continens methodicam explicationem praecipuorum capitum doctrinae Christiana pro Ecclesiis et Scholis orthodoxis Augustanae Confeßionis*. Hof: Pfeilschmidt 1592 [SB Ansbach: SB 110/II i 14].

pearl was a clear description of orthodox Lutheranism and was introduced into the schools, displacing Melanchthon's writings (Vogtherr 1927, p. 36).

His successor *Bartholomäus Welchendorff* (Wolschendorff, Wolschendorfer, 1540–1601) also stemmed from Neustadt on the Oder. He had studied in Jena and held several positions in the area surrounding Bayreuth. In 1563 he returned to Neustadt at the Oder but was discharged in 1570 for being a non-Flacian. At the beginning of the following year, he became the town chaplain of Ansbach, followed by his promotion to dean in Crailsheim in 1578. In 1594 he came to Heilsbronn, where he died in the middle of 1601 (Hocker 1731, p. 153; Simon 1957, p. 560 entry 3347).

The study of astronomy was not taught as an independent subject at the Heilsbronn school. That Marius being able to dedicate himself to it was a rare exception, for as he later remarked, he was "the one Heilsbron student out of so large a number who has been incited, doubtless by Heaven, to these sublime studies of Astronomy."[55]

For his autodidactic studies, Marius lacked, in particular, reference books. A catalog with the inventory of the Heilsbronn library, published by Ludwig Hocker in 1731, lists, with the exception of Marius's own translation of Euclid from the year 1610, no mathematical or astronomical work printed after 1600. Some of these books were possibly only purchased for Marius. But many items Marius referred to—for example, the main work of Nicolaus Copernicus (1473–1543)—cannot be found in Hocker's list. Between Marius' stay in Heilsbronn and the cataloging by Hocker, more than 200 years had passed; books could have been lost or deaccessioned. Most of the books used by Marius were probably provided by his teachers "whose other courses of lectures did not allow them to follow it out themselves, but who helped me greatly with books."[56]

In May 1598 Marius wrote to the margrave that he "with particular pleasure [. . .] in the fifth year [. . .] had a knowledgeable handling of the study of astronomy and astrology."[57] As of 1593 he seems to have been more occupied with this topic, because according to his booklet about the comet of 1596, "an eclipse of the sun came about on May 20th in the year 1593, a good deal earlier than predicted by the calculations of Stadius."[58] Instruments for observations were hardly available to

[55]"[. . .] me unicum ex tanto Alumnorum Heilsbronnensium numero, ad haec sublimia studia Mathematica, divinitus procul dubio excitatum" (Marius 1614, sig.)(4^{v}–)()(1^{r}).

[56]"Insuper Praeceptores meos charissimos testor, qui quod ob alias lectiones ispsis non licebat, me tamen libris plurimum in hoc studio juvarunt" (Marius 1614, sig. C3^{v}).

[57]"[. . .] was sonderlichem Lust [. . .] in das fünfte Jahr [. . .] mit dem Studio Astronomico und Astrologico versiret und umbgangen" (Büttner 2 (1813), p. 78; Hocker 1739, p. 43).

[58]"[. . .] eine Finsternuß der Sonnen [. . .]/anno 1593. den 20. Maii nach Mittag/ein guten theil langsamer als der Calculus Stadii gesetzt," Marius 1596, sig. A4^{v}. In the letter to Bergrat Vicke in Wolfenbüttel, Marius wrote regarding his printed *Tabulae Directionum* from 1599: "I practiced astronomy barely for two years without having any teacher for (astronomy-) mathematics" (Kepler XIV, 1954, p. 383; Translation by J. Schlör). Klug (1906, p. 403) saw a contradiction to the

him; he only mentions a brass quadrant in 1601, which he had used the year before.[59] At the latest since 1594, he performed meteorological observations, because in his calendar for 1601 (with the preface dated June 29th 1600), he described a weather rule by Johann von Glogau[60] (Jan Glogowczyk, Johannes Glogoviensis, 1445–1507)[61]:

> When the Sun runs in Aquarius and the Moon stands against her in the Lion/this means a change in the air/and damp rainy weather in many places.

He "found this to be true for the sixth year in a row." In the *Prognosticon for 1603*, he considered this rule to be confirmed (*Prog. 1603*, sig. B2^r), and even in the *Prognosticon for 1627*, he came back to it (*Prog. 1627*, sig. A4^r). But in principle "the doctrina *Meteorologica* was imperfect and questionable."[62]

According to his own statements, Marius read Copernicus for the first time in the winter of 1595/1596 (Marius 1614, sig. C3^r). This seemingly encouraged him to think about different world systems. In doing so he claimed to have discovered the Tychonic system. The first time he saw a sketch of this world system was in the following autumn together with town pastor Franziskus Raffael[63] (1533–1604), which had been sent it to him by a student from Wittenberg.[64] He had already shown "my hypotheses" (also Tychonic?) to the Ansbach consistory at Easter 1596[65]:

statement above. The *Tabulae* was printed in 1599, but Marius seemingly had worked on them at the latest since 1596 (see section 0 below). Viewed in this way leaves us with no contradiction.

59"[. . .] in the last year 1600, where I found it with a good quadrant made of brass" ("voriges 1600/ Jars/da ich durch einen gerechten messigen quadrantem befunden [...]," *Prog. 1601*, sig. A6^v, cf. Zinner 1942, p. 47).

60For Johann von Glogau, see *NDB* 10 (1974), p. 552 (author: Felix Schmeidler).

61"[. . .] wann die Sonn im wassermann laufft und der Mond ihr entgegen stehet im Löwen/so bedeut es enderung der Lufft/vnnd an vielen orten Feucht Regen wetter" (*Prog. 1601*, sig. A5^v). Marius refers to *Tractatus Preclarissimus in Judicijs Astro[rum] de mutatio[n]ibus aeeris*. Krakau: Florianum Wolgangum 1514 [UB Erlangen-Nürnberg: H61/INC 316].

62*Prog. 1609*, sig. A3^r. In his Prognosticon for 1616 (sig. E2^r), Marius refers to a quadrature of Saturn and Mars, "ist gewesen Anno 1558. den 26. Sept. da findt ich von H. Johan Fischern seligen auffgezeichnet." Johannes Vischer (1524–1587) had studied in Wittenberg, Tübingen, and Italy. In 1555 he was called to be the town's Physicus in Nördlingen: from 1562 to 1568, he was the personal physician of Margrave Albrecht von Brandenburg-Ansbach. Subsequently he was a professor for medicine in Tübingen until his death; Vocke 2 (1797/2001), p. 374f.

63For Markus Franz Raffael, see the entry about the consistory below.

64"[. . .] quae delineation ipsi á quodà studioso Witeberga transmissa fuerat" (Marius 1614, sig. C3^{r-v}).

65"Preter enim modo dictum Eruditissimum virum, omnes etiam tunc temporis Consistorij illustris Assessores quib. post festum paschatis anni 1596. hypotheses meas cum explicatione praesentavi, quorum eta consilio, ab Illustriss. Principe Georgio Friderico March. Brandenburgense laudatissimae memoriae, peculiaris habitation in supradicto monasterio concessa est, ut eò commodius hoc stadium tractere possem" (Marius 1614, sig. C3^v).

> [. . .] under whose advice, a special lodging was assigned to me in the monastery mentioned above by the Illustrious Prince George Frederick, Margrave of Brandenburg, of most honoured memory, for my greater convenience in pursuing this study.

Besides his "honored teachers," all of whom had died by 1614 with the exception of principal Codomann (not mentioned by Marius), Marius called "all members of the then renowned consistory" as witnesses for his draft—see the following overview. Even the consistory councilors of 1596 were all dead in 1614 with only one exception. Only Nikolaus Falk still preached in Crailsheim, although it is still not completely clarified if he was even a consistory councilor. But the consistory was a prestigious institution. It would have been a huge audacity to refer to it, had he no authorization to do so. In addition there were vast personal entanglements between Heilsbronn and Ansbach—see, e.g., the councilor Hohenstein. Marius complained about personal hostilities toward him in the Ansbach scene. There would have been a huge risk were he to have been convicted of dishonesty. Therefore his statements must be considered credible.

The Consistory of Ansbach During Marius's Time

The consistory was an authority, installed by the local ruler, that exercised the church regime in the whole margraviate; "also matrimonial matters and the complete higher and lower education system were in its area of responsibility" (Vogtherr 1927, p. 38). It emerged in Ansbach from the former matrimonial court; its transformation was primarily influenced by the reformer and chancellor of the University of Tübingen, Jacob Andreae[66] (1528–1590). Usually there were two meetings a week (Hausmann 1989, p. 173f.). The order issued in 1594 stipulated that[67]:

> [. . .] the aforementioned should not only be staffed with theologians, or only political persons, but instead with an equal number of both stands (the erudite, pious, honest and honorable theologians as superintendants, pastors and sermonizers (to which also counts our court sermonizer) and likewise three politician from our courts, with one of them being a secretary to the consistory).

President of the Consistory

The "director and president" should be part of the "Politicis"; he was, respectively, the highest-ranking political councilor (Brunner 1746, p. 257; cf. Hausmann 1989, p. 173).[68]

(continued)

[66]For Jacob Andrae see *NDB* 1 (1953), p. 277 (author: Peter Meinhold).

[67]"[. . .] dasselbige nicht allein mit Theologen, oder allein politischen Personen, sondern in gleicher Anzahl zu mahl aus beeden Ständen (nemlich mit deren gelehrten, gottesfürchtigen, aufrichtigen und ehrbaren Theologen, als Superintendenten, Pfarrherren und Stiftpredigern (denen doch alle weg Unser Hofprediger zugeordnet sein soll) desgleichen auch mit drey Politicis aus Unsern Räthen, deren einer zugleich des Consistorii Secretarius seyn [. . .] soll) besetzt und bestellet werde" (Brunner 1746, p. 256; cf. Vogtherr 1927, p. 37f.).

[68]For the succession of the consistory's president, see Geret 1738, sig. 4^r.

From 1588 until 1601 the lawyer *Stephan Mummius* (1532–1601) was president of the consistory. He came from Zwolle and had studied in Paris, Cologne, Mainz, and Basel. He spoke both Hebrew and Greek fluently and had concentrated his efforts on the church fathers. In 1570 he converted to Protestantism in Speyer and became the chancellor of Pfalz-Lüneburg. In 1578 he followed a call to be the court and church councilor in Heidelberg, and as of 1587, he worked for the Margrave of Ansbach (Drüll 2002, p. 399; Geret 1738b; Vocke 1 (1796), p. 200; 2 (1797), p. 323).

His successor was *Nicolaus Stadtmann* (1531–1607). "He was born in 1531 as the son of a council member in Schwäbisch Hall. After a very varied life—in 1548 he had to go to Basel along with the Württembergian reformer Johann Brenz[69], who was exiled because of the Augsburg Interim, moved to Tübingen in 1549, and traveled to Italy in 1559, where he became a doctor of law in Ferrara. In 1561 he became court councilor in Kulmbach and obtained in 1577 the same position in Ansbach, where he died in 1607."[70]

Simon Eisen von Haymen (1560–1619), born on September 20th 1560 in Crailsheim, was the successor to Stadtmann. He had studied law in Straßburg and took his doctorate on December 17th 1582 in Tübingen. In the following year, he married Barabra Heerbrand (?–1604), the daughter of the highly respected Jakob Heerbrand[71] (1521–1600), who was a professor for theology at Tübingen for 40 years. In 1589 Eisen was employed as a court councilor by the margrave. In 1597 he became privy counsilor and vice-chancellor. He was responsible for the Ansbach Amtsordnung of 1608. At the time he even represented the margrave on the Reichstag in Regensburg. In Ansbach he owned a manor in the Platenstraße 17 (Geret 1739b; Layritz 1795, p. 495; Vocke 2 (1797), p. 131, 188f.; Vogtherr 1927, p. 118).[72] He mediated, when Marius complained about Kepler.

Johann Hohenstein (1567–1631) was born in Crailsheim; in 1582 he was one of the first students accepted to the "Fürstenschule." He studied in Tübingen, was crowned poet in Heidelberg in 1596, and took his doctorate in law the following year in Basel. As of 1598 he was an advocate for the poor in Ansbach, 4 years later assessor of the district court and president of the matrimonial court. In 1608 he became vice-president of the consistory and followed the deceased Simon Eisen as president (Vocke 1 (1796), p. 375f.). He seemed to have been close to Marius; at least he wrote a short article in the congratulatory letter for Marius's marriage.

(continued)

[69]For Brenz see *NDB* 2 (1955), p. 598f. (author: Heinrich Hermelink).

[70]Vogtherr 1927, p. 38; cf. Vocke 1 (1796/2001), p. 141; 2 (1797/2001), p. 79

[71]For Heerbrand see *NDB* 8 (1969), p. 194f. (author: Heinrich Fausel).

[72]For his employment with the margrave, a lot of information can be found in Herold 1973.

Vice-President of the Consistory

Until his death in 1604, *Andreas Frobenius* (1532–1604) may have been the vice-president. He was the son of the reformer Volckmar Frobenius (around 1490–1551/2) from the town Stadtilm south of Erfurt. His mother was the converted Jew, Christina Mandel (1503–1602), whose godfather allegedly was Martin Luther. Together with his brother Bonifatius Frobenius (1537–1584), he had studied law in Jena. He started his employment under Margrave Georg Friedrich in 1574 at the latest. The "mighty councilor" lived in Ansbach in the Uzstraße 10 (Geret 1739a; Vogtherr 1927, p. 51).

His successor as vice-president was Johann Hohenstein (1567–1631), who became president of the consistory in 1619 (see above).

Secretary of the Consistory

Secretary from 1601 until his death in 1608 was Wenzeslaus Gurkfelder. Before that, he was deputy head at the "Fürstenschule" (see above). His predecessor is unknown.

His successor was *Moritz Cnod*[73] (Mauritius Cnodius, ?–1631). For the year 1687, the existence of a Wolfgang Cnodius from Kleinlangheim east of Würzburg in the "Fürstenschule" is verifiable (Dannheimer 1959, p. 164 entry 217). This could have been a brother of Moritz Cnod, who called himself only "Cnodius Francus." Under Christoph Pelargus[74] (1565–1633), he had disputed *De Justificatione*.[75] In 1605 he became the cantor at the "Fürstenschule." Salomon Codomann (1590–1637) of Bayreuth had been a student of the "Fürstenschule" from 1602 to 1608 (Dannheimer 1959, p. 164 entry 228). In October 1609 he held a festive speech[76] in Gießen, dedicated to his former teachers, including Cnodius. At the time he had probably already been called to Ansbach. For the death of Abdias Wickner, he composed a song for three voices in 1608.[77] The *Hymni sacres*, published by Johann Neser, the former cantor's son of the same name, incorporates some of the Melodies by Cnod.[78]

(continued)

[73]For Cnod see Fuchs 1837, p. 48; Hausmann 1989, p. 176; Stübner 1690, p. 54.

[74]For Pelargus see *ADB* 25 (1887), p. 328–330 (author: R. Schwarze).

[75]Christoph Pelargus; Mauritius Cnodius: *De Justificatione*. Frankfurt a. d. O.: Voltz 1601 [Evangelisches Predigerseminar Wittenberg: diss203/24].

[76]Codomann, Salomon: *De Pusig. Virgilio Marone Oratio Poetica: In floridissima Giessena Ann. MDCIX. Idibus Octobr. Quae ipsi Maroni Natales, Publice Recitata.* Gießen: Chemlinianis 1610, sig. A1^{v} [HAB Wolfenbüttel: A: 990.11 Theol. (2)].

[77]Cnodius, Mauritius: *Cantio Funebris In Obitum Reverendi, Pietate [...] Dn. M. Abadiæ Wickneri, Abbatis Monasterii Heilsbronnensis fidelissimi.* S.L. 1608 [HAB Wolfenbüttel: A: 386.34 Theol. (10)]. Cnod was also a contributor to the printed funeral sermon: Bermuth, Michael: Christliche Leichpredigt/Bey Bestattung des weiland Ehrwürdigen/Achbarn und Hochgelarten Herrn M. Abdiae Wicknern/Fürstl: Brandenb: Raths. Hof: Pfeilschmidt 1609 [HAB Wolfenbüttel: A: 386.34 Theol. (9)].

[78]A dedication by Cnod can also be found in Schweigger, Salomon: *Ein newe Reyßbeschreibung auß Teutschland Nach Constantinopel und Jerusalem.* Nürnberg: Lantzenberger 1608, sig. d1^{v}–d2^{r} [SB Berlin: Uk 2990]. The work had several reissues.

Court Chaplain

Court chaplain was *Nikolaus Falk* (1540–1616) until 1594. He came from Saalfeld and had studied in Jena and Tübingen. In 1590 he became court chaplain in Ansbach; in 1594 he moved to Crailsheim as a pastor and dean, where he died in 1616 (Simon 1957, p. 114 entry 692).

Falk's successor was *Abdias Wickner* (1560–1608). He originated from Rothenburg and had studied in Wittenberg. After his first positions in Colmberg and Leutershausen, he became court chaplain in Ansbach in 1594. In 1601 he transferred as titular abbot to Heilsbronn where he died in December 1608 (Hocker: Antiquitäten-Schatz 1731, pp. 153–155; Ulshöfer 1991). His position wasn't filled until 1621 (Simon 1957, p. 590).

Town Pastor of Ansbach

Markus Franz Raffael (1533–1604) came from Hettstadt, northwest of Halle, and had studied in Wittenberg under Melanchthon. In 1564 he was the principal of the Latin school in Ansbach. In 1573 he was given the deanship of Feuchtwangen, and as of 1582 for 2 years, he was principal and professor in the Heilsbronn monastery. In 1584 he transferred to Lehrberg; finally in 1587 he held the position as general superintendent and town pastor in Ansbach. In this capacity he also was a councilor of the consistory (Hocker: *Antiquitäten-Schatz* 1731, p. 195; Lang 3 (1811), p. 344; Simon 1957, p. 383f. entry 2305; Vocke 1 (1796), p. 16). Raffael and Marius viewed a sketch of the Tychonic world system together in the autumn of 1596.

His successor as of 1605 was *Laurentius Laelius* (1572–1634). He originated from Kleinlangheim east of Würzburg and had studied in Jena and Wittenberg. In 1598 he became town chaplain of Ansbach, 1602 principal of the "Fürstenschule." In 1605 he returned as town pastor and councilor of the consistory to Ansbach, where he died in 1634 (Hausmann 1989; Hocker: Antiquitäten-Schatz 1731, p. 197; Simon 1957, p. 272 entry 1654; Vocke 1 (1796), pp. 284–287; 2 (1797), p. 56) (Fig. 2.5).

Collegiate Church Preacher[79]

Preacher as of 1579 was *Michael Stieber* (1533–1602), who stemmed from Schwabach and had studied in Wittenberg. Since 1557 he had been active in several Franconian communities, and in 1579 he became the monastery preacher in Ansbach (Simon 1957, p. 488 entry 2927).

His successor was *Johann Meelführer* (1570–1640) of Kulmbach. He had studied in Jena and Wittenberg and took up his first employment at Kulmbach in 1600. In 1602 he became preacher and councilor of the consistory in Ansbach but left in 1608 to be the titular abbot of Heilsbronn. In 1634 he returned to Ansbach, where he carried out his former duties (Simon 1957, p. 314f. entry 1899) (Fig. 2.6).

(continued)

[79]Cf. Hausmann 1989, p 176.

He was succeeded by *Johann Heinrich Priester* (1579–1633) from Feuchtwangen. He had studied in Wittenberg and Frankfurt on the Oder and took his first position at Crailsheim in 1607. From 1611 to 1616, he was the preacher in Ansbach, before he moved to be the pastor and dean in Crailsheim, where he died 1633 (Simon 1957, p. 377 entry 2272).

Johann Hippolyt Brenz (1572–1629) originated from Tübingen, where he studied and graduated. As of 1594 he held several positions in Württemberg. In 1616 he came to Ansbach as preacherr and councilor of the consistory, where he also died in 1629 (Simon 1957, p. 52f. entry 323).

In 1596 Marius's first printed work was issued by the printer Paul Kauffmann of Nuremberg[80] (1568–1632): *Short and actual description of the comet or wonder star/that was seen in this year of Christ our Redemmer/1596 in the month of July/ near the feet of the Great Bear/in the midnight sky.*[81] The city of Gunzenhausen honored him with two guilders for this.[82]

Fig. 2.5 Portrait of Laurentius Laelius (1572–1634), private property of the author

[80]For P. Kauffmann see Grieb 2 (2007), p. 753.

[81]*Kurtze und eigentliche Beschreibung des Cometen oder Wundersterns/So sich in disem jetzt lauffenden Jar Christi unsers Heilands/1596. in dem Monat Julio/bey den Füssen des grossen Beerens/im Mitnächtischen Himmel hat sehen lassen.*

[82]In 1618 Marius also received a financial donation of 5 guilders sent to Ansbach (Mühlhäußer 1993, p. 15, 19).

Fig. 2.6 Portrait of Johann Meelführer (1570–1640). Wikimedia Commons

Marius had seen the comet for the first time on July 12/22, even though he heard from others that it was already visible at an earlier date. He could not determine the parallax of the comet, but his observations persuaded him that this heavenly body stood far beyond the moon.[83] He also engaged in astrological interpretations, relying heavily on the works[84] of the Bohemian astronomer Cyprianus Leovitius[85] (1524–1574).[86]

Augustin Lanius from Ansbach delivered a short contribution to this comet booklet. Three years later Marius dedicated his *Tabulae Directionum Novae* to him: this he had "given to his loyal friend and employee, the talented young man

[83]Marius 1596, sig. B4^{v}, C1^{r}.

[84]Leovitius, Cyprianus: *Grundliche, Klerliche beschreibung, vnd Historischer bericht, der fürnemsten grossen zusamenkunfft der obern Planeten, der Sonnen Finsternussen, der Cometen, vnd derselben wirckung, so sich in der vierden Monarchien erzeigt und begeben*. Lauingen: Emanuel Saltzer 1564 [HAB Wolfenbüttel: H: T 447.4° Helmst. (2)]. A Latin version of this work exists. None of Leovitius' writings can be found in Hocker's bibliography.

[85]For Leovitius see *ADB* 18 (1883), p. 417f. (author: Siegmund Günther).

[86]Further information can be found in the chapter by Jürgen Hamel in this volume.

Aug. Lanius of Ansbach to calculate, after I showed him how to calculate it."[87] During Marius's times Lanius was the organist in Heilsbronn and was[88]:

> [...] an extremely scholarly and well read man who now lives in Halle in Saxonia as a private man [...]. He worked as organist in Heilsbronn in those days and because we had been neighbors and good friends for a long time, he had constant access to my work.

Presumably he also had a room under the roof next to Marius's. Besides Lanius, Georg Ziegelmüller[89] (?–1614) from Wassertrüdingen and Christian Gochsemius[90] from Kitzingen also appear as contributors. Both were students of the "Fürstenschule" (Dannheimer 1959, p. 168 entry 500, p. 180 entry 1195) and friends with Marius.

Simon Marius in Königsberg?

There is a widespread rumor that Marius studied in Königsberg from 1597 to 1599 (Hocker 1739, p. 43; Vocke 2 (1797), p. 415; Muck 3 (1880), p. 40; Junger 1971, p. 227f.),[91] but his name cannot be found in the matriculation register there (Cf. Erler 1917). In 1813 Heinrich Christoph Büttner[92] (1766–1816) published documents that made "Mayr's stay in Königsberg [...] very questionable." Of course Marius wanted to study at a university. He explained the meaning of the comet of 1596 "as well as

---

[87]"Tabulas vero domorum supputandas dedi amico meo singulari & collabroatori fideli juveni ingenioso Augustino Lanio Onoldino monstrat a prius calculi ratione, cui etiam gratia aliqua debes, ppter laborem hunc quem voles subiit, & q tibi, candide Philomathes, profuturus est" (Marius 1599, sig. C1^{r}, cf. Klug 1906, p. 402).

[88]"Inter alios autem non postremum locum occupat doctissimus & multae lectionis vir, Dominus Augustinus Lanius, nunc Halae Saxonum privatam vivens, qui eo tempore organedum Heilsbronnae agebat, & propter habitationes vicinas & amicitatem dudum inter nos initiam, fere perpetuus mearum actionum inspector erat" (Marius 1614, sig. C3^{v}). In *Prognosticon auf 1607* (sig. C4^{r}; cf. Zinner 1942, p. 53), Marius reports about observations of Jupiter, made in 1596 by "me and my good friend Mr Augustinus Lanius, organist of the monastery Heilsbonn in that time" ("ich vnd mein guter freundt Herr Augustinus Lanius, damals Organist im Closter Heylsbronn").

[89]Ziegelmüller already started his first employment in Feuchtwangen in 1596; he therefore hadn't studied at university. In 1601 he became a chaplain and held the position as cantor in the Latin school of Feuchtwangen. In 1606 he became pastor in Gräfensteinberg and in 1612 finally in Berolzheim, where he died in 1614. Schaudig 1927, p. 99; Simon 1957, p. 569 entry 3404.

[90]Gochsemius seemingly studied in Straßburg as of 1601, where he disputed in 1602 and 1603 presided over by the jurist Paul Graseck (1562–1604). In 1608 he presided over a disputation in Brünn in Mähren. Verifiable is a disputation under Paul Graseck: Analyticae Tractionis de Emptione venditione. Disputatio secunda. Respondent: Christian Gochsemius. Straßburg 1602 [BSB München: 4 Diss. 3227,7].

[91]According to Christianson 2000, p. 320, he supposingly stayed at the court in Königsberg.

[92]For Büttner see *ADB* 3 (1876), p. 661 (author: von L.).

I've learned it until God Almighty gives me a better opportunity with proper funds to take up these wonderful studies."[93]

The preface to this booklet is dated September 11, 1596 (Marius 1596, sig. A2^{v}). In May of the following year, Marius again petitioned the margrave[94]:

> About half a year ago your Princely Highness was offered and deliverd my description of the bright comet of the last year with enclosed subservient supplication, wherein I humbly asked from your Princely Highness as a rich and mild father and enabler as well as the most gracious promoter of my studies, that your Princely Highness will consider me in mercy and take into your hands the necessary conditions for the admirable and useful study of astronomy and let graciously provide and follow.

Marius had attached a *supplication* to his comet treatise, with the request to enable him to study astronomy. It is said that the margrave viewed this request benevolently, but nothing happened. This is why Marius reminded him again. The original manuscript of 1596 is untraceable, and in the petition of 1597, printed in the *Franconia* of 1813, Marius mentions no university in particular.

On May 20, 1597, the councilors Nicolaus Stadtmann, Stephan Muhr,[95] Andreas Frobenius, Streuberger,[96] and Johann Gümbelein[97] signed a consideration, in which they recommended sending Marius to Königsberg.[98] Attached was a letter of recommendation to be signed by the margrave. Initially it only says that Marius "submissively asked for the continuation of his studies at other accademies, where he could put his newly invented Tabulas directionum in print and then publish."[99] That doesn't sound like Marius expressed the wish to go to Königsberg. The argument was that Marius "could correct his opus with better opportunities in our printing house in Königsberg on his own [. . .] So that he could better continue his studies began at our

[93]"[. . .] so gut ich es gelernet biß Gott der Allmächtig durch ordentliche mittel mir andere und bessere gelegenheit/dises herrliche Studium recht für die hand zunemen/verschaffen wirdt," Marius 1596, sig. C1^{v}.

[94]"Es ist ungefehr vor einem halben Jahr E.[uer] F.[ürstlichen] D.[urchlaucht] eine Beschreibung vorigen Jahrs leuchtenden Cometen von mir unterthenigst offerirt und übergeben worden, mit beigelegter unterthänigen Supplication darin von E. F. Durchlaucht als reichen und milden Vater und Verläger wie auch gnedigsten Beförderer meines Studirens ich unterthenigst gebeten, E. F. Dchl. wollen mich in Gnaden bedenckhen und nothwendige Verlägung das herrliche und nüzliche Studium astronomiae recht vor die Hand zu nehmen und tractirn Gnädigst verschaffen und folgen lassen" (Büttner 2 (1813), p. 71).

[95]This probably refers to Stephan Mummius (Mum, Muhme, 1532–1601). For him see the list of councilors of the consistory above.

[96]Presumably Viktorin Streitberger, "ein vertrauter Rat des Markgrafen Georg Friedrich" (who was a trusted councilor to the margrave). He had been court councilor in Ansbach since 1590; Vogtherr 1927, p. 51.

[97]Johann Gümbelein from Ammerndorf had been secretary to the court council since 1589.

[98]The *Bedenken* is printed in Büttner 2 (1813), p. 74f.

[99]"[. . .] unterthenigst angelangt zu Vortstellung seiner Studien uf andern academiis, dann seine neuerfundenen Tabulas directionum in Truckh zu geben und hiezu verlag zu thun" (Büttner 2 (1813), p. 75).

and at other universities, which his teachers would always recommend"[100]; he should be paid 80 guilders annually by the Heilsbronn scholarship fund (Büttner 2 (1813), p. 76). But the letter of recommendation was never signed by the margrave.

With another letter from May 1st 1598,[101] Marius again brought himself to the margrave's attention. He did not ask to be allowed to study at a university anymore; instead he only wanted to print his *Tabulae Directionum*. As an alumnus "he doesn't have the money to pay the printers,"[102] so he asked to be reimbursed for the printing cost of an estimated 30 guilders. With the letter of May 16th 1598, the Ansbach councilors referred to their previous year's evaluation "which they couldn't improve any further."[103] Again they suggested Marius should be contented with the travel expenses to Königsberg (Büttner 2 (1813), p. 78, 80).

The relevant Königsberg professor for mathematics at the time was Mathias Menius[104] (Meinius, Meyne, Mävius, Maine, 1544–1601). In 1578 he had published a cometary tract,[105] and it seems that after 1580 he also regularly issued calendars.[106] Menius did not really distinguish himself very much, and there would have been other universities better suited for studying mathematics (or astronomy) at that time. Johannes Praetorius[107] (1537–1616) taught at the nearby Altdorf,

---

[100]"[. . .] sein opus in Unserer Trukherey zu Königsberg mit mehrerer Gelegenheit selbst corrigiren kan [...] Damit er aber solch ahngefangen Studium uf unserer und andern Universitäten dazu ihme seine praeceptores jedesmaln werden gerathen seyn desto besser khönne fortsetzen" (Büttner 2 (1813), p. 76).

[101]This letter had already been printed without date by Hocker 1739, p. 43.

[102]"[. . .] solches und anders mehr bei den Buchdruckrrn[!] und sonsten zu verlegen nit im Vermögen."

[103]"[. . .] welches wir denn auch nicht wisssen zu verbessern" (Büttner 2 (1813), p. 78).

[104]Meinius hailed from Danzig and initially studied in Wittenberg, where he heard the lectures of Melanchthon. In 1571 he departed to Görlitz but became principal of the Johanniter school in the same year and 1 year later professor for astronomy at the high school. As of 1579 he was professor for mathematics in Königsberg. In 1585 he also became royal librarian. On the "zweyten Osterheiligentage des Jahres 1601" (second day of Easter in 1601), he had a stroke and died on June 3rd in Königsberg; Buck 1764, pp. 45–47.

[105]*Von aller geschlecht der Cometen, jeder zeit, wan die erscheinen zugebrauchen, und von dessen wirkungen, der uns zu Dantzigk den 12. Novembris dieses 1577. Jar erschienen ist.* Danzig: Rodhe 1578 [UB Erlangen-Nürnberg: H00/4 MTH-I 99 dg].

[106]Menius, Matthias: *Prognosticon Astrologivm/Super Reuolutionem Mundi anni M.D.LXXXI. In Qvo iuxta Doctrinam Cl. Ptolemaei Alexandrini ordine recensentur discrimina quatuor temporum anni, cum influentijs firmamenti, hoc est, viribus et effectionibus coeli superiorumq; corporum coelestium, in haec inferiora corpora, homines et animalia, etc.* s.l. 1580 [UB Erfurt/Gotha: Math 4°—*Calendarium et Ephemeris sive Diarivm ad Annvm à natiuitate salutifera Domini & Redemptoris nostri Iesv Christi M.D.LXXXI. In Nova et optima formae conscriptvm, una cvm ortv et occasv lvnae et Solis.* Leipzig: Beyer, Johann, 1580 [UB Erfurt/Gotha: Math 4° 00177/01 (01)]. According to Buck 1764, p. 45, he had authored a manuscript in 1576 *De ortu et occasu lunae* that was never published. There could have been a confusion with the quoted calendar [00177/01 (02)].

[107]For Praetorius see Gaab 2011, pp. 10–16.

Melchior Jöstel[108] (Jöstelin, 1559–1611) in Wittenberg, David Origanus[109] (1558–1628) in Frankfurt on the Oder, and finally Michael Mästlin[110] (1550–1631) in Tübingen. At the time Margrave Georg Friedrich was also vice-regent of the Duchy of Prussia; thus Königsberg was under his control (J. Meyer 1890, p. 90; 1892, p. 55). Furthermore in 1558 Hans Daubmann (?–1573) had been appointed university printer. After his death the print shop passed into the possession of his Frankonian son-in-law Georg Osterberger (?–1602), who also maintained two bookshops and a paper mill as well as a bookbindery. In 1585 he received an extended printing privilege (Benzing 1982, p. 260f.). These are probably the real reasons why the Ansbach councilors recommended Königsberg for Marius and the printing of his work.

The *Tabulae Directionum* was printed in 1599 by Christoph Lochner[111] (?–1614) in Nuremberg; its preface was signed by Marius on "pridie Andreae Anno 1598" (November 29).[112] Marius received money to pay for the printing but he never visited Königsberg. Astrologers were the principle users of these tables. Johannes Kepler (1571–1630) was asked several times about his opinion of them. In July 1611 he finally wrote[113]:

> I have no desire to argue any further about his tables (Marii). What I have already said is sufficient, that they are inconvenient to use, which, as I see, the author admits. [. . .] I do not intend therefore to agitate against Marius.

From 1601 until 1629, Johann Lauer[114] (1560–1641) published Marius's calendars year for year in Nuremberg.[115] As indicated on the title pages, they were calculated for Heilsbronn until 1608. They display the coordinates of Ansbach only from 1609 onward (Matthäus 1969, col. 1097). The preface of the first calendar

[108]For Jöstel see Christianson 2000, p. 297f.; Friedensburg 1917, p. 513; Helfricht 2001, p. 34. Marius mentions Jöstel in his *Prognosticon auf 1602*, sig. D2^{v}.

[109]For Origanus see *ADB* 24 (1887), p. 422 (author: Siegmund Günther); Thorndike 6 (1941), p. 60f.; 7 (1958), p. 145f.

[110]For Mästlin see Betsch, Hamel 2002.

[111]For Lochner see Grieb 2 (2007), p. 932f.

[112]November 30th is the day of the apostle and martyr Andreas. Cf. Steinbeck 1795, sig. Y1^{r}.

[113]"Sed de ipsius tabulis disceptare ulterius animus mihi non est. Sufficit hoc quod dixi incommodas esse usu, quod video authorem fateri. [. . .] me publicas adversus ipsum suscepisse inimicitias" (Kepler XVI, 1954, p. 388; cf. Klug 1906, p. 403). Johann Georg Herwart von Hohenburg (1553–1622) asked Kepler in a letter from March 18th 1600 for his opinion; he answered on July 12th. David Fabricius (1564–1617) expressed himself on the topic on April 28th 1602 (Kepler XIV, 1949, p. 111, 131f., 231). In a letter from February 23rd 1610, Nikolaus Vicke requested from Kepler to explain him the directions with three examples "und alle drey figuras uia Ptolemaica ex tabulis Simonis Marij zu erigiren." On March 25th he asked Kepler for the differences in Marius's and Magini's tables (Kepler XVI, 1954, p. 290, 376).

[114]For Lauer see Grieb 2 (2007), p. 891. Marius later married a daughter of the publisher.

[115]For Marius as calendar maker, see Matthäus 1969, col. 1096–1099 as well as his chapter in this volume.

is signed June 29th 1600 (*Prog. 1601*, sig. A3^{r}). In it he mentions that for the last 2 years, one[116]:

> [...] inconvenience followed the other, and those who envy me, whoever they are, turn all my actions to the worst, denigrate and belittle me

A "Fürstenschüler" attracting attention with his own work possibly appeared as boasting for some, who accordingly provoked a negative atmosphere at the margravial court. This calendar was dedicated to the widow Maria von Eyb. The chronicle of the Lords of Eyb, compiled by Wenzeslaus Gurkfelder, claims[117]:

> Hanns Martin vonn Eyb zue Bruckberg was born on St. Martin's Eve in the year 1536, [...] he was Master of Ceremonies for her Highness, the noble born Princess and Lady, Lady Aemilia, born Dutchess of Sachsen and widow of Margrave Georg of Brandenburg, his wife was Maria of Crailshaim, daughter of Wolf of Crailshaim to Summersdorf and Neunhaus, who was first Count Palatine Magistrate to Haideck, afterwards Margravial Bailiff to Kitzingen, whom he married in the year 1560 on Divine Mercy Sunday in Bruckberg, but had no children with her. He died 52 years old on February 28th in the year 1588 and was buried next to his mother in Großenhaslach.

As a widow Maria von Eyb together with Georg Ernst von Eyb (1579–1626) twice donated 100 guilders for the church in Großhaslach as well as for the poor in Bruckberg. The money should remain with the respective lords of Bruckberg, who had to pay an annual interest of 5 guilders to the church (Eyb 1984, p. 217).

The families von Eyb and von Crailsheim "hold executive positions in the central- and provincial administration, the highest offices at court as well as highest military ranks" (Schuhmann 1980, p. 534). In his calendar's preface, Marius talks about "the various honors and good deeds he received"[118] from Maria von Eyb. He appears to have seen her as a patron of his projects, but even her influence wasn't enough to obtain a place at university for Marius. So Marius remained an autodidact, who could claim that he never "had the assistance of any living man as tutor."[119]

---

[116]"[...] vngelegenheit der andern die Hand reichet/vnnd meine mißgönner/wer sie nun auch sein/ mir ohne daß mein thun vnd lassen nur zum ärgsten außlegen/mich verleumbden und verkleinern" (*Prog. 1601*, sig. A2^{v}).

[117]"Hanns Martin vonn Eyb zue Bruckberg, wardt geboren Anno 1536 an S. Martins Abent, [...] ist gewesen der Durchleuchtigen hochgebornen Fürstin und Frauen, Frauen Aemilia, gebornen Hertzogin inn Sachsen, Marggraf Georgen zue Brandenburg wittiben, Hofmeister, Seine Hausfraue war Maria von Crailshaim, Wolffen von Crailshaim zue Summersdorff unnd Neunhauß, so erstlich Pfaltzgräfischer Pfleger zu Haideck, hernach Marggrävischer Amptmann zu Kitizingen gewesen, Dochter, mit welcher er Anno 1560 am Weißen Sonntag zue Bruckberg hochzeit gehalten, hat aber keine Kinder mit ihr gezeuget, Starb Anno 1588 den 28. February seines Alters im 52 Jahr und würdt zue seiner Mutter zue Großenhaslach begraben." The chronicles were published by Ottmar Friedrich Heinrich Schönhuth in 1860, there p. 255f.; cf. Eyb 1984, p. 216, 245.

[118]"[...] vielfeltigen mir erzeigte Ehr und gutthaten."

[119]"[...] nullo unquam vivo praeceptore usus sum" (Marius 1614, sig.)()(1^{r}).

Georg Friedrich von Eyb[120] (1563–1620) was the brother of Georg Ernst von Eyb. For his son of the same name (1596–1633), born in 1596, Marius cast a birth horoscope that is held in the State Archives in Nuremberg.[121]

Simon Marius in Prague

In his *Prognosticon* for the year 1601, Marius repeatedly highlighted the accomplishments of Brahe. The date of the Sun's entry into Capricorn had been determined (*Prog. 1601*, sig. A4^{r})[122]:

> [. . .] after conscientious/proper and infallible calculation/through the protracted observations/incredible efforts/diligence and work/and with great costs by the noble and widely renowned Lord Tycho Brahe/who through God's help has been put to work and conveyed for this purpose

Marius calculated the entrance of the Sun into Aries—and thus the beginning of spring—with the tables of Brahe and Origanus,[123] the *Prutenic Tables*, as well as those of Johann Stadius[124] (1527–1579) and Martin Everatus[125] (Marten Everaert, Everartho, approx. 1540–1601), on which he commented[126]:

---

[120]For Georg Friedrich von Eyb, see Schönhuth, Gurkfelder 1860, pp. 258–260.

[121]National Archives Nuremberg: nobility archive of the Freiherren von Eyb zu Neuendettelsau A 1465; cf. Schott 2015.

[122]"[. . .] nach einer gewissen/eigentlichen vnd vnfehlbarn rechnung/so durch langwiriges observirn/vngleublich mühe/fleis vnd arbeit/vnnd vber grossen vnkosten dermal eines von dem Edlen und weitberümbten Herrn Tychone Brahe durch Gottes hilff ist in das werck gesetzt und herfür bracht worden" (*Prog. 1601*, sig. A4^{r}).

[123]Marius probably used the following edition: Origanus, David: *Ephemerides Novae Annorvm XXXVI, Incipientes Ab Anno Christogennias 1595, quo Ionnis Stadii maxime aberrare incipiunt, & desinentes in annum 1630.* Frankfurt an der Oder: Eichhorn 1599 [UB Tübingen: Bd. 178.4]

[124]For Stadius see Holden 2006, p. 169. His Seine ephemerides were *Ephemerides secundum Antverpiae longitudinem. Ab anno 1554 usque ad annum 1606.* Köln: Arnold Birckmann 1581 [HAB Wolfenbütel: N 132. 40 Helmst. (2)].

[125]Not very much is known about Everatus. He was seemingly born around 1540 in Brügge and died 1601 in Leiden. Besides his ephemerides he published an edited version of *De Astrolabo Catholico* [SB Regensburg: 99/Philos. 2606] by Gemma Frisius with Withagius in 1583 in Antwerpen. His ephemerides are from 1597: *Ephemerides novae et exactae: ab anno 1590 ad annum 1610; ex novis tabulis Belgicis authoris supputatae. Ad longitude. 24.0. latid. 51.30 gradum.* Leiden 1597 [SUB Göttingen: 8 ASTR I, 3013]. Further information about Everatus online at http://www.personenencyclopedie.info/E/Ev/EVERAERTMartin. Marius's opinion of him was not good: "Deß *Euerardi* mag ich kaum gedencken/sintemahl seine *correctio in motu Mercurii* nicht einer faulen Birn werth ist" (Prog. 1611, sig. B4r; cf. Zinner 1942, p. 58). In the *Prognosticon auf 1612* (sig. B1^{v}), he refered to him as "malignus Martini Everhardi."

[126]"Unter disen vier widerwertigen rechnungen [...] ist die beste vnd gewisseste Tychonis Brahe [...] sonderlich mit dem aequinoctio verno voriges 1600/Jars/da ich durch einen gerechten messigen quadrantem befunden/das die Sonn den ersten punct deß Widers erreichet hat/den 10 tag Martij zwischen 6. vnd 7. vhr vor mittag/damit gentzlich übereinstimmet calculus Tychonis/deme ich dißmals vnd vorthin folgen wil" (*Prog. 1601*, sig. A6^{r-v}).

> amongst these four contrarious calculations [. . .] the best and truest is that of Tycho Brahe [. . .] especially for the spring equinox for the previous year 1600, for which I found with the help of a good brass quadrant that the sun reached the first point of Aries on March 10th between 6 and 7 o'clock before midday, in total accordance with the calculus of Tycho, which I followed this time and will in future.

Marius was very familiar with the works of Brahe. Apparently through an acquaintance, he already had tried to determine whether Brahe would employ him as an assistant. He received a letter of recommendation, dated on May 12, 1601, "to our dearest Tycho Brahe, Roman Imperial Councilor to Prague," after which Brahe "on recommendation of several of his acquaintances was inclined" to employ Marius for some time.[127]

He immediately hit the road and arrived in Prague at the end of the month[128] at the palace of the deceased Vice-Chancellor Jakob Kurz of Senftenau (1533–1594) on the Loretoplatz (Christianson 2000, p. 320).[129] The city impressed him very much: He, who strolls around the Royal Saloon "will hear or maybe even see wonders/that are not commonplace/and which, I have heard and seen Anno 1601."[130] Marius only recalled his stay in Prague in a few other notes in his calendars without providing any details about his personal contacts. For example, he remembers "the downpour/which also happened anno 1601 on August the 10th or 11th/when I was in Prague."[131]

Tycho's active observer then was Johannes Eriksen from Hamburg,[132] who probably introduced Marius to Brahe's observational methods (Christianson 2000, p. 320). Eriksen was the nephew of Simon von Petkum (?–1620), who was Brahe's agent in Hamburg around 1600 and very likely placed him in Brahe's service. In 1599 he stayed in Wittenberg together with Brahe; in July he helped with the observation of the solar eclipse in Prague. In spring 1601 he became Brahe's personal secretary. He traveled with Franz Gansneb Tengnagel von Camp[133] (1576–1622) to the Netherlands, whose service he eventually entered (Christianson

[127]"[. . .] an den vesten, unsern besondern lieben Tycho Brahe, Röm. Kaiserlichen Rath zu Prag [. . .] uf Commendation etlicher seiner Bekannten genaigt" (Büttner 2 (1813), p. 81). The report by Doppelmayr (1730/1972, p. 90, footnote y) that Marius went to Brahe in Denmark in 1590 is untenable. This may be going back to Giovanni Antonio Magini (1555–1617), who mentions Marius in a remark in his *Supplementum ephemeridum* (p. 297) of 1614: "Simon Marius, & in motuum obseruatione accuratissimus ob familiaritatem, quam cum Tychone olim in Dania habuit."

[128]Johannes Eriksen wrote on May 27th 1601 to Kepler that Marius's arrival could be any day now, cf. Hansch 1718, p. 176.

[129]The palace doesn't exist anymore. Nowadays the Kepler school stands in its place; in front of it, two big statues are dedicated to Brahe and Kepler.

[130]"[. . .] der wird wunder hören oder gar wol sehen/nicht wie täglich geschicht/und ich selbsten Anno 1601 gehöret unnd gesehen" (*Prog. 1619*, sig. A4^{v}–B1^{r}).

[131]"[. . .] platzregen/dergleichen ist gewesen Anno 1601. den 10. oder 11. Augusti/da ich zu Prag gewesen" (*Prog. 1612*, sig. D1^{r}; cf. Zinner 1942, p. 60).

[132]For Erikson see Christianson 2000, p. 272f.

[133]Tengnagel was a royal from Westphalia, who had married one of Brahe's daughters. Occasionally he also worked as Brahe's assistant. See Christianson 2000, pp. 366–372.

2000, p. 272f.). In a letter of May 27, 1601, Eriksen told Kepler about the arrival of Marius[134]; 4 days later Barbara Kepler (ca. 1574–1611) wrote to her husband in Graz: "Tycho has taken in a mathematician from Ansbach, who is a capable assistant."[135]

In his *Prognosticon* for the year 1606, Marius wrote about the "instruments of the noble and excellent astronomer Tycho Brahe, which I have seen in Prague Anno 1601, as I stayed with him for some time."[136] Not only did he see the instruments "but also used them" (*Prog. 1608*, sig. B4^{v}; cf. Zinner 1942, p. 54). He was "servant and observer [. . .] for Tycho" (Marius 1619, sig. A4^{r}). He does not mention a direct encounter with Brahe, who was sick and bedridden at the time. Marius probably never met him in person, but was able to work with his instruments. He also copied his fixed star tables, which he later took to Italy.[137]

Even though a personal contact never took place, that did not prevent Marius from praising Brahe's work until the end of his life. In the *Prognosticon* for the year 1612, he talks about the[138]:

> [. . .] restitution of Tycho, which is the best in my opinion and that of other diligent and famous astronomers/and it should be reasonable to stop the handwork of the lazy and careless calendar makers [. . .]

In 1624 he still mentions the "Tycho Brahe's good and improved corrections."[139]

The preface of his *Prognosticon for 1602* was signed by Marius without a location on September 21st 1601 (Julian calendar). In it, he states, "the journey I undertook with the merciful permission of E. F. D (Your Princely Highness)."[140] He

[134]"Marggrauij Anspachensis Mathematicus, Simon Mayer, post vnum vel alterum diem familiae nostrae numerum adaugebit, et vti confido me liberabit, obseruationibus quantum ex colloquijs mutuis intelligere potui aliquomodo assuefactus, alias in Astronomicis Haereseos non condemnabitur" (Kepler XIV, 1949, p. 168).

[135]"Der Diho hat ein Matematiguß aufgenumen von annspach es ist ein lötiger gesöll" (Kepler XIV, 1949, p.170).

[136]"Instrumenten dess Edlen unnd vortrefflichen Astronomi Tychonis Brahe, wie ich sie anno 1601. zu Prag gesehen, da ich mich ein zeitlang bey jm auffgehalten habe" (*Prog. 1606*, sig. A2^{r}; cf. Zinner 1942, p. 49).

[137]"[. . .| ich hatte etlicher Fixstern veras distantias von Tychone auss Prag mit in Italiam gebracht" (*Prog. 1618*, sig. A2^{v}; cf. Zinner 1942, p. 65).

[138]"[. . .] restitution Tychonis, welche nach meiner vnd anderer fleissiger und berhümbter Astronomorum observation, die beste ist/vnd solte billich den faulen vnd groben Calendermachern/das Handwerck verbotten werden" (*Prog. 1612*, sig. B1^{v}). In the *Prognosticon auf 1610* (sig. E2^{v}), he speaks about the noble "and really great astronomer Tycho Brahe, whose celebrated name would merit to be mentioned together with the heaven" ("vnd vere Magus Astronomus Tycho Brahe, cuius celebre nomen merito cum mundo coaeuum erit"). In the *Prognosticon auf 1619* (sig. A4^{r}), he calculated the new moon "according to Tycho's teachings and calculations, which was the best in this time" ("nach Tychonis lehr oder rechnung/welche dem nach zur zeit die beste ist").

[139]"[. . .] guten und verbesserten correction Tychonis Brahe" (*Prog. 1625*, sig. A3^{v}; cf. also sig. C2^{r}, C4^{r}).

[140]"[. . .| die Reis/welche auß E. F. D. gnedigstem erlaubnuß ich verrichtet hab" (*Prog. 1602*, sig. A3^{r}, A4^{v}). "E. F. D." means "Eure Fürstliche Durchlaucht," your princely Highness.

had obviously already finished his journey. In the middle of August, he must still have been in Prague, but he was in "Znaim in Mähren" on September 15th 1601 (*Prog. 1613*, sig. C3^r; cf. Zinner 1942, p. 61). According to his *Prognosticon for 1625* on September 16th, a certain planetary constellation brought "good and warm weather in Mähren and Austria [...], where I have been at the time."[141] Furthermore in Austria 1599, an "extraordinarily and delightful wine has been grown, like the one I tasted Anno 1601 in Vienna."[142] Since it is impossible that he traveled from Znaim to Vienna and from there back to Ansbach in only 6 days, he had to have been in Vienna at the beginning of September.

Brahe died on October 24, 1601; therefore Marius had definitely departed from Prague before his death. On his arrival in the city, he might have met Ambrosius Rhodius[143] (1577–1633), who later became professor for higher mathematics in Wittenberg. He resided in Prague from November 1600[144] until at least the end of May 1601.[145] Soon thereafter he must have left Prague and wandered through Bohemia, Moravia, and Styria. At the beginning of September 1601, at the latest, he was in Wittenberg again.[146] However Marius nowhere mentions Rhodius, and they obviously never got to know each other. Marius also probably did not get to know Kepler personally when he was in Prague (Fig. 2.7).

[141]"[. . .] in Mährn vnd Osterreich auch gut warm wetter [. . .], allda ich damals gewesen" (*Prog. 1625*, sig. D1^r, cf. Zinner 1942, p. 70).

[142]"[. . .] aussbündiger herrlicher Wein gewachsen, wie ich denn solchen Anno 1601. zu Wien wohl versucht hab" (*Prog. 1611*, sig. C4^v; cf. Zinner 1942, p. 59).

[143]For Rhodius see Schöneburg 2007, pp. 67–75. For Rhodius there is also a claim that he departed from Prague after Brahe's death, which is verifiably false, as it was in Marius's case.

[144]"Venit ad me Desda hisce diebus M. AMBROSIUS RHODIUS, et nonnulla, quae circa Hypotheses tentasti, clarißime et praestantißime D. D. IÖSTELI, secum a te attulit. Utrumque mihi gratum fuit" (Brahe 1925/1972, p. 391).

[145]"Der Amprosius der stutent fragt mih stäz ob jer mier nihts geschriben habt ob jer jm jm steirmarch khönt jnn diensten unterhelfen." Barbara Kepler in Prague wrote to her husband in Graz, May 31st 1601 (Kepler XIV, 1949, p. 170).

[146]"His spatiis, his itineribus cum progrederetur in studiis, accidit, ut magnus ille coeli Vates, TYCHO BRAHEUS, ad se mittendum hominem literis postularet, quo ejus opera in observatio[n] ibus, quib' incubebat novus Hercules, uteretur. Qua provincia ut neminem alium functurum rectius, quam RHODIUM, existimabat nobis ille Mathematicus Acad. nostrae MELCHIOR JOSTELIUS, sic ei auctor erat ac suasor, ne tam praeclaram occasionem co[n]sulendi studiis suis praetermitteret. Nec defuit ille fortunae suae, vel auctoritatem optime de se meriti Praeceptoris defugit; sed obsecutus ejus consilio Pragam ire, & magno illi Astronomiae Instauratori addicere sese strenue porrexit, cum Philosophie laurea ante esset ornatus: qua publici testimonii loco Kalendis Martiis donatus fuit. Cum Pragam venisset, non modo BRAHEO probavit se facile, sed & decus harum artium ingens ac seculi nostri JOHANNEM COPLERUM, eruditio suo ingenio mirum in modum sibi adjunxit divinxitque. Altero anno, cum Bohemiam, Moraviam, Styriam, maximam partem peragrasset, Mense Septembri huc rediit, privatasque scholas habere instituit" (Reusner 1634, p. 346f.). Since he wrote back to Kepler from Wittenberg on January 1st 1601, he returned back to Wittenberg in this year. The entry in Friedensburg 1917, p. 514, as well as in Kathe 2002, p. 230, according to which Rhodius only came back because of Brahe's death, is untenable. This probably goes back to Röbner 1634 (sig. a3^r), where September 8th 1602 is mentioned as the date of arrival in Wittenberg.

Fig. 2.7 Memorial for Tycho Brahe and Johannes Kepler in Prague. Wikimedia Commons, User Mohylek

With David Fabricius (1564–1617), it was different. In retrospect, Marius wrote to Mästlin on March 29th 1612; this friendship had started when they resided together with Tycho Brahe for 1 month in 1601.[147] Fabricius was an East Frisian pastor with a deep interest in astronomy. He corresponded with Brahe, who valued his observations. On May 1st 1601, he traveled to Prague, where he arrived on May 28th and stayed with Brahe's familiy for about 3 weeks (Christianson 2000, p. 274). On July 3rd, Fabricius was back home in East Frisia. He didn't meet Kepler, but they were soon engaged in a lively correspondence, which Kepler broke off in 1608 because Fabricius didn't support the Copernican system, but rather the Tychonic one. Kepler's later derogatory remarks about Marius might have had a similar background. David Fabricius was slain by a farmer, whom he had accused in a sermon of stealing a goose.[148]

Marius stayed in contact with Fabricius throughout his life. In the *Prognosticon for 1616*, he called the "astronomer David Fabricius my especially favored sir and

[147]"Inter me et illum [D. Fabricius] inita est primitus amicitia, quando Anno 1601 Pragae in aestate per mensem q: aput Tychonem Una eramus" (letter to Mästlin from March 3rd 1612, HAB Wolfenbüttel: Hs. 2174, sig. 283; printed in Zinner 1942, p. 42).

[148]Christianson 2000, p. 274f. In the *Prognosticon* for the year 1621 (sig. C1^r, cf. Zinner 1942, p. 69), Marius wrote about Fabricius: "who was despicably killed by his neighbor in the evening in his church yard" ("welcher vnversehener weiß Jämmerlich von seinem Nachbauern/Abends auff seinem Kirchhoff ist ermordet worden").

good friend."[149] In "Mister Fabricius's tables or Prognosticon [. . .] one can find out the reason"[150] for the discrepancies in Origanus' and the *Prutenic Tables*. Fabricius tried to correspond with Michael Mästlin in Tübingen via Marius, but apparently to no effect, perhaps because some letters got lost.[151]

Simon Marius in Padua

"Anno 1601 in December it was very cold during the union of Jupiter and Spica/I traveled through the Italian Alps at that time,"[152] reported Marius in his *Prognosticon for 1625*. He even recorded the weather during his journey and in Padua.[153] On December 8th/18th, 1601, he matriculated at the university in Padua[154]:

> Simon Marius Guntzenhusanus Francus inclytae Germaniae nationi nomen meum dedi, expositis pro more 6 libris Venetis, 18 decembris anno 1601

The beginning of the solar eclipse on December 14th/24th 1601 "happened just here in Padua Italy/about a quarter/or 20 minutes past 1 p.m./however ended at 10 past 4 p.m." Hereby the "Calculus prutenicus missed the beginning of the eclipse by a whole hour, and the duration by more than 20 minutes."[155]

According to Zedler's *Universal-Lexikon*, Marius resided "in particular for several years in Padua and Venice."[156] But Marius himself never mentioned staying in Venice. From the *Prognosticon for 1614*, one can only extract that on January 1st/2nd, 1602, the weather in "Padua and Venice was damp and rainy."[157] If he were ever in Venice, then only for a visit. In those days, Padua belonged to the dominion of Venice which might explain the rumor that Marius had resided there.

149"Astronomus David Fabricius, mein sonders günstiger Herr und guter Freund/vor diesem vertrewlich communicirt hat" (*Prog. 1616*, sig. C2^r; cf. sig. C4^v).

150"Herrn Fabricij tabulij oder Prognostico [...] den Grund erfahren können" (*Prog. 1616*, sig. B3^v).

151"I recently got a letter from him [David Fabricius], in which he assumes I didn't send his writings to E.E., which however was done long ago" ("[. . .] hab neülich wider brieff von Ihme [David Fabricius] gehabt, darinnen er mich gleichsam will in verdacht ziehen als wenn Ich sein schreiben E. E. nit Zugeschickt hette, welches doch vorlengst geschehen ist" (Zinner 1942, p. 44)).

152"Anno 1601. im Decemb. ist es bei solcher vereinigung Jupiters und spica auch gar kalt gewesen/ Ich reisete damals durch die Alpes in Italiam" (*Prog. 1625*, sig. D2^r).

153According to *Prog. 1626*, sig. A4^v, his brother noted warm and foggy weather for December 8th 1602, while it was "summer weather in Padua, Italy."

154Rossetti 1986, p. 122 entry 1026. Cf. also Klug 1906, p. 398, 418.

155"[. . .] ist allhie in Italia zu Padua geschehen/ohngefehr ein viertel stund/oder 20. minuten nach 1. uhr nach Mittag/das ende aber umb 4. uhr 10. minut. [Dabei habe der] Calculus prutenicus gefehlet umb eine gantze stund im anfang der Finsternuß/in duratione umb anderthalb viertel" (*Prog. 1603*, sig. D1^r). He observed also the lunar eclipse on May 25th/June 4th 1602 in Padua; on May 28th 1602, Marius experienced a minor earthquake in Padua at 3 a.m. (*Prog. 1603*, sig. A3^r; D4^r).

156Zedler 19 (1739), col. 1588; cf. also Vocke 2 (1797/2001), p. 415.

157"Padua und Venedig feucht unnd regnisch erzeigt" (*Prog. 1614*, sig. B2^v).

Fig. 2.8 View of Padua in the World Chronicle of Schedel 1493. Wikimedia Commons

Padua (Fig. 2.8) was a popular place of study for German students. Like Marius, many were Protestants, which often brought them into contact with the Inquisition. Therefore the Doge granted them a special immunity privilege in 1587, which protected them from persecution.[158] Between 1553 and 1630, "not less than 8672

[158]Rüegg 1996, p. 241. Philipp Camerarius (1537–1624) is an example of a Nuremberg citizen, who got in trouble with the Inquisition—however in Rome, not in Padua. Will (1 (1755/1997), p. 176) reports about this: "he was unfortunate, that he was committed to the inquisition in Rome by a Jew, Mararius, had to stay in prison for two months together with his traveling companion, Peter Rieter, Patricius and later caregiver in Hersburck from Nuremberg, and got into danger of being burned because of being accused of heresy. In this temptation he kept his faith, which the Jesuits tried to crumble with guile and malice, and finally he was freed together with his companion by a miraculous cause by the Envoy of the Emperor and other high pleaders. He got cured in Rome of an illness acquired in the prison and returned to his homeland in 1566." ("unglücklich aber war er, daß er in Rom durch einen Juden, Mararius, in die Inquisition kam, über 2 Monate mit seinem Reisgefehrden, Peter Rieter, einem Nürnbg. Patricius und nachmaligen Pfleger in Hersbruck, im Gefängnisse liegen muste, und wol gar in Gefahr war, der beschuldigten Ketzerey wegen verbrennt zu werden. In dieser Versuchung hielte er standhafft an seinem Glauben, worinnen ihn die List und Bosheit der Jesuiten wankend machen wollte, und wurde endlich aus einem wunderbahren Grund

law students and 1864 students of the faculty of arts, that is physicians, theologians and philosophers" studied there. "In Padua alone about 60 Germans matriculated every semester."[159] But there was high fluctuation: For many, their visit was only part of their Grand Tour and so they often only stayed 3 months; the teaching courses in Italy were orientated to that period of time (Stölzel 1 (1872), p. 66f.).

In 1553 a distinct "German Nation" (Natio Germanica) was established in the faculty of arts in Padua, whereby also Baltics, Bohemians, Dutch, and Scandinavians were admitted (Döhnel 1996, p. 298). From 1565 on their meeting place was the church of Santa Sofia (Matschinegg 1999, p. 28). The Natio Germanica had numerous privileges; among other things they were allowed to matriculate their own students (Friedl 1994, p. 16). About 30 people are listed as paying members of the library for each year between 1602 and 1605. Marius himself payed an annual fee of 10 Venetian lira for the library,[160] the equivalent of 2 guilders,[161] which amounted after all to 2% of the money provided by Ansbach for his annual expense.

On March 5th, 1604 (Gregorian calendar) for the first time, Marius shows up as procurator of the German Nation.[162] Caspar Hoffmann[163] (1572–1648, Fig. 2.9) had become procurator before him on July 14th, 1603 (Friedl 1994, p. 123). It seems they were friends, because Hofmann called him "collega meus amicissimus" (Favaro II 1912, p. 225). Both were of the same age—meaning they were significantly older than the average German students in Padua; like Marius, Hofmann came from poor circumstances (Will 2 (1756), p. 162).

The procurators's duties were the execution of the matriculation and management of the Nation's finances (Matschinegg 1999, p. 27). The procurators had a very high fluctuation: At the beginning of August 1604, Philipp Hoechstetter[164] (?– ca. 1635) from Augsburg and Melchior Sebitz[165] (Sebiz, 1578–1671) were appointed. In November they had already been superseded by Paul Clauss[166] (1576–1651)

durch den Kaiserl. Gesandten und andere hohe Vorbitten mit seinem mitgefangenen Gefehrden befreyet, lies sich noch in Rom von einer durch das Gefängnis verursachten Krankheit curiren, und kehrte 1566 in sein Vatterland zurück").

[159]Seelbach 2003, p. 138. Between 1550 and 1599, over 6000 Germans were in Padua (Premuda 1963, p. 98; Printing error in Rüegg 1996, p. 241: 1559 instead of 1599).

[160]Favaro 2 (1912), p. 189 (Contribution for 1602), 195 (1603), 214 (1604), 231 (1605).

[161]Conversion online: http://www.pierre-marteau.com/currency/converter.html

[162]"5 Martii Dominus Georgius Furenus, quem alter Procuratorum Dominus nimirum Fridericus Guntherus 23. Decembris Patavii discendens in suum locum substituerat, re sua ita ferente, discessum etiam hinc parabat, ac in suum locum suffecit Dominum Simonem Marium Francum" (Favaro II 1912, p. 211; cf. also p. 214). In Rossetti 1986, pp. 128–131, Marius isn't mentioned as procurator. For Georg Fuiren (Jorgen Furenius, 1581–1628), who later made his name as botanist in Scandinavia, see Caroe 1913, p. 208.

[163]For Caspar Hofmann see Grieb 2 (2007), p. 686; Will 2 (1756/1997), p. 162–168.

[164]Hoechstetter became a physician in Augsburg (Hirsch 3 (1886), p. 230).

[165]Sebitz is said to have studied at 27 different universities. In 1610 he took his doctorate in Basel; 2 years later he became the successor to his father as professor for medicine in Straßburg (Hirsch 5 (1887), p. 335; Zedler 36 (1743), col. 836–838).

[166]Clauss became a rural physician in upper Austria. In 1628/1629 he emigrated to Ortenburg and later to Regensburg (Matschinegg 1999, p. 253 entry 479).

Fig. 2.9 Portrait of Caspar Hofmann (1572–1648). Private property of the author

from Vienna and Johann Caspar König[167] from Rottweil, followed by Wilhelm Männer[168] (1580–1645) from Linz and Wolfgang Weiss[169] from Ybbs in Lower Austria. In May Männer was to be found in Basel; he was replaced by Christian Rosian[170] (?–1617) in April (Rossetti 1986, pp. 131–133).

The management of the Nation's businesses was carried out by two councilors, who were elected annually like the procurators. They were responsible for the most important university affairs, such as the election of the chancellor or appointments

[167]König registered with the German Nation in Padua on April 23rd 1604. He hailed from Rottweil and took his doctorate in the nearby Freiburg on July 21st 1605 under Georg Marius (1533–1606). In 1613 he was town physician in Hagen. Marius, Georg: *Universales De Febrium Essentia, Causis, Differentiis Et Curatione Assertiones*. Rottweil: Helmlin 1605 [HAB Wolfenbüttel: Mx 247 (6)]; cf. Rossetti 1986, p. 130 entry 1097.

[168]Wilhelm Männer came from Linz and became a medical doctor in July 1614 in Basel. He became a physician in Linz (Matschinegg 1999, p. 481 entry 1630; Speta 2002, pp. 98–102).

[169]No further information is known about Weiss (Matschinegg 1999, p. 606 entry 2306).

[170]Christian Rosian (Roscian) from Meißen matriculated with the German Nation in November 25th 1604 as a student of pharmacy and medicine. He died 1617 as apothecary in Vienna. Rossetti 1986, p. 132 entry 1111; Senfelder 1904, p. 76.

(Matschinegg 1999, p. 27). In 1603/1604 Matthias Untzer[171] (1581–1624) from Halle and Martin Lucken[172] from Lemgo in Westphalia held these positions. From July 28th 1604, they were replaced by Caspar Hofmann and Simon Marius. Marius stayed in office until his departure in July 1605; Hofmann left Padua in April.[173] In early May he matriculated in Basel,[174] where he graduated under Johann Niklaus Stupanus[175] (1542–1621) on September 23rd.[176] Marius and Hofmann seemed to have lost contact, although Hofmann later became a respected professor for medicine in Altdorf, which is easily reached from Ansbach.

In Padua Heinrich Trigel[177] from Heidelberg became Hofmann's successor as councilor on April 15th. In June he had already been replaced by Martin Paelanius[178] from Harlem. On August 4th 1605, new elections were held; Paulo

[171]Untzer had already studied medicine in Leipzig and Tübingen. In 1605 he took his doctorate in medicine in Basel under the Swiss medic Felix Plattner (1536–1614) with his work *De Mola Matricis* (Mola Matricis translates literaly as a uterus millstone; in modern medical terminology, it is an anembryonic gestation or blighted ovum. It is a fertilized ovum that attaches to the womb, but does not develop any further. Usually this leads to a miscarriage). On the cover Untzer called himself "Hallensis" (Platter, Untzer 1605). He later became a famous physician in his hometown and published several medical writings. He is said to have belonged to the "Sekte der Chemiatriker" (meaning he was a disciple of Paracelsus) (Hirsch 6 (1888), p. 47).

[172]On June 27th 1606, Lucken became doctor for medicine in Basel with his work on circulatory disorders (De Apoplexia). On the cover he refers to himself as "Lemgouiensis Westphalus" (Lucken 1606). In an occasional paper in 1612, he calls himself "Poliater Hanno," which means he became a town physician in Hannover (H. Müller 1612, sig. A1^{v}).

[173]Premature replacements were not unusual: In April 1603 Lorenz Hofmann (1582–1630) from Halle at the Saale became successor to the Silesian Heinrich Nagel, who left Padua toward Basel, where he took his doctorate in the same year under Caspar Bauhin (1560–1624). Cf. Grasser, Johann Jacob: *Epincium in Honorem Cl. & Praestantiß. Virorum, M. Valentini Nitneri Mysnici. M. Joachim. Stadtmann. Hallens. Suev. Henrici Nagel Vratislav. Sil. M. Petri Hoffmann. Rothag. Franci. Bartholomaei Crugeri Sax. M. Georgi Meindeli Ratisponens. Cum Athenis Rauracis, forte sic locati, ab Excellentiß. & Cl. Vero D. D. Casaro Bauhino summa in arte medica digitate ornarentur*. Basel: Konrad von Waldkirchen 1603 [StaBi Berlin: Xc 583/3 (10)].

[174]Wackernagel 1962, p. 50 Nr. 131. Wills' statement about Hofmann going to Basel in autumn 1605 is a confusion with the date of his doctorate (Will II 1756/1997, p. 162).

[175]For Stupanus see Hirsch 5 (1887), p. 574.

[176]Stupanus, Johann Niklaus: *Cum Deo, et Consensu Collegii Medici Basileensis. Praeside D. Joh. Nicolao Stupano, Lectiones Suas Caniculares. De Febribus Malignis, Publico Examini Subiicit Caspar Hofman, Gothanus. Ad XXIII. Septembr. Ann. M.DC.V.* Basel: Excertier 1605 [HAB Wolfenbüttel: Mx 115 (10)].

[177]For Trigel see Metzger 1986, p. 18. He had matriculated at the German Nation in Padua on May 24th 1603. Rossetti 1986, p. 127 entry 1067.

[178]Martin Paelanius had become "medicinae baccalaureus" in Montpellier and matriculated at the German Nation on November 20th 1604 in Padua. He became a physician in the Netherlands. Rossetti 1986, p. 132 entry 1110.

von Hullegarde[179] became first councilor, Christian Rosian second, replacing Marius, who had already left Padua (Rossetti 1986, p. 133).

In the middle of April 1605, Marius occupied another important position in the Nation. He became the librarian (Favaro 2 (1912), p. 231, 238). In 1586 the German Nation had started building its own library, supervised by an elected librarian. The Nation's assembly decided that every departing member should donate at least one book to the library, steadily increasing the inventory (Matschinegg 1999, p. 29). When Marius left Padua, Christian Rosian became the librarian. At his farewell the Nation unanimously decided to give Marius a small monetary gift for his loyal services.[180]

It is not known when Marius formed the wish to study medicine. In the *Prognosticon for 1615*, he merely states that "20 years ago [. . .I] had no thoughts about medicine" (*Prog. 1615*, sig. A2^{v}).[181] The idea could have only emerged in the 1590s, maybe even in 1601 in Prague. As we know from his letter to Kepler (Kepler XIV, 1949, p. 257) before he left Ansbach, Marius was in contact with the personal physician Johann Pape[182] (Papius, 1558–1622), who had studied in Tübingen, Straßburg, and Basel, not Padua. Without a doubt he conversed with him about his planned medical studies. Julius Meyer suspected that the decision for Padua was made because Fuchs von Bimbach auf Möhren[183] (around 1567–1626) had studied in Padua in 1587 (J. Meyer 1892, p. 56). After his return to Ansbach, Marius had very much to do with Fuchs; however there is no information about them having been in contact before 1608. Furthermore Fuchs is not mentioned in the matriculation registers of the Nation in Padua.[184] But it is also possible that Marius simply went to Padua at the margrave's direction.

Usually medical studies focused on the medical classics. In Padua, by contrast, it had become customary to hold lectures at the patient's bedside since the middle of the sixteenth century and even to allow students to examine the patients. This practice was only slowly adopted at the other European universities; Caspar Hofmann, for example, did not adopt these methods from Padua and still lectured in the old way (Landau 1902, p. 14).

[179]Hullegarde matriculated at the German Nation on July 5th 1605 in Padua, cf. Rossetti 1986, p. 133f. entry 1120. Together with others he published in 1641 a Pharmacopoeia Bruxellensis (Brussels: Mommart 1641 [UB Erlangen-Nürnberg: H61/4 TREW.P 538]).

[180]"Eodem tempore, consensu totius Nationis, pedello nostro ex aerario nostro decem coronatus dono dedimus, quia longo tempore nobis fideliter servierat et parum lucri propter paucitatem nostrorum ex Natione habuerat" (Favaro 2 (1912), p. 238).

[181]"vor etlich 20. Jahren [...] noch im geringsten keine Gedancken zur Medicina hatte."

[182]For Pape see Drüll 2002, p. 432f.; Vocke 1797/2001, p. 45f. After 1603 Pape became professor for medicine in Königsberg, not as Vocke writes, in Heidelberg.

[183]For Fuchs von Bimbach, see the chapter by Wolfgang Dick in this book.

[184]Cf. Rosetti 1986. Fuchs probably only stayed in Padua for a short time, cf. the chapter by Wolfgang Dick in this book.

The most well-known professor of medicine during Marius time was Hieronymus Fabricius ab Aquapendente (1537–1619). On his initiative an anatomical theater in Palazzo del Bo was built in 1594. For 300 years anatomical lectures were held there and of course Marius would often have been encountered here. The anatomical theater has been maintained in top condition to this day (Rossetti 1985, p. 29) (Figs. 2.10 and 2.11).

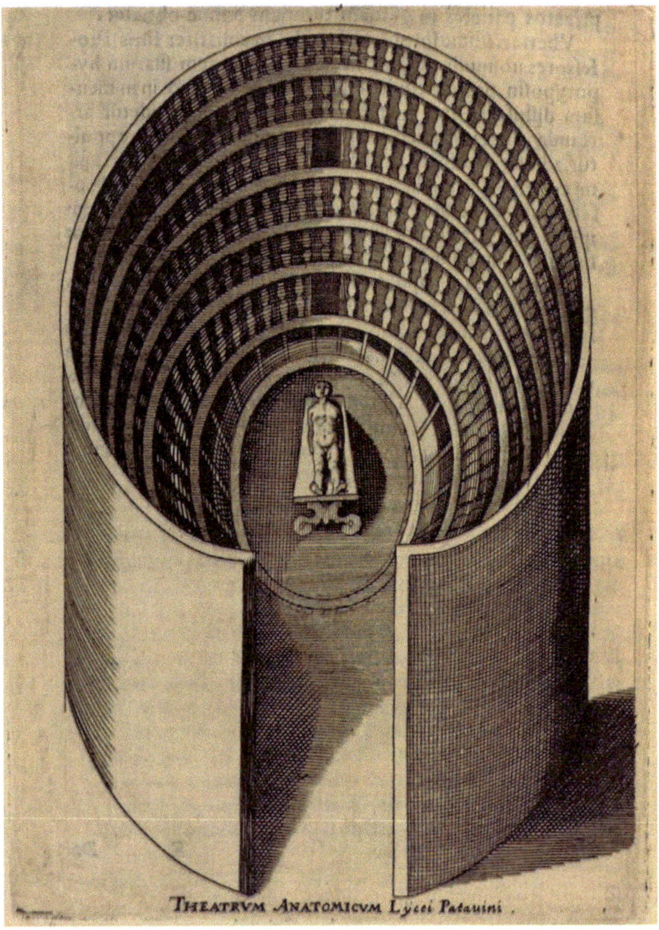

Fig. 2.10 The anatomical theater in Padua by Tomasini 1654. Courtesy of the BSB München: 4 H. lit.p. 216, S. 74, urn:nbn:de:bvb:12-bsb10679955-2

Fig. 2.11 Todays view of the anatomical theater in Padua. Wikimedia Commons, picture taken by Marco Bisello

Medical Professors in Padua During Marius' Study Period

The professor of medicine *Hieronymus Fabricius ab Aquapendente*[185] (Girolamo Fabrizio, 1537–1619) attracted students from all over Europe (Eckart 2000, p. 177). As of 1565 he held lectures on surgery, as of 1589 on anatomy (Drake 1978, p. 437). He also was clearly interested in embryology (Ackerknecht 1979, p. 93). His best-known student was William Harvey (1578–1657), who was motivated to do his own experiments on blood circulation by Fabricius's description of the venous valves (Eckart 2000, p. 145). Fabricius was Galileo's personal physician (Fig. 2.12).

Julius Casserio (Giulio Casserius, ca. 1552–1616) had studied under Fabricius but made a name for himself as anatomist, leading to friction between him and his former teacher.[186] In 1600 and 1601, he published his two volumes *De Vocis Auditusque Organis Historia Anatomica* with numerous

(continued)

[185]For Fabricius see Drake 1978, p. 437, there, however, with the wrong date of birth 1533.

[186]http://galileo.rice.edu/Catalog/NewFiles/casseri.html. Author: Richard Westfall.

illustrations.[187] In 1604 he substituted for Fabricius due to illness. He only got his own regular professorship in 1609 (Thorndike 5 (1941), p. 44 footnote 16).

In Padua there was also a garden for medicinal herbs and a model garden for spices (Rossetti 1985, p. 28, Fig. 2.13). The professor of botany had been *Prospero Alpini* (1553–1617) since 1594; he cultivated numerous exotic plants. He also offered botanical excursions. In 1603 he became the successor of the director of the botanical garden, *Melchiore Guilandino*.[188] The successor for his professorial chair may have been *Giacomo Antonio Contuso* (1513–1603) (Friedl 1994, p. 124).

Since 1598 *Andrighetto Aldrighetti*[189] (Andrighetti, 1573–1631) lectured on the *Canon medicinae* of Avicenna (980–1037), emphasizing the third book.[190] Because of its consistent and cohesive presentation, the canon is considered to be *the* standard reference of the Middle Ages. The five books "pursued the theoretical medical science (I), pharmacology (II), special pathology and therapy (III), surgery (IV) as well as the pharmacology of antidotes.[191]" (Eckart 2000, p. 102) *Antonio Negro* (Antonius de Nigris, ?–1658) lectured on general pathology in 1602 and 1603 (Friedl 1994, p. 122, 124).

Orazio Augenio[192] (1527–1603) and *Emilio Campolongo*[193] (1550–1604) lectured on clinical medicine in 1602 (Friedl 1994, p. 122). Augenio held the view that a person trained in astrology could foresee the future, but it would be very difficult to accomplish. Accordingly he had little faith in prognostications about the plague (Thorndike 6 (1941), p. 212). After his death in 1603/1604, his professorial chair was left vacant. Emilio Campolongo held his professorial chair from 1578 and at the same time was chief physician at the Franziskus hospital (Friedl 1994, p. 124f.; Vendova 1 (1831), p. 193f.).

For practical medicine, lectures were held by *Eustachio Rudio*[194] (1551–1611) and *Ercole Sassonia*[195] (1551–1607) in 1602. Both were intensively occupied with the study of syphilis, among other things. Sassonia

(continued)

[187]Casserio, Julius: *De Vocis Auditusque Organis Historia Anatomica*. Ferrara: Baldinus 1600–1601 [SUB Hamburg: C 1946/662].

[188]http://galileo.rice.edu/Catalog/NewFiles/alpini.html. Author: Richard S. Westfall.

[189]For Aldrighetti see Vendova 1 (1831), p. 40f.

[190]Vendova 1 (1831), p. 40; cf. Friedl 1994, p. 122, 124f.

[191]Antidotarium is a designation for old recipe collections and pharmacopoeias. In its modern meaning, it is a directory for antidotes.

[192]For Augenio see Thorndike 6 (1941), p. 211f.

[193]For Campolongo see Vendova 1 (1831), pp. 193–196.

[194]For Rudio see Thorndike 5 (1941), p. 43f.

[195]For Sassonia seeThorndike 6 (1941), p. 237.

blamed the sickness on promiscuity and prostitution. During intercourse the vulva would heat up, resulting in syphilis. Rudio recommended the use of a linen condom to avoid infection (Bergdolt 2004, p. 182; Schonlau 2005, p. 57).

This general curriculum, obligatory for all students, was complemented with extraordinary presentations not conducted by the professors. In 1602, for example, presentations on clinical medicine were held by *Annibale Bimbioli*[196] (Hannibal Pimbiolus, 1577–1613), *Nicolaus Trevisanus*, and *Tarquinio Carpanedo*[197] (Carpinetus, ?–1616) and on practical medicine by *Giovanni Tomasso Minadoi* (1545–1618), *Alessandro Vigenza* (Viguntia), and *Giovanni Pietro Pellegrini*. In 1603/1604 Benedictus Sylvaticus (1575(?)–1658) lectured on Galen.

To finance his studies, Marius received an exceptional annual stipend of 100 guilders; most other students only received 50 (Muck 3 (1880), p. 38, 40). However, Margrave Georg Friedrich died after a brief illness without any children on April 25th 1603 in Ansbach (Schuhmann 1980, p. 105). This passing (Schuhmann 1980, p. 127):

> [. . .] marks a significant incision in the history of the margraves of Brandenburg-Ansbach. Its dominion became smaller due to the independence of the Principality Kulmbach-Bayreuth, the cessation of Prussia and the Silesian Duchy Jägerndorf [. . .] Henceforth its sovereignty was limited to the small Franconian territory of the sub-burgraviate principality.

In 1598 the succession in Franconia had already been regulated, with the candidates being two younger brothers of the Brandenburg Elector Joachim Friedrich[198] (1546–1608). The lot decided that the older brother Christian (1581–1655) received the uplands of Kulmbach and the not yet 20-year-old Joachim Ernst[199] (1583–1625) was acceded the provincial government in Ansbach on July 23rd 1603 (Schuhmann 1980, p. 127). These changes also led to significant distortions in the administration; then for 8 months, Marius received no money (J. Meyer 1892, p. 56; Vocke 2 (1797), p. 415; Zedler 19 1739, col. 1588). But it seems he bridged this period well by teaching astronomy to wealthy students.[200] He mentions a "discipulum [. . .] in astronomy, whose name was Paul Böym von Löwenburg in Reussen,[201] who on my account, because he was very rich, ordered a sextant to be made."[202]

[196]For Bimbioli see Vendova 1 (1831), p. 114.

[197]For Carpanedo see Vendova 1 (1831), p. 230.

[198]For elector Joachim Friedrich, see *NDB* 10 (1974), p. 438f. (author: Johannes Schultze); Schuhmann 1980, pp. 101–106.

[199]For elector Joachim Ernst, see *NDB* 10 (1974), p. 439 f. (author: Hans-Jörg Herold); Schuhmann 1980, pp. 127–130.

[200]According to Vocke (2 (1797/2001) p. 415), he lacked funding in Padua. According to Zedler (19 (1739), col. 1588), it was very hard for him in Padua because of the missing subsistence payments. Marius himself never mentions financial difficulties for his time in Padua.

[201]Reussen is an old name for Russians or Russia. "Löwenburg inn Reussen" therefore probably means Lemberg (Lwiw) in the todays Ukraine. Cf. Zedler 18 (1738), col. 237.

[202]"[. . .] discipulum [. . .] in Astronomicis, dessen Name Paul Böym von Löwenburg inn Reussen, welcher auf meine instruction, als er sehr reich war, liesse einen Sextanten machen" (*Prog. 1618*, sig. A2^v; cf. Zinner 1942, pp. 25f., 65).

Fig. 2.12 Portrait of Hieronymus Fabricius ab Aquapendente (1537–1619). Wikimedia Commons, User Valérie75

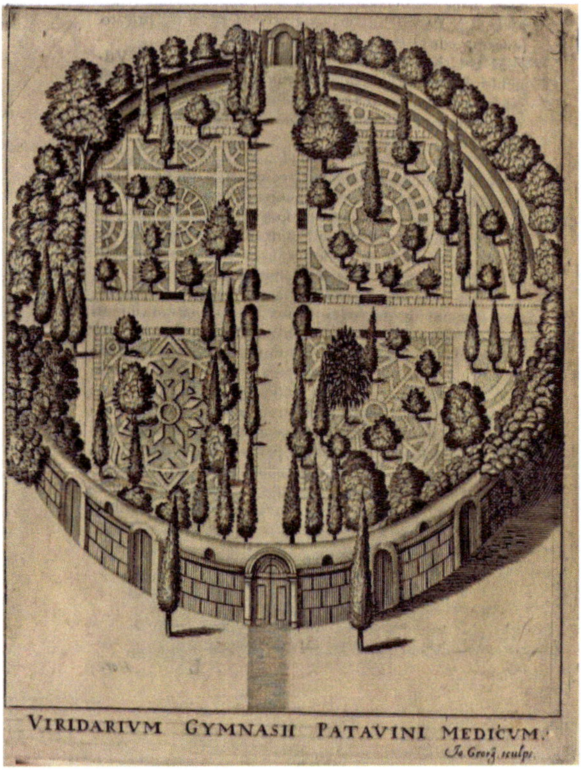

Fig. 2.13 The botanical garden in Padua, 1656. Courtesy of the BSB München, 4 H.lit.p. 216, S. 82, urn:nbn:de:bvb:12-bsb10679955-2

He appears to have tutored him in 1603 and later from1604 also *Baldessare Capra* (1580–1626), the son of Aurelio Capra.[203] Capra had started his studies in mathematics in 1602 (Drake 1978, p. 122). He was also persuaded by Marius "to make astronomical instruments on his own cost," namely, a quadrant and a "mediocre" sextant.[204] A great conjunction in Sagittarius was forecasted for October 8th 1604,[205] whereby Jupiter and Saturn meet in the sky, which occurrs roughly every 20 years. If one marks the positions in the zodiac, where three consecutive great conjunctions occur, an equilateral triangle is formed. Astrology divides the 12 signs of the zodiac into four categories, each forming an equilateral triangle: Aries, Leo, and Sagittarius are the fire signs; Taurus, Virgo, and Capricorn are the earth signs; Gemini, Libra, and Aquarius are the air signs; and finally Cancer, Scorpio, and Pisces are the water signs. The great conjunctions each occur for a period of about 200 years in the fire, earth, air and water signs. Kepler commented to that[206]:

> Fiery triangle [. . .] is a period of 200 years, within which the two upper planets Saturn and Jupiter start to collide, nowhere else as in the heated signs Aries, Leo and Sagittarius.

The particular aspect of the conjunction of 1604 was that for the first time since 600 (or rather 800) years, it had entered the fiery signs. Additionally Mars likewise entered Sagittarius on September 13th and on October 9th was only 16 arcminutes from Jupiter, between Saturn and Jupiter.[207] Astrologers and astronomers had good reason to pay particular attention to this conjunction.[208]

During their joint observation on October 10th 1604, Marius and Capra in the presence of the nobleman Camillo Sasso discovered a new star.[209] It was located on

---

[203]Drake 1978, p. 83. Marius and Capra manufacturing telescopes together is, of course, nonsense (Liesenfeldt 2003, p. 52).

[204]*Prog. 1606*, sig. A3^{r}; cf. Zinner 1942, p. 49.

[205]In fact the conjunction occurred the following day: Jupiter and Saturn were in Sagittarius at 19°9′ or 10°51′, only being 8°18′ apart.

[206]Feuriger Triangel [. . .] ist eine Zeit von 200 Jahren, innerhalb deren die 2 obersten Planeten Saturn und Jupiter anfangen, nirgends anderswo als allein in den drei hitzigen Zeichen Widder, Löwe und Schütze [. . .] zusammenzustoßen" (Kepler 1926, p. 77).

[207]http://www.astro.com/swisseph/ae/1600/ae_1604.pdf; 17.12.2014; cf. Westman 2011, p. 382, according to whom Mars only entered Sagittarius on September 29th.

[208]Kepler (2006, p. 8) wrote about this: "whatabout the astrologers have written so much because the fiery triangle starts exactly in the month, when Mars also meets these two highest planets" ("[. . .] darvon die Astrologi so viel geschrieben/das der fewrige Triangul drinnen angehe/gerad in den Monat/drinnen auch Mars zu baiden höchsten Planeten khommen").

[209]"dico che secondo il costume mio (che era di osservare ogni giorno si le Stelle erranti come le fisse) volendo ridurmi con il Sig. Simon Mario Alemano mio carissimo Maestro in questa professione, & il Sig. Camillo Sasso gentil'huomo Calabrese, il giorno dieci di Ottobre, ad osservare marte, Giove & Saturno, mentre io preparavo un mio quadrante per pigliare certe altezze d'alcune stelle fisse, per havere l'elevatione del Polo di Padova, li Sig.nori sopradetti si conferirono a vedere li sopradetti Pianetti, & mentri il Sig. Simone fra di se sospeso stava mirando la nova Stelle" (Capra 1891, p. 293). Marius described his discovery a bit differently in the *Prognosticon for 1606* (sig. A3^{v}; cf. Zinner 1942, p. 49): "When we didn't realise anything about this new star. On the 29th it was very foggy because of the conjunction of ♃ and ♂ the following day, September 30th

the base of Ophiuchus with its coordinates being "17 degrees/43 minutes of Sagittarius" (Kepler 2006, p. 7), only a little more than 1 degree distant from Jupiter and Mars. Since the following nights were cloudy, they could only confirm their observations on October 15th.[210] This star was visible for about 1 year. In a text authored in the end of August 1605 for his *Prognosticon* of the following year, Marius mentions the new[211]:

> [...] wonder star anno 1604, September 30th in the old calendar or October the 10th N.C., seen for the first time, and still at the same place without having moved/but very small.

Marius had anticipated the appearance of a comet because of the conjunction of the upper planets (Zinner 1942, p. 36), which was confirmed by Caspar Hofmann in the protocols of the German Nation: "On October 10th this star appeared, predicted eight years ago by my dear colleague."[212] In fact Marius had announced the appearance of a comet for the period after 1603 in his comet tract of 1596, but he didn't predict a definite date.[213]

At this point Galileo comes in play. During Marius's time as a student, Galileo was a professor for mathematics in Padua who had not yet made a significant name for himself. He held lectures on the *Sphaera* of Sacrobosco, the *Elements* of Euclid, as well as an introduction to calculating planetary positions (Drake 1978, p. 35), topics

or October 10th N. C. [New Calender], which was a Sunday. In the evening I went to the aforementioned garden to observe, and I went alone, meanwhile it was quite light, but afterwards I saw ♂ and ♃ very close together, but I found between the two another star, which never ever stood there before, and it was similar to Mars in its redness and size, but it sparkled, which the planet Mars does not do. So I stood there and was wondering about the new light, when my discipulus came by and also found, that there is a newly generated light in this place." ("Da wir dann im geringsten nichts von dem neuen Stern vermercket/den 29. ward es sehr nüblicht von wegen der vereinigung ♃ vnnd ♂ folgendes tags/als den 30. 7ber oder 10. October N.C. welches war ein Sontag/kame ich auff den Abend wider in benandten Garten zu observiren, vnd gehe also allein/dieweil es zimblich hell war/hernach aber/besehe ich ♂ vnnd ♃ die nahe beysammen waren/befinde aber zwischen jhnen beeden einen andern Stern/der zuvor niemals allda gestanden/vnd ware durchauss dem Marti gleich an röte vnnd grösse/allein dass er funckelte/welches Mars damals ein Planet nicht thate. Stehe also vnnd verwunder mich vber diesem neuen Liecht/in mittels kompt mein discipulus auch darzu/vnd befinden/dass es ein neu generites Liecht allda sey").

[210] "[...] continue pioggie fu impossibile il vederla, alli 15 di Ottobre finalmente si fecce serenità" (Capra 1891, p. 293; cf. Drake 1978, p. 104).

[211] "Wunderstern Anno 1604. den 30. Septembris altes Calenders oder 10. Octobris N.C. erstlich gesehen worden, vnd noch an solchem Ort fix vnd vnverruckt/doch sehr klein gesehen wird," *Prog. 1606*, sig. B1^r; cf. Zinner 1942, p. 51. In the *Prognosticon for 1608* (sig. B4^r; cf. Zinner 1942, p. 54), he attacked Johann Krabbe (1553–1616) from Wolfenbüttel for his claim, "that the new star of the years 1604 and 1605 had had a motion on its own [...] this Mr. Johann Krab must be a comical, diligent and undaunted talent, though up to now I don't understand his instruments or his method of observation" ("dass der neue stern Anno 1604. vnd 1605. einen proprium motum hab gehabt [...] so muss dieser H. Johann Krab/ein lustig fleissig vnnd vnverdrossen ingenium sein, wiewol ich seine instrumenta wie auch sein modum observandi noch derzeit nit verstehe").

[212] "10. Octobris apparuit stella illa, quam hinc octennium praedixerat collega meus amicissimus, et de qua tot etiam nun quotidie eduntar chartae" (Favaro 2 (1912), p. 225).

[213] Cf. Marius 1596, sig. B1^v–B2^r, as well as his *Prognosticon auf 1606*, sig. A3^r.

Marius probably was not much interested in since he already mastered them.[214] The news about the new 1604 star also reached Galileo, who gave three big lectures about it in December and January to a total of 1000 listeners (Westman 2011, p. 386). They were never published in print and only one page of a manuscript is preserved (Westman 2011, p. 582 footnote 11). It can hardly be assessed what he said.

Capra and presumably Marius attended these lectures.[215] Capra published an astronomical paper[216] about the findings in Padua 1605, about which Marius later claimed:[217]

> During the foreseen great conjunction in Sagittarius, a majestic new star appeared in Sagittarius in the autumn of 1604/much has been written about this/I have also dictated in Italian a tract to my mathematical student Balthasar Capra of Mailand nobility/which he also published in his name with credits to me in the Italian language/in which I disproved a noble professor of philosophy, who had published inept things against the astronomical observations.

The "noble professor" could have been Antonio Lorenzini, who published a *Discorso* about a new star in January 1605 in Padua (Lorenzini 1605). Capra strove to refute the claims in the *Discorso*. For example, the nova could not be evaporation from Earth because of its uniform movement with the stars (Westman 2011, p. 387). However, identifying this professor as Lorenzini is problematic; Capra never mentions this name in his work and Lorenzini was not a Paduan professor.

---

[214]However Marius occupied himself intensively with astronomy during his stay in Italy, as he writes in his *Prognosticon for 1606* (sig. A4^{v}; cf. Zinner 1942, p. 51): "nevertheless I prosecuted in foreign countries beside my studies in medicine even those in astronomy, and have taught it with great hounour to several Italiens" ("gleichwol in frembden Landen, neben meinem studio Medico auch dz Astronomicum ich getriben/vnd mit sondern ehren bey etlichen Italis fortgepflantzet habe").

[215]"havendo veduto che l'Eccellentissimo Sig. Galileo, nelle sue dotte lettioni, che di questa Stella alli giorni passati publicamente fece" (Capra 1891, p. 291).

[216]Capra, Baldessare: *Consideratione astronomica circa la nova, & portentosa Stella che nell anno 1604. adi 10. Ottobre apparse. Con un breve guidicio delli suoi sig.nificati.* Padua: Nella Stamparia Di Lorenzo Pasquati 1605 [Florence: Biblioteca nazionale centrale: FI0098 CFICF]. The Consideratione astronomica is quoted from the edition from 1891; see further the list of references.

[217]"Dieweil auf vorgedachte grosse vereinigung im Schützen folgents 1604. Jar im Herbst der Herrliche schöne Newe Stern im Schützen erschienen ist/davon viel schreibens gewesen/ich auch zu Padua im Welschland meinen in Mathematicis discipulo Balthasar Capra, einem Meylandischen vom Adel einen Tractat in die Feder dictirt/welchen er auch vnter seinem Namen/mir zum besten/in welscher Sprach hat trucken lassen/dieweil ich in sol¬chen einen vornehmen Professorem Philosophia daselbsten/welcher gantz vngeschickte sachen wider die observationes astronomorum hette in truck publicirt nach nohtturfft widerleget habe" (*Prog. 1623*, sig. A2^{v}). In the *Prognosticon for 1606* (sig. A2^{v}; cf. Zinner 1942, p. 48), he wrote about this: "though this was already done by my beloved student in Italy named Balthasar Capra, a nobleman from Milan, but in south European and so in a, in this country, unknown language, who because of his inclined will and his faithfull heart againgst me, did this for the sake of this art in Italy against my despisers and enemies as a thankful student for the sake of truth and my innocence" ("Wie wol solches allbereit von meinem geliebten discipulo in Italia mit Namen Balthasar Capra einem Meylendischen vom Adel, aber in Welscher vnd dieser Landen vnbekandtenr Sprach, ist verrichtet worden, der auss Antrieb seines geneigten Willen und getrewen Hertzen gegen mir, zur Rettung der Wahrheit, vnd meiner vnschuld, wider meine Verächter und Feinde wegen dieser Kunst in Italia solches als ein danckbarlicher discipul gethan hat").

This leaves Galileo as the most likely candidate, although he never published about the nova then. One could understand Marius to mean that Galileo abstained from publishing because of Capra's publication. Galileo seemingly claimed that he saw the great conjunction on October 9th and "the new light" for the first time on the following day. Capra attacked him because of this statement.[218]. Capra compared its color and brightness to Mars.[219] A few days later, he apparently talked to the Venice nobleman Giacomo Cornaro (1483–1542), who in return informed Galileo.[220] If Galileo withheld Cornaro's role, then he indirectly did not recognize the observation made by Capra and Marius. Furthermore, identifying the professor as Galileo is not completely unproblematic because there are no "inept things against astronomical observation" known of him. However, Marius' statement was made 20 years after the event; possibly Marius did not remember it correctly.

In 1607, 2 years after Marius left Padua, Capra published a tract about a proportional compass, dedicated to the Elector Joachim Ernst of Brandenburg (Capra 1607, sig. A2^{r}). In the foreword Capra mentions Marius as his mathematics teacher.[221] That Marius knew nothing about this tract is highly unlikely against this background. It was plagiarized; Capra merely made a Latin translation of Galileo's work. Justifiably, Galileo objected, although he made Marius responsible for Capra's work only much later in Il *Saggiatore*. Marius never reacted to this accusation (Fig. 2.14).[222]

[218]According to Klug (1906, p. 404), Galileo presented Marius and Capra to the public as the discoverers of the new star, rendering Capra's statements in his scripture hardly comprehensible. It is probably only the matter of another example of Klug's bias against Marius.

[219]"[. . .] vidi una Stella nel colore, & grandezza in tutto simile a Marte" (Capra 1891, p. 293).

[220]"Doppo dunque che alli 15 fu di novo osservato il predetto portento, andando un giorno a visitare l'Illustriss: Sig. Iacomo Aloysio Cornaro [. . .] Da questo cavasi una conclusione necessaria, cioè che l'Eccellentiss: Galileo habbi havuto il tempo, & il loco di questo novo portento dall'Illustriss: Cornaro" (Capra 1891, p. 294; cf. Westman 2011, p. 386).

[221]"[. . .] inter alios Germanos quos mei amantissimos esse non semel expertus sum, accessit Simon Marius Guntzenhusanus, is illa qua praeditus est humanitate, & rerum mathematicarum cognitione, quae animus meus maxime desiderabat adeo concinne & miro ordine exposuit" (Capra 1607, sig. A3^{r}).

[222]Together with the *Tyrocinia astronomica* (*Astronomical exercises*), Capra published another, 14 pages long, astronomical work, signed in the preface with April 30th 1606. In it he explains the calculation of the solar eclipse according to Tycho Brahe as well as the setup of a horoscope according to Ptolemy. Marius isn't mentioned at all. He refers rather to Giovanni Antonio Magini (1555–1617), whose ephemerides were very popular, since they were the first using the Gregorian calendar. Furthermore he used the trigonometry textbook by Thomas Finck (1561–1656) from Flensburg: *Geometria rotundi*. Basel: Heinrichpeter 1583 [BSB München: 4 Math.p. 124]. As an example Capra used the new star of 1604, however, without mentioning any reference that he had found it together with Marius in Padua. In May 1606 he published another two disputes: *Disputationes duæ*. Vna De Logica, & eius partibus, Altera De Enthymemate. Padua: Pietro Paulo Tozzi 1606 [Houghton Library: IC6.C1748.606d], a work about logic and logical deductions.

Fig. 2.14 Cover of Capra's script about the proportional compass of 1607. BSB München: 4 Math. a.64, sig. A1^{r}, urn:nbn:de:bvb:12-bsb10052795-9

In 1604 Marius again received 100 guilders from the margrave, in 1605 another 150, "so he can discharge himself and travel back."[223] The call for his return to Ansbach seemed to be surprising for Marius. Only in June 1605, he had been confirmed as the councilor of the German Nation (Rossetti 1986, p. 133); 1 month

[223]Quoted according to Muck 3 (1880), p. 40.

later he departed.[224] Before leaving he signed the register[225] of Heinrich Hartmann (1577–1625) from Wolfhagen near Kassel: "Padua, 1605, Simon Marius Francus."[226]

In the middle of July 1605, Marius was already back in Southern Germany: "Anno 1605 on July 15th/25th [. . .] there was a heavy storm early in the night, when I lay in a village, one mile from Donauwörth near Augsburg."[227] At the latest by the end of August, he appears to have been back home.[228]

The reason for the call to return might have been related to the death of the pastor Georg Caesius[229] (1543–1604), who had died on September 4th 1604. He had acquired a good name as a calendar writer (Zedler 65 (1748), col. 1709f.). Since April 1577 he received an annual payment of 25 guilders from the margrave as appointed astronomus—even though Caesius never called himself that. Apparently he got the money when he delivered his calendar and practica each New Year.[230] Following Hocker, Oertel claims that Marius became the successor of pastor Johann Schülin (Schulin, 1561–1606) from Gnodstadt[231] (Oertel 1775, p. XIII.), who died in 1606. However, Klaus Matthäus doubts that Schülin had ever been the appointed court astronomer.[232] Marius's return from Padua in 1605 further reinforces this doubt.

[224]"In July 1605 [...] I traveled from Italy through the Alps, there was a severe heat and tremendous thunder" ("sonderlich Anno 1605. im Julio [. . .] Ich bin damals eben auff der Reiss auss Italia in den Alpibus gewesen, war grosse Hitz, vnd gewaltig gedonnert," *Prog. 1628*, sig. B3^r, cf. Zinner 1942, p. 72).

[225]This family register is located in the state library in Kassel (Lehsten 1 (2003), p. 369 footnote 599).

[226]Stölzel 2 (1872/1964), p. 36. Hartmann completed a remarkable grand tour: 1600 London, Oxford; autumn 1601 Paris, Bourges, Orléans; 1603 again in London; July–Aug 1603 s'Gravenhage, Gent, Leyden; autumn 1603 Heidelberg; March 1604 Straßburg; summer 1604 Venice, Padua, Bologna, Florence; 1605 Pisa, Venice, Verona, Rome, Padua; autumn 1605 again in Germany; 1609 Lübeck, Hamburg, Bremen, Magdeburg, Goslar (Lehsten 1 (2003), p. 369 footnote 955). Hartmann stayed from the beginning of May until the end of August in Padua. He later became the mayor of Wolfhagen (Lehsten 1 (2003), p. 369; cf. also Stölzel 1 (1872/1964), pp. 66–68).

[227]"Anno 1605 den 15./25. Juli [...] da hat es die Nacht zu frühe ein hefftig Gewitter gehabt und eingeschlagen, als ich in einem Dorff lag ein meilwegs von Donnawerth gegen Augspurg" (*Prog. 1612*, sig. B4^v; cf. Klug 1906, p. 398 footnote 2).

[228]"[. . .] for I'm writing this, after my journey from Italy, at the end of August" ("denn, da ich dises schriebe, nach meiner Reiss aus Italia, als zu Ende des Augustmonats," *Prog. 1606*, sig. E1^r; cf. Zinner 1942, p. 51).

[229]For Caesius see Barnes 1988, p. 151, 160; Leppin 1999, p. 183; Matthäus 1969, col. 1087–1092; the funeral sermon by Michael Lochner (Nürnberg 1604) as well as the chapter by Dieter Kempkens in this volume.

[230]Matthäus 1969, col. 1089f. Marius seemingly liked the practicas by Caesius very much, because in his *Prognosticon for 1602* (sig. A3^v; cf. Zinner 1942, p. 47) he wrote: "and finally because I am not equipped with old observations (which are important in these things and are well noticed in the practica of Herr Caesius)" ("vnd endlich auch mit alten obseruationibus (die viel in solchen sachen thun/vnnd in deß Herren Caesij Järlichen Pratiken[!] wol gespürt wirdt) nicht gerüstet bin").

[231]Gnodstadt is located a few kilometers southeast of Ochsenfurt.

[232]For Schulin see Matthäus 1969, col. 1093–1096.

Caesius also recorded the weather, because in the *Prognosticon for 1612* Marius refers to those "who have the storm register of the blessed Herr M. Georg Caesius to hand" (*Prog. 1612*, sig. B2^{v}).

Court Mathematicus in Ansbach

After his return from Italy, Marius seemingly stayed with his relatives in Gunzenhausen, since his *Prognosticon* for the year 1606 is signed "Guntzenhausen, September the 12th 1605."[233] In the following year a feast was held in his honor, with all 24 members of the council in attendance.[234]

As of 1606 Marius was employed by the Margrave of Ansbach (Fig. 2.15) and received an annual salary of 150 guilders[235] until his death—a relatively sparse

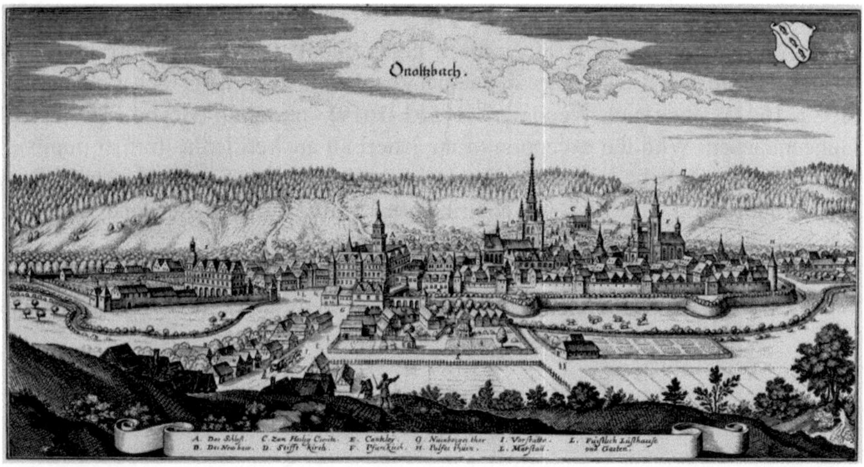

Fig. 2.15 Ansbach in the seventeenth century. Copper engraving. Private property of the author

[233]*Prog. 1606*, sig. A4^{v}. Already in the *Prognosticon for 1610* (sig. B1^{v}), he made remarks about the weather, which "can be found in the register of the blessed Georg Caesius" ("in Herrn Georg Caesij seligen verzeichnuß zu finden").

[234]"In the Mayor's office accounts for 1606 one can find under the category expenditure for meals, which is so to say the representation fund of the aldermen, the expense of 8 fl. 2 ort, which by Georg Bauer the complete honest city councilor consumed, as Herr Simon Marius was the special guest" ("In der Bürgermeisteramtsrechnung von 1606 erscheint unter der Rubrik Außgeben uff Zehrung, quasi dem Repräsentationsfond der Stadtoberen, die Ausgabe über 8 fl. (Gulden) 2 ort bey Georg Bauer ein ganzer Ehrbarer Rath verzehrt, alß man Herrn Simon Maiern zu Gast gehabt," Mühlhäußer 2012, p. 42, by reference to Municipal archives Gunzenhausen, Repertorium I Fach 76/1, Nr. 2).

[235]Muck 3 (1880), p. 40. For comparison, Fuchs von Bimbach had an annual salary of 2581 guilders, the vice-chancellor Simon Eisen however only 463 guilders. Other councilors like Johann Strebel also only got—like Marius—50 guilders; Herold 1973, p. 46, footnote 66, his source: National Archives Nuremberg: Rep. 103 a I AGR-Akten Nr. 16.

remuneration: "I own neither money nor gold, and the hardship of almost all mathematicians is so to speak an inevitable side effect" (Marius 1614, sig.)(3^{v}). He was active not only as a mathematician but also as a physician. His "appointment" was graciously ordered "so that besides the study of medicine, I can also execute the study of mathematics" (*Prog. 1607*, sig. A4^{r-v}). It also seems that he was in demand as a physician, because he sometimes complained about the "peasants/up early in the morning/seeking advice from me, because of illness" (Marius 1619, sig. A4^{r}).

Family Life in Ansbach

On May 8, 1606, Marius married the 16 year old Felicitas Lauer, the daughter of his Nuremberg publisher Johann Lauer. The following people congratulated in writing: the Ansbach Stift preacher Balthasar Bernhold, whom Marius probably got to know as his teacher in Heilsbronn; Kaspar Finck[236] (1574–1632) and Johann Löser[237] (1569–1635), both teachers at Ansbach high school; the Brandenburg councilor Johann Hohenstein[238]; the later municipal chaplain Johann Christoph Lohbauer[239] (1582–1641); and Paul Weniger[240] (1552–1619), pastor of Beyerberg west of Gunzenhausen. With the exception of the latter, all are verified as former pupils of the "Fürstenschule."[241] Kaspar Finck himself married Sarah Merklein from Leutershausen on July 8, 1606; contributions from Marius, Löser, and Lohbauer can be found in the congratulatory letter.

Felicitas Lauer and Marius had seven children. The two first born were the sons Johann Balthasar and Johann Samuel. Balthasar died after a quarter year, Samuel

[236]In 1606 Kaspar Finck became the teacher of the third grade in Ansbach. From 1610 on he taught the fourth grade. In 1611 he became pastor in Insingen, 1619 in Obernbreit. In 1628 he was dishonorably discharged; cf. Simon 1957, p. 123 entry 735.

[237]As of 1600 Johann Löser was the teacher of the second grade in Ansbach. As of 1603 he taught the third grade, as of 1605 the fourth. In 1607 he finally became pastor in Dornhausen southeast of Gunzenhausen; cf. Muck 3 (1880), pp. 38–40; Simon 1957, p. 294f. entry 1777.

[238]For Hohenstein see the list of consistory councilors above.

[239]In 1612 Johann Christoph Lohbauer became assistant professor at Wassertrüdingen. In the same year he transferred to the rectorate in Seeheim, 1614 to the same position in Uffenheim. In 1616 he became the town chaplain in Ansbach, 1619 pastor in Schmalfelden; cf. Simon 1957, p. 295 entry 1781.

[240]Paul Weniger became pastor in Bofsheim in 1591. In 1598 he transferred to Markt Breit, 1601 finally to Beyerberg; cf. Simon 1957, p. 546 entry 3260.

[241]Bernold 1582–1588; Finck 1595–1601; Löser 1582–1587; Hohenstein 1582–1588; Lohbauer 1595–1602; cf. Dannheimer 1959, p. 161, 167, 171f.

after 3 weeks.[242] All the other five children were daughters.[243] Marius referred to this, when he wrote to the margrave in 1614[244]:

> Herein, for myself and my descendants, if there shall be any, I gratefully acknowledge the extreme generosity of the so noble prince

If at all, one daughter may have survived him.[245] He did not count her as a descendant.

Simon Marius died on the evening of December 26, 1624 (Julian calendar).[246] He repeatedly stressed his poor health,[247] which was supposedly worsened by "a

---

[242]Johann Balthasar was baptized on June 7th 1607 and buried in September 17th. Johann Samuel was baptized on June 16th 1608 and already buried on July 6th (Landeskirchliches Archiv, Nürnberg: Kirchenbuch Ansbach, St. Johannis: LAELKB_46_6 Sig. 69 entry 102, Sig. 65 entry 111; LAELKB_46_70 Sig. 9 entry 179, Sig. 12 entry 65).

[243]Anna Margaretha * June 28th 1609; Maria Magdalena * May 7th 1611; Margaretha Elisabeth * August 23rd 1612; Margaretha Barbara * July 18th 1614; Helena Susanna * October 10th 1615 (Landeskirchliches Archiv, Nürnberg: Kirchenbuch Ansbach, St. Johannis: LAELKB_46_6 Sig. 7 entry 14, Sig. 97, entry 160, Sig. 100 entry 172, Sig. 172 entry 162, Sig. 58 entry 82). According to Zinner 1942, p. 26, Marius had five sons and five daughters, with only the daughters surviving the father. Zinner gives no source for his claim.

[244]"Qua in re ego cum posteris meis, si qui erunt summam liberalitatem Tantorum Principum ex tam Illustrissima familia Oriundorum, gratus agnosco" (Marius 1614, sig. B1^{v}).

[245]The death records for St. Johannis in Ansbach show a big gap as of 1620. In the Simon Marius high school, a family tree was found, printed in Wolfschmidt 2012, p. 160f. Here the seven children of Marius are listed correctly with their names but without their dates of birth. Also, there is a Michael Marius (1593–1660/61) listed, who apparently died on February 21th 1660 or 1661 as a Royal Scribe in Lobenhausen near Crailsheim. The existence of this Michael Marius is confirmed by a family register listing of him in the album of Johann Balthasar Bernhold (1618–1652) (Schnabel 1995, Nr. 82/9) that he is the son of Simon Marius is not verifiable. Furthermore on November 7th 1621, Michael Marius from "Guntzenhusanus Altmülensis" matriculated in Straßburg (Knod 1897, p. 592). Regarding his age it would be a good fit to the Michael from Lobenhausen. He married a Catharina Barbara Strebel. Their son Theodor (1640–1690), born in Lobenhausen, became pastor in Württemberg (Haug 1981, p. 285 entry 1671). Simon Marius's brother Michael baptized a son on January 9th 1596 on the name Michael, who died a few weeks later (e-mail on February 13th 2015 from Claudia Heuwinkel of the Municipal Archives Creglingen). He therefore is definitively not the father of Michael Marius from Lobenhausen. The Michael Marius, who matriculated in Straßburg, could also be a Mayr from Gunzenhausen, who wasn't related to the family of Simon Marius. For example, on December 21st 1598, a Michael Mayer was baptized, son of Nicolaus Mayr from the anterior suburbs. Furthermore on February 15th 1590, a son of Hans Mayr from Oberasbach and in July 29th 1592 a son of Hans Mayer from Unterwurmbach were baptized with the name Michael (information by municipal archivist Werner Mühlhäußer from Gunzenhausen). Therefore the Michael Marius from Straßburg is not necessarily related to the family of Simon Marius.

[246]Oertel 1775, p. XXVI: "MARIVS ipse secunda feria natiuitatis Christi, A. 1624. sub meridiem inter horam X. et XI. breui defunctus morbo, anno aetatis [...]."

[247]"I will keep silent now about the weakness of my body and especially that of my head" ("wil jetz geschweigen meiner Leibs sonderlich aber deß Haupts Schwachheit," *Prog. 1609*, sig. A3^{r}); "with such effort as made possible by my lengthly disease of my head" ("mit solchem fleiss als mir in dieser meiner langwirigen Hauptkranckheit müglich gewesen," *Prog. 1616*, sig. A4^{v}); "though I incline from nature, to find more with the help of God, than to learn from others, because of which I

perilous fall from an impressive height" in Italy.[248] In a letter to Mästlin, he apologized for his late response because of "cerebri imbecillitatem" (Zinner 1942, p. 45); he may have suffered from migraine. In 1623 he complained[249]:

> [. . .] and the defamation was my biggest misfortune since my youth/for the rest of my life I won't get rid of it.

The last work of Marius—apart from the calendars—was published posthumously in 1625 by Daniel Mögling[250] (1596–1635) in Frankfurt a. M. The dedication was signed by Mögling on December 10, 1624 (Marius 1625, p. 4)[251]:

> For not only the Lord [to whom the work was dedicated]/but also the author himself (that he still in life) may be content and satisfied

The preface for his *Prognosticon for 1625* was finished by Marius on June 8, 1624, without mentioning his illness (*Prog. 1625*, sig. A3^r). But he appears from the foreword in the tract published by Mögling no longer able to see the tract through the press. He was thus probably seriously ill in his last year of life.

In the *Ansbacher Todten-Almanach*, Johann August Vocke (1750–1810) praises above all the piety of Marius: "He was an eager admirer of the religion and had read the Bible 19 times" (Vocke 2 (1797), p. 415). His widow apparently possessed a property in Schlauersbach near Neuendettelsau, which she sold for 550 guilders (Muck 3 (1880), p. 40f.).

Life at Court and Publications

Initially Marius's most important contact at the Ansbach court was Colonel Hans Philipp Fuchs von Bimbach, who was in the margrave's services as of 1601 at the latest.[252] From 1607 to 1610, he was the director of the privy and the court councils, making him the most powerful official at the court. "Because of his arrogant and coarse nature he did not get along with the other councilors. After it came to

afflicted and suffered the loss of a good part of my life and my health, the ones who know me since my very youth, they can give witness to this" ("wiewol ich von Natur geneigt bin/mehr selbst durch Gottes gnad zuerfinden/als von andern zu lernen/darüber ich auch guten theils meines Lebens und gesundheit zugesetzet vnnd eingebüsset hab/die mich von Jugent auff gekennet/die können mir dessen warhafftig zeugnuß geben," *Prog. 1619*, sig. A3^r); "while I at night, as an ill and weak man did not stay sleeping" ("dieweil ich ja bey Nacht/als ein Krancker schwacher Mann nicht schlaffen geblieben," Marius 1619, sig. A2^r).

[248] "Quod malum in Italia casus ab alto lethalis plurimum auxit" (Marius 1614, sig.)()(1^r).

[249] *Prog. 1627*, sig. D3^r. The preface is dated on 1623.

[250] Mögling was a Rosicrucian (Edighoffer 2002, p. 12). For him see BBKL 5 (1993), col. 1582–1584 (author, Ulrich Neumann); Seck 1 (2002), p. 40f. For his mechanical art chamber of 1629, a lot of references can be found in Stöcklein 1969.

[251] "Damit nicht allein der Herr [dem die Schrift gewidmet ist]/sondern auch der Author selbsten (da er noch im Leben) wol content vnnd zufrieden seyn" (Marius 1625, p. 4).

[252] See the chapter by Wolfgang Dick in this volume.

confrontations with the prince in 1614, the Colonel entered imperial service in 1616 and even Danish service in 1625" (Schuhmann 1980, p. 129). He was killed in action on August 27, 1626, at the Battle of Lutter am Barenberge (J. Meyer 1892, p. 62).

"On order of the noble and strict Lord/Hans Philipp Fuchs von Bimbach," Marius published a German translation of the first six books of Euclid's *Elements* in 1610 (Fig. 2.16); the dedication to the Margraves Christian and Joachim Ernst were signed by him on January 6 (Marius 1610, sig. A2^{v}). The edition seems to have been 300 copies (Doppelmayr 1730/1972, p. 170, footnote qq). In his own dedication, Fuchs von Bimbach defended his decision to "order the appointed mathematicus to make such a translation into German."[253] The *Elements* had already been "48 years ago brought out in the High German language in the year 1562."[254] Back then the first six books of the *Elements*[255]:

> [. . .] were translated from Greek into High German language for the first time by the erudite Wilhelm Holtzmann[256] [Xylander, 1532–1576]/at the time professor for Greek at the university of Heidelberg/with many beautiful artistic illustrations, brought to light.

Already in 1532 the Bamberger practical mathematician (Rechenmeister) Wolfgang Schmid had published *Das erste Buch der Geometria*, classified as the oldest German edition of the *Elements*, but this text was only nominally based on Euclid (Reich 1996, p. 189). In 1555 Johannes Scheubel[257] (1494–1570) translated the books VII, VIII, and IX (Schönbeck 2003, p. 233). In 1562 followed Holtzmann, whose work was also known to Kepler, which is why he asked what Marius had changed.[258]

According to Fuchs Holtzmann's work was very hard to obtain and the translation was unclear.[259] He defended the use of the German language, even though new technical terms had to be invented. Kepler mentions in this context that Marius was also speaking of parallels in the German version. He hoped for an implementation of

[253]"[. . .] bestalten Mathematicum, zu solcher umbsetzung vnd Verdeutschung zubewegen verursachet" (Marius 1610, sig. A3^{v}).

[254]"[. . .] für 48. Jahren/als Anno 1562 auch in Hochdeutscher Sprach außgangen" (Marius 1610, sig. A4^{r}).

[255]"[. . .] durch den hochgelärten Herrn Wilhelm Holtzmann [Xylander, 1532–1576] see/derzeit Griechischen Professorn bey der Vniversitet zu Heydelberg/erstmals auß der Griechischen in unser Hochteutsche sprach übergesetzt/und mit vielen schönen künstlichen anhengen illustriert, anß licht gebracht [worden]" (Kurz 1618, preface, sig. **2^{r}. Cf. Schönbeck 2003, pp. 232–234).

[256]For Holtzmann see *ADB* 44 (1898), pp. 582–593 (author: Fritz Schöll); Drüll 2002, p. 562f.

[257]For Scheubel see Reich 1996.

[258]Kepler too pointed to the works of Holtzmann and Scheubel: "Nihil per has turbas ad nos importatur, cuperem inspicere librum, si quid in versione Xylandrina [...] mutatum sit. Extat et Schebelij, ni fallor, versio trium VII. VIII. IX" (Kepler XVI, 1954, p. 388; cf. Klug 1906, p. 420).

[259]The famous first sentence of the *Elements* reads in the translation by Xylander as follows: "Ain punct oder tipfflein/wirtt das genant/so khain thail hatt." The diameter describes Xylander as "des zirckels Diameter," the triangle as "Triangel" (Xylander 1562, pp. 1–3). This is not an unclear translation into German.

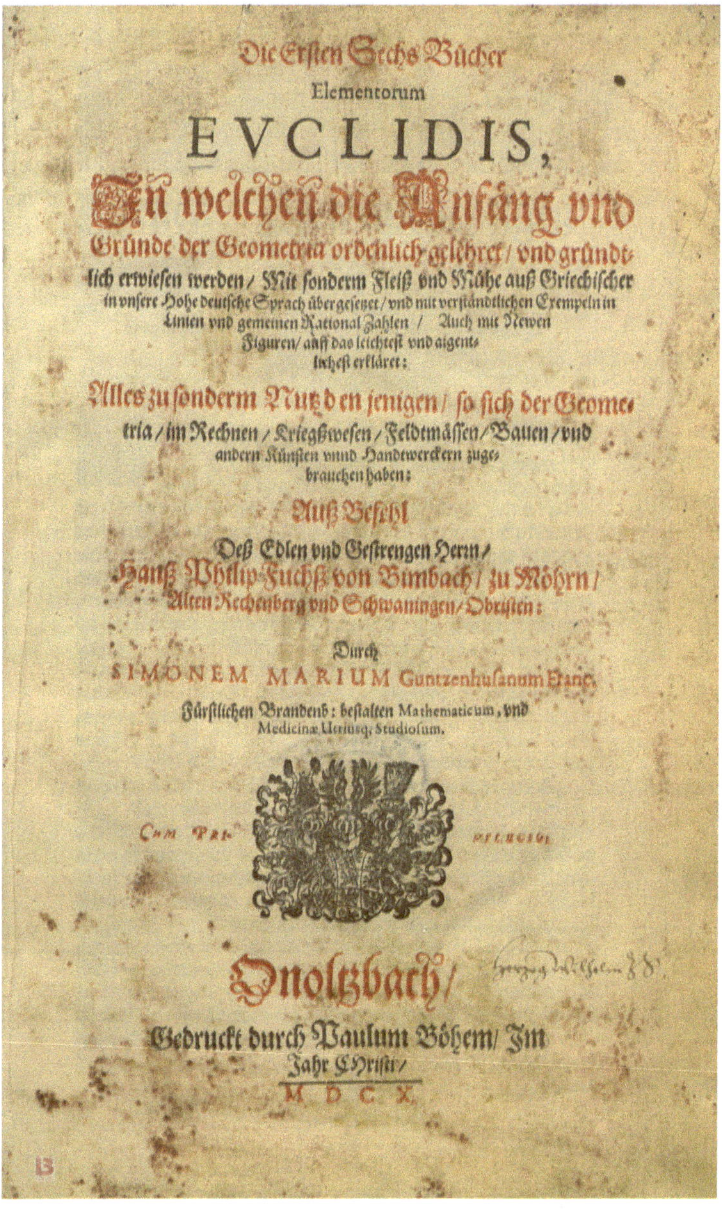
Die Ersten Sechs Bücher
Elementorum
EVCLIDIS,
In welchen die Anfäng vnd Gründe der Geometria ordentlich gelehret/ vnd gründtlich erwiesen werden/ Mit sonderm Fleiß vnd Mühe auß Griechischer in vnsere Hohe deutsche Sprach übergesetzet/ vnd mit verständtlichen Exempeln in Linien vnd gemeinen Rational Zahlen / Auch mit Newen Figuren/ auff das leichtest vnd aigentlichest erkläret:
Alles zu sonderm Nutz den jenigen/ so sich der Geometria/ im Rechnen/ Kriegßwesen/ Feldtmässen/ Bauen/ vnd andern Künsten vnnd Handtwerckern zugebrauchen haben:
Auß Befehl
Deß Edlen vnd Gestrengen Herrn/
Hanß Philip Fuchß von Bimbach/ zu Möhrn/ Alten Rechenberg vnd Schwaningen/ Obristen:
Durch
SIMONEM MARIUM Guntzenhusanum Franc.
Fürstlichen Brandenb: bestalten Mathematicum, vnd Medicinæ Utriusq; Studiosum.
Cum Privilegio
Onoltzbach/
Gedruckt durch Paulum Böhem/ Im Jahr Christi/
M D C X.

Fig. 2.16 Cover of Euclid's Elements in Marius's translation. Courtesy of the LB Coburg: Cas A 970, sig A1r, urn:nbn:de:bvb:70-dtl-0000027017

a standardized linguistic usage and not that different terms would be used everywhere.[260]

According to Fuchs, a good German translation was urgently necessary, because[261]:

> In what great errors are the land surveyors caught? In my opinion there can rarely two be found who agree in their conclusions (facit) [. . .]

In Fuchs' sense, Marius referred in his preface to the "huge and coarse errors," "in which a common land surveyor together with all the others is caught/and therefore all purchases and sales occur in huge and evident error."[262] For land surveying then, the *Geometrey*[263] by Jacob Köbel[264] (1460–1533) was often used, which had evident deficiencies (Hergenhahn 1996, p. 78f.):

> The [. . .] formulas for the irregular quadrilateral, the triangles and the regular polygons are wrong in principle, however the rule used for the irregular quadrilateral delivers a solution that is more and more true, the more the form to be determined approaches a rectangle.

The art of surveying was in fact in a pitiable condition. Probably from the same motivation as Marius, the mathematician Ambrosius Rhodius[265] (1577–1633) from Wittenberg had published a cheap version of Euclid[266]; however, it was in Latin and addressed to students.

260Kepler in a letter to Vicke: "Pudet Parallelas germanice non aliter dici posse quam parallelas. Hic vero communi cura opus esset, ut termini transirent in usum publicum, neque aliter hic loqueretur aliter ille. Et puto me in decimo libro aliquid hic profecisse. Eodem enim iure quo Euclides in Greca lingua, nova constituit nomina, constitui ego in vernacula" (Kepler XVI, 1954, p. 389).

261"Inn was grossem Irthumb stecken die Landtmässer? Deren meiner Mainung nach gar nimmer/ oder selten zwen gefunden/die in dem facit überein-stimmen [...]" (Marius 1610, sig. A4r, wrong counting in the book, there sig. A3r). In the *Prognosticon for 1610* (sig. A3r), Marius defends the use of the German language with reference to the preface by Fuchs von Bimbach.

262"[. . .] gewaltigen und groben Irrthumb [. . .] darinnen gemeine Feldtmässer alle mit einander stecken/und daher in Kauffen und Verkauffen grosser und augenschlicher jrrthumb vorgehet" (Marius 1610, sig. A5r). Apart from this Marius also defends the use of German in his comet script of 1619 (Marius 1619, sig. A3r).

263Köbel, Jacob: *Geometrei: Von künstlichem Messen vnd absehen, allerhand höhe, fleche, ebene, weite vnd breyte [...] mit künstlich zubereyten Jacob-stab, Philosophischen Spiegel, Schatten vnd Meßruten/Durch schöne Figurn vnd Exempel.* Frankfurt a.M.: Christian Egenolff 1535 [SUB Göttingen: MC 95-200: E2020].

264For Köbel see Hergenhahn 1996, 1997.

265For Rhodius see Schöneburg 2007, here the biography about Rhodius on pp. 67–75.

266Of this book three editions were printed in Wittenberg. The first was *Euclidis Elementorum Libri XIII/Succinctis & perspicuis demonstrationibus comprehensi a M. Ambrosio Rhodio.* Wittenberg: Paul Helwig; Johann Gormann 1609, 593 p. (although the page count jumps from p. 447 to 458) [BSB: A.gr.b. 1458; HAB: N 50.8° Helmst. (1)]. With Paul Helwig and Ambrosius Rothe, a "Editio postuma, priore Correctior & Emendatior," meaning an improved and corrected reprint, was issued in 1634 with 594 pages [HAB Xb 1724]. A third and unchanged edition, compared to the second one, was issued in 1661 with Hiob Wilhelm Fincelius [HAB Xb 1769 (1)]. Accordingly the statement in *ADB* (XXVIII, 1889, p. 329; author: Siegmund Günther) is to be corrected, that a complete edition of Euclid with commentary was issued only after his death.

To meet the demands of von Fuchs, Marius emphasized trigonometry, especially the measurement of areas, with many examples which he took "partially from Clavio[267]" (Marius 1610, sig. A5^{v}) and partially from his own devising.

In 1618 the practical mathematican Sebastian Kurz[268] (1576–1659) from Nuremberg provided a new version.[269] In the preface he recalled the first German translation by Holtzmann, continuing[270]:

> Hereafter, however, the first six books were translated anno 1616 by Johann Peter Dou the appointed surveyor of Leyden in the Netherlands from German and French to the Dutch language[271] [. . .] beyond this they were finally translated by the appointed astronomer and medicus Simon Marius into German anno 1610/who promised to also unveil the following 9 books in the same fashion/because these are no longer obtainable/the others however are also a hard nut to crack, so to speak [. . .]

Since Kurz especially liked the work of Dou and spoke Dutch, he provided the new German edition. In the preface he addressed the reader with the problems regarding the coining of new terms[272]:

> [. . .] because in nearly all mathematical books, words like Basis, Cathetus, Hypotenusa, perpendicular, parallel, parallelogram, Quadrat, Centrum, Diameter, circumferentia, and other vocabulary of this art, are common and so to speak issued with German civil liberty, therefore they are all here also kept in their meaning.

---

[267]Christoph Clavius (1537–1612) had already published a Latin version of the *Elements* in 1574. "Enzyklopädisch angelegt enthielt diese Edition, die mehr als zwanzig Auflagen erfuhr, schon alles, was seinerzeit über Text und Textkritik zu den Elementen bekannt war [...]. Charakteristisch ist ihre bewusst pädagogische Ausrichtung, die sie als 'Handbuch über Euklid' für den Schulunterricht empfiehlt." ("Encyclopedically structured, this edition, running for more than 20 reprints, already contained everything known about text and text criticisms of the Elements [. . .]. Characteristic is its conscious educational orientation, making it suitable as a "manual of Euclid" for school instruction" (Schönbeck 2003, p. 228, 231)). On Clavius see Lattis 1994.

[268]For Kurz see Hawlitschek 1999.

[269]A reprint of this work was issued 1634 in Amsterdam by Johannes Janssonius (HAB Wolfenbüttel: 55.3 Astron. (3)).

[270]"Hernach aber/seindt die 6 ersten Bücher anno 1606. durch Herrn Ioann Petersz Dou, der statt Leyden inn Hollandt bestelten Landtmesser vnd Visirer/auß Teutscher vnd Frantzösischer sprachen/in die Niderlendische transferirt worden [...] Vber diß/seindt sie endtlich auch durch Herrn Simon Mayrn F. Br. bestelten Astronomorum und Medicum, anno 1610. ins Teutscht gebracht/und die 9 folgenden Bücher gleicher gestalt ans licht zubringen versprochen worden/ weilen dan diese nicht mehr zu bekommen/die andern aber gleichsam also hardt in der nuß stecken [...]" (Kurz 1618, preface sig. **2^{v}).

[271]The *Elements* was translated into Italian by Niccolo Tartaglia (1499–1557) already in 1543, 1564/1565 into French by Pierre Forcadel, 1570 into English by Henry Billingsley (ca. 1538–1606) with a preface by John Dee (1527–1608), and 1576 into Spanish by Rodrigo Zamorano. The Arabic translation by Nasiraddin at-Tusi (around 1200–1274) was printed in 1594. In 1603/1607 a translation into Chinese followed by Matteo Ricci (1552–1610) [and Xu Guangqi (1562–1633)], only then followed the translation by Dou (Schönbeck 2003, p. 234).

[272]"daß weil fast in allen Mathematischen Buchern/diese wörter Basis, Cathetus, Hypotenusa, perpendicular, paralell, paralellogram, Quadrat, Centrum, Diameter, circumferentia, vnd andere vocabula artium, gemein vnd gleichsam mit dem Teutschen Burgerrecht begabet/darumb so seindt sie auch alhie in ihrem wesen behalten werden" (Kurz 1618, preface, sig. **3^{r}).

In September 1613 Marius was visited by the mathematician Lucas Brunn[273] (1572–1628), to whom he showed the Andromeda Nebula, which he had discovered the previous December.[274] Brunn issued a *Euclidis Elementa Practica*[275] in 1625. According to its title, it contains "an excerpt of all problems and handworks [. . .] for useful application of the compass and the quill" (Schönbeck 2003, p. 323). This was the second German edition of Euclid, following Marius's translation.[276]

As conclusion to his translation of the six books, Marius wrote[277]:

> So far I understand it/you, the kind-hearted reader will be served with this/and will have a special fill of it/so I hope I have invested my effort and diligence properly/therefore it hopefully will lead to/a faster translation of the remaining books/As long as my health and God make it possible.

In fact he occupied himself with the translation of the remaining books of Euclid,[278] but this plan, as hinted at in the quote from Kurz above, was never realized.

The most important event in Marius's and Fuchs von Bimbach's cooperation was the purchase of a telescope, with which Marius "discovered" the moons of Jupiter.

---

[273]For Brunn see Doppelmayr 1730/1972, p. 97f.; Zinner 1979, p. 266f.

[274]Marius 1614, sig.)()(4^r. On December 15th 1612, Marius directed his telescope to the Andromeda nebulae: "Without the instrument the same is seen as some sort of little cloud; and with the instrument no distinct stars are seen [. . .], but rather only white rays, which the closer to the center the brighter they come out. [. . .] About the same brilliance appears when a bright candle is observed through a clear lantern from a long distance" ("Absque instrumento cernitur ibidem quaedam quasi nebecula: at cum instrumento nulla videntur stellae distinctae, [. . .] sed saltem radij albicantes, qui quo propiores sunt centro eò clariores evadunt [. . .]. Similis fere splendor apparet, si à longinquo candela artens per cornu pellucidum e nocte cernatur," Marius 1614, sig.)()(4^r). This nebula is visible with the bare eye and was already listed in the star charts of the Arabic astronomer As-Sufi (903–986) (cf. Strohmaier 1984, p. 50). When Marius is called the discoverer of the Andromeda nebulae, this refers to him being the first one to examine this nebula with a telescope and introducing this subject to the astronomical sciences. According to the old Aristotelian definition, the nebula belonged to the realms of meteorology. I thank Jürgen Hamel for helpful tips.

[275]Brunn, Lucas: *Euclidis Elementa Practica, Oder Außzug aller Problematum und Handarbeiten auß den 15. Büchern Euclidis/Allen und jeden/deß uhralten Geometrischen nutzlichen gebrauchs/ deß Circkels Liebhabern zu gut in Teutsche Sprach dargegeben*. Nürnberg: Simon Halbmaier 1625 [StB Nürnberg: 1 an Math. 4. 12].

[276]It seems as if the first comprehensive German translation of the Elements was only published in 1651. Apparently only the second edition of 1653 is preserved: Teutscher Euclides, Das ist: Geometrische Beschreibung und zwar Furnehmlich Von derselben Elementen. The author, Heinrich Hofmann (1576–1652), was professor for mathematics from Jena (Schönbeck 2003, p. 233).

[277]"So ferne ich nun verstehe/daß dir guthertziger Leser hiermit wird gedienet seyn/vnnd ein sonderlich wolgefallen darob habest/so verhoffe ich meine mühe und fleiß wol angelegt zuhaben/ auch vrsach dahero zunemen/die übrigen Bücher desto ehe zuverfertigen/So viel mir durch gesundtheit deß Leibs/und verleyhung Gottes zuthun müglich seyn" (Marius 1610, p. 165).

[278]On July 6th 1611, the Bergrat Vicke from Wolfenbüttel wrote to Kepler, adding a letter by Marius to it, which states: "The first six books by Euclid are now translated into German and it has been printed one year ago, now I concentrate on the remeining ones" ("Priores sex libri Euclides in linguam germanicam iam et translati et ante annum impressi sunt, in residuis nunc versor," Kepler XVI, 1954, p. 382. Many thanks for help with this translation to Joachim Schlör).

But this story is told elsewhere in this book. Ernst Zinner (1886–1970), the historian of astronomy, doubted that Marius made his observations from a tower in the Ansbach palace (Zinner 1942, p. 36):

> Marius tells us, that at first, he made his observations from the residence of his patron Fuchs von Bimbach and was allowed to take the telescope home in autumn 1609, where he observed the stars in his observatory, an attic.

In the acquisition of a second telescope, the margravial privy councilor Johann Baptist Lenck (Lenccius) was of assistance[279]:

> Meanwhile, two extremely well polished glasses, convex and concave, were sent from Venice by that most distinguished and accomplished man, Lord John Baptist Leuccius, who had returned from Belgium to Venice after peace was made, and who was already thoroughly acquainted with the instrument. These glasses were fitted into a wooden tube [. . .].

Lenck was of the reformed faith (Herold 1973, p. 29) (Fig. 2.17). He had probably studied law in Strasbourg; in any case his *Observationes Politicae* was published there in 1601.[280] Early in the seventeenth century, he had started to work for the margrave in Ansbach.[281] Joachim Ernst himself had participated "until 1706 in the field and diplomatically" in the war of independence between the Netherlands and Spain (Schuhmann 1980, p. 128). On April 12, 1609, an armistice was established in Antwerpen, which was kept for 12 years. Lenck took part in the negotiations as an adviser to the margrave. Consequently he returned to Ansbach in April and was immediately sent by the margrave to Venice, where he arrived in August. His official assignment was to observe "how the leadership of Venice resisted the Roman Catholic Church's presumptuous intervention." He should scout for reformatory aspirations and eventually initiate new relationships. He stayed until his recall in October 1610 (Herold 1973, p. 124; Rein 1904, p. 119, 148).[282] During this time, around the turn of the year, he provided Marius and Fuchs von Bimbach with a telescope.[283]

---

[279] "Interim etiam mittebantur è Venetijs duo vitra egregrie polita, convexum & concavum, à clarissimo & prudentissimo viro Domino Johanne Baptista Lenccio, qui è Belgio post factam pacem reversus Venetias concesserat & cui instrumentum hoc jam notissimum fuerat. Haec vitra tubo ligneo coaptata fuerunt [. . .]" (Marius 1614, sig.)()(3^r). Lencks portrait from 1608 is preserved in the Germanisches Nationalmuseum (Graphical collection: Inventory-Nr. P 7914, capsule-Nr. 825). He's referred to as "IOAN BAPT. LENCKIUS. MARCG. BRAND. CONSIL. ET LEGATUS." According to the French inscription, he was the advisor of the margrave of Brandenburg and ambassador for the Dutch-Spanish ceasefire of 1608. It was made by the copper engraver and cartographer Henrik Hondius (1573–1650).

[280] Lenccius, Johannes Baptista: *Observationes Politicae, ex variis Historiarum & civilis doctrinae Scriptoribus praeterpropter conciliatae*. Straßburg: Rihel, Rietsch 1601 [BSB München: Pol. G. 488]. Reprint 1606.

[281] He was intermittently at the 1608 Reichstag in Regensburg (Herold 1973, p. 89).

[282] For Lencks activity in Venice, see Herold 1973, pp. 124–127.

[283] On March 17th 1611, Lenck represented the margrave in Jüterbogk, to protest, that the emperor Rudolf II (1552–1612) had declared the Duchy Jägerndorf reverted and denied the house of Brandenburg any rights to it. In 1622 he received a domicile in Knoblochsdorf in Lehrberg as a knight's fiefdom, where he started to build a small castle in 1628; J. Meyer 1892, p. 63.

Fig. 2.17 Portrait of Johann Baptist Lenck. Courtesy of the German National Museum Nuremberg: Graphic collection: Inventory-Nr. P 7914, Capsule-Nr. 825

Despite the support of high-ranking administrators,[284] it appears that Marius did not feel comfortable at the Ansbach court[285]:

> I don't really care for a life or a position at the court, which a lot of people are striving for with a lot of effort and greed; much more I enjoy the pleasures of a private life and philosophical studies, and so I fulfill my duty.

[284]In the *Prognosticon for 1607* (sig. B3^v, B8^v), Marius refers to the chronic of Mansfeld twice when relating to the meteorological forecasts. It was published in 1572 in Eisleben (see Spangenberg 1572). Mansfeld today is a small locality in the southern Harz Mountains (Saxony-Anhalt). This somewhat digressive literature was probably brought to his attention by the count Wilhelm von Mansfeld (1555–1615), the lord marshal of the margrave.

[285]"Non ego nullam laborum compensationem, aut honoris sive officij alicujus auctionem à Celstitudine Vestra pero siquidem ea omnia, quae hac in parte à me proficisci possunt, Vestre sunt, Vestris sumptibus parta & procurata" (Marius 1614, sig.)(4^v).

It seems that he later had been the target of hostility, because he complained that "in my name false notes were written/even the letters sent to me from foreign places were opened without hesitation/and from all side I have been vilified."[286] This could have been a consequence of his relationship with Fuchs from Bimbach, who was very unpopular in the Ansbach palace.

Despite the court, Ansbach was neither a center for literary production nor for scientific research. Even though Marius was in contact with a few scholars like Fabricius or Mästlin,[287] this could never completely replace a direct exchange.[288] It seems that he did not receive visitors from afar very often. Lucas Brunn has already been mentioned. On August 3, 1611, Ahasver Schmitner[289] (Schmidner) from

[286]"[...] unter meinem Namen falsche Zettel geworffen/auch mein von frembden orten geschickte Brieff ohne scheu geöffnet/unnd allerseyts bin verunglimpfft worden" (*Prog. 1619*, sig. A3^r).

[287]The following letters by Marius are preserved: To Michael Mästlin: December 6th 1609, December 29th 1611, March 29th 1612, August 1st 1613, June 15th 1614 (Zinner 1942, pp. 40–45; Original in the HAB Wolfenbüttel: Cod. Guelf. 15.3 Aug. 2°, sig. 268, 270, 283, 290; Cod. Guelf. 10.2 Aug. 2°, sig. 310). To Johann Caspar Odontius: December 19th 1620 (Zinner 1942, p. 46; Original in the HAB Wolfenbüttel: Cod. Guelf. 15.3 Aug. 2°, sig. 256). From Kepler: November 10th 1612 (Kepler XVII, 1955, pp. 33–37). To Kepler: August 16th 1613 (Kepler XVII, 1955, pp. 72–74). Furthermore Marius wrote to Nikolaus Vicke in Wolfenbüttel in 1611. Vicke added this letter as a copy to a letter to Kepler on July 6th 1611 (Kepler XVI, 1954, p. 382f.).

[288]The last known letter from Marius was sent on December 19th 1620 to Johann Caspar Odontius (1580–1626), the late professor for mathematics in Altdorf. Marius thanked him for his two letters that are not preserved (Zinner 1942, p. 46). For Odontius see Gaab 2011, pp. 22–24. Furthermore Kepler (XVII, 1955, p. 34) wrote to Marius: "Dixi te imperitum hominem. Nullum probrum: Tu enim in Franconia sedens, necessariò minorem habes eorum notitiam, qui Pragam sub Rudolfo ex omnibus orbis partibus confluxerant, quàm Ego." In the translation by Joachim Schlör: "I have said, you are an inexperienced human. This is not an insult; you, as someone who lives in Franconia, have necessarily fewer knowledge about the things, that were collected under Rudolph in Prague from all directions, than I have."

[289]Assuerus Schmitnerus matriculated on October 29th 1608 in Wittenberg (Weissenborn 1 (1934), p. 77). In 1610 he disputed under Tobias Tandler (1671–1617) in medicine—*Diaskepseon aeirourgikon dekas*. Wittenberg: Gormann 1610 [SUB Göttingen: DISS MED COLL MAX 587 (6)] as well as on October 24th under the physician Bartholomäus Krüger (1579–1613) a *Disputatio Medica Prima De Morbo Morbique Essentia In Genere*. Wittenberg: Rüdinger 1610 [SUB Göttingen: DISS MED COLL MAX 523 (20)]. In 1612 he disputed in Basel under Caspar Bauhin (1560–1624) to achieve his doctorate *De phrenitide*. Basel: Genath 1612 [WLB Stuttgart: Med. Diss. 5211]. He studied medicine and visited Marius in Ansbach on his way from Wittenberg to Basel. From Basel he visited Italy; on October 30th 1612, he matriculated in Padua with the German Nation (Rossetti 1986, p. 160 entry 1317). He later settled as physician in Königsberg. He probably died before 1634 (Komorowski 2008, p. 32, 40). From the year 1654, the *Einfältige Klag- und Trost-Reime*. Königsberg: Reusner 1654 [StaBi Berlin: Yi 851-3 (14)] by Simon Dach (1605–1659) is preserved, in which the death of the physician Ahasver Schmittner (1618–1654) is mourned. It is assumed that he was the son of the Schmitner, who was visited by Marius. The sunspots were probably shown to him by Johann Fabricius (1587–1617), when he published his book about his discovery in Wittenbergin: Fabricius, Johann: *De Maculis in Sole observatis, et apparente earum cum Sole conversione, Narratio*. Wittenberg: Seuberlich 1611 [BSB München: Res/4 Astr.p. 516,33].

Königsberg showed him sunspots.[290] In July 1615 the later Altdorfer lecturer for mathematics Petrus Saxonius (1591–1625) stopped in Ansbach during his journey through southern Germany.[291] Apparently Marius showed him his sunspot observations from November 17/27, 1611.[292] Nothing is known about further visits.

In 1613 Paul Böheim[293] (1561–1641) issued the *New Invention, through which in Arithmetic many beautiful examples can be solved shortly and effectfully* (*Newe Invention, Dadurch in der Arithmetic viel schöner Exempla sehr kurtz und gering auffgelöst seind)*, dedicated to Fuchs von Bimbach and published by the Royal Brandenburger National Chamberlin (Landtpfennigmeister)[294] Matthias Wilhelm. All we know about Böheim is that he originated from Ulm and had published *A New Reckoning Booklet* (*Ein Newes Rechenbüchlein)*[295] in 1596 as "German school and reckoning master" (Weyermann 1829, p. 618f.). Marius wrote a short preface for the *New Invention*: "A good wine does not need a sign post," rendering his preface actually redundant, when on top of this his "special profession does not include mercantile reckoning." But he had "examined the invention himself and found it with pleasure and wonder reasonable," which is why he recommends this booklet to the readers.[296]

Aside from his calendars Marius only published one further work during his lifetime, the tract on the comet of 1618,[297] with preface signed April 16, 1619. It could have been a work commissioned by his publisher and father-in-law, because

---

[290]In the *Prognosticon for 1613* (sig. A4^{v}), Marius wrote: "Concerning the sunspots [...] I saw them for the first time last year in August 1611, by the help of Ahasver Schmidnero from Prussian Königsberg, who was visiting me" ("Die maculas in sole belangt, [...] die hab ich voriges Jahr 1611 im Augusto zum erstenmal gesehen, monstrante Ahasvero Schmidnero Regiomontano Borusso, der damals mich visitiert hat," cf. Klug 1906, p. 524). According to the *Mundus Iovialis* (Marius 1614, sig.)()(3^{v}), he wanted to present everthing about the sunspots, "setting out all my observations upon them from August 3, 1611 to the present time" ("Acturus nunc eram de maculis in Sole, uti ante hac proposueram, quidquid etiam in eis à 3. Augusti 1611. usque huc observavi manifestare." Cf. also Marius 1619, sig. C3^{r}).

[291]For Saxonius see Gaab 2011, pp. 17–22, for his journey through Southern Germany p. 17, 20.

[292]Gaab, Leich 2014, p. 13. For Marius and the sunspots, compare the chapter by Neuhäuser in this volume.

[293]For Böheim see Benzing 1982, p. 9f.

[294]In early modern period, a "Landpfennigmeister" was the head of the local financal administration.

[295]Wilhelm, Matthias: *Ein Newes Rechenbüchlein: mit vilen schönen gesellschafften Wächsel vnd ander dergleichen Kauffmans Rechnungen, so zuvor in truck nie außgangen, durch die Wälsch Practick, mit mancherley Müntz fortten soluiert vund auffgelößt.* Augsburg: Manger 1596 [BSB München: 4 Merc. 36].

[296]"Ein guter Wein bedarff keines außgehenckten Zeichens [...] Vornemlich/weil meine sonderliche Profession nicht inn dergleichen Kauffmans-Rechnungen bestehet [...] dieweil ich solche Invention selbsten examinirt, und mit lust und verwundern gerecht befunden" (Wilhelm 1613, sig. A3^{v}–A4^{r}).

[297]In 1618 three comets appeared, which is not discussed here. For this see the chapter by J. Hamel in this volume. For the 1607 comet, Marius did not—contrary to Kepler—publish an autonomous essay.

such works were very salable.[298] The title page is adorned with a sketch of a geocentric system. The Earth A lies in the center of the world, surrounded by five additional spheres. The spheres B and C belong to Earth: B is reserved for rain, wind, thunder, and hail. In this watery air[299]:

> [. . .] next to the soil/betimes we can find unusual phenomena, weather/rain/hail/living frogs/ and other things/who wants to tell it all.

Konrad von Megenburg[300] (1309–1374) had already written in his *Book of Nature* (*Buch der Natur*) that it is often "raining small frogs and small fish" (Megenberg 1994, p. 82 line 16f.). This belief probably came into being by the fact that young frogs often emerge from water in droves (Bächtold-Stäubli 3 (1987), col. 125).[301] Anyway this was still a widespread belief in Marius's times.

The sphere C is responsible for "the upper, thin, watery air, in which rainbows are found."[302] Sphere D is the firmament, wherein the planets are moving—without Marius settling for the Tychonic system here. Sphere E is reserved for the fixed stars, the outer sphere F for the "waters above the firmament."

It should be noted, at this point, that Marius warned against drawing too hasty conclusions, as was not unusual in comet tracts[303]:

> One thing needs to be considered/that many come to the disreputable thought/and also publish in writing/as if this comet was several thousand times bigger than the land/and had only been exhalations from the Earth/which I found very strange/particular from those who have been supporters of the Tychonic system/But this happens like the old saying/one rarely benefits from rushing. And so it happened to them/since they were very fast with their works wanting to achieve or contribute great honor.

The last work of Marius—apart from the calendars—the *Complete Refutation of the Position Circle Claudius Ptolemy* (*Gründliche Widerlegung der PositionCirckel/ Claudii Ptolomaei*), was published posthumously, in Frankfurt on the Main by

[298] Lauer published two other writings about comets:– Odontius, Johann Caspar: Komētakribographia ; das ist: eygentliche, gründliche Beschreibung deß im November und December erschienenen Cometen, im 1618. Jahr Jesu Christi. Nürnberg: Lauer 1619 [SB Regensburg: 999/Philos.2041/2046]– Herlicius, David: Kurtzer Discurs, vom Cometen, und dreyen Sonnen, so am Ende deß 1618. Jahrs erschienen sindt. Nürnberg: Lauer 1619 [UB Erlangen-Nürnberg: H61/4 TREW.X 34].

[299] "[. . .] nechst dem Erdboden/haben wir zu zeiten ungewöhnlich phaenomena, Wetter/Regen/ Hagelstain/lebendige Frösch/und anders/wer wollte es alles erzehlen" (Marius 1619, sig. C1^{v}–C2^{r}).

[300] For Konrad von Megenberg, see *NDB* 12 (1980), p. 546f. (author: Sabine Krüger).

[301] As a matter of fact, it rained frogs sometimes, which can be explained with a tornado picking up a pond and later dropping it at another place (Simons 1997, p. 21f.).

[302] Cf. Klug 1906, p. 400.

[303] "Eines muß ich allhier gedencken/daß etliche in die ungereumbten gedancken kommen/auch in öffentlichen schreiben publicirn lassen/als wenn dieser Comet viel tausendmal grösser als der Erdboden/vnd doch nur ab exhalationibus von der Erden gehabt haben/gewesen/welches mir sehr wunderlich vorkommen/sonderlich von denen so Schole Tychoniame participes gewesen/Aber es ist solchen geschehen nach dem gemeinen Sprichwort/Eylen thut selten gut. Also ist diesen auch geschehen/da sie so geschwind mit ihren Schrifften herfür gewischt und grosse ehr erlangen oder einlegen wöllen" (Marius 1619, sig. B3^{r}, cf. Gindhart 2006, p. 260).

Daniel Mögling in 1625[304] (1596–1635). Mögling had traveled during his studies to Nuremberg in January 1616 and had matriculated in medicine shortly after in Altdorf. From 1621 he worked as court mathematician for Landgrave Philipp von Hessen Butzbach[305] (1581–1643), where he carried out an extensive scientific correspondance.[306]

Mögling dedicated the volume to "the honorable and respected Mister Philip Eggebrechten/noble merchant in Nuremberg/my favored and valued friend." Philipp Eckebrecht[307] (Eckenbrecht, Eckeprecht, Eggebrecht, 1594–1667) was a merchant and hobby astronomer in Nuremberg, who was very interested in astrological questions.[308] Marius had sent him his manuscript, which Eckebrecht forwarded to Mögling. In the *Prognosticon for 1624*, Marius only mentioned that "printing the tract was too expensive" (*Prog. 1624*, sig. A2^r). Obviously his father-in-law Johannes Lauer did not want to print this volume.

In it Marius took a stand against the house division of Regiomontanus.[309] Four points divided the zodiac into four equal parts: the ascendant, as the point on the horizon at the time of the birth rising sign, opposite the descendant, as well as the midheaven (midday, Medium Coeli) and the midnight (Imum Coeli). In the classical astrology, these quarters were again divided into three parts, called the "houses." In order they're named life, gain, brothers, parents, children, illness, marriage, death, piety, work life, good deeds, and captivity.

Among astrologers it is disputed, to this day, how one approaches the division of the houses. Widespread at that time was the "rational method" made popular by Regiomontanus, who wanted to achieve more reliable astrological calculations through this (Strauß 1926, p. 53). The zenith is the point on the firmament that is exactly above the spectator (especially at the time of his birth). Extending the axis zenith-Earth's center, this gives a second intersection point with the firmament, the nadir. The plane through Earth's center, vertically lying on the axis zenith-nadir, is the true horizon. The meridian is the largest circle in the sky, defined by the north and south poles as well as the zenith. It intersects the true horizon at two points, the north and the south point, whereby the north point is closer to the north pole, the south point closer to the south pole. The meridian's intersection point with the equator (near the south point) is called culmination point. From there the equator is divided

[304]For Mögling see footnote 251.

[305]For Philipp von Hessen-Butzbach, see *NDB* 20 (2001), p. 379f. (author: Katharina Schaal).

[306]*BBKL* 5 (1993), col. 1582–1584 (author: Ulrich Neumann).

[307]For Eckebrecht see Grieb 1 (2007), p. 318.

[308]Cf. Gaab: Trew 2011, pp. 58–60.

[309]He emphasized in the *Prognosticon for 1624* (sig. A2^r) that he formulated his critique for Regiomontanus' method already in 1599 in his *Tabulae*: "I printed a Latin treatise already 24 years ago, wherein I discarded the method of Regiomontanus concerning his astrological houses, against it I have used the old process with the tempora horaria" ("Nun hab ich aber vor 24. Jahren einen Lateinischen Tractat trucken lassen/darinnen ich den modum Regiomontani mit seinen circulis positionum verworffen/hergegen den alten proceß duch die tempora horaria wider herfür gesucht").

into 12 equal parts 30 degrees each. These 12 points together with north and south point define 2 semicircles (Surya, Sindbad 1980, p. 42):

> This practically gives us the twelve houses on a globe, when we first start from the north and the south point A and B, further divide the equator (beginning with culminated point B') into twelve equal parts and then by connecting the points A and B with this equator sections into the biggest circles.

These circles divide the ecliptic into 12 sections, identified in the horoscope as houses. But since the ecliptic is tilted toward the equator, these houses are not of the same size (Surya, Sindbad 1980, pp. 41–46); Marius polemized against this.

Discursion: Simon Marius and Johannes Kepler

Johannes Kepler left Prague heading toward Graz at the end of April 1601 and only returned at the end of August.[310] It is therefore highly unlikely that Marius and Kepler had already met in Prague. If they had met, it was not an intensive encounter. This is supported by a letter from the personal physician Johann Pape from Ansbach that he had sent to Kepler on August 22, 1608, according to which Pape didn't hear about Kepler from Marius either in writing or during his stay in Ansbach at the end of 1601.[311] A brief exchange of letters took place in 1612/1613, prompted by a 1611 letter from Marius to Nikolaus Vicke in Wolfenbüttel, wherein he talked about his endeavors. Vicke forwarded an excerpt of it to Kepler[312]:

> Firstly, I claimed the immobility of the earth, leaving out the personal issues[a)], only examining the arguments against Copernicus's reasoning, which is accepted in our times as serious and true by Kepler and the mathematician Galileo from Padua[b)]. I take the arguments for my claim Holy Scripture[c)], which also agree with the physics[d)] and the astronomy. Then the view of those will be disproved, who attributed an excessive mass to the celestial bodies. And I will give a new and more probable measure of their size, whereby the Belgian, usually called Perspicill, instrument was especially helpful. Thirdly, I will prove that Venus as well [as the Moon] is illuminated by the Sun and it becomes horned and half,[313] as it has been most accurately

[310] Dating according to the Julian calendar; Caspar 1995, p. 139; Christianson 2000, p. 303; Klug 1906, pp. 418–425; Thoren 1990, p. 460.

[311] "Rogauerem quidem Simonem Meierum; ut me de rerum tuarum statu erudiret; sed nihil ex homine, necque per literas, neque cum ad nos ipse redijsset, cognoscere potui" (Kepler XIV, 1949, p. 257; cf. Klug 1906, p. 419).

[312] Quoted according to Wohlwill 1926, p. 380f.; the Latin original can be found in Kepler XVI, 1954, p. 382f. Kepler and Vicke had already known each other longer. One of Vicke's family registers contains an entry by Kepler from April 25th 1600 (Stralsund Museum, Kulturhistorisches Museum der Hansestadt Stralsund, Inv.-Nr. A 1993:160, Sig. 137).

[313] Marius carefully observed the phases of Venus. For example, he wrote in the *Prognosticon for 1614* (sig. B7^{v}; cf. Zinner 1942, p. 63): "Venus [. . .] will stay morning star until January 1615. It is seen now again crescentshaped, and its horn will show to the West" ("Die Venus [. . .] wird Morgenstern bleiben biss in den Jenner des 1615. Jahrs, wird jetzt wider falcata [sichelförmig] gesehen, vnd ihre spitzen gegen Nidergang wenden."). A comprehensive reflection on this topic can be found in the *Prognosticon auf 1612* (sig. A2^{v}–A3^{r}).

observed by myself several times with a Belgian Perspicill from the end of the last year[e)] until April in the present one, when Venus was close to Earth, both as the west and east star. Fourthly, I will talk about the new Jovian planets, orbiting Jupiter, as the other planets do around the Sun, although with varying distances and periods. I have already examined and created tables for the periods of the two outer planets, so that it can find out at any time by how many minutes they stick out to the left and right of Jupiter, and both of these final chapters are unheard of for all times. Maybe I will encounter other things during my work.

Kepler answered Vicke that he was looking forward to[314]:

[. . .] someone in Germany competing with the Italian Galileo, to uncover the secrets of the sky, and I ask you, noble Herr, to admonish Marius, that he refrains from the tendency to diminish, as it is usual with nations, with the same diligence, like he had done before, to avoid personal issues; because it is all about truth.

Kepler phrased his criticism rather more harshly in the introduction of his *Dioptric*, which was published in Augsburg 1611[315]:

Since the nations never lack competitiveness nor diminishment in the sciences and a lot of people in Germany will demand proof of the Germans, I will inform them about the same things written in a German's letter, from which one can recognize, that Galileo was not wrong, in that he worried about his rights at least communicated his invention through letter riddles to us in Prague.

Kepler printed the letter from Marius to Vicke in his preface and commented in the margins the five sections noted above[316]:

(a) Regarding the remark, personal issues should stay aside[317]:

[. . .] he had freed Kepler from an apprehension, because if Marius had advocated the Earth's motion mentioning his name, he would have feared for his reputation.

(b) Regarding the remark, Galileo and Kepler are Copernicans[318]:

[. . .] a first good sign for victory, because Marius through ignorance limits the number of followers to two, whereas they are already widespread.

(c) Regarding the remark, Marius has taken his arguments against the Earth's movement from Holy Scripture[319]:

---

[314]Quoted according to Wohlwill 1926, p. 381f.

[315]Quoted according to Wohlwill 1926, p. 383.

[316]Kepler 1611, preface p. 27f. (the preface has its own pagination).

[317]Translation by Klug 1906, p. 422. In the Latin original: "Liberavit Kepler metu: qui valde scilicet, honori suo metuebat, si Marius motui Terrae intercessisset cum sui nominis mentione" (Kepler 1611, preface p. 27).

[318]Translation by Klug 1906, p. 44. In the Latin original: "Primum victoriae omen ante pugnam, quod Marius imperitia hominum, sectae hujus amplitudinem intra duos restringit, quae jam pene publica est: nisi flos omnis doctorum hominum intra Academiarum septa sit conclusus" (Kepler 1611, p. 28).

[319]Translation with the help of Joachim Schlör. In the Latin original: "Obsistite Theologi, rem impertinentem aggreditur; authoritatem Scripturae abusum ist" (Kepler 1611, preface p. 28).

> Contradict, theologians, he is doing something very improper; he wants to abuse the Bible's authority.

(d) Regarding the remark, the physics is in agreement.[320]

> May we be judged by our deeds.

(e) Regarding the observations of Venus at the end of the previous year[321]:

> [. . .] just at the time, as Galileo wrote about Venus from Florence to Prague, and already before Marius predicted, that this will appear in this manner.

Kepler created a propaganda for the Copernican system. As mentioned above, he broke off his correspondence with Fabricius, because he did not vouch for this system. The remarks (a) to (d) should be seen in this light. If Marius, for example, writes that Galileo and Kepler make a case for Copernicus, then this does not necessarily reflect the view that only these two did. That this opinion was widespread at the time is also unsustainable. For most, the competition to Ptolemy during that period was Brahe, not Copernicus. Kepler loses himself here in cheap polemics. That he wants to keep the Bible out of this discussion is honorable, but for the majority of people, it was still the basis for decision-making.

For Marius the last remark that he was accussed of misusing Galileo's observations must have been offensive. Apparently Marius had spoken about this issue with Fuchs von Bimbach, who requested mediation from the Imperial Councilor Eisen[322, 323] which prompted Kepler to write directly to Marius from Prague on November 10, 1612: His remark in the *Dioptricis* was neither unfair nor dishonorable.[324] The issue probably wasn't that important to Marius, because he only answered him on August 16th the following year, after Kepler had already sent a second letter.[325] He accepted his apology but defended the[326]:

[320]Translation with the help of Joachim Schlör. In the Latin original: "Cernamur agendo" (Kepler 1611, preface p. 28).

[321]Translation by Klug 1906, p. 422. In the Latin original: "Quo ipso tempore Galilaeus Florentia Pragam scripsit de Matre amorum, et haec Mario sic ordine apparitura jam tunc praedixit" (Kepler 1611, preface p. 28).

[322]For Simon Eisen von Haymen, see the list of the consistory councilors above.

[323]"Haec cum antea Lincio ad te scribere concupivi, tum vero maxime a Nob. et Ampl. D. Doctore Eisen Caesaris ad Appellationes Bohemicas Consiliario nuperrime jussus, quin et exoratus (ita enim volebat), me scripturum recepi: dum mihi Dn. Capitanei Fuchsij, Patroni nostrarum artium, ut audio, maximi solicitudinem in nobis conciliandis exposuit" (Kepler XVII, 1955, p. 35).

[324]"Caeterum ipsum fero arbitrum: fatebitur scio, et his literis, et loco in dioptricis pellecto, te nec injuriose nec inhoneste a me tractatum" (Kepler XVII, 1955, p. 35).

[325]"Doctissime Domine Keplere [. . .] Tuae literae superioris anni, sub finem ejusdem recte ad manus pervenere meas" (Kepler XVII, 1955, p. 72; cf. Klug 1906, p. 424). The second letter from Kepler is not preserved.

[326]The answer is printed in Kepler XVII, 1955, pp. 72–74; here the summary of the final report (p. 462) is quoted.

Earth's immobility with reference to the first chapter of Genesis; the world system here was different from the Copernican one. Also the orbiting satellites of Jupiter cannot count as a testimony for Copernicus.

Finally Marius hoped for an imminent opportunity to speak directly to Kepler.[327] This opportunity occurred in October 1613 in Regensburg, where Kepler proposed to call the Jupiter moons after liasons of Zeus or Jupiter. This led to the names Io, Europa, Ganymede, and Callisto[328]:

> This fancy, and the particular names given, were suggested to me by Kepler, Imperial Astronomer, when we met at the Regensburg Fair in October 1613. So as a jest, and in memory of our friendship then begun, I hail him as joint godfather of these four stars [. . .].

Marius and Kepler had no further personal encounters. We only know from another exchange of letters that Marius' *Prognosticon for 1622*: He complained of having not seen the recent great solar eclipse because of bad weather and continued:[329]

> Herr Johannes Kepler also complained about this in a letter to me, that we could have had a good astronomical discussion, if there had been clear air.

Kepler mentioned Marius again in print in 1618. In his *Epitome Astronomiae Copernicanae*, he used the orbital period times of the Jupiter moons found by Marius and recorded in his *Mundus Iovialis* to verify his third law. In brackets he also mentions Galileo, but he seems to have had a positive attitude toward Marius.[330]

But Kepler's tone changed the following year. In August 1619 he wrote to the calendar maker Johannes Remus in Vienna[331]:

> As to the sunspots, Marius agrees with you, but besides that he is an unpopular and brazen seer who only can interpret signs, as he also acknowledges. He should keep his opinions to himself and stop annoying his friends.

[327]"[. . .] sed in proximis caetera persequar, nisi forsan coram de his et similibus nostrum studium concernentibus conferendj occasio dabitur" (Kepler XVII, 1955, p. 74; cf. Wohlwill 1926, p. 392).

[328]"Huic figment & propriorum nominum imposition occasionem praebuit Dominus Keplerus Caesareus Mathematicus, quando mense octobri Anni 1613. Ratisbonae in Comitijs una eramus. Quare si per jocum & per amicitiam inter nos tunc initium, illum compatrem horum quatuor siderum salutavero, haud male fecero" (Marius 1614, sig. B2^{v}).

[329]"[. . .] darüber auch Herr Johann Kepler in einem schreiben an mich sich beklaget, da es denn sonsten ein schönen Astronomischen discurs hette geben, wenn helle lufft gewesen were" (*Prog. 1622*, sig. A2^{v}–A3^{r}; cf. Zinner 1942, p. 69f.).

[330]"Intervalla enim quatuor Jovialium a Iove prodit Marius in suo mundo Iovialia ista 3.5.8.13 (vel 14. Galileo) [. . .] Periodica vero tempora prodit item Marius ista [. . .]" (Kepler 1620, p. 554). Vgl. auch p. 537: "Deprehendit enim Marius in suo mundo Ioviali restitutiones satellitum Iovialium circa Jovem, nequaquam regulares esse ad lineas, quas ex centro Terrae in Iovem ejicimus; esse vero regulares, si comparentur ad lineas ex centro Solis per Iovem eductas" (cf. Drake 1978, p. 275). Klug (1906) didn't know about these passages or he ignored them in his prejudice against Marius.

[331]"De maculis assentitur tibi Marius, caetera vates invisus et audax et plus quam prognostes, ut quidem et fatetur. Habeat sibi res suas seorsim; ne gravis sit amicis" (Kepler XVII, 1955, p. 376; cf. Wohlwill 1926, p. 404).

Kepler signed his *Prognosticon for 1620* on November 10, 1619, in which he also took a stand against Marius in connection with the fact that the Sun, Venus, Mars, and Jupiter met in Aries during spring[332]:

> If I had formed a well-founded opinion about politics, and had surrounded it with necessary circumstances of the persons, the time and the location, like Marius does in his interpretation of this quarter, so it would be easy for me to apply this figure of revolution and hide my opion within it. But because I never was of the opion, that the circumstances of certain earthly actions were praedestined by the stars and even if it were so, then this figure, which you draw for the beginning of an quarter, would not matter; that is why I don't mention this theme in the horoscope and try to concentrate on natural things.

Marius wrote in his *Prognosticon for 1620* of which he had already finished the preface on April 13, 1619[333]:

> [...] this gathering in Aries means something special in the regions under Aries. I think about a huge meeting of potentates in Germany/what will be decided there/time will show us/but it will not be in vain.

Marius seemingly let slip some rumors circulating in the Ansbach court. For Kepler this went too far. Perhaps it was in reaction to Kepler when Marius wrote in his *Prognosticon for 1622*[334]:

> Whatever may be the general meaning with the world and with the leaders, this time I will report either nothing or just little [...] Never again will I give a statement so clear as happened before, since my speech and writing is obviously held against me, and I have not caused anything else but friends becoming enemies and getting hostility for my hard work instead of rewards.

Since the meeting in Regensburg, Marius had called Kepler his good friend, which did not change even after this small dispute.[335] Worth mentioning is another

332"Wann ich mir von dem Politischen Wesen einen gemessenen Casum formirt/und mit notdürfftigen Umbständen der Person Zeit und Ort umbzircket hette/wie Marius in außlegung dieses Quartals von sich zuverstehen gibt/so wäre es mir unschwer diese Revolutions-Figur zu appliciren/unnd meine Maynung drein zuverklaiden. Weil ich aber nie der Mainung gewest/daß einige irrdische Handlungen nach ihren Umbständen im Gestirn praedestiniert/und da es schon wäre/daß doch dergleichen figurae so man auff den Eintritt deß Quartals zu stellen pflegt/nichts bey der Sachen thun würden/also laß ich diß Thema mit seinem Horoscopo fahren/und betrachte das Natürliche." (Kepler in the *Prognosticum auf 1620*, quoted according to Kepler XI/2, 1993, p. 202).

333"[...] diese Zusammenkunfft in Wider bedeut etwas besonders in Landen dem Widder unterworffen. Ich halte auff eine grosse Versamblung mächtiger Potentaten in Teutschland/was alldar geschlossen wird werden/wird die zeit wol geben/vergebens geschicht sie gewiss nicht" (*Prog. 1620*, sig. B5^{v}).

334"Was nun die allgemeinen bedeutung sey in der vnteren Welt, vnter Hohen-häuptern, so melde ich dissmals entweder gar nichts oder doch wenig davon [...] Ich gehe nimmer so deutlich herauss wie vor disem geschehen, dieweil ich augenscheinlich vermercke, dass man mir mein reden vnd schreiben zum ärgsten ausslegt, vnd ich damit nit mehr aussgericht, als mir dadurch freund zu feinden worden vnnd mit meiner sauren arbeit, an stat einer gebührlichen belohnung, nur feindschafft verdient hab" (*Prog. 1622*, sig. B5^{r}; cf. Zinner 1942, p. 70).

335"Herr Johann Kepler/Imperial Mathematician/as my good friend" ("Hern Johann Kepler/ Keyserlicher Mathematicus/als mein guter freund," Marius 1619, sig. B2^{v}); "Herr Kepler Imperial

dispute between the two about the moon, although it certainly was not of vital importance. In the *Prognosticon for 1620*, Kepler argued against Marius's opinion that the moon emits his own light[336]:

> What Marius states in his description about the Moon's own light/when it already darkened/ still making it appear to be like glowing iron/Also what he remarked in August 1617 about the same lunar eclipse/that is not happening right know in the shadow of earth/and instead I have seen in 1596 the moon to be darkened/that it could only be perceived a little bit pale/or iron-colored.

In the *Prognosticon for 1621*, Marius reports about the lunar eclipse of April 14/24, 1595, and of August 6/16, 1617. During the latter, he saw the moon[337]:

> Beautifully rounded and red [. . .], like a glowing iron, about which I was very amazed. [. . .] Now I'm telling you, that I am very certain about the moon emitting its own light, which exactly resembles a glowing iron [. . .].

Of course Kepler was right about the moon not emitting its own light. Marius however could not be deterred from his opinion. In his *Prognosticon for 1628*, he again speaks about "the moon is going to be red, like a glowing iron, which is its own light" (*Prog. 1628*, sig. D3^{r}; cf. Zinner 1942, p. 72). After 1619 Kepler never again expressed an opinion on Marius.

Mathematicus also my good friend" ("Herr Kepler Keyserlicher Mathematicus auch mein guter freund," *Prog. 1620*, sig. B5^{v}); "Herr Johann Kepler/the excellent Astronomus" ("Herr Johann Kepler/der vortreffliche Astronomus," *Prog. 1626*, sig. C1^{v}); "even Herr Kepplerus/my good friend/remembers this in August of this 1623. year. It would be desirable, that his labores in motibus planetarum would be published. But where are the patrons?" ("auch H. Kepplerus/mein guter freund/im Augusto dieses 1623. Jahrs erinnerung thut. Were wol zu wünschen das seine labores in motibus planetarum möchten publicirt werden. Aber wo sein die Maecenates?," *Prog. 1626*, sig. D2 ^{r}); "the excellent Astronomus Johannes Kepler [...] because he is my very good friend, I wish him to find a rich patron for all his efforts, so that the unbelieveable effort, industriousness and expenses are not destroyed again" ("der vortreffliche Astronomus Johannes Kepler [...] Wollte ihm als meinem sehr guten Freund ein reichen Patron seiner gehabten mühe wünschen/damit die unglaubliche Mühe/Fleiß und Unkosten nicht wider zu grund giengen," *Prog. 1627*, sig. D3^{r}).

336 "Dann was Simon Marius in seiner Beschreibung meldet von des Monds eignem Liecht/ vermittelst dessen er/wann er schon gantz verfinstert/noch anzusehen seyn solle wie ein glüend Eysen/Item was er Anno 1617. im Augusto an derselben Monds=Finsternussen dergleichen angemerckt/das findet jetzo mitten in dem Schatten der Erde keine statt/und hab ich hingegen Anno 1596. und sonsten den Mond sogar verdunckelt gesehen/daß er kaum ein wenig mit einer blaichen/oder Eysenfarb zu mercken gewest" (Kepler in the *Prognosticum for 1620*, quoted according to Kepler XI/2, 1993, p. 205f.).

337 "Schön rund vnnd roth [. . .], wie ein hoch glüenteisen, darüber ich mich hoch verwundert. [. . .] Nun sag ich, das ich glaube vnnd nun mehr gantz vergwisert bin, das der Monn sein eigen Licht hab, welches sich gäntzlich vergleicht einem glüenten eisen [...]" (*Prog. 1621*, sig. A3^{r-v}; cf. Zinner 1942, p. 68).

Honors

Without claiming to be exhaustive, here are some honors awarded to Marius. During his lifetime he did not experience many of them. In 1612 his hometown Gunzenhausen presented him with a small beaker worth 6 and a half guilders. The goldsmith Lienhart Heckel had been commissioned with its manufacture (Clauß 1922, p. 19; Mühlhäußer 1993, p. 17). He probably received this honor for his discovery of the Jupiter moons (Clauß 1922, p. 19; Zinner 1942, p. 26).

In the conference center Onoldia, a hall is named after him. The Lions Club commissioned the Munich artist Friedrich Schelle to design a memorial for Marius that since 1991 can be seen on the Kleinen Schlossplatz (Figs. 2.18 and 2.19).[338]

Fig. 2.18 The Simon-Marius-Fountain on the Kleinen Schlossplatz in Ansbach. Image taken by the author

[338]Information according to http://surfan.de/rundgang/rundgang.php4?station=Simon-Marius-Denkmal, viewed on February 8th 2009.

Fig. 2.19 Inscription on the Simon-Marius-Fountain in Ansbach. Image taken by the author

In 1969, a gymnasium in Gunzenhausen was named after Marius. In 2014, a memorial plaque was unveiled at the Sparkasse on the Hafnersmarkt near his former birthplace. Already in December 1924, a relief was mounted for Marius on the spot in the Ansbach palace (Fig. 2.20) where the former palace tower had been.

A crater on the moon is named after him. Its coordinates are 11.9° N and 50.8° W and its diameter is 41 kilometers (Fig. 2.21).[339]

[339]Cook 1999, Index of Named Formations at the end of this book.

Fig. 2.20 Memorial plaque for Marius in the palace of Ansbach. It is hanging at the place, where in former times the castle tower stood. If Marius had made observations from here, is questionable. Image taken by the author

This designation can be found for the first time on the famous lunar map by Giovanni Battista Riccioli[340] (1598–1677), which had been bound with his *Almagestum Novum*[341] in 1651 (Whitaker 1999, p. 61f., 213). In the anniversary year 2014, a minor planet was also named after Marius.[342]

---

[340]For Riccioli a lot of references can be found in Grant 1984 as well as in Siebert 2006.

[341]In the *Almagestum Novum*, Marius is mentioned three times, always in connection with his discovery of the Jupiter moons (Riccioli 1651, Praefatio p. XII; Chronici Pars I p. XXVII, Chronici Pars II p. XLV).

[342]See the chapter by Thomas Müller in this volume.

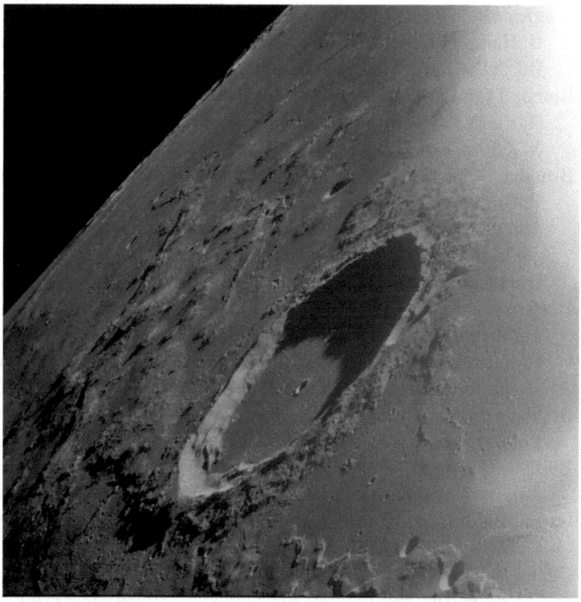

Fig. 2.21 The moon crater Simon Marius, recorded by Apollo 12. Wikimedia Commons, National Aeronautics and Space Administration

Acknowledgments I have to thank Werner Bürger, city archivist of Ansbach (letter of April 15, 2009), for his friendly information. To Claudia Heuwinkel from city archives Creglingen, I express my thanks for information concerning the brother Michael Marius. Klaus-Dieter Herbst from Jena gave me important hints concerning the calendars of Marius. Wolfgang Marius from Graz provided me with materials in the possesion of his family. My former colleague Dr. Thomas Lambertz as well as Joachim Schlör helped with translations from Latin.

References

Ackerknecht, Erwin H. (1979): *Geschichte der Medizin*, 4. Aufl. Stuttgart: Ferdinand Enke 1979

Allgemeine Deutsche Biographie (ADB, 1877-1912), 56 Vol. Leipzig 1877–1912. Reprint Berlin: Duncker & Humblot 1968

Bächtold-Stäubli, Hanns (Ed.) (1987): *Handwörterbuch des deutschen Aberglaubens*, 10 Vol. Berlin/New York: Walter de Gruyter 1987

Benzing, Josef (1982): *Die Buchdrucker des 16. und 17. Jahrhunderts im deutschen Sprachgebiet*, 2., verb. u. erg. Aufl. Wiesbaden: Otto Harrassowitz 1982

Barnes, Robin Bruce (1988): Prophecy and Gnosis. Apocalypticism in the wake of the Lutheran Reformation. Stanford: University Press 1988

Bergdolt, Klaus (2004): *Das Gewissen der Medizin. Ärztliche Moral von der Antike bis heute*. München: Beck 2004

Betsch, Gerhard; Hamel, Jürgen (Eds.) (2002): *Zwischen Copernicus und Kepler – M. Maestlinus Mathematicus Goeppingensis 1550–1631*. Vorträge auf dem Symposium, veranstaltet in Tübingen vom 11. bis 13. Oktober 2000 von der Fakultät für Physik der Universität Tübingen (= *Acta Historica Astronomiae*, Vol. 17). Frankfurt a. M.: Harri Deutsch 2002

Biographisch-Bibliographisches Kirchenlexikon (BBKL, 2001–2003), 22 Vol. 1990–2003. Bd. I, II: Hamm: Friedrich-Wilhelm Bautz 1990; Bd. III–XVIII Herzberg: Traugott Bautz 1992–2001; Bd. XIX–XXII: Nordhausen: Traugott Bautz 2001–2003

Brahe, Tycho (1925): Brief an Melchior Jöstel vom 18. November 1600. In: Dreyer, John L. E. (Ed.): *Tychonis Brahe Dani opera omnia*, Bd. 8. Kopenhagen 1925. Reprint Amsterdam: Swets & Zeitlinger 1972, pp. 391–393

Brunner, Johann Caspar (1746): *Corpus Constitutionum Brandenburgico-Culmbacensium, Oder Vollständige Sammlung Der Vornehmsten so wohl allgemeinen als besondern in dem Marggrafthume Brandenburg-Culmbach in Ecclesiasticis und Politicis Theils einzeln gedruckten, Theils noch nicht gedruckten Landes-Ordnungen und Gesetze*, Vol. 1. Bayreuth: Brandenburgisches Zeitungs-Comptoir 1746 [UB Erlangen-Nürnberg: H00/4 JUR-I 564[1]

Buck, Friedrich Johann (1764): *Lebensbeschreibungen derer verstorbenen Preußischen Mathematiker überhaupt und des vor mehr denn hundert Jahren verstorbenen großen Preußischen Mathematikers P. Christian Otters insbesondere*. Königsberg, Leipzig: Johann Heinrich Hartungs Erben und J. d. Zeise 1764 [SUB Göttingen: 8 H L BI I, 7110]

Buddeus, Johann Franz (1709): *Allgemeines Historisches Lexicon*, Dritter und Vierdter Theil: H–Z. Leipzig: Thomas Fritsch 1709 [ULB Sachsen-Anhalt: Na 1404 (3/4)]

Büttner, Heinrich Christoph (1813): Franconia. *Beiträge zur Geschichte, Topographie und Litteratur von Franken*, Bd. 2. Ansbach: Gassert 1813 [UB Augsburg: 02/IV.15.8.607]

Capra, Baldessare: Consideratione Astronomica circa la Stella Nova dell'anno 1604. In: Galilei, Galileo: *Le opere di Galileo Galilei*, Vol. II. Florenz: Barbera 1891

Capra, Baldessare (1606): *Tyrocinia astronomica. In quibus non solum calculus eclipsis solaris, ab astronomo magno Tychone Brahe restitutus clarissime explicatur; sed etiam facillima methodus erigendi, & dirigendi cœleste thema ad ipsius Ptolemæi mentem traditur*. Padua: Pietro Paolo Tozzi 1606 [National Library Florence: RARI.B.R.176.1]

Capra, Baldessare (1607): *Usus et fabrica circini cujusdam proportionis*. Padua: Petrus Paul Tozzius 1607 [BSB München: 4 Math.a. 64]

Caroe, Kristian (1913): Danske eg norske studerende i Padua 1559–1615. *Personalhistorisk Tidsskrift* 6/4 (1913), pp. 201–210

Caspar, Max (1995): *Johannes Kepler*. Stuttgart: Verlag für Geschichte der Naturwissenschaften und Technik 1995

Christianson, John Robert (2000): *On Tycho's Island. Tycho Brahe and His Assistants, 1570–1601*. Cambridge: Cambridge University Press 2000

Clauß, Hermann (1922): Zum Lebensbild des Simon Marius. *Gunzenhäuser Heimat-Bote* (5.2.1922), p. 18f.

Cöler, Philipp Albert (1664): *Simonis Marii Brandenburgica Sidera, Planetas Circumjoviales*. Autor und Respondent: Johannes Christoph Meelführer. Wittenberg: Röhner 26.3.1664 [UB Erlangen-Nürnberg: H00/HIST 601s; H00/HIST 727g (2)]

Cook, Jeremy (Ed.) (1999): *The Hatfield Photographic Lunar Atlas*. London u.a.: Springer 1999

Dannheimer, Wilhelm (1952): *Verzeichnis der im Gebiete der freien Reichsstadt Rothenburg o. T. von 1544 bis 1803 wirkenden ev.-luth. Geistlichen*. Nürnberg: Verlag Die Egge 1952

Dannheimer, Wilhelm (1959): Die Heilsbronner Fürstenschüler von 1582–1631. *Zeitschrift für bayerische Kirchengeschichte* 28 (1959), pp. 154–183

Diefenbacher, Michael; Fischer-Pache, Wiltrud (Eds.) (2003): Das Nürnberger Buchgewerbe. Buch- und Zeitungsdrucker, Verleger und Druckhändler vom 16. bis zum 18. Jahrhundert. Bearbeitet von Manfred H. Grieb. Mit einem Beitrag von Peter Fleischmann. Aus Archiven zusammengestellt von Lore Sporhan-Krempel und Theodor Wohnhaas (= *Quellen und Forschungen zur Geschichte und Kultur der Stadt Nürnberg*, 31). Nürnberg: Selbstverlag des Stadtarchivs Nürnberg 2003

Döhnel, Andreas (1996): Medizinstudium in Padua. In den Fußstapfen großer Ärzte. *Deutsches Ärzteblatt* 93/6 (1996), p. 298

Doppelmayr, Johann Gabriel (1730/1972): *Historische Nachricht von den Nürnbergischen Mathematicis und Künstlern*. Nürnberg: Peter Conrad Monath 1730. Reprint Hildesheim/New York: Georg Olms 1972

Drake, Stillman (1976): Was Simon Mayr Galileo's "Ancient Adversary" in 1607? *Isis* 67/238 (September 1976), pp. 456–459

Drake, Stillman (1978): *Galileo at Work. His Scientific Biography*. New York: Dover Publications 1978

Drake, Stillman (1979): Galileo and Satellite Prediction. *Journal for the history of astronomy* 10/2 (Juni 1979), pp. 75–95

Drake, Stillman; O'Malley, Charles Donald (Ed. and translater) (1960): *The Controversy on the Comets of 1618*. Philadelphia: University Press 1960

Drüll, Dagmar (2002): *Heidelberger Gelehrtenlexikon 1386–1651*. Berlin/Heidelberg/New York/ Barcelona/Hongkong/London/Mailand/Paris/Tokio: Springer 2002

Eckart, Wolfgang U. (2000): *Geschichte der Medizin*, 4. Aufl. Berlin: Springer 2000

Edighoffer, Roland (2002): *Die Rosenkreuzer*, 2. Aufl. München: Beck 2002

Erlangische gelehrte Anmerkungen und Nachrichten. 1746–1789 [UB Erlangen-Nürnberg: H00/ LTG-VIII 5]

Erler, Georg (Ed.); Lehmann, Clara; Joachim, Erich (Bearbeiter) (1917): *Die Matrikel der Albertus-Universität zu Königsberg i. Pr.*, Vol. III: Personenregister und Heimatsverzeichnis. Leipzig: Duncker & Humblodt 1917

Eyb, Eberhard Freiherr von (1984): Das reichsritterschaftliche Geschlecht der Freiherrn von Eyb (= *Veröffentlichungen der Gesellschaft für Fränkische Geschichte*, IX, 29). Neustadt a. d. Aisch: Degener 1984

Favaro, Antonio (1912): *Atti della nazione germanica artista nello studio di Padova*, Bd. 2. Venedig: Societá 1912 [StB Nürnberg: D 1066 = HL 1, 635]

Fikenscher, Georg Wolfgang Augustin (1801–1805): *Gelehrtes Fürstenthum Baireut; oder biographische und literarische Nachrichten von allen Schriftstellern, welche in dem Fürstenthum Baireut geboren sind und in oder ausser demselben gelebet haben und noch leben; in alphabetischer Ordnung*. Vol. 1–4: Erlangen: Palm 1801; Vol. 5–12: Nürnberg: Lechner 1803–1805

Flood, John Lewis (2006): *Poets Laureate in the Holy Roman Empire: A Bio-bibliographical Handbook*, 4 Vol. Berlin: De Gruyter 2006

Foerstemann, Karl Eduard; Hartwig, Otto; Naetebus, Gotthold (1894): *Album Academiae Vitebergensis: 1560–1602*. Leipzig: Tauchnitz 1894 [BSB München: 6971241 Hbl/885.444-1]

Friedensburg, Walter (1917): *Geschichte der Universität Wittenberg*. Halle (Saale): Niemeyer 1917

Friedl, Gabriele (1994): Deutsche Medizinstudenten an der Universität Padua im 15., 16. und 17. Jahrhundert und deren Einfluss auf die Entwicklung der Medizin im deutschen Sprachraum. Universität Saarbrücken, Dissertation, 1994

Fuchs, Johann Michael (1837): *Einige Notizen zur Schul-Geschichte von Heilsbronn und Ansbach. Bekannt gemacht bei der Säcular-Feyer des Ansbacher Gymnasiums am 12. Juni 1837*. Ansbach: Brügel 1837 [SB Ansbach: SB 110/X e 16]

Gaab, Hans (2011a): Astronomie in Altdorf. *Altnürnberger Landschaft e.V.* Sonderheft 2011. Neuhaus: Altnürnberger Landschaft 2011

Gaab, Hans (2011b): Der Altdorfer Mathematik- und Physikdozent Abdias Trew (1597–1669). Astronom, Astrologe, Kalendermacher und Theologe (= *Acta Historica Astronomiae*, Vol. 42). Frankfurt a. M.: Harri Deutsch 2011

Gaab, Hans; Leich, Pierre (2014): Marius' Replik auf Scheiner. Der Anhang zum Mundus Iovialis von Simon Marius. *Globulus* 18 (2014), pp. 11–14

Geret, Johann Georg (1738a): *de die nominali, Carolo in fastis sacro, ipso die Caroli*. Ansbach: Johann Valentin Ludersius 1738 [BSB München: 2 Diss. 1 Beibd. 85]

Geret, Johann Georg (1738b): *Illustris Carolini Onoldini natalem secundum [...] sollemnibus orationibus celebrandum indicat* [Inest vita Stephani Mummii]. Ansbach: Messererus 1738 [BSB München: 2 Diss. 6 Beibd. 94]

Geret, Johann Georg (1739a): *Praemisso excellentissimi Andr. Frobenii elogio*. Ansbach: Messererus 1739 [BSB München: Res/ 2 P.o.lat. 82, 52]

Geret, Johann Georg (1739b): *Praemisso perillustris Simonis Eisenii elogio*. Ansbach: Messererus 1739 [BSB München: 2 Diss. 14 Beibd. 35]

Gindhart, Marion (2006): *Das Kometenjahr 1618. Antikes und zeitgenössisches Wissen in der frühneuzeitlichen Kometenliteratur des deutschsprachigen Raumes*. Wiesbaden: Dr. Ludwig Reichert Verlag 2006

Goercke, Ernst (1986): Mediceische Sterne kontra Brandenburgisches Gestirn: Das Leben des Simon Marius. *Die Sterne* 62/4 (1986), pp. 223–231

Grant, Edward (1984): In Defence of the Earth's Centrality and Immobility: Scholastic Reaction to Copernicanisum in the Seventeenth Century. In *Transactions of the American Philosophical Society* 74/4 (1984), pp. 1–69

Grieb, Manfred (Ed.) (2007): *Nürnberger Künstlerlexikon. Bildende Künstler, Kunsthandwerker, Gelehrte, Sammler, Kulturschaffende und Mäzene vom 12. bis zur Mitte des 20. Jahrhunderts*, 4 Vol. München: Saur 2007

Groß, Johann Matthias (1727): *Großes Historisches Lexicon Evangelischer Jubel-Priester*, Vol. 1. Nürnberg: Endter und Engelbrecht 1727 [BSB München: 2120406 4 Biogr.c. 56-1]

Hansch, Michael Gottlieb (Ed.) (1718): *Epistolae ad Joannem Kepplerum Mathematicum Caesareum scriptae*. o.O. [Leipzig] 1718 [UB Erlangen-Nürnberg: H61/2 TREW.C 509]

Haug, Otto (Bearbeiter) (1981): *Baden-Württembergisches Pfarrerbuch*, Vol. 2: Pfarrerbuch Württembergisch-Franken, Teil: 2. Die Kirchen- und Schuldiener. Karlsruhe: Verlag Evangelischer Presseverband für Baden 1981

Hausmann, Gerhard (1989): Laurentius Laelius. Stadtpfarrer von Ansbach und Generalsuperintendent des Fürstentums Brandenburg-Ansbach (1572–1634). Universität Erlangen-Nürnberg, Dissertation, 1989

Hawlitschek, Kurt (1999): Sebastian Kurz (1576–1659). Rechenmeister und Visitator der deutschen Schulen in Nürnberg. In: Gebhardt, Rainer (Ed.): Rechenbücher und mathematische Texte der frühen Neuzeit (= *Schriften des Adam-Ries-Bundes Annaberg-Buchholz*, Bd. 11). Annaberg-Buchholz: Adam-Ries Bund 1999, pp. 257–266

Helfricht, Jürgen (2001): *Astronomiegeschichte Dresdens*. Dresden: Hellerau-Verlag 2001

Herbst, Klaus-Dieter (2009): Galilei's astronomical discoveries using the telescope and their evaluation found in a writing-calendar from 1611. *Astronomische Nachrichten* 330/6 (2009), pp. 536–539

Hergenhahn, Richard (1996): Jakob Köbel 1460–1533. Stadtschreiber zu Oppenheim, Feldmesser, Visierer, Verleger, Druckherr, Schriftsteller und Rechenmeister. In: Gebhardt, Rainer (Ed.): *Rechenmeister und Cossisten der frühen Neuzeit. Beiträge zum wissenschaftlichen Kolloquium am 21. September 1996 in Annaberg-Buchholz* (= *Freiberger Forschungshefte*, D 201 Wirtschaftswissenschaften, Geschichte). Freiberg 1996, pp. 63–82

Hergenhahn, Richard (1997) Jakob Köbel, seine Bedeutung als mathematischer Schriftsteller. *Oppenheimer Heft* 15 (1997), pp. 2–73

Herold, Hans-Jörg (1973): *Markgraf Joachim Ernst von Brandenburg-Ansbach als Reichsfürst*. Göttingen: Vandenhoeck & Ruprecht 1973

Hirsch, August (Ed.) (1888): *Biographisches Lexikon der hervorragenden Ärzte aller Zeiten und Völker*, 6 Vol. Wien: Urban & Schwarzenberg 1884–1888

Hocker, Johann Ludwig (1731a): *Hailsbronnischer Antiquitäten-Schatz*. Ansbach: Lüders [ETH Zürich: Rar 658 q]

Hocker, Johann Ludwig (1731b): *Bibliotheca Heilsbronnensis sive Catalogus librorum omnium tam manuscriptorum, quam impressorum, qui in monasterii Heilsbronnensis bibliotheca publica adservantur*. Nürnberg: Monath [StB Nürnberg: Solg. 1189 2°]

Hocker, Johann Ludwig (1739): *Supplementa zu dem haylßbronnischen Antiquitäten-Schatz*. Nürnberg: Monath 1739 [StB Nürnberg: Solg. 1189 2°]

Holden, James Herschel (2006): *A History of Horoscopic Astrology*. Tempe, AZ: American Federation of Astrologers 2006

Iselin, Jacob Christoph (1729): *Neu-vermehrtes historisch- und geographisches allgemeines Lexicon*, Vol. 3. Basel: Johann Ludwig Brandmüller 1729 [BSB München: 11732084 2 H.un. 150 m-3]; Neuauflage Basel: Brandmüller 1747 [FU Berlin: 32/2013/ 35435-3]

Jöcher, Christian Gottlieb (1751): *Allgemeines Gelehrten-Lexicon: Darinne die Gelehrten aller Stände sowohl männ- als weiblichen Geschlechts, welche vom Anfange der Welt bis auf ietzige Zeit gelebt, und sich der gelehrten Welt bekannt gemacht, Nach ihrer Geburt, Leben, merckwürdigen Geschichten, Absterben und Schrifften aus den glaubwürdigsten Scribenten in alphabetischer Ordnung beschrieben werden*; 3. Leipzig: Gleditsch 1751 [BSB München: 5947975 Hbh/4 N.libr. 52-3]

Johnson, J. H. (1930): The Discovery of the First Four Satellites of Jupiter. *The Journal of the British Astronomical Association* 41 (1930), pp. 164–171

Junger, Karl (1971): *Die Fürstenschule zu Heilsbronn*. Erlangen: Hespelein 1971

Kepler, Johannes (1611): *Dioptrice sev Demonstratio eorum quae visui & visibilibus propter Conspicilla non ita pridem inventa accidunt: Praemissae Epistolae Galilaei de ijs, quae post editionem Nuncij siderij ope Perspicilli, nova & admiranda in coelo deprehensa sunt. Item Examen praefationis Ioannis Penae Galli in Optica Euclidis, de usu Optices in philosopia*. Augsburg: Francus 1611 [HAB Wolfenbüttel: 43 Astron. (1)]

Kepler, Johannes (1620): *Epitome Astronomiae Copernicanae, Usitata forma Quaestionum & Responsionum conscripta, Liber Quartus*. Linz: Johannes Plancus 1620 [SB Regensburg: 999/Philos.1655]

Kepler, Johannes (1603): Gutachten über das feurige Trigon. 1603. In: Strauß, Heinz Artur; Strauß-Kloebe, Sigrid (Eds.): *Die Astrologie des Johannes Kepler. Eine Auswahl aus seinen Schriften*. München: Oldenbourg 1926, pp. 77–83

Kepler, Johannes (1937–2009): *Gesammelte Werke*, 21 Vol. München: Beck

Kepler, Johannes (2006): *Über den Neuen Stern im Fuß des Schlangenträgers*. Nachwort von Eberhard Knobloch. Übersetzung von Otto und Eva Schönberger und Eberhard Knobloch. Würzburg: Königshausen & Neumann 2006

Klug, Josef (1906): Simon Marius aus Gunzenhausen und Galileo Galilei: ein Versuch zur Entscheidung der Frage über den wahren Entdecker der Jupitertrabanten und ihrer Perioden. *Abhandlungen der Mathematisch-Physikalischen Klasse der Königlich Bayerischen Akademie der Wissenschaften* 22 (1906), pp. 385–526

Knod, Gustav C. (Bearbeiter) (1897): *Die alten Matrikel der Universität Strassburg 1621 bis 1793*. Vol. 1: Die allgemeinen Matrikeln und die Matrikeln der philosophischen und theologischen Facultät. Strassburg: Karl J. Trübner 1897

Komorowski, Manfred (2008): Basel als Promotionsort Königsberger Akademiker vor 1700. In: Marti, Hanspeter; Komorowski, Manfred (Eds.): *Die Universität Königsberg in der frühen Neuzeit*. Köln/Weimar/Wien: Böhlau 2008, pp. 24–41

Kronk, Gary W. (1999): Cometography. *A Catalog of Comets*, Volume 1: Ancient - 1799. Cambridge: University Press 1999

Kurz, Sebastian (1618): *Die sechs ersten Bücher Euclidis, Deß Hochgelärten weitberümbten / Griechischen Philosophi und Mathematicis: Von den anfängen und fundamenten der Geometriae. Dabey dann mancherley ajß disen Büchern gezogene nutzbarkeiten angefüget seind: sampt den Speciebus inn Geometrischen figurn / als machen / verändern / zusammenfügen / abziehen /vielfältigen unnd theilen: Per Demonstrationes Lineales*. Amsterdam; Willem Janszon Blaeu 1618 [StB Nürnberg: Math. 4. 3]

Landau, Richard (1902): *Die Universität Altdorf und ihre medizinische Fakultät*. Sonder-Abdruck aus der Festschrift zur Feier des 50jährigen Bestehens des ärztlichen Vereins. Nürnberg 1902

Lang, Karl Heinrich von (1811): *Neuere Geschichte des Fürstenthums Baireuth*. Vol. 3, welcher die Regierungszeit des Markgrafen Georg Friedrich von 1557 bis 1603 begreift. Göttingen: Schneider 1811 [UB Erlangen-Nürnberg: H00/G.E.O 78[3]

Lattis, James M. (1994): *Between Copernicus and Galileo. Christoph Clavius and the Collapse of Ptolemaic Cosmology*. Chicago/London: University of Chicago Press 1994

Layritz, Johann Christoph (1795): Genealogie der Eisen im Nürnbergischen. In: Siebenkees, Johann Christian; Kiefhaber, Johann Carl Sigmund: *Materialien zur Nürnbergischen Geschichte*, Vol. 4. Nürnberg: Schneider 1795, pp. 491–500

Lehsten, Lupold von (2003): *Die hessischen Reichstagsgesandten im 17. und 18. Jahrhundert*, Bd. 1: Prosopographische Untersuchung. Quellen und Forschungen zur hessischen Geschichte 137. Darmstadt, Marburg: Selbstverlag der Hessischen Historischen Kommission Darmstadt und der Historischen Kommission für Hessen 2003

Leich, Pierre (2012): Die Copernicanische Wende bei Galilei und Kepler und welche Rolle Simon Marius dazu einnimmt. In: Wolfschmidt 2012, pp. 163–193

Leppin, Volker (1999): Antichrist und Jüngster Tag. Das Profil apokalyptischer Flugschriftenpublizistik im deutschen Luthertum 1548–1618 (= Quellen und Forschungen zur Reformationsgeschichte; 69). Gütersloh: Gütersloher Verlagshaus 1999

Liesenfeld, Cornelia (2003): Die Astronomie Galileis und ihre Aktualität heute und morgen (= *Augsburger Schriften zur Theologie und Philosophie*, Bd. 2). Münster: Lit 2003

Lorenzini, Antonio (1605): *Discorso intorno alla nuova stella*. Padua: Pietro Paolo Tozzi 1605 [UB Leiden: 1397 C 13]

Lucken, Martin (1606): *Positiones De Apoplexia*. Basel: Schroeter 1606 [UB Erlangen-Nürnberg: H00/DISS.A.S 1039]

Mägerlein, Fritz (1989): *Familienregister der Stadtpfarrei Creglingen*, Bd. I–VII. Neustett 1989. Bound typescript, available in the Creglingen City Archive

Magini, Giovanni Antonio (1614): *Supplementum ephemeridum, ac tabularum secundorum mobilium*. Venedig: Zenarij 1614 [UB Graz: I 4090]

Major, Johann Daniel (1683): *See-Farth nach der Neuen Welt, ohne Schiff und Segel. Anno 1670. zu erst, und nu wiederumb, der gelehrten Welt vorgestellet*. Hamburg: Wolff 1683 [UB Erlangen-Nürnberg: H61/Trew.Yx 190]

Marius, Simon (1596): *Kurtze und eigentliche Beschreibung des Cometen oder Wundersterns / So sich in disem jetzt lauffenden Jar Christi unsers Heilands / 1596*. Nürnberg: Paul Kauffmann 1596

Marius, Simon (1599): *Tabulae Directionum Novæ*. Nürnberg: Christoph Lochner 1599

Marius, Simon (1610): *Die Ersten Sechs Bücher Elementorum Evclidis*. Ansbach: Paul Böhem 1610

Marius, Simon (1614): *Mundus Iovialis Anno M.DC.IX. Detectus Ope Perspicilli Belgici*. Nürnberg: Johann Lauer 1614

Marius, Simon (1614/1916/2019): *The World of Jupiter*, translated by A.O. Prickard and Albert van Helden, in *Simon Marius and his Research*, Hans Gaab and Pierre Leich (ed.), Cham: Springer 2019

Marius, Simon (1619): *Astronomische vnd Astrologische beschreibung deß Cometen so im November vnd December vorigen 1618. Jahrs ist gesehen worden*. Nürnberg: Lauer 1619

Marius, Simon (1625): *Gründliche Widerlegung der PositionCirckel*. Franckfurt a. M.: Lukas Iennis 1625

Matschinegg, Ingrid (1999): Österreicher als Universitätsbesucher in Italien (1500–1630). Regionale und soziale Herkunft – Karrieren – Prosopographie. Universität Graz, Dissertation, 1999

Matthäus, Klaus (1969): Zur Geschichte des Nürnberger Kalenderwesens. Die Entwicklung der in Nürnberg gedruckten Jahreskalender in Buchform. *Archiv für Geschichte des Buchwesens*. Frankfurt a.M., Buchhändler-Vereinigung 9 (1969), col. 965–1396

Megenberg, Konrad von (1994): *Das Buch der Natur. Die erste Naturgeschichte in deutscher Sprache*. Ed. by Franz Pfeiffer. Hildesheim/Zürich/New York: Olms 1994

Mencke, Johann Burckhardt (1715): *Compendiöses Gelehrten-Lexicon*. Leipzig: Gleditsch 1715 [BSB München: 1603869 N.libr. 203 m]

Metzger, Wolfgang (2006): *Quellen zur Gartenkultur des Heidelberger Humanismus: Gedichte und Pflanzenkatalog zum Garten des Hofapothekers Philipp Stephan Sprenger von 1597*. Heidelberg: Universitätsbibliothek 2006

Meyer, Christian (1893): Aus dem Lebensgang eines evangelischen Geistlichen und Gelehrten im 17. und 18. Jahrhundert [Johann Ludwig Hocker]. Nach eigenhändigen Aufzeichnungen mitgeteilt von Christian Meyer. *Zeitschrift für deutsche Kulturgeschichte*, Folge 3, 3 (1893), pp. 339–354, 428–448, 488–521

Meyer, Julius (1890): *Erinnerungen an die Hohenzollerherrschaft in Franken*. Ansbach: Brügel 1890 [StB Nürnberg: Hist 3151 8°]

Meyer, Julius (1892): Osiander und Marius. *Jahresbericht des Historischen Vereins für Mittelfranken* 44 (1892), pp. 51–71

Muck, Georg (1880): *Geschichte von Kloster Heilsbronn von der Urzeit bis zur Neuzeit*, Vol. 3. Nördlingen: Beck 1880

Mühlhäußer, Werner (1993): *Gunzenhausen – Geschichte durch Jahrhunderte*. Gunzenhausen: Stadt Gunzenhausen 1993

Mühlhäußer, Werner (2012): "Mensch Mayer!" Oder, wie man zum berühmten Sohn von Gunzenhausen wird. In: Wolfschmidt 2012, pp. 34–49

Müller, Heinrich (1612): *Carmen Epicum Encomiastice Conscriptum*. Stadthagen 1612 [HAB Wolfenbüttel: H: 139 Helmst. Dr. (16)]

NDB (1953–2005): *Neue Deutsche Biographie*. Berlin: Duncker & Humblot

Oertel, Eucharius Ferdinand Christian (1831): *Taubmanniana, oder des launigen Wittenberger Professors, Friedrich Taubmann, aus Wonsees, Leben, Einfälle und Schriftproben*. München: Fleischmann 1831 [BSB München: Biogr. 1147 d]

Oertel, Georg Christoph (1775): *De vita et fatis Simonis Marii mathematici quondam Brandenburgici / non nihil praefastus*. Erlangen: Camerariae 1775 [UB Tübingen: Kg 284.4]

Oertel, Georg Christoph (1780): *Animadversiones quaedam in vitas M. Petri Kolbii, Simonis Marii ac Eliae Levitae*. Nürnberg: Bileing 1780 [UB Erlangen-Nürnberg: H00/G.N.A. 914]

Pagnini, Pietro (1930): Galileo and Simon Mayer. *The Journal of the British Astronomical Association* 41 (1930), pp. 415–422

Platter, Felix; Untzer, Matthias (1605): *Disputatio Medica de Mola Matricis*. Basel: Schroeter 1605 [UB Erlangen-Nürnberg: H00/DISS.A.S 1039]

Premuda, Loris (1963): Die Natio Germanica an der Universität Padua. Zur Forschungslage. *Sudhoffs Archiv für Geschichte der Medizin und der Naturwissenschaften* 47 (1963), pp. 97–105

Reich, Ulrich (1996): Johann Scheubel (1494–1570), Wegbereiter der Algebra in Europa. In: Gebhardt, Rainer: *Rechenmeister und Cossisten der frühen Neuzeit. Beiträge zum wissenschaftlichen Kolloquium am 21. September 1996 in Annaberg-Buchholz*. Freiberg: Technische Universität Bergakademie 1996, pp. 173–190

Rein, G. (1904): *Paolo Sarpi und die Protestanten: Ein Beitrag zur Geschichte der Reformationsbewegung in Venedig im Anfang des siebzehnten Jahrhunderts*. Helsingsfors: Lilius und Hertzberg 1904

Reusner, Jeremias (1634): Programma in funere Cl. Viri Ambros. Rhodii, Medicinae Doctoris & Mathem. Super. In *Academia Wittenbergensi Profess. Publici*. Wittenberg 1634. In: Witte I, 1677, pp. 345–349

Riccioli, Giovanni Battista (1651): *Almagestum Novum Astronomiam Veterem Novamque Complectens: Observationibus Aliorum, Et Propriis Nouisque Theoreamtibus, Problematibus, ac Tabulis promotam*. Bologna: Victor Benati 1651

Röber, Paul (1634): *Geistlicher Rohrstab zur Abmessung des Tempels / Altars und derer / so darinnen anbeten / Auß der Offenbarung Johannis/cap.11v.1.2. Bey Christlicher Hinsetzung Deß Ehrenvesten / Großachtbarn und Hochgelahrten Herrn Ambrosii Rhodii, der Artzney Doctoris, der Mathematum Superiorum Professoris Publici in der löblichen Universitet Wittenberg / nunmehr Seligen / Welcher in recht Christlicher Andacht von dieser Welt abgefordert Anno 1633. d. 24. Augusti, Abends nach sechs Uhr / Seines Alters 56. Jahr 13.Tage*. Wittenberg: Johann Haken 1634 [StaBi Berlin: Ee 700-2733]

Rossetti, Lucia (1985): *Die Universität Padua. Ein geschichtlicher Querschnitt*. Triest: Edizioni Lint 1985

Rossetti, Lucia (1986): *Matricula Nationis Germanicae Artistarum in Gymnasio Patavino (1553–1721)*. Padua: Ed. Antenore 1986

Rüegg, Walter (1996): *Geschichte der Universität in Europa*, Bd. 2: Von der Reformation zur Französischen Revolution (1500–1800). München: Beck 1996

Schaudig, Wilhelm (1927): *Geschichte der Stadt und des ehemaligen Stiftes Feuchtwangen*. Feuchtwangen: Sommer & Schorr 1927

Schlichtegroll, Friedrich (1791): *Nekrolog auf das Jahr 1790*, Vol. 1. Gotha: Justus Perthes 1791 [UB Erlangen-Nürnberg: H00/HIST 875 a]

Schlund, Hans (1987): Heilsbronner Fürstenschüler aus dem Gunzenhäuser Land (1582–1631). *Alt-Gunzenhausen* 43 (1987), pp. 28–37

Schnabel, Werner Wilhelm (1995): Die Stammbücher und Stammbuchfragmente der Stadtbibliothek Nürnberg (= *Die Handschriften der Stadtbibliothek Nürnberg*, Sonderband), 3 Vol. Wiesbaden 1995

Schönbeck, Jürgen (2003): Euklid (= *Vita Mathematica*, Bd. 12). Basel/Boston/Berlin: Birkhäuser 2003

Schöneburg, Silvia (2007): Zur mathematischen Lehrtätigkeit an der Universität Wittenberg im 16. und frühen 17. Jahrhundert, dargestellt unter besonderer Berücksichtigung des Wittenberger Mathematikers Ambrosius Rhodius (1577–1633). Universität Halle (Saale), Dissertation, 2007

Schönhuth, Ottmar Friedrich Heinrich; Gurkfelder, Wenzel (1860): Chronik der Herren von Eyb. *Zeitschrift des Historischen Vereins fuer das Wuerttembergische Franken* 5,2 (1860), pp. 242–265

Schonlau, Anja (2005): *Syphilis in der Literatur*. Würzburg: Königshausen & Neumann 2005

Schott, Herbert (2015): Das Horoskop des Ansbacher Hofmathematicus Simon Marius für Georg Friedrich jun. von Eyb-Neuendettelsau 1596. *Altfränkische Bilder*, Neue Folge 10 (2015), pp. 10–13

Schreibmüller, Hermann (1928): *Das Ansbacher Gymnasium 1528–1928*. Ansbach: C. Brügel & Sohn 1928

Schuhmann, Günther (1980): Die Markgrafen von Brandenburg-Ansbach. Eine Bilddokumentation zur Geschichte der Hohenzollern in Franken (= *Jahrbuch des Historischen Vereins für Mittelfranken*, 90). Ansbach: Selbstverlag des Historischen Vereins 1980

Seck, Friedrich (Ed.) (2002): *Wilhelm Schickard – Briefwechsel*, 2 Vol. Stuttgart-Bad Cannstatt: frommann-holzboog 2002

Seelbach, Ulrich (2003): Humanistische Bildungsreise in die antike Welt: Roland von Waldenburg in Italien (1566–1567). In: Ertzdorff, Xenja von; Giesemann, Gerhard (Eds.): *Erkundung und Beschreibung der Welt. Zur Poetik der Reise- und Länderberichte. Vorträge eines interdisziplinären Symposiums vom 19. bis 24. Juni 2000 an der Justus-Liebig-Universität Gießen*. Amsterdam/New York: Editions Rodopi B. V. 2003

Selle, Götz von (1956): *Geschichte der Albertus-Magnus-Universität zu Königsberg in Preußen*, 2. Aufl. Würzburg: Holzner 1956

Senfelder, Leopold (1904): *Öffentliche Gesundheitspflege und Heilkunde*. Wien: Gilhofer & Ranschburg 1904

Siebert, Harald (2006): Die große kosmologische Kontroverse. Rekonstruktionsversuche anhand des Itinerarium exstaticum von Athanasius Kircher SJ (1602–1680) (= *Texte und Abhandlungen zur Geschichte der Mathematik und der Naturwissenschaften*, Bd. 55). Stuttgart: Franz Steiner 2006

Simon, Matthias (1957): *Ansbachisches Pfarrerbuch. Die Evangelisch-Lutherische Geistlichkeit des Fürstentums Brandenburg-Ansbach 1528–1806*. Nürnberg: Verein für Bayerische Kirchengeschichte 1957

Simons, Paul (1997): Froschregen, *Kugelblitze und Jahrhunderthagel. Warum das Wetter verrückt spielt*. München: Knaur 1997

Sommer, Eduard (1842): Das Vaterland der Sachsen; *Mittheilungen aus Sachsens Vorzeit und Gegenwart*. Vol. 2. Dresden: Blochmann 1842 [SLUB Dresden: Hist.Sax.A.188.b-2]

Spangenberg, Cyriakus (1572): *Mansfeldische Chronica* / Der Erste Theil. Von Erschaffung und Austheilung der Welt, und insonderheit von der Graueschafft Mansfelt, und den alten und ersten Königen und Fürsten, der Schwaben und Marckmannen, Cherusken, Francken und Sachsen und von gemeinen Politischen und Weltlichen hendeln / so sich in Friede oder Kriegsleufften in dieser Landart / Sachsen / Thüringen und am Hartz / auch etwan anderswo zugetragen / dabey dieser Landart / Oberkeiten oder Unterthanen mit gewesen. Eisleben: Petri 1572 [UB Erlangen-Nürnberg: H00/2 HIST 785 k]

Speta, Franz (2002): *Joachim BURS[CH]ER und seine Sammelreise in Österreich im Jahre 1615.* Stapfia 80 (2002), pp. 25–123

Steinbeck, Christoph Gottlieb (1795): *Chronologischer Handkalender für die Vorzeit Gegenwart und Zukunft.* Gera: Gottlieb Heinrich Illgen 1795

Stöcklein, Ansgar (1969): *Leitbilder der Technik. Biblische Tradition und technischer Fortschritt.* München: Heinz Moos 1969

Stölzel, Adolf (1872): *Die Entwicklung des gelehrten Richtertums in den deutschen Territorien,* Bd. 1. Stuttgart: Cotta 1872. Reprint Aalen: Scientia Verlag 1964

Strauß, Heinz Artur (1926): *Der astrologische Gedanke in der deutschen Vergangenheit.* München/ Berlin: Oldenbourg 1926

Strohmaier, Gotthard (1984): *Die Sterne des Abd ar-Rahman as-Sufi.* Hanau/Main: Müller & Kiepenheuer 1984

Stübner, Johann (1690): *Das Alt-Berühmte Closter Heilsbronn / Samt dessen edelsten Kleinod und besten Landes-Schatz, der Löblichen Fürsten-Schul / Ehbevor / Auf hochgebietenden Befehl / mit Poetischer Feder entworffen / bey daselbstigen Schul-Jubilaeo, auch Anno 1682. den 5. April. und dessen solemnen Feyer / in so genanntem Capitolio, offentlich vorgestellet; Anjetzo aber mit einem historischen Anhang / des Closters und der Fürsten-Schul Ursprung / Aufnehmen / Zustand / und Bediente betreffend / vermehret und erkläret / auch der werthen Posterität zu Lieb / aus heiliger Gemüths-Neigung gegen das liebe Closter / zum Druck ausgefertiget.* o.O. 1690 [StB Nürnberg: Mel. Bav. 287]

Surya, G. W.; Sindbad (1980): *Astrologie und Medizin*, 6. Aufl. Bietigheim: Rohm 1980

Thoren, Victor E (1990): The Lord of Uraniborg. A Biography of Tycho Brahe. Cambridge: University Press 1990

Thorndike, Lynn (1923–1958): *History of Magic & Experimental Science.* 8 Vol. New York: Columbia University Press 1923–1958

Tomasini, Giacomo Filippo (1654): Gymnasium Patavinum. Udine: Nicolai Schiratti 1654 [BSB München: 4 H.lit.p. 216]

Trew, Abdias (1643): *Discursus Von Grund und Verbesserung der Astrologiae und was durch Anleytung derselben auß dem Gestirn / Conjunctionibus, Finsternussen und andern Constellationibus, sowol von natürlichen / als WeltHändeln / ohne Aberglauben könne geurtheilet / und prognosticirt werden.* Nürnberg: Wolfgang Endter 1643 [GNM Nürnberg: Nw. 2837 4°; UB Erlangen-Nürnberg: H61/4 TREW.Q 110]

Ulshöfer, Kuno (1991): Das einfache Leben des Rothenburger Gymnasialrektors Abdias Wickner (1528–1564). *Jahrbuch des historischen Vereins für Mittelfranken*, Ansbach 95 (1991), pp. 125–145

Vendova, Giuseppe (1831): *Biografia degli Scrittori Padovani*, Bd. I. Padua: Minverva 1831

Vocke, Johann August (1796 and 1797): *Geburts- und Todten-Almanach Ansbachischer Gelehrten, Schriftsteller, und Künstler; oder: Anzeige jeden Jahrs, Monats und Tags, an welchem Jeder derselben gebohren wurde, und starb, nebst ihrer kurz zusammengedrängten Lebens-Geschichte und dem Verzeichnis ihrer Schriften und Kunstwerke.* Augsburg: Georg Wilhelm Friedrich Späth, Theil 1 1796, 2. Theil 1797. Reprint Neustadt a. d. Aisch: Verlag für Kunstreproduktionen 2001

Vogtherr, Friedrich (1908): *Geschichte der Familie Vogtherr im Lichte des Kulturlebens*, 2. Aufl. Ansbach: Seybold 1908

Vogtherr, Friedrich (1927): *Geschichte der Stadt Ansbach.* Ansbach: Brügel & Sohn 1927

Wackernagel, Hans Georg (Ed.) (1962): *Die Matrikel der Universität Basel*, III. Vol.: 1601/02–1665/66. Basel: Verlag der Universitätsbibliothek 1962

Walch, Johann Georg (1726): *Philosophisches Lexikon*. Leipzig: Gleditsch 1726 [BSB München: Ph.u. 533]

Weidler, Johann Friedrich (1741): *Historia astronomiae sive de ortu et progressu astronomiae liber singularis*. Wittenberg: Schwartz 1741 [SUB Göttingen: 8 HLU IV, 4105]

Weissenborn, Bernhard (1934): *Album Academiae Vitebergensis Ältere Reihe*, Bd. I: Textband; Bd. II: Registerband. Magdeburg: Selbstverlag der Historischen Kommision 1934

Westman, Robert S. (2011): *The Copernican Question. Prognostication, Skepticism, and Celestial Order*. Berkeley/Los Angeles/London: University of California Press 2011

Weyermann, Albrecht (1829): *Neue historisch-biographisch-artistische Nachrichten von Gelehrten und Künstlern, auch alten und neuen adelichen und bürgerlichen Familien aus der vormaligen Reichsstadt Ulm*. Ulm: Stettinische Buchhandlung 1829. Reprint Neustadt a.d. Aisch: Verlag für Kunstreproduktionen 2000

Whitaker, Ewen Adair: *Mapping and Naming the Moon. A history of lunar cartography and nomenclature*. Cambridge: University Press 1999

Will, Georg Andreas (1755–1758): *Nürnbergisches Gelehrten-Lexicon*, 4 Vol. Nürnberg u. Altdorf 1755–1758. Reprint Neustadt a. d. Aisch: Christoph Schmidt 1997

Witte, Henning (1677–1679): *Memoriae philosophorum, oratorum, poetarum, historicorum et philologorum nostri seculi clarissimorum renovatae decas prima (- nona)*. Königsberg, Frankfurt a.M.: Hallervord, 2 Vol. 1677–1679 [UB Mannheim: Sch 071/335]

Wolfschmidt, Gudrun (Ed.) (2012): Simon Marius, der fränkische Galilei, und die Entwicklung des astronomischen Weltbildes (= *Nuncius Hamburgensis – Beiträge zur Geschichte der Naturwissenschaften*, Vol. 16). Hamburg: tredition 2012

Wohlwill, Emil (1926): *Galilei und sein Kampf für die Copernicanische Lehre*, Vol. 2. Leipzig: Leopold Voss 1926

Xylander (= Holtzmann, Wilhelm, 1562): *Die Sechs Erste Bücher Euclidis Vom Anfang oder grund der Geometrj*. Basel: Oporinus 1562 [UB Erlangen-Nürnberg: H00/2 PHL-VII 137 c]

Zedler, Johann Heinrich (1732–1754): *Grosses vollständiges Universal-Lexicon Aller Wissenschaften und Künste*, 68 Vol. Halle, Leipzig: Zedler

Zinner, Ernst (1942): Zur Ehrenrettung des Simon Marius. *Vierteljahrsschrift der Astronomischen Gesellschaft* 77/1 (1942), pp. 23–75

Zinner, Ernst (1979): *Deutsche und niederländische astronomische Instrumente des 11.–18. Jahrhunderts*, Nachdruck der 2. Aufl. München: Beck 1979

Chapter 3
Hans Philip Fuchs von Bimbach (ca. 1567–1626), Patron of Simon Marius

Wolfgang R. Dick

The colonel, later general Hans Philip Fuchs von Bimbach (ca. 1567–1626) was for some years the most influential official at the court of the Protestant Margrave Joachim Ernst of Brandenburg-Ansbach and was engaged as a diplomat beyond the state's borders. However, later he fell out with the Margrave and entered into imperial (i.e., Catholic) service. He changed sides again and finally fell at the Battle of Lutter in the service of the Danish crown. Fuchs von Bimbach informed Simon Marius about the invention of the telescope already in the autumn of 1608 and as patron helped him to obtain one of the first copies. Thus, he seems to have played a considerable role in the early history of the telescope. This article gives biographical data about Fuchs von Bimbach. It attempts to define more precisely his role in the first astronomical applications of the telescope and also presents new insights into Simon Marius's work.

Introduction

The name of his sponsor Fuchs von Bimbach zu Möhren arises several times in the literature about Simon Marius, without mentioning his biographical dates and without examining the relationship between this general and politician and Simon Marius. Certainly, without his help Marius would only have received and used a telescope much later—maybe even not at all—and may have been in the history of astronomy just one of many calendar makers; additionally, this concerns one of the earliest ever mentions of telescopes, only weeks or months after its invention.

W. R. Dick (✉)
Potsdam, Germany

Frankfurt am Main, Germany
e-mail: wdick@astrohist.org

H. Gaab, P. Leich (eds.), *Simon Marius and His Research*, Historical & Cultural Astronomy, https://doi.org/10.1007/978-3-319-92621-6_3

Who was this man with the exceptional name? Which business had a military man and politician at the Frankfurt fair, where he saw one of the first telescopes in 1608? What is secure knowledge about Simon Marius, Fuchs von Bimbach, and the telescope and what is only speculation? The present study undertakes an initial examination of these questions.

A scientific biography of Hans Philip Fuchs von Bimbach using archival sources is still a desideration of research; this is also desirable for a biography of Marius. For general reasons I could use only secondary literature and couldn't consult all relevant publications for the present study. Using new or previously unused literature gives, nevertheless, a more comprehensive picture of Marius's sponsor, compared to previous biographies. Some incorrect statements about Fuchs von Bimbach and Simon Marius that were perpetuated in popular as well as in scientific literature are debunked (see section "Errors and Speculations about Fuchs von Bimbach in the Literature"). Yet I can't verify that all biographic dates mentioned in the other sections are correct. Much of the following information should in general be treated with caution, as they originate from particularly unreliable sources. The review with original sources, as far as this is possible, remains a task for future historians.

The following description is more detailed in terms of family origins and the early years of his life, because there has been almost no information in Fuchs's biographies until now. There is a lot of material about the Ansbach years and the later period that can only be bundled together for a brief characterization. The main focus is on the cooperation with Simon Marius. Concerning this, all known facts will be named and reinterpreted. The final section outlines approaches for further research.

Biographical Data About Hans Philip Fuchs von Bimbach

The Fuchs von Bimbach Family

The family of the Fuchs von Bimbach with the manor Bimbach belonged to several cantons of the Franconian Ritterkreis (Knight's Company), especially to the canton Baunach from the end of the tenth century until 1806 (Köbler 1999, p. 184).[1] Several other families named Fuchs existed until the seventeenth and eighteenth centuries in Franconia (e.g., Fuchs von Dornheim, von Neidenfels, and von Wiesentheid). Almost all of them died out (Tittmann 1998).[2] The Fuchs von Bimbach belonged in the seventeenth century to the lower or middle untitled Franconian nobility. In 1699 they were awarded the title baron ("Freiherr") (Fuchs von Bimbach 1975; B [ressensdorf] 1988). It is significant that the Franconian Fuchses, among them the Fuchs von Bimbach, held the rank of Franconian Reichsritter (imperial knights) and so were subordinated only to the emperor, not to the sovereigns. Nevertheless there

[1] Also contains leads to literature on the Franconian knights and the Fuchs families.

[2] Köbler 1999, p. 184, lists only four of these lines.

were tight-woven, partly symbiotic bonds to fiefdoms, patronates, and court employments but also conflicts resulting from struggles for independence from the liege lords.

Founding father of the family Fuchs von Bimbach was the Franconian nobleman Dietrich Fuchs who bought Bimbach in 1404. The local castle, destroyed during the German Peasants' Revolt and later rebuilt, was in the family's possession until 1913. The little town Bimbach belongs to Prichsenstadt (county of Kitzingen) in Unterfranken (Lower Franconia) since 1972.[3] The Fuchs we are interested in had little to do with Bimbach.[4] In his lifetime five family lines of the Fuchs von Bimbach existed. They were differentiated by their family seats. One of them resided in Bimbach, the others in Burgpreppach, Gleisenau, Eltmann, Bischofsheim, and Möhren.[5]

Hans Philip's great uncle Dietrich as the eldest son received the dominion Bimbach and founded the line Fuchs von Bimbach-Bimbach. His younger brother Christoph, grandfather of Hans Philip, purchased the castle of Mehren (later written Möhren) in the Duchy of Palatinate-Neuburg in 1522. Christoph's sons Endres and Sigmund called themselves from 1545 on Fuchs von Bimbach zu Mehren (= Möhren) or Fuchs von Bimbach-Möhren (Tittmann 1998, p. 79, note 385). The community of Möhren is a part of Treuchtlingen (county of Weißenburg-Gunzenhausen) in Mittelfranken (Middle Franconia) since 1972,[6] while Neuburg today belongs to the Bavarian administrative district of Oberbayern (Upper Bavaria). That Möhren belonged to Palatinate-Neuburg, at that time, was a decisive factor for the life of Hans Philip Fuchs von Bimbach.

The male line of the Fuchs von Bimbach has expired today but the name is continued in the female line.[7] The present seat of the Fuchs von Bimbach is the castle of Burgpreppach, where the family archives is also stored.

After the reformation most family members probably became Protestants, though Hans Philip's uncle Hans Fuchs von Bimbach (b. 1562) was a Catholic clergyman, Domkapitular (canon) in Bamberg and Würzburg (Biedermann 1747, Tabvla LIX). For Hans Philip no special faith preferences are known; he served on the Protestant side as well as the Catholic. His nephew, last of the Bimbach-Möhren line, converted to Catholicism to take possession of his heritage with the Emperor's help. Today the family Fuchs von Bimbach und Dornheim is Catholic (Fuchs von Bimbach 1975).

[3]https://de.wikipedia.org/wiki/Bimbach_%28Prichsenstadt%29 (accessed on May 2, 2016).

[4]However, he must have known his relatives there, as in 1581 his father became one of the custodians of mentally ill Hans Dietrich Fuchs von Bimbach zu Bimbach (ca. 1522–1586) and his children (Müller 2001, p. 11).

[5]An extract of the family's genealogy over eight generations can be found in Flurschütz da Cruz 2014, p. 385; compare the not clearly structured but complementary genealogical tabloids in Biedermann 1747, Tittmann 1998, p. 93, und Müller 2001, p. 13.

[6]https://de.wikipedia.org/wiki/M%C3%B6hren_%28Treuchtlingen%29 (last accessed on May 17, 2016).

[7]https://de.wikipedia.org/wiki/Fuchs_%28Adelsgeschlecht%29 (last accessed on May 7, 2016), with further references. To the later history of the family line, see also Rößner/Hammerich 2011.

Family, Birth, and Youth in Palatinate-Neuburg[8]

His father Endres (ca. 1519–1599)[9] is believed to have already come to the court of Palatinate-Neuburg in 1537 (Ludwig 1968, p. 42) then under Otto Henry, Elector Palatine (Ottheinrich von der Pfalz), from 1557 on under Wolfgang of Palatinate-Zweibrücken. In 1556 he became Landrichter (state judge) of Grailsbach and Pfleger (governor) of Monheim and in 1561 Statthalter (governor) of Neuburg/Donau.[10] After Wolfgang's death in 1569, the Duchy of Palatinate-Neuburg was separated and became independent under the new duke Philip Ludwig. As a governor Endres was some kind of representative in his absence and consultant of the sovereign; besides he was chief of protocol when foreign nobleman had to be welcomed. He also conducted negotiations, e.g., about marital contracts. His activities as a diplomat, administrator, and judge are documented in detail (Schöndorf 2006). His biographer writes about his "gradlinigen, etwas raubeinig wirkenden Art" ("straight, somewhat roughnecked character") and describes his efforts to increase and protect his possessions.

In 1546 Endres married Margaretha von Seckendorff-Aberdar who died in 1564; the marriage remained childless.[11] His mother-in-law, also Margaretha, was his cousin and was brought up in his father's household.[12] On November 28 or December 8 or 18, 1566, he re-married to Anna von Zeiskam (Zaiskam) from the Electoral Palatinate, the daughter of another governor of Wolfgang of Palatinate-Zweibrücken.[13] It would be interesting to know the exact date for the earliest date of birth of the eldest child, as it has to be excluded that Endres and Margaretha had "in

[8]Some biographic dates were, if not otherwise noted, extracted from Buchner/Mavridis (2009). See section "Source Situation and Approaches for Further Researches" for a discussion of this biography.

[9]Also called Endriß, Andreas, or Andrä. Endres stated he was 79 years old in 1598 (Schöndorf 2006, note 6); Schöndorf concludes, "somit dürfte das Geburtsjahr 1519 feststehen" ("so 1519 as the year of birth should be certain"), what is not correct as his birth could just as well have been at the end of 1518. Müller 2001, p. 13 (family tree), indicates without proof 1522 as the year of birth, and Flurschütz da Cruz 2014, p. 385 (genealogy), names none. The year of death 1599 is given in both family trees and also by Biedermann 1747, Tabvla LIX. Ludwig 1968 gives 1519 to 1592 as the dates of his life but 1592 is verifiably wrong.

[10]Ludwig 1968, p. 42, and Rechter 1997, p. 124, both refer to two different archival sources.

[11]Rechter 1997, p. 124, with archival references. Schöndorf (2006) calls her Magdalena.

[12]Rechter 1997, p. 108; he cites an archival source according to which she was the sister of Heinrich Fuchs von Bimbach. Heinrich existed in the line Bimbach-Gleisenau (Flurschütz da Cruz 2014, p. 385). According to Biedermann 1747, Tabvla LVIII, she was the sister of Hans Diet[e]rich from the line Bimbach-Bimbach. This genealogical confusion shows how insufficiently the family has been studied to date.

[13]Rechter 1997, p. 124, names November 28 with reference to an archival source. Ludwig 1968 sets December 18 as the date of marriage with reference to a Neuburg parish register. Schöndorf 2006, p. 6, names December 8. Between each of these three dates lie 10 days; it should be investigated whether one of the authors converted from the Julian to the Gregorian calendar though 1566 was before the Gregorian calendar reform of 1582.

Unehren zusammengekrochen" ("dishonorably crept together"), as the pastor of Bimbach would have called such cases.[14] In this marriage eight sons and two daughters were born. Four survived, Hans Philip, Lud[e]wig Veit, [Hans] Carl, and Anna Maria.[15] Without proof Hans Philip is considered to be the eldest, but I doubt this (see below).[16]

His brother Ludwig Veit served as a Hofmeister (court tutor/master of ceremonies) in Palatinate-Neuburg. He seems to be the only brother who had children, Hans Carl and Anna Maria.[17] Ludwig Veit died accidently in 1607,[18] and his brother Carl fell in Hungary in a battle against the Turks in 1604.[19]

The first names of the son, who interests us, are written very variably in the literature: Hans/Han[n]ß/Johan[nes] Philip[p][s], latinized Iohannes Philippus (by Marius; Marius 1614/1988, p. 36) or Iohan Philip (in the cartridge of his portrait; see Fig. 3.4). Even finding the baptismal register wouldn't give clarification about the "correct" spelling, as there were no fixed orthographical rules for names at those times. I presume that today's widespread notation "Johann Philipp" results from the assumption (in admissible analogy to modern use) that Hans could only be a modified version of the baptismal name Johann. Hans (in different spellings) instead of Johann obviously was the contemporary form, for Fuchs von Bimbach, as well as for other persons. The parish register of Bimbach from 1576 to 1588 (Schmidt/Müller 2001), for example, lists in the entries of marriages and baptisms only "Hans" as first name of several persons in various spellings (occasional three spellings for

[14]See Schmidt/Müller 2001 for his parish register with numerous cases of premarital or illegitimate relationships among commoners.

[15]Schöndorf 2006, p. 6; Biedermann 1747, Tabvla LIX. In 1587 Anna Maria married a Kämmerer (chamberlain) of Palatinate-Neuburg. Zwanziger 1919, p. 23, gives 1557 as the year of her birth which can't be correct. This date is from Biedermann, ibidem, and obviously refers to the husband's year of birth.

[16]This assumption in the literature has no source and presumably dates back to Biedermann (ibidem), who lists the brothers in this order. Apparently he had no birth dates so the order could well be random. A comparison of his Tabvla LVIII with the genealogical table in Müller 2001, p. 13, shows that Biedermann made the list not in the order of birth but on graphical principles in order to better illustrate the family lines. Therefore the widespread view that Hans Philip was the eldest son seems unsubstantiated to me.

[17]Biedermann 1747, Tabvla LIX: Here the son is called Johann Carl but he might have been called Hans Carl.

[18]Biedermann, ibidem, "kam an[no] 1607. im Wasser ums Leben" ("died in water in 1607"); Zwanziger 1919, p. 23, "ertrank 1608 in der Donau" ("drowned in the Danube in 1608"); Veh 1984–1985, p. 150, "war der bedrohten Stadt zu Hilfe geeilt und hatte durch Sturz vom Pferd das Leben verloren" ("rushed to help the threatend town [Donauwörth that was occupied by Catholic troops in December 1607] and lost his life falling from his horse"); ibidem, p. 151, note 36, "dem in Donauwörth 1607 verunglückten Ludwig Veit" ("the in Donauwörth in 1607 lethally injured Ludwig Veit"); Flurschütz da Cruz 2014, p. 385 (genealogy), "† 1607."

[19]Zwanziger 1919, p. 23; Veh 1984–1985, p. 151, note 36: "hatte nach 1603, im kaiserlichen Dienst gegen die Türken in Ungarn kämpfend, bei Kaschau den Tod gefunden" ("lost his life near Kaschau after 1603 fighting in imperatorial service against the Turks in Hungary"). Košice (German: Kaschau) today lies in Eastern Slovakia near the Hungarian border.

one and the same man), but I couldn't find a "Johann." The introduction to Marius's translation of Euclid he signed "Hanß Philips Fuchs von Bimbach" (Marius 1610, sig. A4^{v}); a letter written short before his death he signed "Hannß Philip Fuchsen" (Lichtenstein 1850, p. 145); also the nobleman today known as Johann Ernst von Anhalt he called "Hanß Ernst" (see below), which indicates his own use of Hans instead of Johann. Also other representatives of the family before and after him were called Hans; a Johann can be found only in the eighteenth century ([Gotha] 1924, p. 253). So I decided to use "Hans" instead of "Johann" and to write both first names in their shortest form, though "Philipp" would be as correct as the today unknown "Philips" or a spelling of Hans in other, today unusual forms.

Nothing is known directly until now about the birth of Hans Philip; conclusions can only be drawn from other data. If he really was the eldest son, as it is assumed, he could have been born in the second half of 1667, however not earlier, as second eldest son in 1668 but not much later.[20] The most likely birthplaces are Möhren or Neuburg.

In 1580 Hans Philip is verifiable in the Fürstliche Schule (Princely School) in Lauingen, where he held two reported speaches as a student; the school's historian assumes a stay from at least 1579 to 1583 (Ludwig 1968, p. 42).

The next biographical date given in the literature is a stay in Padua in October 1587 at the age of not more than 20.[21] It has been concluded that Fuchs studied there from this date alone. But he is not included in the registers of the German Nation in Padua,[22] so studies there seem very unlikely. There's nothing known about possible university studies until now; all information concerning this are nothing but speculations. From his further work and from his and Marius's writings, we can only conclude that he had extensive military experiences and some knowledge of military history, which he also recorded in writing (see section "Fuchs von Bimbach as a Military Author"). He had acquired at least basic skills in optics (see section "Fuchs, Marius and the Telescope") and concerned himself with geometry and its applications (see section "The Translation of Euclid").

The stay in Padua may have been during a "Kavalierstour" (educational tour), as was usual for young noblemen (often after finishing their university studies). About 100 years later, this is documented in detail for some of his relatives (Rößner 2003; Flurschütz da Cruz 2014, Sect. 1.3.1). There are indications that he might have stayed in Lyon in his younger years (Zwanziger 1919, p. 24).

Certain conclusions about his education might possibly be drawn by considering the education of 16-year-old Ludwig Reinhold Fuchs von Bimbach in 1682, who was expected to follow his famous relative Hans Philip in his military career

[20]Lent 2006 names "ca. 1567," Buchner/Mavridis 2009 "about 1568," Flurschütz da Cruz 2014, p. 385 (genealogy), "ca. 1567." Ludwig 1968, p. 42 and note 118, calculated a birth around 1567 from the average age of the Tertiani of the "Gymnasium illustre" in Lauingen.

[21]Biedermann 1747, Tabvla LIX, without further details about this stay.

[22]See section "Simon Marius in Padua" in Chap. 2. According to Zwanziger 1919, p. 24, he couldn't be found in the university registers of either Altdorf or Heidelberg.

(cf. section "Attempt to Assess of Hans Philip Fuchs von Bimbach"). His godfather recommended his mother that the son "eine gute Wißenschafft, in der Rechen=, Meß=, Kriegs= und Friedens=Bau=Kunst, auch etwa eine Verständnüß in Ernst= und Lust=Feüern, dann in der Geographie bey zubringen wäre."[23] "Die frantzösische Sprache, Reiten, Fechten und Tantzen"[24] would also be important. Ludwig Reinhold was sent to the University of Tübingen and afterward to France.

From his own records, it is clear that Fuchs von Bimbach had been taken part in military campaigns since the 1580s (Jähns 1890, p. 922; cf. section "Fuchs von Bimbach as a Military Author"). If he was born in 1567/1568, he would have been 17–18 years old in 1585.

In 1596 he became Hauptmann (captain) of Neuburg and was assigned to bring the contingent of the Protestant estates to the war against the Turks in Hungary.[25]

In 1599 Hans Philip inherited together with his two brothers the property Möhren and the family properties near Gerolzhofen (Buchner/Mavridis 2009). His brother Ludwig Veit became Lord of Möhren.[26] So one could assume that Ludwig Veit was the older brother and Hans Philip's year of birth would be 1568 at the earliest.[27] Possibly he added "auf Möhren" to his name only after his brother's death; that means from 1607/1608 on.

His father's inheritance was obviously not insignificant. For example, in 1582 Endres bought several properties and rights in Sulzfeld and five other places for 3700 guilders; in 1594 he sold the castle Rauenbuch that he had inherited from his mother-in-law with all belongings and rights to Margrave Georg Friedrich of Brandenburg-Ansbach and Bayreuth for 16,000 guilders (Rechter 1997, p. 112 and 125).

The connections of the family to their liege lord, the Duke of Palatinate-Neuburg, seemed to have been close. For example, in a chronicle of Möhren, the following is

---

[23]"A good knowledge, in the arts of calculating, measuring, war and peace architecture and also some knowledge of serious and joyful celebrating and in geography should be achieved." Rößner 2003, p. 106, s.a. p. 105 and the introduction to this book.

[24]The French language, riding, fencing and dancing.

[25]Veh 1984–1985, p. 146. Veh refers to information from the Österreichisches Staatsarchiv (State Archives of Austria) about the military career of Hans Philip.

[26]Veh 1984–1985, p. 146. Biedermann 1747, Tabvla LIX, calls Hans Philip and Carl only "Fuchs zu Bimbach," Ludwig Veit instead "Fuchs von Bimbach zu Mehren anno 1599." A chronicle about Möhren (Boller 1834, p. 8) placed Ludwig Veit at the beginning: "1599 folgten dem Andrä Fuchs seine drei Söhne—Veit Ludwig, Hans Philipp und Hans Karl im Besitze von Möhren [. . .]" ("In 1599 the three sons of Andrä Fuchs followed him in the possession of Möhren—Veit Ludwig, Hans Philipp and Hans Karl"). Note that the third brother is called "Hans Karl" here—maybe a confusion with Ludwig Veit's son.

[27]Veh 1984–1985, p. 146, claims without reference that Hans Philip, "obschon ältester der drei Söhne" ("though the eldest of the three sons"), was "nicht dazu bestimmt worden, als Majoratsherr das Rittergut zu übernehmen, sondern sollte im gehobenen Hofdienst eines Landesherrn auf militärischem oder diplomatisch-verwaltungsmäßigem Gebiet sein Glück machen." ("not designated to assume responsibility of the manor as lord but should find his luck at the court of a sovereign in military or diplomatic-administration service"). But Veh claims several demonstrably wrong or very doubtful facts about Fuchs von Bimbach and Simon Marius, so that his statements are not to be trusted.

reported in connection with the death of Carl in Hungary in 1604: "Seine beiden Brüder feierten seine Leichengottesdienste dahier, und luden zu dieser Feier den Herzog Philipp Ludwig von Neuburg ein, welcher in höchster Pracht in Person erschien, und den Leichengottesdiensten beiwohnte."[28] About 1601 Hans Philip quit the service in Palatinate-Neuburg but stayed connected to the dukes there, partly because of the fief Möhren but certainly also through a, then usual, patronage.

Nothing is known about a marriage or a family of his own. After his brother's, Ludwig Veit's, death, he became guardian of his son Carl.[29] Later this nephew succeeded him. As Hans Carl died in 1662, the family line Fuchs von Bimbach-Möhren expired.

Serving the Brandenburg Margraves in Ansbach

In 1599 Hans Philip led Ansbach troops in the so-called Straßburger Fehde (Strasbourg Feud). This started his career at the Ansbach court.[30] Whether he had already entered this service in that year or was just "lent out" from Palatinate-Neuburg for this military campaign has still to be investigated. Generally, all of his military service was interim, as was then usual. In 1601/1602 Fuchs von Bimbach fought again in the "Long Turkish War" (1593–1606) (Veh 1984–1985, p. 146) in which his brother would fall 3 years later. Later he remembered one episode where "wir selbsten anno 1601 vor Wienn vnd Preßburg, da wir des Obersten Fürsten Hanß Ernst von Anhalt Oberstlieutenant gewesen, gesehen."[31]

Also in 1601 he became Kriegsrat (military advisor) to Margrave Georg Friedrich. One year later he joined the Spanish-Dutch war as some kind of custodian for the margrave's 19-year-old relative and designated successor, Joachim Ernst, which led to a close relationship between them.

Joachim Ernst inherited the Margravate of Ansbach in 1603. In the fall of this year, he gave Schwaning and Rechenberg to Fuchs von Bimbach and his brothers as fiefs to very favorable conditions. Fuchs von Bimbach immediately started to build his own castle in Schwaning, today's Unterschwaningen in the Middle Franconian

[28]Boller 1834, p. 8: "His two brothers celebrated his funeral service here and invited Duke Philipp Ludwig of Neuburg who joined the funeral services dressed resplendently."

[29]Zwanziger 1919, p. 23, who calls the nephew Karl Johann. Biedermann, ibidem, gives "Johann Carl." "Johann Karl" is indicated in the genealogy at Flurschütz da Cruz 2014, p. 385.

[30]For details about his time at the Ansbach court, see Herold 1973; for the numerous references to Fuchs von Bimbach, see the personal register or search for "Fuchs" in the digital copy.

[31]"We met in the year 1601 outside of Vienna and Preßburg [Bratislava], when we were the lieutenant colonels of the Archduke Hanß Ernst von Anhalt." Cited after Jähns 1890, p. 925; cf. section "Fuchs von Bimbach as a Military Author". This was certainly Johann Ernst von Anhalt-Zerbst (1578–1601), who died in December 22, 1601, in Vienna and who was significantly younger than his inferior Fuchs von Bimbach; compare his criticism of too young and unexperienced colonels (section "Fuchs von Bimbach as a Military Author").

administrative district of Ansbach, halfway between Ansbach and Möhren.[32] From 1604 to 1606, he dwelled in the Netherlands again with the young Margrave, from 1605 on as an "Obrist" (colonel). At the inducement of Joachim Ernst, he assembled a battalion of soldiers for the Dutch States General.[33]

Simon Marius called Fuchs von Bimbach in 1614 "a man of the highest celebrity, not only for his ancient and noble lineage, but also and chiefly for his great deeds, his heroic exploits, and his consummate skill in war throughout France, Hungary, Belgium, and Germany."[34] "France" could mean the Strasbourg feud; "Belgium" was the name for the Netherlands at that time.

After his return from the Netherlands, his short political career at the Ansbach court began. From 1607 to 1610, Fuchs von Bimbach was the director of the Geheimer Rat (Privy Council), as well as of the Hof- und Kammerrat (Court and Chamber Council). He was the most powerful court official and highly paid.[35] He also worked outside the margravate, e.g., through participation in the founding of the Protestant Union in 1608, and was often underway in diplomatic missions across the Empire, e.g., in Frankfurt am Main.

From 1610 on Fuchs withdraw bit by bit from the Ansbach court. This was related to quarrels between him, other court officials, and also later the Margrave. First he quit his position as director of the Court and Chamber Council but continued leading the conferences of the Privy Council (Herold 1973, p. 209). His full income was paid until 1614.

In 1610 the building of his castle was almost finished but he seems not to have retired to his estates. Instead he became an artillery general in the Jülich-Klevische Erbfolgestreit (War of the Jülich Succession, an inheritance dispute). This was in the interest of his Palatinate-Neuburg liege lord Philip Ludwig, who was supported by the dukes of the Protestant Union (Jähns 1890, p. 922).

In 1616 the break with Margrave Joachim Ernst was definitive when Fuchs von Bimbach approached the Catholic side. There were also financial claims by Fuchs, who started litigation at the Reichskammergericht (Imperial Superior Court of Justice) against Joachim Ernst, as well as other conflicts (Zwanziger 1919, p. 27; Herold 1973, p. 46 and pp. 209–212).

It might have been an unfavorable coincidence for Simon Marius that his *Mundus Iovialis*, in which he praised Fuchs von Bimbach, was published just when Fuchs's conflicts with the Ansbach court escalated. Nothing is known about tangible impacts on Marius, but he complained about intrigues to his detriment just at that time.[36]

---

[32]To the history of the castle, see especially Veh 1984–1985, also [Unterschwaningen] 2009 and https://de.wikipedia.org/wiki/Unterschwaningen (last access May 17, 2016).

[33]Jähns 1890, p. 922, who used Fuchs von Bimbach's own report (cf. section "Fuchs von Bimbach as a Military Author").

[34]Marius 1614/1916/2019, Preface.

[35]See the comparison with Marius in Chap. 2.

[36]See Chap. 2.

Fuchs von Bimbach as a Military Author

The state library of Württemberg holds a handwritten military tract, whose author is not mentioned, but who is obviously Fuchs von Bimbach.[37] This manuscript also includes autobiographical notes. It was written at the earliest in 1610, military historian Jähns assumed it originates from around 1612.

After a short historical introduction, the essay expands on many organizational grievances in the military, based on the author's own experiences, e.g., as the following:

> Unter Kaiser Rudolf[38] habe man geradezu die jungen unerfahrenen Herren, zumal wenn es hohe Standespersonen gewesen, den erfahrenen Obersten grundsätzlich vorgezogen, weil sie sich mit geringer Besoldung begnügt und den oft fehlerhaften Anweisungen aus Wien nicht widersprochen hätten.[39]

As an evil custom of the foot soldiers he reprimanded "das Mitschleppen eines übermäßig großen Weibertrosses" ("that they dragged an abundantly large train of women") but admitted:

> Wiewol die Teutschen weiber den Soldaten beuorab in Ungarn mit tragender notturfft sowohl in wartung in kranckheiten denen Soldaten sehr nützlich sein. Zum tragen findet man selten eine, die vnder 50 oder 60 Pfund tregt; da etwan der Soldat mit Victualien oder ander dergleichen tragende wahren nit versehen, so ladet er ihr Stroh oder Holz daruor auf, zu geschweigen, daß manche ein, zwei oder mehr Kinder uf dem Ruckhen tregt.[40]

Then he listed in detail the clothing and tableware that a woman carried for a man along with their tent.

After this introduction the following three chapters primarily apply to the training of soldiers, especially the infantry (in today's parlance) with a lot of drawings. Chapter II presents for the most part a "in ganz unwesentlichen Punkten geänderte Abschrift der 'Instruction' des Landgrafen Moriz von Hessen v. J. 1600, die jedoch nicht genannt wird." "Das IV. Kapitel gibt einen kurzen Abriß der Feuerwerkerei ohne besonderen Wert."[41]

---

[37] According to Herold 1973, p. 67, note 79: Württembergische Landesbibliothek, Stuttgart, Handschriftenabt., Cod. milit. 2° 65. Herold was the first to point to Fuchs von Bimbach as the author. A summary can be found in Jähns 1890, pp. 922–925 and 1034.

[38] Rudolf II (1552–1612).

[39] Jähns 1890, p. 923: "Under Emperor Rudolf the young and unexperienced men, especially persons of high rank, were preferred to experienced colonels, because they were satisfied with low salaries and didn't contradict the often wrong orders from Vienna." This quote presents not Fuchs von Bimbach's own words but an analogous rendition by Jähns.

[40] Quoted after Jähns 1890, p. 924: "The German women were of use to the soldiers in Hungary for carrying their gear and caring for them in case of illness. One rarely finds one that carries less than 50 or 60 pounds; as the soldier has no grocery or other goods, he loads her up with straw or firewood, not to mention that some carry one, two or more children on their back."

[41] Jähns 1890, p. 925: "in insignificant details modified copy of the 'Instruction' by Landgrave Maurice of Hesse from the year 1600, which however is not named." "Chapter IV gives a short summary of fireworks without special value."

Jähns discussed the unpublished work of Fuchs von Bimbach within the framework of *Geschichte der Wissenschaften in Deutschland* ("History of science in Germany"—a history of astronomy by Rudolf Wolf was also published in this series). One can't refer to this work as "scientific" in the narrow sense whereby the general level of military lore at that time must be taken into consideration. It contains a lot of empirical findings as well as pragmatic conclusions and might have been intended as an instructional and textbook.

It is remarkable that Fuchs von Bimbach probably worked on his book during the same years in which Simon Marius wrote *Mundus Iovialis*. Did they perhaps inspire each other to compose a longer publication?

In the Thirty Years' War

From the following years until Fuchs's death, no more contacts to Simon Marius are known. Therefore this period of time will be handled very briefly, though a lot of material exists.[42]

Fuchs von Bimbach entered imperial service after lengthy negotiations in 1618. As a reason for his change of station, Johann Ernst's biographer cites the insults at the Ansbach court, from which Fuchs suffered as an imperial knight (Herold 1973, p. 46). As background it has to be taken into consideration that his new Palatinate-Neuburg liege lord and patron, Wolfgang Wilhelm, had converted to the Catholic Church in 1614, shortly before his father's death and against his will. In particular he hoped for the Emperor's support in the War of the Jülich Succession. Thirdly, as an imperial knight, Fuchs was formally only subordinated to the emperor, so serving him was normal rather than scurrilous. Confessional concerns seem unimportant to him. And finally he was principally an officer and therefore always on search for new appointments.

Emperor Matthias appointed Fuchs von Bimbach on July 7, 1618, to his "Obristen, Hofkriegsrat und Obristfeldzeugmeister" (colonel, court councilor of war and colonel gun master).[43] ("Feldzeugmeister," literally "battlefield ordnance master," was the name of the artillery officers; they were subordinated to a colonel.[44]) It is disputed whether he was involved in the Battle of White Mountain in 1620. Afterward he was accused that, as commander of the artillery, he had willfully ordered too short bombardments and was discharged without full payment (Zwanziger 1919, p. 27). So he also had to put up with an insult in imperial service.

[42]Detailed, partly contradictory data inter alia in Zwanziger 1919–1920 and Veh 1984–1985.

[43]Veh 1984–1985, p. 151, with reference to a note from the Österreichisches Staats-Kriegsarchiv (Austrian State Military Archive).

[44]Concerning the artillery officers in the foot soldier troops of the fifteenth and sixteenth centuries, see https://de.wikipedia.org/wiki/Landsknecht#Artillerie (accessed May 2, 2016) and the literature given there. The statements might also be basically valid for the beginning of the seventeenth century.

From 1621 on we find him again on the Protestant side as an officer and diplomat. The relationship to his (meanwhile Catholic) liege lord Wolfgang Wilhelm of Palatinate-Neuburg however remained very close. The latter lobbied for him against the Emperor for Fuchs's dominion Möhren and assigned to him the mediation between Emperor Ferdinand II and the Danish king Christian IV, which remained unsuccessful. In 1625 Fuchs was urged by Christian IV to join his service as an infantry general and later artillery general (Lichtenstein 1850, p. 143), which resulted in a condemnation by Ferdinand II and a threat to confiscate his estates.

Like other officers, Fuchs had deposited his most valuable movable possessions at the company of Samuel Rademacher in Hamburg during the war (Zwanziger 1920, p. 15).

Death in the Battle of Lutter

On August 27, 1626 (on the Julian calendar, this was August 17), one of the biggest and most momentous battles of the Thirty Years' War took place on a plane west of the Harz and south of Salzgitter near the village Lutter am Barenberge (Fig. 3.1).

Fig. 3.1 View from road B 248 in direction Nauen (Fuchs's dying place) to a part of the Lutter battlefield; under the tree the memorial stones for Fuchs von Bimbach on a rest area. Photo by the author, March 15, 2008

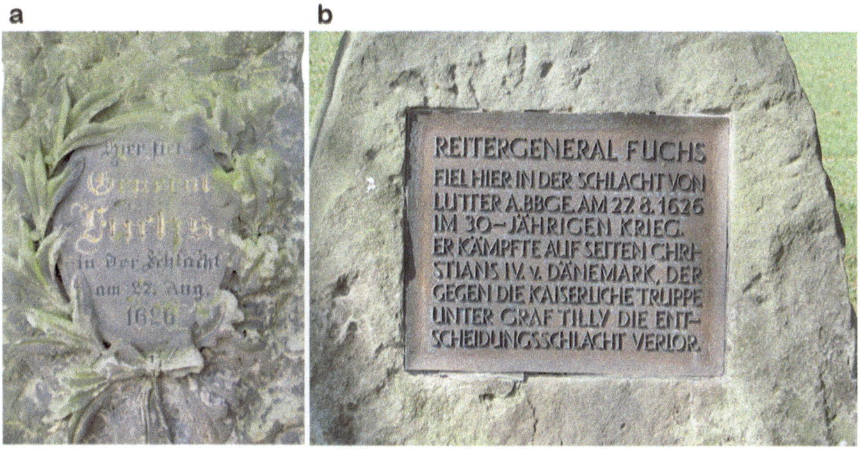

Fig. 3.2 (**a, b**) Memorial stones for Fuchs von Bimbach at road B 248 (details). To the left the memorial stone of 1908. Photo by the author, March 15, 2008

Instead of defeating the troops of Tilly and Wallenstein, as intended, the Danish king suffered a disastrous defeat. The battle finally ruined his imperial aspirations and after the war he possessed less than before. Fuchs von Bimbach, the highest ranking officer next to the King, is said to have warned him about entering the battle.

Whether the following description of Fuchs von Bimbach's death is authentic or was elaborated later has still to be researched:

> Groß und stark beleibt war ihm an dem heißen Schlachttage die Rüstung zu unbequem, er trug dafür eine weiße seidene Aermelweste (Wamms) und über diese einen kurzen leichten Oberrock (Casake) von grauer Farbe, so daß die hohe Figur überall leicht zu erkennen war. Verwundet nahm er, der anfänglich von einigen Ligisten für den König gehalten, den ihm angebotenen Pardon nicht an und wurde, von noch mehren Streichen tödtlich getroffen noch lebend nach Nauen in des Königs gewesenes Quartier, den riemschneiderschen Hof gebracht und auf die Bank hinter den Ofen niedergelegt. Er befahl hier, ihn an der Stelle, wo er gefallen, zu begraben, auch sein Grab zu respectieren und starb dann.[45]

So Hans Philip Fuchs von Bimbach died on August 27, 1626 in the small village Nauen near Lutter at the most 59 years old. Today two memorial stones stand on a parking area near to his former grave (Fig. 3.2). The oldest dates from 1908 (Melzner 1982). The Fuchs von Bimbach family paid an annual amount to the owner of the field to maintain the grave until the end of the eighteenth century. Around 1800 a road from Lutter to Seesen was built, today's B 148. The ditch went through Fuchs

[45]Lichtenstein 1850, p. 143: "Tall and corpulent as he was, his body armor was uncomfortable on hot battle days, instead he wore a white, silk doublet and over this a short, light, grey tunic (Casake), so that his tall figure was easily recognizable everywhere. Wounded he, who first was thought to be the King by some Catholic League soldiers, did not accept the offered pardon and was, lethally injured by several blows, brought still alive to Nauen in the King's former quarters, the Riemenschneider Court, and laid on a bank behind the oven. He ordered them to bury him here where he had fallen, to respect his grave and then he passed away."

von Bimbach's grave, which was opened. It contained a remarkable tall skeleton and a valuable sword (Lichtenstein 1850, pp. 143–144). To the facts about the grave and the memorial stones, more exact researches are desirable because the dates in literature and in the Internet are inaccurate, contradictory, and partially grossly incorrect.

In passing it is noted that before the Battle of Lutter, a nightly luminous effect was reported in the form of a sword that pointed from the imperial to the Danish troops and inspired the former to fight. This could be interesting for astronomers and meteorologists who deal with reports about noctilucent clouds and similar phenomenon. However, it can't be excluded that this was only a rumor put into the world by Tilly as psychological warfare (Lichtenstein 1850, pp. 134–135).

Fuchs's Financial Circumstances

Through inheritance and purchase in the form of fiefs, Fuchs von Bimbach owned several estates and castles. Besides the main property of Möhren, this included the large, richly endowed castle Schwaningen (Fig. 3.3), Rechenberg manor, and the castle of Cronheim near Gunzenhausen. From these he received income from the farms and payment in kind such as fish, wood, and the hunt bag.

Fig. 3.3 Castle Schwaningen. Etching of Matthaeus Merian, in Zeiller 1648 (Digitized version of the original print: http://bildsuche.digitale-sammlungen.de/?c=viewer&bandnummer=bsb00065888&pimage=00218), printed facsimile around 1960. Collection of the author

After his departure from Ansbach, he was criticized for not having paid for the fiefdom of Schwaningen and through abuse of office to have used margravial material and workers for the building of the castle (Veh 1984–1985, p. 148 and pp. 151–152). The truth of these accusations might be difficult to verify.

The worth of the fiefdoms can be assessed by the amount the widowed Margravine Sophie paid in 1630 to the heir Hans Carl Fuchs von Bimbach for the return of Schwaningen and Rechenberg, namely, more than 75,000 guilders (Veh 1984–1985, p. 153).

His annual income in Ansbach was 2581 guilders, additionally a large payment in kind of wine, cereals, and fish (Herold 1973, p. 46, note 66). In 1612 he lent the margrave 20,000 guilders for his marriage. For repayment Joachim Ernst used all the incoming taxes (Herold 1973, p. 191). His income in earlier and later assignments, as an officer, has not yet been determined.

The inheritance deposited in Hamburg included cash, silverware, precious clothing, jeweled harnesses, canons, horses, and other things amounting to 10,000 thalers (= 240,000 guilders).[46]

To roughly estimate these amounts in today's currency, we set Marius's annual payment of 150 guilders[47] as today equal to 10.000 € as a lower limit. We receive for the value of both fiefdoms not less than five million euros, for Fuchs's annual pay about 170,000 €, for the credit to the Margrave 1.3 million euros, and for his disposable inheritance 16 million euros.

Fuchs's Physical Appearance

We are mostly informed about Hans Philip Fuchs von Bimbach's appearance by his portrait (Fig. 3.4). A comparison to the portrait of Simon Marius from about 1614 (see Fig. 5.2) shows a great similarity stemming from the beard style, the haircut, and the clothing, whereby Fuchs's one is of course more splendid. Portraits of other contemporaries (see Chap. 2) show less similarities. It remains speculation as to whether Marius adapted his appearance to match his patron.

Fuchs is described as "groß und stark beleibt" (tall and corpulent); later an "auffallend große[s] Skelett" (remarkable tall skeleton) was found in his grave (Lichtenstein 1850, p. 143 and 144). In his inheritance in Hamburg was "eine Stadtliche, fürstliche, ja Königl. Kleidung" (splendid, princely, even royal clothing; quoted after Zwanziger 1920, p. 15), so he attached value to a representative appearance.

[46]Zwanziger 1920, p. 15; Buchner/Mavridis 2009, p. 77; in both cases without naming sources. Buchner/Mavridis call this inheritance a spoil of war which is inaccurate.

[47]Cf. Chap. 2.

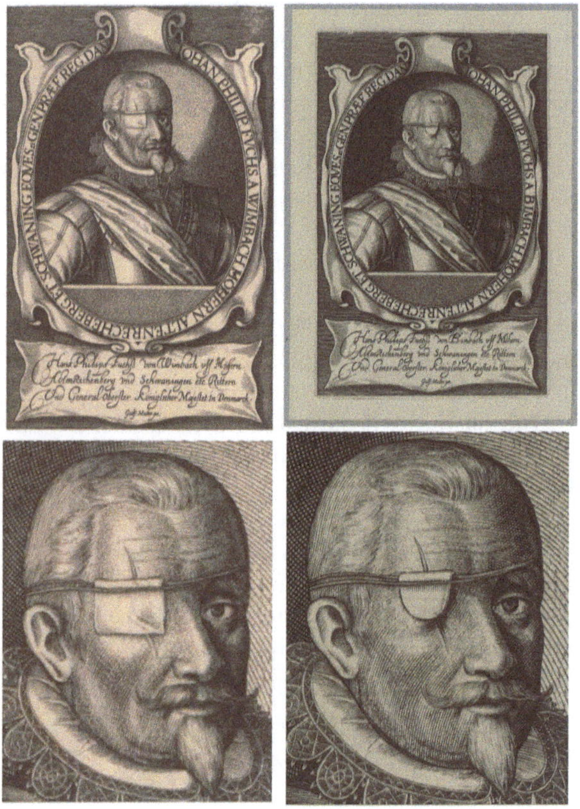

Fig. 3.4 (**a**) Portrait of Fuchs von Bimbach in an earlier version with mistakes in the epigraphs. Source: Wien, Österreichische Nationalbibliothek (see footnote 48). (**b**) Portrait of Fuchs von Bimbach in a later, corrected version. Source: Staatsbibliothek Berlin—Preußischer Kulturbesitz (see footnote 49). (**c**, **d**) Details of the Portraits (**a**, **b**)

The eye patch seen in the portrait indicates a severe injury or even the loss of his right eye, maybe in the battle, but I couldn't find anything about that.[48] He has a vertical scar across his right eye, apparently from a sword strike. If the portrait was painted posthumously, it could be a wound from one of the last battles before his death.

The form of his eye patch obviously results from artistic freedom of expression; an earlier version (or draft?)[49] shows the patch bigger and rectangular (Fig. 3.4a, c);

[48]Veh 1984–1985, p. 151, note 36: "Johann Philip had—it is not known when—lost his right eye." This statement might have made on the basis of the portrait and so have no validity.

[49]Wien, Österreichische Nationalbibliothek, Bildarchiv und Grafiksammlung, Porträtsammlung, Inventar-Nr. PORT_00099848_01, Digitized version with high resolution: http://www.portraitindex.de/documents/obj/oai:baa.onb.at:7847095; this version contains the typing error "Wimbach" instead of "Bimbach" in the cartouche and in the sign of the image, what indicates a draft or a very early version.

the final version[50] shows it smaller and semicircular (Fig. 3.4b, d).[51] The first version shows clearer that it could be a provisional covering of the wound with a piece of cloth that was attached to a string around his head and fixed with a second string on top of the cloth. For a permanent eye patch after the loss of an eye, I would expect another material (leather), another form, and a more comfortable fitting, but only a medical historian could give more precise information about that. One can at least assume that he wore this eye cover only at the end of his life. The engraving was made around 1626, maybe only posthumously, as the signature shows, but possibly also from an earlier template.

Fuchs von Bimbach as Sponsor of Simon Marius

Introduction

> All that I have observed, developed and already published in this regard, I owe to this great and most noble gentleman, my protector and patron, who holds all my reverence.[52]

The Latin word Marius uses here, translated as "protector," is "Patron"—again we come across patronat, very usual in those times. Protection and sponsoring by Fuchs von Bimbach, who was the highest official in the margravate with significant influence on the margrave, made him at least temporarily feel more secure. Contrary to Kepler, who found patrons in Rudolf II and later Wallenstein only for short periods, Marius was secured by the patronage of the Ansbach margraves in a long term, and so he was more or less independent of his second protector Fuchs von Bimbach.

John Robert Christianson assumed that Marius did not hurry to publish his discoveries due to his secure position, differing from Galileo who was looking for a good position.[53] This could have been a reason but just one among others such as uncertainty as discoverer, inexperience with respect to priority claims, relative isolation from other scientists, lack of time because of his calendar production, health problems, etc.

The most important events in the relationship between Simon Marius and Fuchs were the acquisition and use of early telescopes. For the history of their invention, Marius's report about Fuchs's visit to the Frankfurt fair in 1608 is very important

---

[50]Berlin, Staatsbibliothek zu Berlin—Preußischer Kulturbesitz, Handschriftenabteilung, Inventar-Nr. Portr. Slg/Mil. m/Fuchs von Bimbach, Johann Philipp, Nr. 1, b019047, Digitized version: http://www.portraitindex.de/documents/obj/33017232

[51]The later version is held in different archives and has been published several times; see also https://de.wikipedia.org/wiki/Datei:Fuchs_von_Bimbach.png (seen May 26, 2016).

[52]Marius 1614/1916/2019, Preface.

[53]Christianson 2000, p. 320: "Galileo [...] rushed into print with his discoveries [...] to achieve the scientific and patronage triumph of his carreer. Marius was already assured of patronage and did not rush [...]".

because this seems to be one of the earliest recorded dates that we have. The first exact date is September 25, 1608 (Gregorian).[54] It would be desirable to narrow the time span of Fuchs's visit in Frankfurt am Main, as the question of which of the three Dutch inventors could have been to Frankfurt depends on this.[55] Besides the general question arises (though irrelevant to the telescope), why an officer and politician visited a trade fair. Therefore in the following section, general remarks about trade fairs of the period and about Fuchs's visit at the fair of 1608 will be made before we focus on the description given by Marius. Finally we will discuss the cooperation of Marius and Fuchs on a translation of Euclid's writings that was published in 1610.

Possibly Fuchs von Bimbach came closer to Marius only during the telescope episode in 1608 though he must have known him and his calendars earlier. (Contrary to Wallenstein, nothing is known about any interest in astrology that Fuchs might have had.) Unfortunately, we know nothing at all about their relationship before the fall of 1608 and after 1612. Marius's statement that Fuchs "... frequently talked the matter over with me after supper" (see below) indicates that Marius was invited to supper several times and their relationship must have been quite close at least in the fall of 1608. For the year 1612, Fuchs's assistance to Marius is documented.[56] Hans Gaab assumes in his chapter (section "Life at Court and Publications" in Chap. 2) that the problems Marius had in Ansbach were somehow related to his good relationship to the unpopular Fuchs von Bimbach.

The Visit of the Frankfurt Fair in 1608

At that time, the Frankfurt Reichsmessen (Imperial fairs) were not only important economic events but also top-ranking social occasions (Stahl 1991; Brübach 1994). They served trade among merchants as well as the retail sector. What was lacking at ordinary markets could be found here, e.g., gems and books (also antiquarian). It was a place for settlements (often cashless but also with cash) and a financial center for exchange, credits, and investments. As many people met there, the fairs were also used for the exchange of information, not only between the merchants, and also for entertainment; there were theater productions, jugglers performed, and exotic animals were displayed (e.g., an elephant in 1629) or just pictures. Besides, tolerated by the councilmen, prostitution flourished, with which the local women would be protected from the many strangers. The fair replaced that which is matter of course for us today: shopping centers, banks, newspapers, television, theater, varieté, circus, and so on. Noblemen loved to visit the fair to meet their peers, to shop, or to just be entertained.

[54]Van Helden 1977, pp. 35–36; cf. Willach 2007, p. 109. The following statements about the general history of the telescope are based on these two publications. There is also an English version of Willach 2007, 2008.

[55]See the discussion of this question in Van Helden 1977, pp. 21–22.

[56]See Chap. 2.

Fuchs von Bimbach had been to Frankfurt on other occasions. Marius's report about a merchant "whose acquaintance he [Fuchs] had formerly made" suggests that Fuchs had also visited the trade fair in earlier years (Marius 1614/1916/2019, Preface). He might not only have had private motives (shopping, investment) but also business reasons such as purchasing for the Ansbach court together with other officials,[57] negotiating credits for the margrave, repaying his debts, and meeting other noblemen for diplomatic conversation. It might well be that he and other armed travelers guarded money or goods transports between Ansbach and Frankfurt.

When did the trade fair take place in 1608? For this there are earlier considerations in the literature about the history of the telescope, without taking all circumstances into consideration. A detailed history of the Frankfurt trade fair from 1765, evaluated by Albert van Helden (1977, pp. 21–22), reports that originally the trade fair took place between Assumption Day (August 15) and the Nativity of Mary (September 8). Already in the sixteenth century, its start had switched to the Nativity of Mary.[58]

To that another tradition must be considered; if the Nativity of Mary fell on Monday, Tuesday, or Wednesday, the trade fair started on Monday; if the holiday fell on Thursday, Friday, Saturday, or Sunday, it first started on the Monday of the following week.[59]

Finally it must be taken into account that in the imperial city Frankfurt, as well as in Ansbach, the old Julian calendar was still valid until 1700 (Brübach 1994, p. 136). However, in the Republic of the Seven United Provinces (the Netherlands), the Gregorian calendar had been valid since 1582.

Taking all these conditions into consideration, the date of the beginning of the fair, determined by Emil Wohlwill, responds exactly: "Nach Erkundigungen, die ich im Frankfurter Archiv eingezogen habe, wurde die Frankfurter Herbstmesse des Jahres 1608 am 12. September eröffnet [. . .]."[60] On the Julian calendar of 1608, the Nativity of Mary fell on a Thursday; the following Monday was September 12.[61] On the Gregorian calendar, this Monday was already September 22.

---

[57]It has been examined in detail, for the court of Landgrave Maurice of Hesse-Kassel (reigned 1592–1627), how regularly and to what extent royal courts satisfied their needs for consumer goods and luxury at the fairs in Frankfurt and Leipzig: Becker 1991.

[58][Orth] 1765, pp. 64–65 and 544–546; see also the quote in the chapter "Priority, reception and rehabilitation of Simon Marius" by Pierre Leich (Chap. 15).

[59][Orth] 1765, p. 66; Dietz 1910/1970, p. 40. I owe the reference to the last source to Dr. Michael Matthäus, Institut für Stadtgeschichte, Frankfurt am Main.

[60]Wohlwill 1926, p. 347: "According to inquiries I made in the Frankfurt archives, the Frankfurt fall fair of 1608 was opened on September 12 [. . .]." In his researches, Wohlwill does not take the calendar differences into consideration as he continues: "erst drei Wochen später bildet das Patentgesuch des Brillenmachers Johann Lippersher zum erstenmal den Gegenstand der Beratungen der holländischen Generalstaaten." ("It was only 3 weeks later that the patent application of the spectacle maker Johann Lippersher was for the first time part of the consultations of the States General of the Netherlands.")

[61]Compare Marius's own calendar for 1608: two digital copies in libraries are linked on the *Marius Portal*.

In the second half of the sixteenth century, there were 18 fair days in Frankfurt (Brübach 1994, p. 31). This might have been the same in 1608, because the original fair date between August 15 and September 8 also stretched over approximately 3 weeks. I assume that the 18 fair days mentioned excluded Sundays and the fair took place on 3 × 6 days. Thus, the last day would have been Saturday, October 1, on the Gregorian calendar October 11.

The dating by Wohlwill and the calculated end of the 1608 fair are confirmed by the following archival registrations: "Anfang der Herbstenmeß. NB. Ist die Meß ausgeleut worden donnerstags vf Nat. Mar. vor anfang der Meß."[62] So the fall trade fair of 1608 started after Nativity of Mary, in accordance with the rules above on the following Monday, September 12 (Julian date). "Alß man Montags den 3t. 8bris 1608 nach vollend[er] herbstmeß an den Veltpf[orten] vfgeschlossen, Ist seither jungsten Meß vber gefallen [...]."[63] So in 1608 the fair was finished before October 3, according to the calculations above, on Saturday, October 1 (Julian).

A verification of the begin and end of the 1608 fair using contemporary calendars seems impossible, though they also included fair dates (they are missing in Marius's own calendar for 1608). I was unable to consult a calendar for 1608 other than Marius's; however, an example for 1606 only mentions the start of the fair and that only approximately.[64]

As for the presence of the merchants, it must be taken into account that they arrived and left only in large "Geleiten" (convoys) because of the danger of holdups. By imperial order, these convoys had to be protected by the respective sovereign ([Orth] 1765, pp. 75–99; Dietz 1910/1970, pp. 41–44). Less than ten convoys existed that were obviously assembled in some meeting places in larger cities and were composed of mounted merchants, four-in-hand freight carriers, and numerous (certainly armed) escorts. In 1446 the Nuremberg convoy, though quite small in that year, was composed of 250 people, 450 horses, and 69 carriages (Dietz 1910/1970, p. 61). The convoys arrived before the official beginning of the fair, on Wednesday or Thursday of the previous week, and the merchants unpacked their goods (Dietz 1910/1970, p. 40 and 44).

As to the further procedure of the fair, there are different claims; a Frankfurt jurist and (amateur) historian Alexander Dietz claimed, without source references, first a business week and then a week to settle up (Dietz 1910/1970, p. 40). A new academic investigation refers to 5 days of accounting for the repayment of debts,

---

[62]"Begin of the fall fair trade. NB. Fair was rung [i.e. announced by bell ringing] on Thursday of the Nativity of Mary before the beginning of the fair." Diurnal of 1608/09, Institut für Stadtgeschichte, Frankfurt am Main, Call number: Rechneiamt: Bücher 448, fol. 24^{v}; by kind information from Dr. Michael Matthäus, Institut für Stadtgeschichte, Frankfurt am Main, E-Mail of July 5, 2016.

[63]"As on Monday the 3rd October 1608 after completion of the fair, the field gates were opened, is from this time the latest fair finally over [. . .]." Ibidem, fol. 37^{v}.

[64]Krabbe [ca. 1605], p. [62]: "Franckfurt am Mayn helt Meß/[. . .]/Die 2. [d.h. Herbstmesse] auff Marie Geburt."; Transl.: "Frankfurt am Main held a fair/[. . .]/The 2nd [i.e. fall fair] on the Nativity of Mary." I owe this information to Dr. Klaus-Dieter Herbst, Jena. Link to the corresponding page: http://reader.digitalesammlungen.de/de/fs1/object/display/bsb10392756_00062.html

followed by a week of trading for the sale of goods (Brübach 1994, p. 310). The convoys "wurden in der dritten Woche [. . .] großenteils schon am Dienstag abgeführt. Als letztes zog am Samstag Nachmittag das hessendarmstädtische ab."[65] Residual sales to small customers were continued until 3 p.m. on Saturday (Dietz 1910/1970, pp. 40–41).

We don't know how long Fuchs von Bimbach stayed in Frankfurt in 1608. Considering the long distance, about 200 km, between Frankfurt and Ansbach, it would probably have been a 3-day ride,[66] and taking into account the many things, he would have had to accomplish and to enjoy, and he certainly would not have stayed for only a couple of days. Let us suppose that he stayed until the middle of the third week and left, for example, on September 28 (a Wednesday in Julian calendar); he would have arrived at Ansbach not earlier than the evening of September 30 (October 10 in Gregorian calendar). At least, it is very unlikely that he would have returned before the beginning of October (Gregorian). If Frankfurt was only an intermediate stop on a longer journey, his return might well have been significantly delayed.

Dutch merchants had a far longer journey of about 450 km, and with loaded carriages and in a large convoy, the traveling speed would have been much slower than for riders. We can certainly calculate 10 days for the journey if not more. Such a trip was only worth it if one stayed in Frankfurt for the whole fair, so that the Dutchmen would have only returned to Amsterdam or other places at the earliest around October 20 (Gregorian). In individual cases merchants might have traveled without a convoy, though a very short stay also seems unlikely for them, considering the long, arduous, and dangerous route. This also applies to the visitors.

Fuchs, Marius, and the Telescope

Everything in the literature on how Simon Marius got his first telescopes obviously goes back to his own presentation in *Mundus Iovialis* and is only occasionally supplemented with details, which are not to be found by Marius. Though often cited, Marius's account will be repeated literally, because in certain details the exact wording is important. Primarily, Prickard's and van Helden's translation is used. A new analysis of the Latin original with regard to the following considerations would be desirable:

> In the year 1608, when the Frankfurt autumn fair was going on, it happened that there was at the same place the most noble, gallant, and energetic John Philip Fuchs, of Bimbach in Mohr, [. . .] Various things went on there, and among others it chanced that a certain merchant met the nobleman mentioned above, whose acquaintance he had formerly made,

[65] Dietz 1910/1970, p. 44: "The convoys left for the most part already on Tuesday in the third week. The one from Hesse-Darmstadt left finally on Saturday afternoon."

[66] To traveling speed, see, for example, https://de.wikipedia.org/wiki/Reisegeschwindigkeit

> and told him that there was then present in Frankfurt at the fair a Belgian, who had invented an instrument by means of which the most distant objects might be seen as though quite near. Hearing this, he begged the merchant to bring the Belgian to him, which the merchant at last consented to do. Our nobleman had a long discussion with the Belgian first inventor, and felt doubts as to the reality of the new invention.[67]

This "first inventor" ("primus inventor") is unknown. Cornelis de Waard drew in 1906 the conclusion, "dass es sich mit großer Wahrscheinlichkeit um Sacharias Janssen gehandelt hat."[68] Arjen Dijskstra argued that it could have been Jacob Metius, who would have had the best reasons to travel to the Frankfurt fair.[69] To discuss this question, we have to recapitulate, which inventors of telescopes are currently known.

It is recorded that Hans Lipperhey and Jacob Metius applied for patents on their inventions on October 2 and 15, 1608, respectively. They were both refused because neither was the unique inventor. A document from October 14 quotes, in this context, a third, unnamed inventor (Van Helden 1977, pp. 36–40). Earlier in the literature, Sacharias Janssen was supposed to be this third inventor. But Huib Zuidervaart argued convincingly that Janssen was unable to create telescopes in 1608 and proposed Lowys Lowyssen as a more suitable candidate (Zuidervaart 2010).

If the "Belgian" whom Fuchs met was a merchant, stayed until the third fair week, and left with a convoy, so according to the calculation in the previous section, it probably wasn't one of the two known inventors, because they wouldn't have made it home until October 15. Lipperhey can be excluded, and also Metius had to leave significantly before the fair ended, which seems unlikely, however not impossible. Apart from the known inventors, there might have been another Dutch optician, who manufactured a functioning telescope in 1608:

> At last the Belgian produced the instrument, which he had brought with him, and one glass of which was cracked, and told him to make trial of the truth of his statement. So he took the instrument into his hand, and saw that objects on which it was pointed were magnified several times. Satisfied of the reality of the instrument, he asked the man for what sum he would produce one like it. The Belgian demanded a large price, and when he understood that he could not get what he first asked, they parted without coming to terms.[70]

A detailed analysis of this report shows that the demonstration of the first telescope was a private, nonpublic matter. Fuchs von Bimbach didn't meet the "Belgian" himself but heard about him through an acquaintance. The owner didn't demonstrate his instrument straight away but only after a lengthy discussion. Therefore it can't be concluded that this telescope was offered for sale at the fair. Without

---

[67]Marius 1614/1916/2019, Preface.

[68]"that in all probability it must have been Sacharias Janssen." Willach 2007, p. 111; Wohlwill 1926, p. 347, calls this "a very weak proof." I couldn't examine the original publication.

[69]Dijkstra 2012, p. 137. For a summary of Dijkstra's thesis, see the chapter "In the Turmoil of the Early 17th-century Cosmology Debate—Simon Marius as a Supporter of the Tychonic System" by Pierre Leich in the present volume.

[70]Marius 1614/1916/2019, Preface.

the merchant, who he had known for some time, Fuchs von Bimbach probably would not have heard about the telescope. The "first inventor," e.g., a spectacle maker, could have been selling lenses and glasses or in the case of Metius his brother's new book. However, Marius's report nowhere states why he was at the fair; he might as well have been a visitor just like Fuchs and not a merchant.

In the literature about Marius, it is presumed that Fuchs von Bimbach was interested in the telescope for military reasons. This is an obvious assumption as Dutch documents of 1608 mention this and also a desired secrecy; but there are no statements by Fuchs or Marius about this. We also don't know if Fuchs ever used one of the instruments he later bought for military purposes. As he handed them over to Marius, his interest might well have been purely scientific. The image quality of the first telescopes was very low and their field of view very limited, so their military value was low. It would be desirable to find out since when telescopes were actually used for military purposes.

Also the conclusion that there was no trade because of the large price claimed is just a presumption. The main reason may have been that "one glass [...] was cracked." The statement "The Belgian demanded a large price" may not necessarily have been the reason that "they parted without coming to terms." We don't know what sum had been demanded, whether Fuchs eventually would have accepted it, if a usable instrument had been for sale immediately, and how much he paid nearly 1 year later for a Dutch instrument. The repeated mention of the high price as an obstacle for the sale certainly goes back to Klug, whose original translation of *Mundus Iovialis* was truncated and therefore false: "Der Belgier verlangte eine hohe Summe; deswegen zerschlug sich der Handel."[71] Anyway, Fuchs couldn't get a telescope at once, and the "delivery" of another copy under the prevailing conditions would certainly have taken a long time, maybe even until the next fair. The attempt to build his own telescope could have been a question of time rather than costs[72]:

> When he returned to Ansbach, the Nobleman sent for me, and told me that an instrument had been devised by which very remote objects were seen as though quite near. I heard the news with the utmost surprise. He frequently talked the matter over with me after supper, and at last came to the conclusion that such an instrument must necessarily be composed of glasses, of which one was concave, the other convex. He took up a piece of chalk and with his own hand drew a sketch on the table to show what sort of glasses he meant. We afterwards took

[71] Klug 1906, p. 410: "The Belgian demanded a large amount of money; so the trade was abandoned."

[72] Please note that the English and the German translations differ. The English one reads: "The Belgian demanded a large price, and when he understood that he could not get what he first asked, they parted without coming to terms." The German one is: "Der Belgier forderte eine hohe Geldsumme. Als Philipp nun kennengelernt hatte, was er aufs erste Mal nicht erhalten konnte, schied man also unverrichteter Dinge voneinander." (Marius 1614/1988, p. 38 ["The Belgian demanded a large amount of money. As Philipp had seen what he couldn't get at the first time, they left without having achieved anything."] The English translation more clearly suggests a relation between the large price and the fact that Fuchs did not obtain the instrument.

> glasses out of common spectacles, a concave and a convex, and arranged them one behind the other at a convenient distance, and to a certain extent ascertained the truth of the matter.[73]

The calculation above shows that Marius probably wouldn't have met Fuchs and heard about the telescope before October 10 (Gregorian date).

Marius's report shows that he had never before heard or read about the possibility of such an instrument, though the enlarging effect of such a combination of lenses had been known since the sixteenth century (Van Helden 1977; Willach 2007, pp. 105–109). Rolf Willach assesses that at the end of this century, "die Kenntnis des teleskopischen Effektes sicher weithin Allgemeingut unter den Brillenmachern und Gelehrten geworden war."[74] At least for Marius this was not correct; it is not generally known if he had been concerned with optics before or after 1608 and his access to scientific literature was limited. In contrast to Kepler, he just used telescopes and didn't care about their construction and theory. So unfortunately, we have no detailed specifications of his instruments, only a picture together with his portrait (see Fig. 5.2).

It must also be doubted whether Fuchs von Bimbach knew about the telescopic effect of a combination of a concave and a convex lense before. It is therefore all the more remarkable that he recognized the form of the lenses in the telescope he inspected in Frankfurt. As he was probably about 40 years old, he could have been in contact with spectacles and their makers. However he failed to notice something else—the diaphragm, with some certainty, present in front of the objective lens.

Experimenting with two spectacle lenses, Fuchs and Marius persuaded themselves of the magnifying effect of the lens arrangement. Thus they repeated what others had found out decades before; however just like their predecessors, they were unable to construct a telescope with only this knowledge:

> But as the convexity of the magnifying-glass was too great, he made a correct mould in plaster of the convex glass, and sent it to Nuremberg to the makers of ordinary spectacles that they might prepare glasses like it; but it was no good, as they had no suitable tools, and he was unwilling to reveal to them the true principle of the process. No expense was spared, and several months elapsed. If we had been acquainted with the method of polishing glasses, we should have produced excellent spy-glasses immediately after our return from Frankfurt.[75]

With the last sentence, Marius was fundamentally wrong. Rolf Willach, who examined numerous sixteenth-century spectacle lenses, stated that most of them were of insufficient quality to be used in telescopes. He concludes "dass die Erfindung eigentlich nur einem guten Brillenmacher gelingen konnte, dem ein entsprechend großer Vorrat an Linsen für seine Versuche zur Verfügung stand. Gelehrte und Halbgelehrte […], welche vielleicht bestenfalls über ein halbes Dutzend Gläser verfügten, hatten nur eine sehr geringe Chance."[76]

---

[73]Marius 1614/1916/2019, Preface.

[74]Willach 2007, p. 109: "the telescopic effect was certainly common knowledge among the spectacle makers and scholars."

[75]Marius 1614/1916/2019, Preface.

[76]Willach 2007, p. 116: "that only a good spectacle maker with a large reserve of lenses for his attempts could succeed in the invention. Scholars and semi-scholars […] who at the best had half a dozen lenses, had only a very small chance."

For the construction of a telescope, the magnification effect is insufficient; one also needs to produce a sufficiently sharp image. Most lenses didn't provide the necessary result because the quality of the glass was inadequate and/or they were not ground well enough. In 1608 and the following years, even the best lenses needed a modification to meet with the requirements of a telescope, a diaphragm that reduced the aperture of the objective. This improves the quality of the image significantly. Willach found that around 1608 lenses were only a little better than before and it was only the diaphragm, which led to the decisive breakthrough. For example, a lense of 3 cm diameter was stopped down to 1 cm. He proposed the thesis that the telescope only gained such a rapid dissemination by copyists because a craftsman immediately noticed the diaphragm in front of the lens.[77]

But Fuchs and Marius were no experts; the function of the diaphragm obviously remained unknown to them because it is nowhere reported. Therefore and because of their small selection of lenses, they were unable to recreate a telescope. All future purchases were of complete telescopes.

In the Marius literature, there is discussion that the failure of Fuchs and Marius was because Fuchs "was unwilling to reveal to them [the Nuremberg spectacle makers] the true principle of the process." Apart from the fact that Fuchs's understanding of this method was minimal and his knowledge was not sufficient to construct a telescope (which Marius however didn't know), this statement is dependent on the correct translation. In the original it says "veram conficiendi rationem illis revelare noluit." The translator Joachim Schlör pointed to the form of "revelare" in this sentence that is basic to his German translation (Schlör 2012, p. 57 and 59), which is similar to the English one used here. Josef Klug translated the sentence more than 100 years before as "das Geheimnis der Fertigstellung wollte sich ihnen [den Brillenmachern] nicht offenbaren"[78] for which according to Schlör the form in the sentence had to be "revelari." So one letter, e or i, makes a significant difference in the statement. On the other hand, Schlör lists some grammatical mistakes in *Mundus Iovialis*, and, as the chapter by Richard Kremer in the present volume shows, Marius's publications are not always clear in their claims and partially incorrect. As well as an inaccurate formulation or a grammatical mistake made by Marius, a printing error of the typesetter is also possible, who misunderstood the meaning of the sentence or didn't understand it at all. As was discussed above, there is no other proof that Fuchs was interested in secrecy for military reasons. Actually he had to assume that sooner or later others would also receive a telescope from the Netherlands or copy one. The interpretation of Klug, though grammatically wrong, seems more plausible to me. At least one can't clearly decide between the two possibilities. Whatever choice is made, the reason for the failure to construct a telescope was another one. Because of its dubiousness, it is inadequate as proof of Fuchs's pursuit of secrecy; equally through circular reasoning, the assumed secrecy

[77]Willach 2007, pp. 112–117. For new insights in the history of telescope invention, see also Van Helden 2009.

[78]Klug 1906, p. 410: "the mystery of completion wouldn't reveal to them [the spectacle makers]."

cannot be used as proof of the correctness of the new translation—for which the only evidence is the letter "e".

> In the meantime, glasses of the same kind were becoming common in Belgium, and a fairly good one was sent, with which we were highly pleased. This was in the summer of 1609. From this time I began to look into the heavens and the stars with this instrument, whenever I was at the house of the nobleman so often mentioned, at night time; sometimes he used to allow me to carry it home, and in particular about the end of November, when I was observing the stars according to my custom in my own observatory.[79]

In October 1608 Marius was probably the first astronomer outside of the Netherlands who learned about the invention of the telescope, but it took about another 9 months until he could hold one in his hands.

Though Marius wrote "a fairly good one was sent," it was obviously Fuchs's property, who with certainty paid for it. At first, Marius was only sometimes allowed to take it home. As Marius could not expect assistance from the Ansbach court, which was always in financial difficulties, Marius stayed dependent on a private sponsor.

Arjen Dijkstra noted that in June 1609 Adamus Valentinus Fuchs von Bimbach was enrolled in Leiden and could have got the telescope for his relative in Ansbach.[80] But there is no proof for this until now. Adam Valentin Fuchs von Bimbach zu Burgbreitbach (= Burgpreppach), younger son of Georg Fuchs von Bimbach zu Gleisenau, was a generation younger than Hans Philip Fuchs von Bimbach zu Möhren and was related to him in the male line through his great-great-grandfather (Biedermann 1747, Tabvla LX; Flurschütz da Cruz 2014, p. 385).

> Meanwhile, two glasses extremely well polished, a convex and a concave, were being sent from Venice by that most distinguished and accomplished man, the Lord John Baptist Leuccius,[81] who had returned from Belgium to Venice after peace was made, and who had already been thoroughly acquainted with the instrument. These glasses were fitted into a leaden tube, and made over to me by the most noble and active nobleman whom I have mentioned already, in order that I might try what they would show among the constellations and stars near Jupiter. Accordingly, from this time until January 12, I gave my diligent attention to these Jovian stars ....[82]

As the lenses were "fitted into a leaden tube," Marius in this case also received a complete telescope. In Venice particularly high-grade glass was produced so that the lenses in Marius's and Fuchs's second telescope were probably better than those in the first. The receipt of this instrument can be dated around mid-January 1610 (Gregorian calendar), as Marius reports his observations made earlier with the first telescope that he recorded from December 29, 1609 (Julian date) on. He received the new instrument obviously some days before January 12, 1610 (Julian date). This time he didn't have to observe at Fuchs's home but could take the telescope home

[79]Marius 1614/1916/2019, Preface.

[80]Dijkstra 2012, p. 138. I owe the reference to this dissertation to Dr. Günther Oestmann, Bremen.

[81]To John Baptist Leuccius, see section "Life at Court and Publications", Chap. 2.

[82]Marius 1614/1916/2019, Preface.

immediately. From January 13 until February 8, Marius traveled and left the instrument at home. Afterward he continued his observations with this telescope and got it from Fuchs as a gift or "permanent loan," because:

> In order that I might observe the Jovian stars with greater closeness and diligence, the illustrious nobleman whom I have frequently mentioned, out of his special affection towards these astronomical studies, placed the instrument entirely at my disposal. From that time accordingly to the present, I have made continuous observations with this instrument and with others afterwards constructed.[83]

About the "afterwards constructed" instruments, nothing is known, also not how Marius obtained them. Had prices lowered since then, so he could afford them himself or did Fuchs von Bimbach help again? We don't know. It can be excluded that Marius could construct them himself. He lacked the necessary manual skills and knowledge, the indispensable tools, the access to good glass or numerous finished lenses, and the abilities of an experimenter. As opposed to Galilei, there is no information that Marius had carried out experiments. Self-constructing would have taken a lot of time and he certainly would have written about it. The production of the annual calendars and prognostica might have left him with no time for the complex construction of telescopes.

Indirectly Marius makes Fuchs a witness to the correctness of his report about the acquisition of telescopes and the observations made with them:

> This is the exact truth. I should never be allowed thus in a public document to say what was not true about so great a man [. . .].[84]

Marius might have remembered incidents from the past when others wanted to "say what was not true" about Fuchs von Bimbach. It was just around 1614 that Fuchs vented his anger about two court officials who mocked him through indiscretion and finally even addressed the Reichskammergericht (Imperial Chamber Court) (Herold 1973, p. 46).

The Translation of Euclid

Marius's translation of Euclid, which was published in 1610, is discussed by Hans Gaab in more detail (section "Life at Court and Publications" in Chap. 2; see also Fig. 2.14). Here we only outline Fuchs von Bimbach's intentions with this book.

The title page and the introductions show that Marius did the translation at Fuchs's instigation. The title page also mentions the applications: "Alles zu sonderm Nutz denjenigen/so sich der Geometria/im Rechnen/Kriegßwesen/Feldtmässen/ Bauen/vnd andern Künsten vnnd Handtwerckern zugebrauchen haben."

[83] Marius 1614/1916/2019, Preface.

[84] Marius 1614/1916/2019, Preface.

("Everything shall be of special profit to those who have to use geometry, in reckoning, warfare, field surveying, constructing and other arts and crafts.")

Fuchs added an introduction, dated January 1, 1610 in which he, after a detailed explanation why the German language is important for such books, described the applications:

> [. . .] zuforderst aber im Kriegswesen/die jenigen/so Quartier schlagen/Schantzen/Vestungen etc. bawen vnd zervbrechen/mit Zeug: oder Geschützwesen/Werkcken [. . .] Wie viel sind der Wercklеut/deren handtwerck allein auff der Geometria bestehet?[85]

He explained which mistakes architects and builders make because they know too little about geometry and asked:

> Inn was grossem Irthumb stecken die Landtmässer? Deren meiner Mainung nach gar nimmer/oder selten zwen gefunden/die in dem *facit* übereinstimmen [. . .].[86]

He estimated which errors come together in large areas (10,000 acres) when small errors are made in measuring and reckoning.

As an officer Fuchs occasionally must have come into contact with geometry, e.g., when directing a canon or using maps. Ruling over several territories where border conflicts happened, Fuchs had a relationship to surveying. As owner of a castle, he employed architects, builders, and craftsmen.

However, it seems astonishing to use Euclid's rather abstract book for these practical purposes instead of an instruction manual for land surveyors or other textbooks. But at that time Euclid's book was the primary text among the introductions to geometry. Fuchs had probably first heard about it at school in Lauingen. He wrote on the significance of Euclid's "Elements" that they "der gantze grundt und Fundament der Geometria seind" and "in vielen Handthierungen vnvermeidtlich gebraucht muß werden."[87]

He then expresses his hope that the margraves of Ansbach and Bayreuth to whom the book is dedicated:

> nicht allein [. . .] mir zu gut halten/Daß gedachten dero Mathematicum [d.h. Marius] ich dieser Verdeutschung/So gleichwol nicht ohne sondere Mühe und Versäumnuß abgangen/an vnd vielleicht von andern Verrichtungen abgehalten/Sondern auch mit ihm Allergnädig zu frieden seyn [. . .].[88]

[85]Marius 1610, sig. A3^{v}: "First in warfare/those who prepare the quarters/build or deconstruct redoubts/fortresses/with things or ordnance/works [. . .] How many craftsmen are there whose craft is based only on geometry?".

[86]Marius 1610, sig. A4^{r}: "Which large errors are made by land surveyors? In my opinion never or rarely two are found who agree on the facit [result] [. . .]."

[87]Marius 1610, sig. A3^{v}: "are the complete basis and fundament of geometry" and "inevitably must be used in many actions."

[88]Marius 1610, sig. A4^{r}: "not only [. . .] due to me that I requested their Mathematicum [mathematician, i.e. Marius] this translation into German, which couldn't be made without some effort and omissions, and possibly kept him from other duties; however are most graciously content with him [. . .]."

Marius writes in his introduction that his translation:

> geschehen ist auß Befehl deß Edlen vnd gestrengen Herrn Hanß Philips Fuchßen von Bimbach etc. So der Geometrischen sachen nicht allein ein besonderer Liebhaber vnd Beförderer ist/sondern daß der Anfang vnd Grunde denjenigen/so sich darinnen zu üben willens zu wissen für hochnötig geachtet/vnnd durch sein vielfältiges nachfragen experimentiren vnnd außsinnen/den gewaltigen vnd groben Irrthumb vermercket/darinnen gemeine Feldtmässer alle mit einander stecken/und daher in Kauffen und Verkauffen grosser und augenscheinlicher irrthumb vorgehet [. . .].[89]

Afterward he lists further applications of geometry in which ignorance leads to mistakes. If Marius was not exaggerating, Fuchs von Bimbach seems to have engaged intensively with geometry and its applications, through "manifold inquiries, experiments, and cogitation."

Attempt to Assess of Hans Philip Fuchs von Bimbach

To describe Fuchs von Bimbach as a shrill figure (Buchner/Mavridis 2009, p. 78) because of his switching sides and his combativeness is too shortsighted. To a certain extent, the same applies to him, as was written about his liege lord, Margrave Joachim Ernst of Ansbach, "daß auch er von den Tendenzen seiner Zeit zum Abenteurertum nicht unberührt geblieben war, die den Individualismus übersteigerte und das Kondottierentum[90] förderte. Immer haftete seinen politischen Entscheidungen etwas Verwegenes an. Zudem gingen Joachim Ernsts religiöse Bindungen nicht sehr tief."[91] However, Fuchs's military book and his military and political approach show him to be more prudent than Joachim Ernst or King Christian IV. As an imperial knight, officer, and a landlord, he was typical for his times.

Changing sides for actual or supposed advantage was common, also for sovereigns. In this, Fuchs von Bimbach probably followed his main liege lord and patron, Wolfgang Wilhelm of Palatinate-Neuburg.

The tendency to quarrels was also widespread. "Adelige des 16. Jahrhunderts befanden sich anscheinend mit jedermann in Konflikt."[92] In distinction to earlier

---

[89]Marius 1610, sig. A5^r: "was made by order of the most noble and strict gentleman Hanß Philips Fuchßen von Bimbach etc. who is not only a special enthusiast and sponsor of geometry but considers the introduction and fundament for those willing to learn to exercise therein as urgently necessary and recognizes through his manifold inquiries, experiments and cogitation the massive and primitive errors land surveyors have in common in this and therefore in buying and selling proceed with large and obvious errors [. . .]."

[90]Condottieri were Italian leaders of soldiers up to the sixteenth century who for better salary often changed sides, even during a battle.

[91]Herold 1973, p. 65: "that he didn't remain untouched by the tendency to adventurism of his time that overemphasized individuality and supported condottierism. His political decisions were always tainted with something foolhardy. Moreover Joachim Ernst's religious ties were not very deep."

[92]Flurschütz da Cruz 2014, p. 14: "Noblemen of the 16th century seemed to have conflicts with everyone."

centuries, blood feuds had been replaced by confrontations in court, at least on a regional level. In the disputes about power and territories between the emperor and the sovereigns, between realms and religious confessions, war was often the chosen solution.

Fuchs von Bimbach was primarily an officer; he was a politician only for a few years of his life. Naturally, during all the years, he also had to supervise the administration of his territories, but not much is known about this to date.

In the early modern era, it was absolutely common to plunder in order to finance war and also for personal enrichment. A colonel like Fuchs von Bimbach was not only an officer but also a small-scale businessman. He had to employ his subordinated officers and foot soldiers temporarily and to pay them. For this he did not always get money from his employer but also had to acquire credit, and for their repayments sufficient revenue had to be generated.[93] For the spoils of war, there were quite fixed rules, for example, what share of the conquered canons a quartermaster had to give to the colonel. There was also an official (Beutmeister), who was responsible for "fair" distribution. Which part of Fuchs's large assets came from the military expeditions and what was from other sources (inheritances, high income in Ansbach, financial investment, management of goods) can't be answered here.

In Ansbach Fuchs seems not to have been liked—except by Marius and in the beginning by the Margrave. Other court officials complained about him (envy certainly played a role here); when his star began to descend, satirical poems circulated and finally he fell from the Margrave's grace.[94] It is hard to decide what was true about his alleged immoral way of life and his "course" manners, and what was exaggerated or even fictional.[95] Marius must have known the accusations but nevertheless praised him highly.

Christian IV made Fuchs posthumously responsible for the defeat at the Battle of Lutter, but it seems that he was diverting attention from his own unsound decision (Zwanziger 1920, p. 14). In the older military literature, it is said: "Fuchs stand bei Freund und Feind als alter, erfahrener und listiger Kriegsoberster in sehr großem Ansehen und war der ausgezeichneteste Officier in der königlichen [dänischen] Armee."[96]

In the Fuchs von Bimbach family, Hans Philip obviously was held in high esteem. In 1682 a young family member is admonished that he should "in weyland Hannß Philipp Fuchs von Bimbach, Fusstapfen tretten, und eben den Nachruhm hinterlaßen

---

[93]To get an impression of the different matters and people for which Fuchs was responsible as a colonel, read, for example, the detailed Wikipedia article about foot soldiers: https://de.wikipedia.org/wiki/Landsknecht

[94]On the different incidents and reproaches, see Zwanziger 1919, p. 26, Herold 1973, and Veh 1984–1985.

[95]How dubious, at least partly, the accusations were from today's standpoint, one of the reproaches claims: Fuchs would use magic to make himself bulletproof (Zwanziger 1919, p. 26). Compare Herold 1973, pp. 43–44 on the accusations.

[96]Lichtenstein 1850, pp. 144–145: "Fuchs was held in high standing by friend and foe as an older, experienced and cunning colonel and was the most excellent officer of the royal [Danish] army."

möge, daß er in angeführten Wißenschafften wohlerfahren, und Oberster geworden seye [...]."[97]

However, basically Hans Philip failed in his principal occupations. He was not particularly successful either as a politician, a diplomat, or an officer, which however in the given constellation of various parties, interests, and confessions of the period was extremely difficult; Wallenstein has gone down in history and in art as a similarly failed player. In his three most important positions as court official in Ansbach, imperial colonel, and general of the Dutch King, Fuchs suffered extensive slights and died finally as a defeated military commander.

His interest in sciences was less pronounced than, for example, his contemporaries Rudolf II or Maurice of Hesse-Kassel (called "The Learned"). According to Simon Marius, he was "not only benefactor and admirer of the whole of mathematics and other similar sciences but also their greatest patron." A remaining merit for the cultural history of mankind is the support of Simon Marius and one of the first astronomical uses of the telescope. Fuchs von Bimbach would probably have considered it an insult if he had ever imagined that four centuries later he is appreciated for his support of astronomy but not for his service in battle.

Errors and Speculations About Fuchs von Bimbach in the Literature

Unfortunately, all existing biographies about Fuchs von Bimbach contain errors and speculative statements that are formulated as facts, which is also true for the statements about Fuchs in the literature about Marius. Without any claim to completeness, some of these mistakes that are perpetuated again and again without verification will be listed here. Further minor errors have already been discussed in the text above and in the footnotes.

The errors begin with his name. In an eighteenth-century review of people with the family name "Fuchs," he is falsely called "Johann Philip Fuchs von Fuchsberg."[98]

In the *Braunschweigisches Biographisches Lexikon*, he is registered as "Fuchs von Bimbach, Hans Philip Freiherr von" (Lent 2006), what rewritten would result in "Freiherr von Fuchs von Bimbach." Apart from the wrong title "Freiherr" (baron, see below), "Hans Philip von Fuchs von Bimbach" is meaningless. Dijkstra refers to him in short as "Von Bimbach," what is also inadmissible (Dijkstra 2012, p. 137 and 138). The family name was Fuchs; "von Bimbach" was an addition to mark the lineage. Later family members were called "Freiherr Fuchs von Bimbach [und

[97]Quoted from Rößner 2003, p. 105: "follow the footsteps of the erstwhile Hans Philip Fuchs von Bimbach and may leave as posthumous fame that he achieved in the previously mentioned sciences and as a colonel [...]."

[98]Lauterbach 1783, pp. 16–17; the entry gives two older books as sources.

Dornheim]," in no case "Freiherr von Fuchs von Bimbach" or "Freiherr von Bimbach."

In the earlier literature both first names are always used, although in various spellings. It is not known whether there was a preferred first name. Joachim Schlör repeatedly uses only the second name Philipp[99] what seems unreasonable to me because in contemporary documents both names are always used.[100]

Occasionally 1567 is stated as his year of birth,[101] but the exact year is unknown. 1567 is just a plausible but unsubstantiated presumption.

In 1892 Julius Meyer speculated about Marius's choice of Padua as his place of study: "Von Einfluß auf die Wahl der Universität Padua mag auch der Umstand gewesen sein, daß sein Landsmann, der markgräflich onolzbach'sche Geheimrath und Kriegsoberste Freiherr Hans Philipp von Fuchs-Bimbach auf Möhren [...] i. J. 1587 auf derselben Universität seine wissenschaftliche Ausbildung genommen hatte."[102] This sentence is full of mistakes and unproven claims. Fuchs's attendance at the University of Padua is not documented, nor is his influence on Marius already recorded in 1602. Besides the name of the family line (von Fuchs-Bimbach) is wrong. Strictly speaking he wasn't a fellow countryman of Marius, but came from neighboring Palatinate-Neuburg. Fuchs was no "Geheimrat" (privy councilor)—at least not in the meaning of this title as used in the nineteenth century—and not a baron. This title is wrongfully ascribed to him in many publications. This claim of studies in Padua and the title Freiherr were also included in the biographical Wikipedia article[103] and in the record in the "Gemeinsame Normdatei (GND)" of German libraries, which obviously was based on the Wikipedia entry.[104]

A newer essay about Marius and the telescope says: "Janssen stellte bereits 1608 sein Fernrohr auf der Frankfurter Messe vor." "Im Herbst 1608 erfuhr Marius vom Artillerie-Offizier, Freiherr Hans Philip Fuchs von Bimbach, daß auf der Frankfurter Herbstmesse Fernrohre angeboten wurden [. . .]."[105] As explained above, there is no

[99]In his translation of Marius (1614/1988), the first name Philipp repeatedly stands for Fuchs von Bimbach but is missing in the Latin original at these points. Because of the special sentence construction in Latin without personal pronouns, translations have to insert one or a name. See also the use of a singular "Philipp" in Schlör 2012.

[100]Another publication that uses a singular "Philipp" is Ritter 1870, p. 451, in a note by the publisher, not in a document. This might have been Ritter's error; the index p. 744 has "Johann Philipp."

[101]For example, Dijkstra 2012, p. 136.

[102]Meyer 1892, p. 56: "The choice of the University of Padua might also have been the circumstance that his fellow countryman, the Ansbachian margravian privy councilor and colonel, Baron Hans Philipp von Fuchs-Bimbach auf Möhren, [. . .] received his academic education at the same university in 1587."

[103]https://de.wikipedia.org/wiki/Hans_Philipp_von_Fuchs_von_Bimbach (Versions from April 25, 2016 and before, corrected by me on June 23, 2016).

[104]Fuchs von Bimbach, Hans Philipp, http://d-nb.info/gnd/1026798078 (last seen May 17, 2016).

[105]Wolfschmidt 2012, p. 219 and 225: "Already in 1608 Janssen presented his telescope at the Frankfurt fair." "In the fall of 1608 Marius heard from the artillery officer, Baron Hans Philip Fuchs von Bimbach that telescopes were offered for sale at the Frankfurt fair [. . .]."

direct evidence, who had the telescope in Frankfurt; that it was Sacharias Janssen is not a fact, but a theory. Marius's report mentions one telescope, not telescopes in the plural. From Marius's report it can't be concluded that the telescope was "presented" in public and "offered for sale."[106] To name Fuchs von Bimbach, an artillery officer (a modern term) is not accurate for the year 1608; this can only be stated for later years, as he belonged to the "infantry" (also a modern term) before 1610.

Naming the "inventor" who showed Fuchs von Bimbach a telescope at Frankfurt am Main a "merchant"[107] is pure speculation. It might originate from the presumed identification with Janssen who among other things was a merchant.

At this point attention will be drawn once more to a popular error concerning Marius. He didn't construct his own telescopes as is often claimed[108] but used complete ones. That he received his first telescope in October 1608[109] appears much too early; he only got it in summer of 1609.

An extremely absurd statement can be found in a popular book about historical places in Lower Saxony, in this case about the battlefield of Lutter: "Unter den Toten auf protestantischer Seite befand sich auch General Fuchs [. . .] Dieser aus Bayern stammende Haudegen, der mit vollem Namen Hans Philipp Freiherr Fuchs von Rimbach [sic] hieß, war eine recht ungewöhnliche Persönlichkeit. Er war eigentlich Astronom und hatte entscheidenden Anteil daran, daß im Jahre 1610 Galileo Galilei die Entdeckung der vier Jupitermonde gelang. Zudem war Fuchs Mathematiker und Herausgeber der ersten 15 Bücher des altgriechischen Mathematikers Euklid. Was ihn in den Krieg getrieben hat, wußte wohl nur er selber."[110] This needs no comment. However, this "source" was used to create an article for Wikipedia about the Battle of Lutter in which the general was called a "kursächsischer Mathematiker und

[106]Compare, for example, Van Helden (1975) who wrote inaccurately "that a *Belga* was trying to sell a telescope at the autumn fair at Frankfurt in 1608."

[107]For example, by Christianson 2000, p. 320: "peddler."

[108]For example: Van Helden 1974, p. 39, note 3: "It seems thus likely that Galileo constructed his first telescope in much the same way as Simon Marius did"; Dijkstra 2012, p. 137: "received their first working specimen from the Netherlands, which enabled Marius to construct new telescopes"; Riekher 1990, p. 21: "Nach diesen Angaben [von Fuchs] ist es gelungen, ein Fernrohr zu bauen." ("With this information [from Fuchs] it was possible to build a telescope.") But Rolf Riekher could use only inadequate literature about Marius for the first edition of his book in 1957—especially the good translation of *Mundus Iovialis* by Joachim Schlör was lacking and he had to base himself on Klug 1906.

[109]Van Helden 2011 [1997], p. 510: "Simon Marius in Ansbach probably obtained his first spyglass as early as October 1608."

[110]Friedrich 1989, p. 196: "General Fuchs was also among the dead on the Protestant side [. . .]. This Bavarian warhorse, whose full name was Hans Philipp Freiherr Fuchs von Rimbach [sic] was a quite extraordinary personality. He was actually an astronomer and played a decisive role in Galileo Galilei's discovery of the four moons of Jupiter in 1610. Besides Fuchs was a mathematician and publisher of the first 15 books of the ancient Greek mathematician Euclid. Only he knows what drove him to war."

Astronom" ("mathematician and astronomer from the Electorate of Saxony").[111] As a curiosity I remark that it was just this crude mistake that first drew my attention to the Franconian imperial knight. Through an Internet search in 2007 for memorial stones, etc. about astronomers, I found this Wikipedia page about the Battle of Lutter and the "astronomer" Fuchs von Bimbach who was missing in the *Biographical Index of Astronomy* (BIA) (Brüggenthies/Dick 2005). Though he wasn't an astronomer, we have meanwhile included him together with other patrons of astronomy in the second edition of the BIA (Brüggenthies/Dick 2017).

Source Situation and Approaches for Further Researches

Until now there is no academic biography of Fuchs von Bimbach, only popular portrayals and a few encyclopedia entries, as well as a lot of details, spread over numerous publications. Almost all sources cited here in turn refer to older sources that also had to be consulted for a comprehensive biography. The existing printed material would be enough to write a book about Fuchs von Bimbach, but without intensive archival studies, this would be of only small academic value.

The most comprehensive biography until now is by Karl Hermann Zwanziger of 1919/20 that contains much interesting data but unfortunately indicates its sources insufficiently. A newer one can be found in a local history book about Unterschwaningen (Buchner/Mavridis 2009); it cites its sources fairly precisely but is also based on unreliable ones and gives room to speculations that are not marked as such. These local history researches and publications are often very valuable because they exploit local sources—but unfortunately often lack exact references and are not always on an academic level. The latter also applies to many publications of the early twentieth century and before; however, their authors had access to archival documents that have in the meantime been lost. As well as the already quoted biographies, there is a completely useless one of 1899 (Clementi 1899; based on Meyer 1892 and Lichtenstein 1850) that, however, was used repeatedly as a source for others and another one without value from 1982 ([Klay?] 1982).

With one exception everything known about the relationship between Fuchs and Marius, as well as about their acquisition and use of telescopes comes from Marius himself. Independent sources would be valuable but have not been found to date. The extensive literature about the early history of the telescope and about Marius offers at least various approaches as to how Marius's descriptions can be interpreted and classified in the comprehensive history of the telescope. This literature could only be reviewed to some extent here.

[111]http://de.wikipedia.org/wiki/Schlacht_bei_Lutter (Version from February 21, 2016 and before, corrected by me June 23, 2016); the incorrect lineage goes back to Lichtenstein 1850, p. 143: "Aus dem Kursächsischen gebürtig" ("Born in the Electorate of Saxony").

In the literature about the Thirty Years' War and especially about the Battle of Lutter, Fuchs von Bimbach is mentioned repeatedly; this could also only be used very selectively here. The only entry in a big (printed) biographical reference book, the *Dansk biografisk leksikon* (Danish biographical encyclopedia)[112] only deals with his time in Danish military service and is of little biographical use. Additional findings about his military career up to 1610 may be won from his own memories in the Stuttgart manuscript (see section "Fuchs von Bimbach as a Military Author").

Many details to Fuchs von Bimbach at the Ansbach court on an archival basis are presented in the dissertation by Hans-Jörg Herold about Joachim Ernst (Herold 1973). At the same time, it demonstrates how much previously unknown data can be found in archives. The newer dissertation of Andreas Flurschütz da Cruz deals with a conflict between the Franconian knight families Fuchs von Bimbach and Wolf von Wolfsthal in the second half of the seventeenth century but contains also valuable information about previous family history and references to archival holdings especially in Franconia, among them the Fuchs von Bimbach family archives.[113] Numerous other files in state, ecclesiastical and private archives mentioning Hans Philip Fuchs von Bimbach might exist. To him, his ancestors and siblings information might also be found in archival documents about the Neuburg/Donau and Jülich courts, in maybe still existing parish registers of Neuburg and Möhren, in imperial files in Vienna and Prague, in Dutch and Danish archives, maybe also in Padua and elsewhere. And probably there, in a remote place, Simon Marius is also named ....

References

Becker, Horst (1991): Messeeinkäufe für die Hofhaltung des hessischen Landgrafen Moritz. In: Stahl 1991, pp. 328–337

Biedermann, Johann Gottfried (1747): *Geschlechtsregister Der Reichsfrey unmittelbaren Ritterschaft Landes zu Franken Löblichen Orts Baunach*. Bayreuth 1747. Reprint: Neustadt a.d. Aisch: Verlag für Kunstreproduktionen Schmidt 1988. Digitized version of the original: http://www.mdz-nbn-resolving.de/urn/resolver.pl?urn=urn:nbn:de:bvb:12-bsb10328047-2

Boller, Friedrich von (1834): *Notizen über die Burg Möhren und das dazu gehörige Dorf Gundelsheim am Hanenkam, im Rezatkreise, Landgerichtsbezirke Monheim*. Ansbach [1834]. Digitized version: https://books.google.de/books?id=b71AAAAAcAAJ

[112]Fuchs v. Bimbach, Hans Philipp. In: *DBL* 1. Reproduktion: Scandinavian Biographical Archive (SBA). Mikrofiches. London [u.a.]: Saur [1990], I A–89, 192–193. I consulted the online version in World Biographical Information System (WBIS) Online, http://db.saur.de/WBIS/. Later editions: Rockstroh; DBL 3 (the latter couldn't be consulted yet).

[113]Herold 1973, "Ungedruckte Quellen" (unpublished sources) and numerous individual references particularly to H. Ph. Fuchs von Bimbach in the text; Flurschütz da Cruz 2014, "Quellen und Archive" (sources and archives). Zwanziger 1919, p. 24, mentions a "handschriftlichen Nachlaß Hans Philipps im Kreisarchiv zu Nürnberg" (manuscript paper of Hans Philipp in the Nuremberg District Archives).

B[ressensdorf], E[rnst] v[on] (1988): Fuchs v. Bimbach u. Dornheim. In: *Genealogisches Handbuch des in Bayern immatrikulierten Adels*, vol. 17. Neustadt a.d. Aisch: Verlag Degener & Co. 1988, pp. 393–395

Brübach, Nils (1994): Die Reichsmessen von Frankfurt am Main, Leipzig und Braunschweig (14.–18. Jahrhundert) (= *Beiträge zur Wirtschafts- und Sozialgeschichte*; 55; also: Bamberg, Universität, Dissertation 1993). Stuttgart: Steiner 1994

Brüggenthies, Wilhelm; Dick, Wolfgang R. (2005): *Biographischer Index der Astronomie / Biographical Index of Astronomy* (= *Acta Historica Astronomiae*; 26). Frankfurt a. M.: Verlag Harri Deutsch, 2005

Brüggenthies, Wilhelm; Dick, Wolfgang R. (2017): *Biographischer Index der Astronomie / Biographical Index of Astronomy* (= *Acta Historica Astronomiae*; 60). [2nd, enlarged ed.] Leipzig: AVA – Akademische Verlagsanstalt, 2017

Buchner, Siglinde; Mavridis, Carl-Alexander (2009): II. Mittelalter und frühe Neuzeit. In: Unterschwaningen 2009, pp. 57–98. With the chapters: II.5.6 Die Herren Fuchs von Bimbach (1603 bis 1630) (p. 77); II.6 Johann Philipp Fuchs von Bimbach zu Möhren, eine schillernde Persönlichkeit des Dreißigjährigen Krieges (pp. 78–85)

Christianson, John Robert (2000): *On Tycho's Island: Tycho Brahe and His Assistants, 1570–1601.* Cambridge [etc.]: Cambridge University Press 2000

Clementi, Albert (1899): Johann Philipp Freiherr Fuchs von Bimbach. *Das Bayerland* 10 (1899), pp. 197–200, 209–212, 222–226

[DBL 1:] *Dansk biografisk lexikon.* 19 vols. København: Gyldendal 1887–1905

[DBL 3:] *Dansk biografisk leksikon.* 3. udgave. 16 vols. København: Gyldendal 1979–84

Dietz, Alexander (1970): *Frankfurter Handelsgeschichte*, vol. 1. Frankfurt am Main: Verlag von Hermann Minjon 1910. Reprint: Glashütten im Taunus: Verlag Detlev Auvermann 1970. Digitized version of the original: http://tudigit.ulb.tu-darmstadt.de/show/57-1505-01

Dijkstra, Arjen Folkert Benjamin (2012): *Between academics and idiots. A cultural history of mathematics in the Dutch province of Friesland (1600–1700).* Dissertation, Universiteit van Twente, Leeuwarden 2012. Online version: http://doc.utwente.nl/81545/

Flurschütz da Cruz, Andreas (2014): Zwischen Füchsen und Wölfen. Konfession, Klientel und Konflikte in der fränkischen Reichsritterschaft nach dem Westfälischen Frieden (also: Bamberg, Universität, Dissertation 2013). Konstanz, München: UVK-VerlagsGesellschaft 2014

Friedrich, Ernst Andreas (1989): *Wenn Steine reden könnten.* Hannover: Landbuch-Verlag [1989]. Therein pp. 195–196: Eine Schlacht rötete die Erde. Das Blutfeld im Lutterbecken

Fuchs von Bimbach, Dieter (1975): Fuchs v. Bimbach u. Dornheim. In: *Genealogisches Handbuch der Freiherrlichen Häuser*, Freiherrliche Häuser A Band IX (= *Genealogisches Handbuch des Adels*; 59). Limburg a.d. Lahn: C. A. Starke Verlag 1975

[Gotha:] *Gothaisches Genealogisches Taschenbuch der Freiherrlichen Häuser*, Jg. 74. Gotha: Justus Perthes 1924. Therein pp. 252–254: Fuchs von Bimbach und Dornheim

Herold, Hans-Jörg (1973): Markgraf Joachim Ernst von Brandenburg-Ansbach als Reichsfürst (= *Schriftenreihe der Historischen Kommission bei der Bayerischen Akademie der Wissenschaften*; Schrift 10; also: Kiel, Universität, Dissertation 1970). Göttingen: Vandenhoeck und Ruprecht 1973. Digitized version: http://digi20.digitale-sammlungen.de/de/fs1/object/display/bsb00051647_00001.html

Jähns, Max (1740): Geschichte der Kriegswissenschaften vornehmlich in Deutschland. Teil 2, 17. und 18. Jahrhundert bis zum Auftreten Friedrichs des Grossen 1740 (= *Geschichte der Wissenschaften in Deutschland*; 21). München/Leipzig: R. Oldenbourg 1890. Reprint: New York: Johnson; Hildesheim: Olms 1966. Digitized versions of the original: http://resolver.sub.uni-goettingen.de/purl?PPN372096581, https://archive.org/details/geschichtederkr03jhgoog

[Klay, Kurt (?) (1982):] Johann Philipp Freiherr Fuchs von Bimbach. In: Kurt Klay (ed.): *Am 27./17. August 1616: Die Schlacht bei Lutter am Barenberge.* Lutter am Barenberge 1982, [20 unnumbered pp.]; Digitized version: http://www.wolfgangroehl.de/Lutter/General_Fuchs.pdf

Klug, Josef (1906): Simon Marius aus Gunzenhausen und Galileo Galilei: ein Versuch zur Entscheidung der Frage über den wahren Entdecker der Jupitertrabanten und ihrer Perioden. *Abhandlungen der Mathematisch-Physikalischen Klasse der Königlich Bayerischen Akademie der Wissenschaften* 22 (1906), pp. 385–526. Digitized version: https://archive.org/stream/abhandlungenderk22kn#page/n475/mode/2up

Köbler, Gerhard (1999): *Historisches Lexikon der deutschen Länder. Die deutschen Territorien und reichsunmittelbaren Geschlechter vom Mittelalter bis zur Gegenwart.* 6th ed. München: Verlag C. H. Beck 1999

Krabbe, Johannes (1605): *Alt und New Schreibkalender Auff das Jahr nach der Gnadenreichen Geburt unsers lieben Herrn Jesu Christi M. DC. VI.* Erffordt: Beck [ca. 1605]. Digitized version: http://www.mdz-nbn-resolving.de/urn/resolver.pl?urn=urn:nbn:de:bvb:12-bsb10392756-1

Lauterbach, Samuel Abraham (1783): *Historische Nachrichten von einigen Vornehmen und Gelehrten, welche den Geschlechts-Nahmen Fuchs führten.* Breslau 1783. Digitized versions: https://archive.org/details/bub_gb_u1NPAAAAcAAJ , http://www.mdznbn-resolving.de/urn/resolver.pl?urn=urn:nbn:de:bvb:12-bsb10683961-2, https://books.google.de/books?id=u1NPAAAAcAAJ

Lent, Dieter (2006): Fuchs von Bimbach, Hans Philipp Freiherr von. In: Horst-Rüdiger Jarck, Dieter Lent et al. (eds.): *Braunschweigisches Biographisches Lexikon: 8. bis 18. Jahrhundert.* Braunschweig: Appelhans 2006, pp. 242–243

Lichtenstein, G[eorg] (1850): Die Schlacht bei Lutter am Barenberge. Ein Beitrag zur vaterländischen Geschichte (= *Dänemarks Theilnahme an dem dreißigjährigen Kriege bis zum Frieden von Lübeck*). Braunschweig: Oehme & Müller 1850. Digitized version: http://reader.digitale-sammlungen.de/resolve/display/bsb10016441.html, https://books.google.de/books?id=n0EAAAAAcAAJ

Ludwig, Gernot (1968): Zur Geschichte der Fürstlichen Schule des „Gymnasium illustre" in Lauingen (= *Jahresbericht des Albertus-Gymnasiums Lauingen*; 1968). Lauingen 1968. Therein pp. 42–44: Johann Philipp Fuchs von Bimbach zu Möhren

Marius, Simon (1610): *Die Ersten Sechs Bücher Elementorum Evclidis.* Ansbach: Paul Böhem 1610

Marius, Simon (1614/1988): Joachim Schlör, *Mundus Iovialis – Die Welt des Jupiter.* Gunzenhausen: Johann Schrenk 1988

Marius, Simon (1614/1916/2019): Marius Simon, The World of Jupiter, translated by A.O. Prickard and Albert van Helden, in *Simon Marius and his Research*, Hans Gaab and Pierre Leich (eds.), Cham: Springer 2019

Melzner, Fritz-Günther (1982): Die Schlacht bei Lutter am Barenberge. In: Kurt Klay (ed.), *Am 27./17. August 1616: Die Schlacht bei Lutter am Barenberge.* Lutter am Barenberge 1982, [20 unnumbered pp.]. Digitized version: http://www.wolfgangroehl.de/Lutter/Die_Schlacht_bei_Lutter_am_Barenberge-27August1626.pdf

Meyer, Julius (1892): Osiander und Marius. *Jahresbericht des Historischen Vereins für Mittelfranken* 44 (1892), pp. 51–71. Digitized versions: http://www.simon-marius.net/pix/content/4/Meyer_Osiander-und-Marius_1892.pdf, http://opacplus.bsb-muenchen.de/title/6365-4

Müller, Annemarie B. (2001): Die Fuchs von Bimbachische Herrschaft Bimbach und deren Pfarrei im 16. und 17. Jahrhundert. *Blätter für fränkische Familienkunde* 24 (2001), pp. 7–35

[Orth, Johann Philipp (1994):] *Ausfürliche Abhandlung von den berümten zwoen Reichsmessen so in der Reichsstadt Frankfurt am Main järlich gehalten werden [...].* [Frankfurt am Main] 1765. Reprint: [Frankfurt am Main:] Brönner [ca. 1994]. Digitized version of the original: https://books.google.de/books?id=tk2ANzrLFDEC

Pellengahr, Hans-Georg (2012): Simon Marius – die Erforschung der Welt des Jupiter mit dem Perspicillum 1609–1614. In: Gudrun Wolfschmidt (ed.): *Simon Marius, der fränkische Galilei, und die Entwicklung des astronomischen Weltbildes* (= *Nuncius Hamburgensis – Beiträge zur Geschichte der Naturwissenschaften*; 16). Hamburg: tredition 2012, pp. 72–141

Rechter, Gerhard (1997): *Die Seckendorff. Quellen und Studien zur Genealogie und Besitzgeschichte*. Bd. 3., Die Linien Aberdar und Hörauf, Teil 1. Neustadt a.d. Aisch: Kommissionsverlag Degener & Co. 1997

Riekher, Rolf (1990): *Fernrohre und ihre Meister*. 2nd, rev. ed. Berlin: Verlag Technik 1990

Ritter, Moriz (1870): Die Gründung der Union, 1598–1608 (= *Briefe und Acten zur Geschichte des Dreissigjährigen Krieges in den Zeiten des vorwaltenden Einflusses der Wittelsbacher*; 1). München: M. Rieger 1870. Digitized version: https://books.google.de/books?id=KKFBAAAAcAAJ

Rockstroh, [Knud Christian] (1935): Fuchs v. Bimbach, Hans Philipp. In: *Dansk biografisk leksikon*. 2. udgave, vol. 7. København: J. H. Schultz Forlag 1935, pp. 502–503

Rößner, Volker (2003): Studium *und Kavalierstour der fränkischen Reichsritter Christoph Ernst und Ludwig Reinhold Fuchs von Bimbach 1681 bis 1686. Briefe und Dokumente*. Neustadt a.d. Aisch: Verlag Degener & Co. 2003

Rößner, Volker; Hammerich, Helmut (2011): *Die Familie Fuchs von Bimbach und Dornheim im Deutschen Kaiserreich. Ein Lebensbild in Briefen aus dem Nachlass des Reinold Frhr. Fuchs von Bimbach und Dornheim (1845–1903).* Stegaurach: Wissenschaftlicher Kommissions-Verlag 2011

Schlör, Joachim (2012): Mundus Iovialis – Die Welt des Jupiter. In: Gudrun Wolfschmidt (ed.): *Simon Marius, der fränkische Galilei, und die Entwicklung des astronomischen Weltbildes* (= *Nuncius Hamburgensis – Beiträge zur Geschichte der Naturwissenschaften*; 16). Hamburg: tredition 2012, pp. 50–71

Schmidt, Ernst Ludwig; Müller, Annemarie B. (2001): Das Kirchenbuch des Pfarrers Christoph Pfefflein. Teil 2: Bimbach. Taufen und Trauungen September 1576 bis März 1588; aus dem Freiherrlich Fuchs'schen Archiv in Burgpreppach. *Blätter für fränkische Familienkunde* 24 (2001), pp. 36–93

Schöndorf, Kurt (2006): Endres Fuchs von Bimbach zu Möhren. Landrichter von Graisbach-Monheim und Statthalter zu Neuburg. *Mitteilungen des Historischen Vereins für Donauwörth und Umgebung* 2004 (2006), pp. 4–29

Stahl, Patricia (ed.) (1991): Beiträge zur Geschichte der Frankfurter Messe (= *Brücke zwischen den Völkern – zur Geschichte der Frankfurter Messe*; Bd. 2). Frankfurt am Main: Historisches Museum 1991

Tittmann, Alexander (1998): Die ritterschaftliche Familie der Fuchs. Ihre Genealogie und ihr Besitz im Altlandkreis Haßfurt. *Jahrbuch für fränkische Landesforschung* 58 (1998), pp. 37–95. Digitized version: https://www.bayerische-landesbibliothek-online.de/jffl

[Unterschwaningen:] Geschichte der Gemeinde Unterschwaningen (= *Reihe fränkische Geschichte*; 15). Gunzenhausen: Schrenk-Verlag 2009

Van Helden, Albert (1974): The Telescope in the Seventeenth Century. *Isis* 65 (1974), No. 1, pp. 38–58

Van Helden, Albert (1975): The historical problem of the invention of the telescope. *History of Science* 13 (1975), pp. 251–263

Van Helden, Albert (1977): The Invention of the Telescope (= *Transactions of the American Philosophical Society*; 67, No. 4). Philadelphia: American Philosophical Society 1977

Van Helden, Albert (1997): Telescope, Invention of (to ca. 1630). In: John Lankford (ed.): *History of astronomy. An encyclopedia* (= *Garland Encyclopedias in the History of Science*; 1). New York und Abingdon: Routledge 2011 [first published: Garland Publishing, 1997], pp. 509–511

Van Helden, Albert (2009): The beginnings, from Lipperhey to Huygens and Cassini. *Experimental Astronomy* 25 (2009), pp. 3–16

Veh, Otto (1985): Zur Geschichte des Schlosses von Unterschwaningen. *Jahrbuch des Historischen Vereins für Mittelfranken* 92 (1984–1985), pp. 143–175

Willach, Rolf (2007): Der lange Weg zur Erfindung des Fernrohres. In: Jürgen Hamel, Inge Keil (eds.): *Der Meister und die Fernrohre. Das Wechselspiel zwischen Astronomie und Optik in der*

Geschichte. Festschrift zum 85. Geburtstag von Rolf Riekher (= *Acta Historica Astronomiae*; 33). Frankfurt am Main: Verlag Harri Deutsch 2007, pp. 34–126

Willach, Rolf (2008): *The Long Route to the Invention of the Telescope* (= *Transactions of the American Philosophical Society*; 98, No. 5). Philadelphia: American Philosophical Society 2008

Wohlwill, Emil (1926): *Galilei und sein Kampf für die copernicanische Lehre*, vol. 2: Nach der Verurteilung der copernicanischen Lehre durch das Dekret von 1616. Leipzig: Verlag von Leopold Voss 1926. Reprint: Wiesbaden: Sändig 1969. Digitized version of the original: https://archive.org/details/galileiundseinka02wohluoft

Wolfschmidt, Gudrun (2012): Ein neuer Blick ins Weltall – Simon Marius, der fränkische Galilei, und das Fernrohr. In: Gudrun Wolfschmidt (ed.): *Simon Marius, der fränkische Galilei, und die Entwicklung des astronomischen Weltbildes* (= *Nuncius Hamburgensis – Beiträge zur Geschichte der Naturwissenschaften*; 16). Hamburg: tredition 2012, pp. 194–257

Zeiller, Martin (1648): *Topographia Franconiae, Das ist, Beschreibung, Und Eygentliche Contrafactur der Vornembsten Stätte, Und Plätze des Franckenlandes [...].* Frankfurt am Main: Merian [1648]. Digitized version: http://www.mdz-nbn-resolving.de/urn/resolver.pl?urn=urn:nbn:de:bvb:12bsb11215651-6

Zuidervaart, Huib J. (2010): The 'true inventor' of the telescope. A survey of 400 years of debate. In: Albert Van Helden, Sven Dupré, Rob van Gent, Huib Zuidervaart (eds.): *The origins of the telescope*. Amsterdam: KNAW Press 2010, pp. 9–44. Online-Version: www.dwc.knaw.nl/wp-content/HSSN/2011-12-Origins.pdf

Zwanziger, Karl Hermann (1919/20): Hans Philipp Fuchs von Bimbach-Möhren und der Friedensvermittlungsversuch des Pfalzgrafen Wolfgang Wilhelm von Neuburg im niedersächsisch-dänischen Krieg. *Kollektaneen-Blatt für die Geschichte Bayerns insbesondere des ehemaligen Herzogtums Neuburg* 84 (1919), pp. 21–30, 85 (1920), pp. 3–16

Chapter 4
Georg Caesius as Official Court Astronomer of the Margrave Georg Friedrich of Brandenburg-Ansbach

Dieter Kempkens

Abstract Caesius as a court astronomer was able to successfully market his yearly prognostications and writing calendars. In comparison to other authors, his meteorological forecasts were more reliable, and his argumentative defense of the worth of astrology against a large group of astute critics was impressive. He motivated his readers to repent, so that God would reduce the negative prognostications for the harvest, for diseases, and for wars. In the discussion about the comet of 1577, he qualified it as a wonder star and therefore, together with the leading theologians of Wittenberg, opposed the interpretation of astrological naturalism.

Caesius fulfilled Margrave Georg Friedrich's expectations. His texts achieved high circulation numbers, delivered prognostications based on astronomical and astrological justifications, formulated the existential anxiety of the common man, and finally increased their belief in a strict but also merciful God as helper in need. The love of science and the unity of the Lutheran faith—humanism and confessionalization—should strengthen the state.

Georg Caesius was appointed court astronomer by Georg Friedrich of Brandenburg-Ansbach in 1577 and remained in the post until his death in 1604.[1] In 1603 Georg Friedrich decided that the Margraves Christian and Joachim Ernst should rule the territories. In 1606, they appointed Simon Marius as court mathematician, and he thus succeeded Caesius.

In the research, Klaus Matthäus analyzed the prognostications and writing calendars of Caesius and then introduced his book on the history of comets, published initially in Latin and then in German (Matthäus 1969, pp. 1087–1092).

[1]The certificate of appointment is now in the Staatsarchiv Nürnberg, formerly in the Germanisches Nationalmuseum Nürnberg (GMN). Dr. Burger informed the author by e-mail on 13.8.2014 that the old archive numbers of the GMN are no longer registered in the state archive, so that this document is not yet listed in the inventory.

D. Kempkens (✉)
Bergheim, Germany

H. Gaab, P. Leich (eds.), *Simon Marius and His Research*, Historical & Cultural Astronomy, https://doi.org/10.1007/978-3-319-92621-6_4

40 years later, I completed Caesius's biography (Kempkens 2011) and an analysis of why his prognostications and writing calendars were able to hold their position on the book market for more than 30 years (Kempkens 2014). The contemporary reception of his comet book was handled in detail by Marion Gindhart in 2006. According to her, Caesius used Cardano's Tetrabiblos extensively as a database (Gindhart 2006, p. 140). Georg Caesius's library, which is in the Kirchenbibliothek Neustadt a.d. Aisch, has not been studied to date.[2]

I utilize the results of these researches to answer the following three questions:

1. What duties and rights did he have as a court astronomer, and did he fulfill them?
2. How did he collect and use astronomical data?
3. What position did he take in the discussion about the comet of 1577?

What Duties and Rights Had Caesius as Court Astronomer, and Did He Fulfill Them?

In 1543 Georg Grau, Latinized Caesius, was born in Rothenburg o.d. Tauber, where he also went to school and where the teacher Abdias Wickner inspired his interest in astronomy. In 1561, Wickner published a treatise on the determination of the day and night lengths and from 1563, annually, a "prognosticon cum Calendario" (Ulshöfer 1990–1991). Caesius, as a student, gave a speech "de collatione solis et lunae cum filio die et ecclesia" (Ulshöfer 2002, p. 70).

So well prepared, he studied the septem artes liberales in Wittenberg from 1563 to 1565. In retrospect he formulated the essence of his studies (Caesius 1581, p. B2^{v}):

> Und was ich für meine Person in der Astrologia weiß und kan [. . .] das habe ch fürnemlich in der löblichen Uniuersitet Witteberg/Anno 63. 64. und 65. von Herrn Sebastiano Theodorico[3] seligen/der uns die Bücher Ptolemaei de praedictionibus Astronomicis publice für gelesen und erkleret/Studirt und gelernt[4]

The University of Wittenberg, following Philipp Melanchthon's anthropology, made clear that astrology explains the whole of nature of man. For every star or planet has its own radiance, which generates the spiritus vitalis (life force) (Salatowsky 2006, p. 114).[5] The spiritus animalis (mind) is also influenced by it,

[2]This has been known since 1781 (Schnizer 1786, p. 7). According to Reinhold Ohlmann (Neustadt-Aisch), many of Caesius's astronomical books are in this library.

[3]Sebastian Theodoricus (about 1520–1574), born in Windsheim, was Professor of Mathematics in Wittenberg from 1550 onward.

[4]And what I know myself and can in Astrologia [. . .] I have learned it at the praiseworthy University of Wittenberg in the years [15]63 [15]64 and [15]65 from the blessed Herr Sebastiano Theodorico, who read unexplained studies and learnt us the books Ptolemaei de praedictionibus Astronomicis publice.

[5]"Spiritus vitalis" are flames, which arise from the purest blood in the heart, transport the warmth of life to the other limbs, and grant them the strength of activities caused by the warmth of life.

but not the anima rationalis (the will). This is why, in the spirit of Melanchthon, Caesius often urged his readers to repent and to persuade God through prayer to change the impending, astrologically conditioned events such as war, change of rulers, bad weather, and widespread disease. He, on the other hand, understood astronomy as a science, which calculated the laws of the planets and stars, that is, theoria (Dear 2005).[6]

How could he, the son of a poor man, become a court astronomer? The Margrave ruled the territories of Brandenburg-Ansbach and Brandenburg-Kulmbach, was Duke of Jägerndorf, was administrator of the Duchy of Prussia, and was also involved in all the Protestant alliances throughout the empire (Ritter 1895, p. 49). The territories in Franconia were not monolithic but contained numerous free imperial cities and imperial knight dominions. Georg Friedrich, like other rulers, recruited for his official functions burgesses, firstly because during his reign, aristocrats rarely visited a university to qualify for government offices and secondly in order to withdraw the so created, loyal-to-him administration from the influence of the partially co-governing estates (Endres 1997, p. 479).

After studying, Caesius published an annual *Prognosticon Astrologicum* starting in 1567. He also published a *Schreibkalender* (Caesius 1569, p. B4^{v}) from 1569 onward. In 1574 he moved to Ansbach as a priest and was supposed to become court chaplain to the Margrave, which in the end did not happen probably because the accusation that he was a crypto-Calvinist led to the conclusion that he wouldn't represent the pure Lutheran beliefs theologically correctly (Matthäus 1969, p. 1089). However, here at the latest, he came into contact with the Margrave. In 1575 he already published his prognostications with the Margrave's coat of arms. In 1577 he was appointed court astronomer without residential obligation. The position he now entered publicly into was a patronage relationship. His duties consisted of dedicating his Prognostica and Schreibkalender to the Margrave and to present him with several copies each year on New Year's Day. In return, he received 25 guilders a year—as a gift. What did Margrave Georg Friedrich want to achieve with this appointment?

Since ancient times, secular and spiritual rulers have appointed astronomers as their consultants. In humanism, they belonged to the culture of the court, which reflected its significance and the honor of the ruler. The front position of the margravial territories in the religion-political conflict, being a direct neighbor of the counterreformatory bishoprics of Würzburg and Bamberg, necessitated the promotion of the proper faith not only through the extension of controls by consistories and priests signatures under the formula of Concord (Sommer 2002) but also by their dissemination especially in Prognostica. The public should see him as a promoter of science and faith and as a ruler concerned with the daily needs and souls of his subjects. In fact, he supported astrology because—as the dominant position of Melanchthon and his successor Caspar Peucer stated—it predicted both the political and the individual future; yet he was aware that it would not always be the case. For

[6]In the sixteenth century, authors also used the word "theoria" to explain the calculation of cosmic movements.

this reason, he also requested and received horoscopes from Melanchthon, Erasmus Reinhold, Cyprian Leowitz, Christoph Stathmion, Petrus Hoffmann, Balthasar von Rechenberg, and others (GStA PK I. HA Rep. 41. V F1) (Fig. 4.1).[7]

Caesius initially used the *ephemerides* of the mathematician Cyprian Leowitz (Leowitz 1557; Caesius 1581, p. B2^r; Caesius 1594, p. A3^r) for his prognostications and from 1601 on Erasmus Reinhold's *Prutenic Tables* in the improved edition of David Origanus for his astronomical data (Caesius 1601, Title page). Ephemerides provided, among other things, data on the entry, duration, and end of the annual lunar and solar eclipses, the beginning of the seasons, and the daily positions of the planets in the zodiac (Leowitz 1557, p. A). His astrological conclusions were documented in detail with quotations from Ptolemy and the comments on the *Centiloquium* of Girolamo Cardano and Giovanni Pontanus (Ptolemaeus 1553). What did the Margrave expect? Caesius delivered the answer in 1580:

> die Hochlöbliche Churfürstliche Heuser Saxen und Brandenburg/wie auch Hesse/haben je und allweg/neben der reinen lehr des H. Evangeli die freyen Künste/und unter derselben die *Astronomiam* und *Astrologiam*, hoch geliebet und mit grossen unkosten auff den universiteten zu Wittenberg/Leiptzig/Franckfurt an der Oder/Königsberg in Preußen und Marpurg gefördert.[8] (Caesius 1581, p. B2^r)

So the love of science and the unity of the Lutheran faith—humanism and confessionalization—(Kohler 2011, pp. 124–136) strengthen the power of the state.

What did Caesius's texts contribute to this? He regularly defended the value of astrology against the great number of critics. Astronomy, on the other hand, was not disputed (Caesius 1590, Dedication).

1. He could establish his texts as the best weather forecast, which Kepler also confirmed in a letter in 1604 (Kepler 1995, pp. 67–68).
2. Political prophecies triggering political conflicts were omitted, but he pointed to potential peasant uprisings in the wake of inflation.
3. He avoided theological controversies and formulated only brief eschatological predictions.

How Did He Collect and Use Astronomical Data?

Caesius only mentions his own observations in a few passages: he had observed Saturn (Caesius 1579, p. B2) and the comet of 1577, but he does not state which instruments he used. Probably he only made a few of his own observations.[9] He also

[7]This reference was first given by Claudia Brosseder (Brosseder 2004, p. 31).

[8]The highly worthy houses of the Electors of Sachsen and Brandenburg and the house of Hessen, have, as well as the teachings of the Holy Gospel, always loved the liberal arts and astronomy and astrology among them. And at great cost, they have supported the universities of Wittenberg, Leipzig, Frankfurt an der Oder, Königsberg in Prussia, and Marburg.

[9]Note from Klaus Matthäus to the author during the conference on 19.9.2014 in Nuremberg.

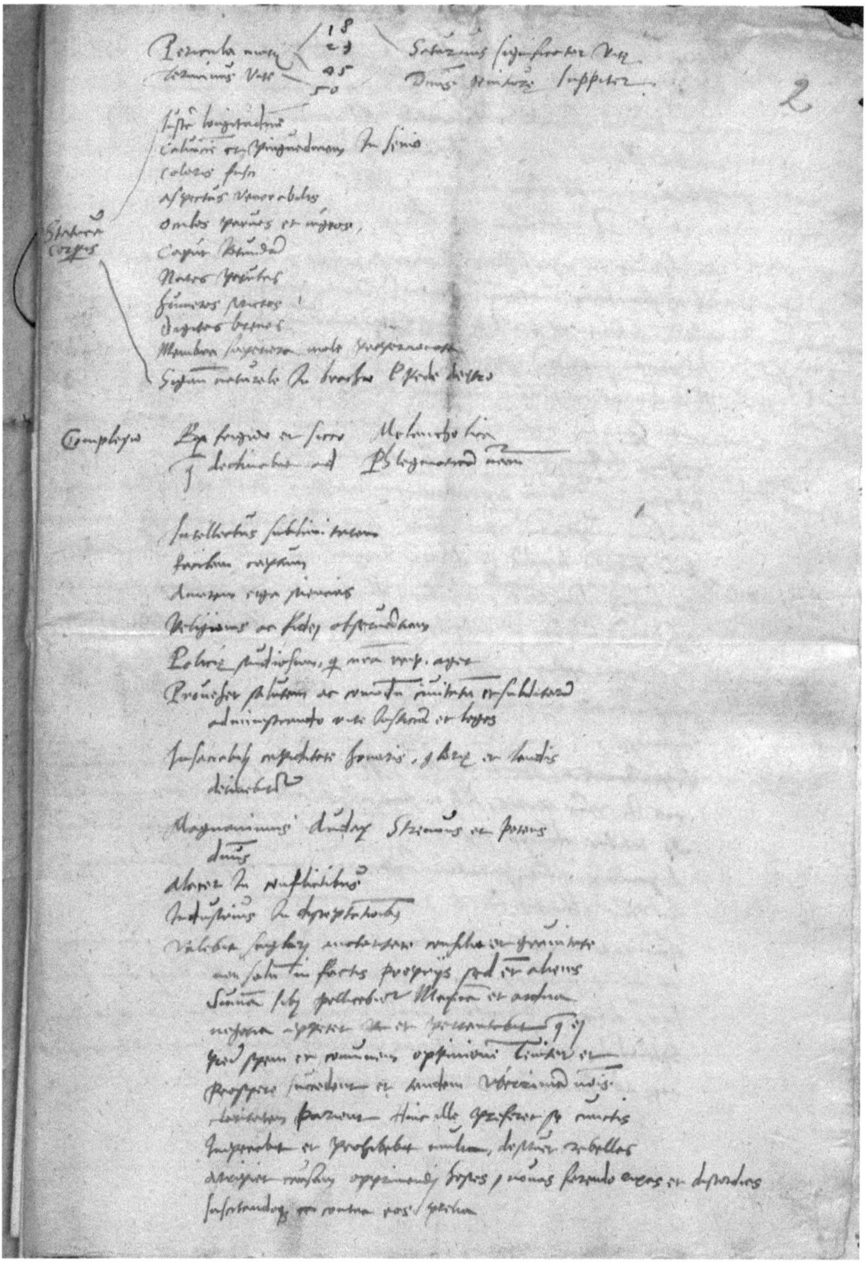

Fig. 4.1 Horoscope for Margrave Georg Friedrich—no date. Geheimes Staatsarchiv Stiftung Preußischer Kulturbesitz, Berlin

does not mention the instruments armillary sphere and Jacob's staff, as was customary for authors of prognostica and calendars. He regularly emphasizes that the new astronomical year starts with the beginning of spring (Caesius 1568, p. A4^{v})[10] and always refers to astronomical data when he describes the first day of a season: "Der Sommer fahet nach Astronomischer berechnung in disem 79. Jar den 12. Junij 1. stund unnd 3. Viertel nach Mittag"[11] (Caesius 1579, p. B3). Autumn starts with the "eingang der Sonnen in den Ersten grad der Wag/Welchs in diesem 1568 jar geschicht Den 13. Septembris: 1 Stunde 45 Minuten/nach Mittag[12]" (Caesius 1568, p. B2). Afterward, he gives a weather forecast for each day, which he establishes with the respective aspect (angular relations between the planets in the zodiac) and which he backed up with historical weather data.

For the description of solar and lunar eclipses, he also uses exact astronomical data because these cosmic events could lead to great political changes according to the astrological doctrine of this period. The darkness "an der Sonnen [...] 25. Februar oder 7. Marz newen calenders/bey dem Drachenkopff/um 10 und 11 uhr gegen hohem Mittag/Da die Sonn 27 Minuten über 11 Zoll/das ist/fast gantz und gar vom Mond verfinstert wird"[13] (Caesius 1568, p. B2). He did not immediately mention the expected effect, which would first occur in 1599 and 1600, but he offered his readers the interpretations of solar eclipses by Heinrich Rantzau, Philipp Melanchthon, Ptolemy, and Hermes Trismegistus Centiloquium Aphorism 53 (Caesius 1598, p. D2). What did he consider to be the use of astrology? For spiritual discipline, it creates order in time (calculation of the Easter feast), for the secular world appointments in court, and for domestic affairs, it offers the correct points in time for seeding, woodcutting, and the taking of medicines[14] (Caesius 1598, p. D2) (Fig. 4.2).

When, in autumn 1577, a new comet was visible to the naked eye in the heavens, a huge number of broadsheets and books were published, so that Caesius, as court astronomer, had to participate with his own evaluation, in order not to damage his reputation and thus also that of the Margrave.

[10]Here recorded for Rothenburg: On the 10th day of March, 11 h and 55 min, the sun reaches the first point of the hot and dry sign, Aries, at which time the winter ends, and according to astronomical calculations, the 1568th year begins.

[11]Summer, according to the astronomical calculation, starts in this 79th year on the 12th of June, 1 h and 3 quarters after midday.

[12]Entrance of the sun in the first degree of Libra, which will happen in this 1568th year the 13th of September: 1 h 45 min after midday

[13]At the sun [...] 25th of February or 7th of March new calendar/at the dragon head/at 10 and 11 o'clock toward high noon/because the sun 27 min over 11 in./that is/almost completely from the moon is darkened.

[14]Moreover, he is well versed in the interpretation of horoscopes. This darkness in the star sign Pisces also means that a high potentate, born in the sign of Pisces or in the counterglow of Virgo, will die, for which he quotes Proclos (Caesius 1598, p. D2).

Prognosticon Astrologicum, oder Teutsche Practick
Auff das Jahr/nach vnsers
HErrn vnd Seligmachers JEsu Christi Geburt
M. D. XCVI. Nach Erschaffung der Welt 5558. Auß der
Planeten Lauff vnd Finsternussen/ mit sonderm fleiß/ doch auff das
kürtzst beschrieben/vnnd zu glückseligem Newen Jar dedicirt:
DEm Durchleuchtigsten
Hochgebornen Fürsten vnd Herrn/Herrn Georg
Friderichen/Marggrafen zu Brandenburg/in Preussen/
zu Stettin/Pomern/der Cassuben vnd Wenden/Auch in Schlesien/zu Jä-
gerndorff vnd ꝛc. Hertzogen/ Burggrafen zu Nürnberg/
vnd Fürsten zu Rügen ꝛc. Meinem Gne-
digsten Fürsten vnd Herrn.
Durch M. Georgium Cæsium Rotenburgensem, Pfarherrn
zu Burckbernheim.
Non falLVabItVr reX In MagnItVDIne VIrtVtIs sVæ. Psal.33.
Mit Röm: Käy: Majest: Freyheiten/nicht nachzudrucken.

Fig. 4.2 Georg Caesius, *Prognosticon for 1596*, Title page. BSB München: Rar 4410 Beibd. 6

What Position Did Caesius Take in the Discussion of the Comet of 1577?

In 1579, Caesius published a book on the history of comets. The Latin version was dedicated to the Council of the City of Nuremberg, the later German edition to the Margrave Georg Friedrich. The discussion about this comet had centered on the question whether it was an unusual phenomenon or one of many. Caesius considered it to be a wonderful star (Caesius 1579a, p. A2^{r}) sent by God to show his anger about the religious and moral wickedness of men. The comet will have negative consequences in the coming years, but these could not yet be predicted. He followed his guarantors Caspar Peucer and Jakob Milich, who are often explicitly mentioned in both versions of the book about the history of comets. They rejected an "astrological naturalism" on the grounds that if planets in certain conjunctions always automatically produce comets, there would be no need for a "caring God" (Weichenhan 2004, p. 409). The comet, according to Caesius, receives its characteristics from two planets in its vicinity or even from one of the two. The present comet had received more from the planet Saturn than from Mars (Caesius 1579b, p. S7^{r}).[15]

In his subsequently characterization of the comets (Caesius 1579b, pp. Q8^{v}–R3^{v}), Caesius elucidates the effects of the present comet: melancholic weather, diseases, as well as drought and inflation (Caesius 1579b, p. Q8^{v}; Weichenhan 2004, p. 410).[16]

Four positions determined the comet discussion:

1. Exhalations of the earth are attracted by the planets and slowly condensed into matter, which is then set on fire by the fire sphere and seen as a comet (Weichenhan 2004, p. 402).
2. Comets are like planets or new creations (Weichenhan 2004, p. 402) (according to Cardanus, Tycho Brahe called the comet from 1577 a "pseudo-planet") (Christianson 1979, p.127).
3. Comets are created by sublunar exhalations of the earth (according to Aristotle) (Weichenhan 2004, p. 243).
4. God alone creates them.

Caesius favored Cardanus's thesis of a new creation. In Cardanus's commentary on Ptolemy (Cardanus in Ptolemy 1553) and in Sigismund Suevus' book (Suevus 1578) on the comets, he found the material to classify it as miraculous (Gindhart 2006, p. 140).[17]

Caesius succeeded in constructing a bridge between theology, astronomy, and astrology; his astrological and astronomical foundations led to the explanation of the "gestirn mit iren krefften" and thus to knowledge of the "allmechtigkeit/güte und unerforschliche weyßheit Gottes."[18] On the other hand, they warned of "zukünftigen unglück und grosse verenderung" and offered the temporal division of the year. He saw his specific contribution as an author in that his "kunst" helped "etliche oerter in der heyligen schrifft zu verstehen und to promote die Lust an solcher Herrlichen und

[15]He could not say what the influence of this would be; the readers should consult Cardanus and other scholars on this (Caesius 1579b, pp. K8$^{r\text{-}v}$, P5^{r}).

[16]Cardanus, on the other hand, despised Pontanus as one who had no idea of the "technical-mathematical demands of the art" of astrology (Grafton 1999, p. 252).

[17]He also recommends to his readers the commentary of Giovanni Pontanus on the Centiloquium and the Cometographia of Mizaldus 1549 (Caesius 1579b, p. R3^{v}). He quotes this book with an extensive catalog of comet phenomena, because it contains many agreeable quotes from Pontanus. Mizaldus himself cited Potanus (Mizaldus 1549, pp. 15, 19, 25).

[18]Stars with their powers. [...] omnipotence benevolence, and inexplicable wisdom of God.

Fig. 4.3 Unknown artist, Epitaph for Georg Caesius, Rhein-Erft-Kreis, Bergheim/Erft

Lieblichen der gestirn erkenntnuß" (Caesius, 1568, pp. $A2^{v}$–$A3^{r}$).[19] Here he followed precisely the Wittenberg school founded by Peucer: "Astrology was in their eyes the key to connect the Book of nature with the Book of Biblical Revelations. Thus—hermeneutics could represent this unity" (Brosseder 2005, p. 575; Westman 2011, pp. 143–145).

The Margrave could consider himself fortunate to have given the office of court astronomer to a man who established the meteorological forecasts astronomically

[19]To comprehend many verses in the holy scriptures and the delight of such glorious and loving knowledge of the stars

and astrologically, justified the existential fears of his subjects, and strengthened their faith in an austere but also benevolent God as a helper in need.

The family of George Caesius created an epitaph in commemoration of him in 1605, showing a man praying to Christ (see Fig. 4.3).[20]

References

Primary Literature

Several authors. *Horoskope für Markgraf Georg Friedrich von Brandenburg-Ansbach*. Geheimes Staatsarchiv Stiftung Preussischer Kulturbesitz Signatur BPH, Rep. 41 V Markgraf Georg Friedrich, F Nr. 1 [. . .] Manuscript.

Caesius, Georg (1568): *Practica oder Prognosticon, auff das Jhar, nach unsers Herren und Seligmachers Jesu Christi Geburt*, 1568. Regensburg: Burger 1567.

Caesius, Georg (1569): *Practica Oder Prognosticon auf das Jar nach vnsers Herrn vnnd sehligmachers Jhesu Christi geburt 1569*. Regensburg: Burger 1568.

Caesius, Georg (1579): *PROGNOSTICVM ASTROLOGICVM* || *Oder* || *Practica Teutsch/Von* || *den vier Zeitten/vnd andern zufellen/Auff* || *das Jar/nach vnsers Herrn vnd Seligmachers Jesu* || *Christi geburt M.D.LXXIX*. Nürnberg: Fuhrmann 1578.

Caesius, Georg (1579a): *Catalogus nunquam antea visus, omnium cometarum secundum Seriem annorum a diluvio conspectorum usq. ad hunc praesentem*. Nürnberg: Fuhrmann.

Caesius, Georg (1579b): *Chronick Oder ordenliche verzeichnuß unnd beschreibung aller Cometen*. Nürnberg: Fuhrmann.

Caesius, Georg (1581): *PROGNOSTICON ASTROLOGICVM* || *Oder* || *Teutsche Practick/von den* || *vier Zeiten/Finsternussen vnd andern zufellen/*|| *dises nach Christi vnsers Herrn vnd Seligmachers Geburt* || *M.D.LXXXI. Jars*. Nürnberg: Fuhrmann 1580.

Caesius, Georg (1590): *Prognosticon Astrologicum, oder Teutsche Practick.*|| *Auff das Jar nach vnsers* || *Herrn vnnd Seligmachers Jesu Christi Geburt/*|| *M.D.LXXXX*. Nürnberg: Fuhrmann 1589.

Caesius, Georg (1594): *Prognosticon Astrologicum, oder Teutsche Practick* || *Auff das Jar/nach vnsers* || *HErrn vnd Seligmachers Jesu Christi Geburt/*|| *M. D. XCIIII*. Nürnberg: Fuhrmann 1593.

Caesius, Georg (1598): *Prognosticon Astrologicum, oder Teutsche Practick* || *Auff das Jahr/nach vnsers* || *HERRN vnd Seligmachers JEsu Christi Geburt* || *M.D.XCVII*. Nürnberg: Fuhrmann 1597.

Caesius, Georg (1601): *Prognosticon Astrologicum, oder Teutsche Practick:*|| *Auff das Jahr/nach vnsers* || *HERRN vnd Seligmachers JEsu Christi Geburt* || *M.D C I*. Nürnberg: Fuhrmann 1600.

Leowitz, Cyprian (1557): *Ephemeridum Novum Atque Insigne Opus Ab Anno Domini 1556 Usque in 1606 accuratissime supputatum*. Augsburg: Uhlhard.

Mizaldus, Antonius (1549): *Cometographia: crinitarum stellarum quas mundus nunquam impunè vidit, aliorumq[ue] ignitoru[m] aeris Phaenomenōn, natura[m]& portenta duobus libris philosophicè iuxtà ac astronomicè expediens*. Paris: Wechelus.

Ptolemäus, Claudius (1553): *De Praedictionibus Astronomicis, cui titulum fecerunt Quadripartitu [m],Grecè &Latinè, Libri IIII. Philippo Melanthone interprete. Eiusdem Fructus librorum suorum siue Centum dicta, ex conversione Ioviani Pontani*. Basel: Oporinus.

Suevus, Sigismund (1578): *Cometen Was sie f[ue]r grosse Wunder vnd schreckliche ding zu bedeuten vnd anzuk[ue]ndigen pflegen. Mit viel gedenckwirdigen Historien vnd Exempeln erkleret. Aus guten Chroniken vnd andern B[ue]chern mit fleiß zusammen getragen vnd menniglich⍰ zur trewen warnung vnd ernsten Bußpredigt [fue]rgestellet*. Görlitz: Fritsch.

[20]The epitaph hung in the church of Burgbernheim until 1840. It was found in Bergheim/Erft in 1940, and today it hangs in the Plenary Hall of the *Kreistag* (Kempkens 2011).

Secondary Literature

Asch, Ronald Gregor (2008): *Europäischer Adel in der Frühen* Neuzeit. Köln/Weimar/Wien: Böhlau.

Brosseder, Claudia (2004): *Im Bann der Sterne*. Berlin: Akademie Verlag.

Brosseder, Claudia (2005): The writing in the Wittenberg sky. *Journal of the history of ideas* 66, pp. 557–576.

Christianson, John Robert (1979): Tycho Brahe's German treatise on the comet of 1577, *Isis* 70, pp. 110–140.

Dear, Peter (2005): What is the history of science the history of? *Isis* 96, pp. 390–406.

Endres, Rudolf (1997): Die Festigung der neuen Lehre. In Andreas Kraus (Ed.), *Geschichte Frankens* (Handbuch der bayerischen Geschichte, 3) (pp. 478–481). München: Beck.

Gindhart, Marion (2006): *Das Kometenjahr 1618*. Wiesbaden: Reichert.

Grafton, Anthony (1999): *Cardanos Kosmos. Die Welten und Werke eines Renaissance-Astrologen*. Berlin: Berlin Verlag.

Jardine, Nicolas (1998): The places of astronomy in early modern culture. *Journal for the history of astronomy* 29, pp. 49–62.

Kempkens, Dieter (2011): Ein Bild mit vielen Geheimnissen – Das Epitaph für Georg Caesius in Bergheim. *Geschichte in Bergheim*, 20, pp. 65–90.

Kempkens, Dieter (2014): Der Erfolg der Prognostica auf dem Buchmarkt in der frühen Neuzeit. *Jahrbuch für Kommunikationsgeschichte* 16, pp. 5–27.

Kepler, Johannes (1995): Briefe 1604–1607. In Max Caspar (Ed.), *Gesammelte Werke, Bd. XV*. (2nd ed.) München: Beck.

Kohler, Alfred (2011): *Von der Reformation zum Westfälischen Frieden* (Oldenbourg Grundriss der Geschichte, 39). München: Oldenbourg.

Matthäus, Klaus (1969): Zur Geschichte des Nürnberger Kalenderwesens. Die Entwicklung der in Nürnberg gedruckten Jahreskalender in Buchform. *Archiv für Geschichte des Buchwesens* 9, pp. 965–1396.

Ritter, Moriz (1895): *Deutsche Geschichte im Zeitalter der Gegenreformation und des Dreißigjährigen Krieges (1555–1648). Bd. 2, 1586–1618.* Stuttgart: Cotta.

Salatowsky, Sascha (2006): *De Anima. Die Rezeption der aristotelischen Psychologie im 16. und 17. Jahrhundert* (Bochumer Studien zur Philosophie, 43). Amsterdam: Grüner.

Schnizer, Georg Matthäus (1786): *Die Kirchen-Bibliothek zu Neustadt an der Aysch, Fünffte Anzeige*. Nürnberg: Bauer.

Sommer, Wolfgang (2002): Theologie und Frömmigkeit. In Gerhard Müller, Horts Weigelt, Wolfgang Zorn (Eds.): *Handbuch der Geschichte der evangelischen Kirche in Bayern*. Bd.1. (419–456). St. Ottilien: EOS.

Ulshöfer, Kuno (1990–1991): Das einfache Leben des Rothenburger Gymnasialrektors Abdias Wickner (1528–1564). *Jahrbuch des Historischen Vereins für Mittelfranken* 95, pp. 125–145.

Ulshöfer, Kuno (2002): Abdias Wickner, der vierte evangelische Titularabt von Heilsbronn (1560–1608). *Zeitschrift für bayerische Kirchengeschichte* 71, pp. 69–96.

Westman, Robert S. (2011): *The Copernican Question. Prognostication, skepticism, and celestial order*. Berkeley/Los Angeles/London: University of California Press.

Weichenhan, Michael (2004): *„ergo perit coelum". Die Supernova des Jahres 1572 und die Überwindung der aristotelischen Kosmologie*. Stuttgart: Steiner.

Wolfschmidt, Gudrun (2012): Ein neuer Blick ins Weltall – Simon Marius, der fränkische Galilei, und das Fernrohr. In: Gudrun Wolfschmidt (Ed.). *Simon Marius, der fränkische Galilei und die Entdeckung des astronomischen Weltbilds.* (pp. 195–258). Hamburg: tredition.

Chapter 5
Simon Marius's *Mundus Iovialis* and the Discovery of the Moons of Jupiter

Jay M. Pasachoff

Abstract Though the details of who was first to see the four major satellites of Jupiter are obscured by the mists of time, it seems that Simon Mayr (Marius) nearly simultaneously and independently discovered them and noted the discovery only 1 day after Galileo similarly discovered and noted it. The twin discoveries were confused by the use of different calendars by Marius and by Galileo, the former using the Julian calendar then still in use in Protestant regions and the latter using the new Gregorian calendar that was adopted in Catholic regions. Galileo was particularly sensitive to his priority, and the use of 1609 by Marius in the title of his *Mundus Iovialis* of 1614 raised particular ire, though adding the required 10 days for the conversion from O.S. to N.S. brought Marius's discovery into early 1610. In the long run, we now use the names that Marius gave—Io, Europa, Ganymede, and Callisto—to what are called the Galilean satellites.

Introduction

Simon Marius's discovery of the moons of Jupiter, which were described in his 1614 book (Fig. 5.1), were not published in wide circulation for years after their discovery. In December 1609, Marius used the telescope of his patron Fuchs von Bimbach and saw three "stars" that moved with Jupiter. He apparently realized that they were satellites of Jupiter but did not write down his observations until December 28 (Julian, i.e., Old Style or O.S.). Though with hindsight, we could wish that Marius had trumpeted his observations to the world, he did not write about them until he put a note and a drawing (Fig. 5.2) in his almanac for 1612, published in 1611, one of a series of annual almanacs that Marius put out for almost 30 years, probably connected with his duties as court astronomer to the Margrave of Ansbach.

J. M. Pasachoff (✉)
Williams College—Hopkins Observatory, Williamstown, MA, USA

Carnegie Observatories, Pasadena, CA, USA
e-mail: jmp@williams.edu

H. Gaab, P. Leich (eds.), *Simon Marius and His Research*, Historical & Cultural Astronomy, https://doi.org/10.1007/978-3-319-92621-6_5

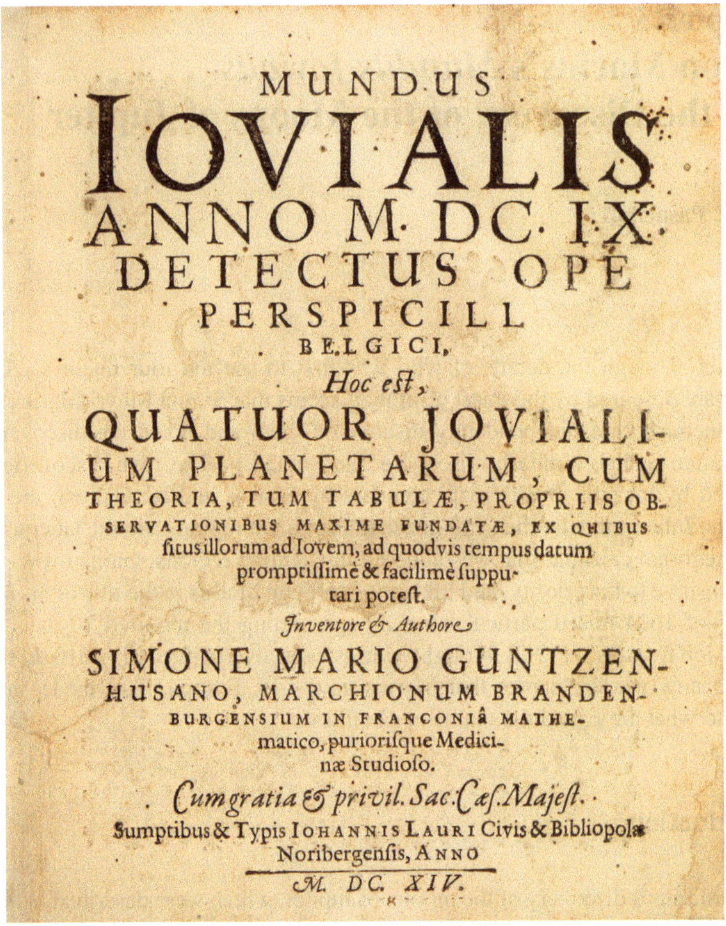

MUNDUS
IOVIALIS
ANNO M. DC. IX.
DETECTUS OPE
PERSPICILL
BELGICI,
Hoc est,
QUATUOR JOVIALI-
UM PLANETARUM, CUM
THEORIA, TUM TABULÆ, PROPRIIS OB-
SERVATIONIBUS MAXIME FUNDATÆ, EX QUIBUS
situs illorum ad Iovem, ad quodvis tempus datum
promptissimè & facilimè suppu-
tari potest.
Inventore & Authore
SIMONE MARIO GUNTZEN-
HUSANO, MARCHIONUM BRANDEN-
BURGENSIUM IN FRANCONIÂ MATHE-
matico, puriorisque Medici-
næ Studioso.
Cum gratia & privil. Sac. Cæs. Majest.
Sumptibus & Typis IOHANNIS LAURI Civis & Bibliopolæ
Noribergensis, ANNO
M. DC. XIV.

Fig. 5.1 The frontispiece of Marius's 1614 book *Mundus Iovialis*. The title page is lacking the final "i" in "Perspicilli" in my copy; I have checked a dozen other copies and none lack that letter. Perhaps that part of the type loosened (other letters in that area of that page seem to be printed weakly) or the letter broke, at one end of the print run or even in the middle, and was immediately or soon fixed (Jay and Naomi Pasachoff Collection)

In the meantime, Galileo Galilei, in Padua, had made his own telescope, improving on the idea that he had heard of from a Dutch traveler. He apparently also saw three "stars" moving with Jupiter, and on January 7, 1610 (Gregorian, i.e., New Style or N.S.), with the first observation being copied from a scrap of paper (Meeus 1962; Gingerich and van Helden 2003). Galileo had started publishing his telescopic discoveries, and following his original descriptions of the additional stars he saw in, for example, the Milky Way, Orion, and star clusters such as the Pleiades and Praesepe, he added page after page of the images of the moons of Jupiter, with asterisks for the moons around a printed letter O for Jupiter itself. He first called the objects "Sidera Cosmica," Cosmic Stars, trying to gain favor with Cosimo de'

Fig. 5.2 The portrait of Simon Marius from his *Mundus Iovialis: Anno M·DC·IX· Detectus Ope Perspicilli Belgici, Hoc est, Quatuor Joviali cum Planetarum, cum Theoria, tum Tabulæ, Propriis Observationibus Maxime Fundatæ [. . .]* Marius used the Julian calendar (Old Style, or O.S.), while Galileo, in his *Sidereus Nuncius* of 1610 and in general, used the Gregorian calendar from 1582 (New Style or N.S.). Four satellites circling Jupiter appear in the upper-left quadrant. A chemical device known as an alembic is in the lower-right quadrant. The content of the circle in the upper-right quadrant is not clear; it has been called a pair of comets and also merely a sky view with the ecliptic and celestial equator (Jay and Naomi Pasachoff Collection)

Medici, but realizing that Cosmica could be confused with the general term "cosmic" instead of the name "Cosimo," a printed slip was placed over most or all copies of the title page changing "Cosmican" to "Medician," ("Cosmica Sidera" to "Medicea Sidera") changing the credit to the family name Medici.

At the time of publication, March 1610, Galileo had no idea that he had a competitor to his north in Marius. Galileo had corresponded with Kepler prior to 1600, and Marius had spent at least a month with Tycho a few years later, but they were not in correspondence. Owen Gingerich of the Harvard-Smithsonian Center for Astrophysics and Albert van Helden of the Netherlands (who had started the Galileo project at Rice University in the United States) have analyzed the sequence of Galileo's observations and publication of *Sidereus Nuncius* in detail (Gingerich and van Helden 2011). They showed that Galileo started writing down his notes about the moons of Jupiter on January 11, 1610 (N.S.), first transcribing his observations of January 7 and 8, the 9th having been cloudy (Gingerich and van Helden 2003), and 10th, this last from an unknown source or from memory.

Galileo's handwritten notes showing the moons are held in the library of the University of Michigan in Ann Arbor, Michigan; I reproduce them in my textbook *The Cosmos: Astronomy in the New Millennium* (Pasachoff and Filippenko 2014).

Marius's 1611 publication, the almanac, showed that he had realized that there were, in fact, four moons instead of the three that he had first seen. The fourth moon had at first, for both Marius and Galileo, been blended with Jupiter's disk or another moon or occulted by Jupiter. (As I write in 2015, a series of mutual occultations and eclipses of Jupiter's moons is going on.) The discovery by Marius and by Galileo (independently, of course) that there were four moons took weeks or months.

Marius's 1610 book was about Euclidean geometry, and his almanac for 1612 apparently had only local circulation, as I described in a January 2015 paper before the Historical Astronomy Division of the American Astronomical Society meeting in Seattle (Pasachoff and Leich 2015).

In 1996, I purchased my copy of Marius's *Mundus Iovialis* (Figs. 5.1 and 5.2) because, as the rare-book dealer described, the frontispiece shows the first image of an astronomical telescope. It is in the foreground labelled "perspicillum"; Rosen (1947) dates the name "telescope" (meaning "far-seeing") to 1611, and the Greek mathematician Giovanni Demisiani is credited with the name that he applied when Galileo presented one of his devices at the Accademia dei Lincei, but whether that was known to Marius is unknown. Also in Marius's frontispiece, Jupiter is in the upper-left quadrant, surrounded by the orbits of the four satellites. Since Galileo used only asterisks to show the moons, we can state that Marius's 1614 diagram is the first to show the satellites' orbits (Fig. 5.3).

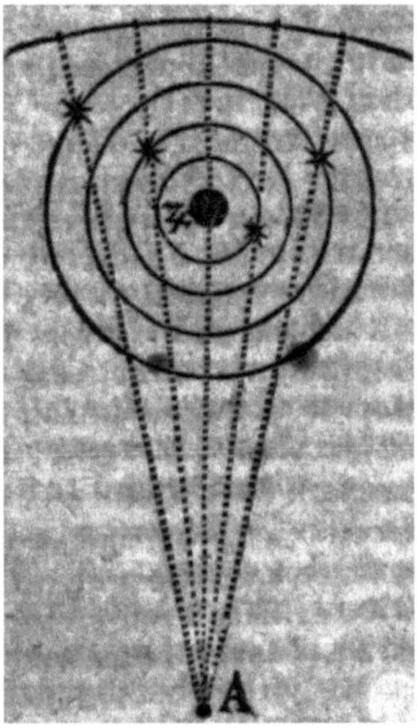

Fig. 5.3 Marius's first printed drawings of the moons of Jupiter, published in 1611 in his *Prognosticon Astrologicum auf 1612*. Kepler in 1611 cited a letter from Marius that he had seen phases of Venus, so probably Galileo found out about Marius's telescopic observations at that time (Staatsarchiv Nürnberg: Fürstentum Brandenburg-Ansbach, Staats- und Schreibkalender (129), Nr. 274, sig. C3r)

Marius's Observations

Much of Marius's *Mundus Iovialis* was translated in the British magazine *The Observatory* for 1916[1] (recalling that all dates are O.S.):

> [. . .] glasses of the same kind were becoming common in Belgium, and a fairly good one was sent, with which we were highly pleased. This was in the summer of 1609. From this time I began to look into the heavens and the stars with this instrument, whenever I was at the house of the nobleman so often mentioned, at night time; sometimes he used to allow me to carry it home, and in particular about the end of November, when I was observing the stars according to my custom in my own observatory. Then for the first time I looked at Jupiter, who was in opposition to the Sun, and made out some tiny stars, sometimes following, sometimes preceding Jupiter in a straight line with him. First, I thought that they were of the number of those fixed stars which cannot be seen without this instrument in other parts, such as those which I was finding in the Milky Way, the Pleiads, the Hyads, Orion, and elsewhere. However, as Jupiter was then retrograding, and still I saw these stars accompanying him throughout December, I was at first much astonished; but by degrees arrived at the following view, namely, that these stars moved round Jupiter, just as the five solar planets, Mercury, Venus, Mars, Jupiter, and Saturn revolve round the Sun. I therefore began to record my observations. The first was taken on December 29, when three stars of this description were visible in a straight line from Jupiter towards the west [. . .] Accordingly, from this time until January 12, I gave my diligent attention to these Jovian stars, and somehow ascertained that there were four such bodies, which themselves revolved around Jupiter.

Adding 10 days for the conversion from the Julian calendar, his observations began on January 8, 1610 (N.S.), though Marius did not draw the configurations.

Marius wrote (and note how he does not rule out his own priority nor does he give Galileo priority anywhere except Italy):

> In recounting all this, I am not to be understood as wishing to lessen Galileo's reputation, or to snatch from him the discovery of these Jovian stars among his countrymen in Italy—far from it. My object rather is, that it may be understood that these stars were not shown to me by any mortal in any way, but were discovered and observed by me, by my own investigation, in Germany, almost at the very time, or slightly before it, at which Galileo first saw them in Italy. The credit, therefore, of the first discovery of these stars in Italy is deservedly assigned to Galileo and remains his. So if this little book of mine shall reach Florence and come into Galileo's hands, I pray that he will receive it in the same spirit in which it is written by me. I am so far from wishing in any way to detract from his authority and his discoveries, that I rather thank him for publishing his 'Nuncius Sidereus,' for in it he has done much to confirm my view.

Marius should get credit and deserves to be better known to the world at large not only for his independent discovery of the moons of Jupiter. He was also one of the first to see sunspots telescopically (his observations of November 1611 are listed in the dedication of June 30, 1612, in *Prognosticon Astrologicum auf 1613*). He is also a discoverer of the Andromeda Galaxy (1612), which Charles Messier in the 1770s listed as 31st in his catalogue (Pasachoff 2014).

---

[1]Arthur Octavius Prickard's incomplete translation has been completed by Albert van Helden and forms the first chapter of this volume.

Who Discovered That Jupiter Had Four Moons?

My student Jake Goldenring (B.A. '15), as part of my course Astronomy 340 = History of Science 340 = Leadership Studies 340, has duplicated the views of Jupiter's moons using the computer-program *Stellarium* (Fig. 5.4). He writes, "it is very easy to pinpoint the time at which Galileo observed four moons but the same cannot be said for Marius. In the *Siderius Nuncius* Galileo writes, 'On the 13th of January, for the first time, I saw four small stars in this arrangement with regard to Jupiter [a figure appears in the book]. Three to the west, and one to the east. They made a line that was almost straight, but the one in the middle of those that were to the west deviated a little from the straight line towards the north'." He has duplicated the view in *Stellarium* and writes that from his calculated image "it is plain to see that Galileo was accurate in his description of the positions of the moons of Jupiter on the night of January 13th (N.S.). Just as he said, there are three moons to the west and one moon to the east. Also, Io deviates slightly north from the plane of orbit, as was expected. Taking the above evidence as one body, we can say, with confidence, that Galileo did see four Jovian satellites on the night of January 13th 1610." Galileo drew one of the moons on that night above the line joining Jupiter's disk with the other moons.

Jake writes, "Marius is very circumspect in describing his timeline of witnessing the fourth major moon of Jupiter. In the *Mundus Iovialis*, Marius reports that, 'At this time [the 29th of December 1609, Julian], as I frankly confess, I thought that there were only three such stars accompanying Jupiter.' (Marius 1614, sig.)(3^{r}). Thus it is clear that Marius did not witness all four of the major moons of Jupiter on the first

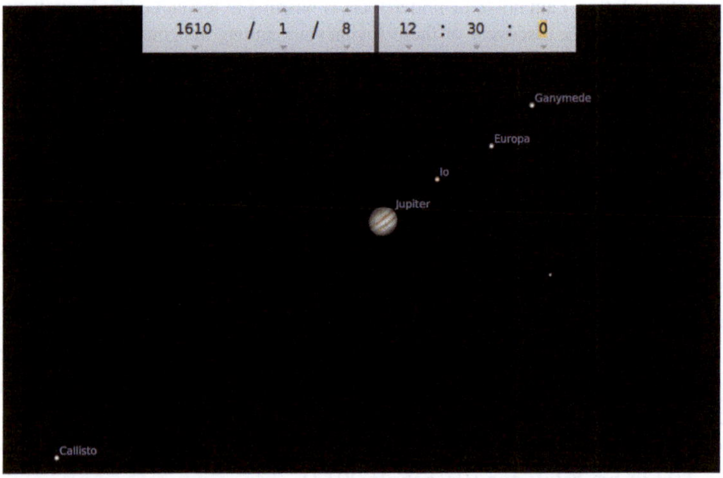

Fig. 5.4 A calculation using the program *Stellarium* showing the configuration of Jupiter and its four major moons on December 29, 1609, Julian, which is Gregorian January 8, 1610, corresponding to Marius's first observation; the calculation matches Marius's description of three stars off to the west; Callisto would have been so far on the east side that it wouldn't have been in Marius's telescope's field of view (Courtesy of Jacob Goldenring)

night that he logged observations. In addition to this conclusion, as one continues to read in the *Mundus Iovialis*, it becomes clear that Marius was limited by his 'telescope.' He describes going through the process of getting a new one and resuming observations once the new instrument was in his possession. Presumably, this new instrument had a wider field of view than the original instrument, which made it more advantageous to observing all four of the moons. In this light, Marius states that, 'About the end of February or beginning of March, I felt entirely confirmed in my view as to the definite number of these stars.' (Marius 1614, sig.)(3^{r}) He later goes on to affirm that this date was March 1610."

Jake continues, "Based on the above, we can now confirm that Galileo did, beyond a shadow of doubt, confirm the existence of four major moons of Jupiter before Marius was able to. However, there is still a gray area surrounding who saw the moons first. Yes, as shown above, Galileo was the first to record his observations of the moons, but there is still the possibility that Marius witnessed them first as his writings in the *Mundus Iovialis* would seem to suggest."[2]

Galileo Was Not Pleased

Galileo was obsessed with his priority for discoveries, perhaps in part related to his job searches and certainly for his internal self-crediting. On the *Sidereus Nuncius* title page, probably written in February 1610 (it was in the last section of the book to be printed), Galileo wrote, about his discovery of the moons, that they were "nemini in hanc usque diem cognitos, novistimè Author depræhendit primus," "known to nobody up to today, the Author most recently discovered for the first time." On March 19, 1610, Galileo wrote the personal secretary of Duke Cosimo II, "I did not want to prolong the publication in case I risked that perhaps someone else might have discovered the same and preceded me."

In *Il saggiatore* from 1623, Galileo was very harsh in accusing Marius of plagiarism. He basically ruined Marius's reputation worldwide for hundreds of years, as I describe in detail in the *Journal for the History of Astronomy* (Pasachoff 2015).

Note that priority disputes were not unique to Galileo. Bertoloni Meli (1993), in a discussion of Newton versus Leibniz, wrote "Priority in the seventeenth century involved a complex series of factors and had different connotations from those with which we are familiar. Partly as a result of such different conventions, the second part of the century has been recently aptly characterized as 'the golden age of the mud-slinging priority disputes'" (Meli 1993, p. 4).

[2]Albert van Helden wrote in 1994: "It seems a hopeless task, after almost four centuries, to find out exactly when Marius first saw what turned out to be satellites of Jupiter through his telescope."

Judged by a Jury

Following a jury convened in Holland, in 1903, Jean Abraham Chrétien Oudemans and Johannes Bosscha published a study in French about the relation of the discoveries of Marius and Galileo with the title "Galilée et Marius." After the death of Oudemans, the lengthy article "Rehabilitation of an astronomer gravely accused" followed from the 4-year-younger Bosscha in 1907.

He concluded (the author's translation):

> As the opinion of the jury involved an error and even a fault of Galileo, Oudemans and I, in order to avoid any suspicion of bias on the part of the jury, have explained the reasons which, in our opinion, are sufficient to conclude that the cruel accusation when the latter suffering from the disease to which he was to succumb [Galileo's *The Assayer* appeared in 1623 and Marius died early in 1625], could no longer defend himself, was completely unfounded and contained assertions of which Galileo himself had to know the falsehood.
>
> [...] had very real merits in astronomy, especially in the theory of Jupiter's satellites, and that, instead of having committed plagiarism toward Galileo, he was given the honor that was he was due.

Clearly, though, writing from 2015, we can see that Marius has not gained the worldwide reputation that he undoubtedly would have had without Galileo's attack.

The Names of the Four Major Satellites of Jupiter

Io, Europa, Ganymede, and Callisto—the names now in general use (van Helden 1990, p. 371f.; 1994), especially since the discovery of Amalthea interior to the four major satellites made Galileo's Roman numbering scheme ambiguous—come from *Mundus Iovialis* (Fig. 5.5a). Marius describes that:

> Jupiter is much blamed by the poets on account of his irregular loves. Three maidens are especially mentioned as having been clandestinely courted by Jupiter with success. Io, daughter of the River Inachus, Callisto of Lycaon, Europa of Agenor. Then there was Ganymede, the handsome son of King Tros, whom Jupiter, having taken the form of an eagle, transported to heaven on his back, as poets fabulously tell [...] I think, therefore, that I shall not have done amiss if the First is called by me Io, the Second Europa, the Third, on account of its majesty of light, Ganymede, the Fourth Callisto [...]

On the following page (Fig. 5.5b), Marius credits Johannes Kepler for the original suggestion:

> This fancy, and the particular names given, were suggested to me by Kepler, Imperial Astronomer, when we met at Ratisbon fair [an assembly of delegates] in October 1613. So if, as a jest, and in memory of our friendship then begun, I hail him as joint father of these four stars, again I shall not be doing wrong.

The table of the positions of the four Jovian satellites that appears at the end of his *Mundus Iovialis* (Fig. 5.6) is one of the pieces of evidence that proved to Bosscha and colleagues that Marius had early, original observations.

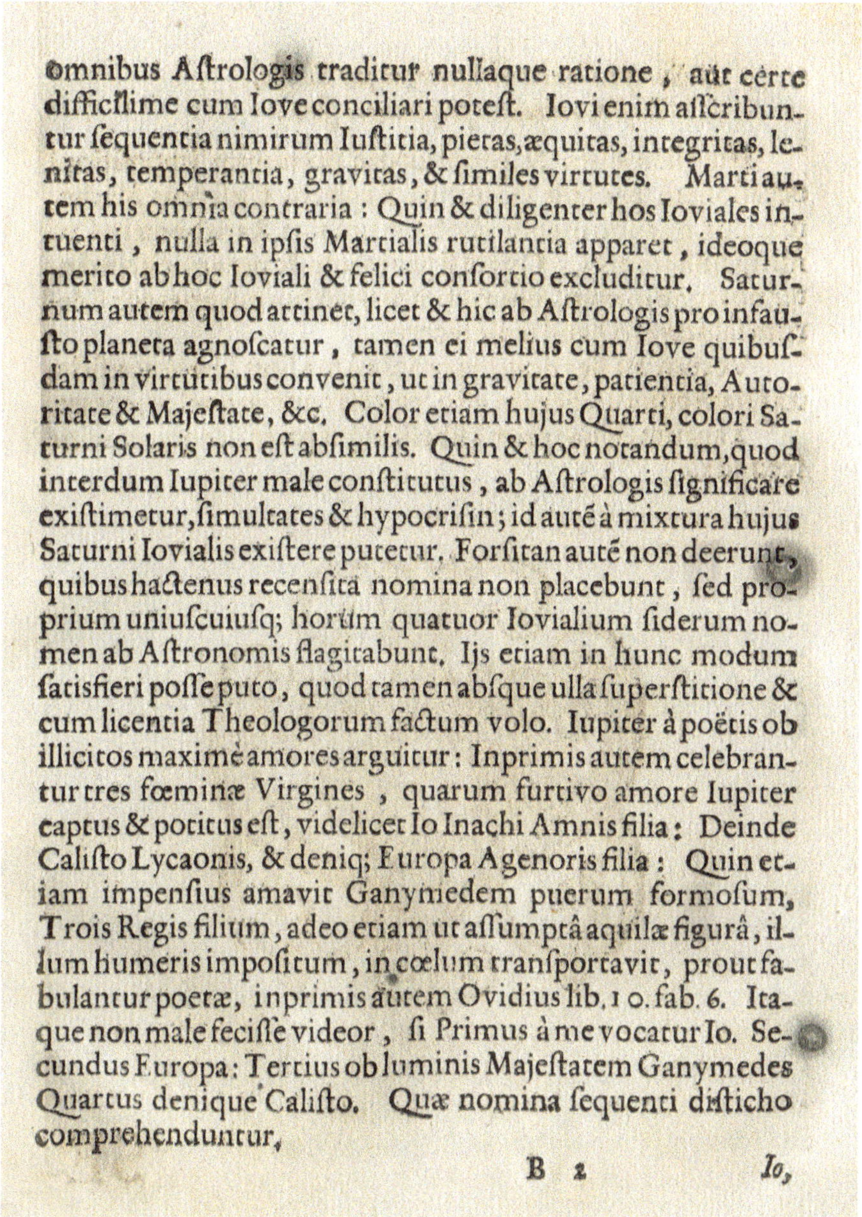

omnibus Aſtrologis traditur nullaque ratione, aut certe difficilime cum Iove conciliari poteſt. Iovi enim aſſcribuntur ſequentia nimirum Iuſtitia, pietas, æquitas, integritas, lenitas, temperantia, gravitas, & ſimiles virtutes. Marti autem his omnia contraria: Quin & diligenter hos Ioviales intuenti, nulla in ipſis Martialis rutilantia apparet, ideoque merito ab hoc Ioviali & felici conſortio excluditur. Saturnum autem quod attinet, licet & hic ab Aſtrologis pro infauſto planeta agnoſcatur, tamen ei melius cum Iove quibuſdam in virtutibus convenit, ut in gravitate, patientia, Autoritate & Majeſtate, &c. Color etiam hujus Quarti, colori Saturni Solaris non eſt abſimilis. Quin & hoc notandum, quod interdum Iupiter male conſtitutus, ab Aſtrologis ſignificare exiſtimetur, ſimultates & hypocriſin; id autē à mixtura hujus Saturni Iovialis exiſtere putetur. Forſitan autē non deerunt, quibus hactenus recenſita nomina non placebunt, ſed proprium uniuſcuiuſq; horum quatuor Iovialium ſiderum nomen ab Aſtronomis flagitabunt. Ijs etiam in hunc modum ſatisfieri poſſe puto, quod tamen abſque ulla ſuperſtitione & cum licentia Theologorum factum volo. Iupiter à poëtis ob illicitos maximè amores arguitur: Inprimis autem celebrantur tres fœminæ Virgines, quarum furtivo amore Iupiter captus & potitus eſt, videlicet Io Inachi Amnis filia: Deinde Caliſto Lycaonis, & deniq; Europa Agenoris filia: Quin etiam impenſius amavit Ganymedem puerum formoſum, Trois Regis filium, adeo etiam ut aſſumptâ aquilæ figurâ, illum humeris impoſitum, in cœlum tranſportavit, prout fabulantur poetæ, inprimis autem Ovidius lib. 1 o. fab. 6. Itaque non male feciſſe videor, ſi Primus à me vocatur Io. Secundus Europa: Tertius ob luminis Majeſtatem Ganymedes Quartus denique Caliſto. Quæ nomina ſequenti diſticho comprehenduntur,

B 2 *Io,*

Fig. 5.5 (**a**) The page of *Mundus Iovialis* describing Io, Europa, Ganymede, and Callisto—their current names (Jay and Naomi Pasachoff Collection). (**b**) The other side of the page, giving credit to Kepler for the original suggestion (Jay and Naomi Pasachoff Collection)

In Europa, Ganimedes puer, atque Caliſto,
Laſcivo nimium perplacuere Jovi.

Huic figmento & propriorum nominum impoſitioni occaſionem præbuit Dominus Keplerus Cæſareus Mathematicus, quando menſe octobri Anni 1613. Ratisbonæ in Comitijs unâ eramus. Quare ſi per jocum & per amicitiam inter nos tunc initum, illum compatrem horum quatuor ſiderum ſalutavero, haud male fecero.

Verum uti hęc nomina omnia à me ſunt liberè conficta, ita etiam cuique liberum eſto, ea vel repudiare vel acceptare.

Tantum de hac primâ libelli hujus parte, ſequitur nunc ſecunda.

SE-

Fig. 5.5 (continued)

EPOCHÆ

QUATUOR PLANETA-
RUM JOVIALIUM IN ANNIS
COMPLETIS.

	Primi			*Secundi*			*Tertij*			*Quarti*		
	sig.	*gr.*	*m.*	*sig.*	*gr.*	*m.*	*sig.*	*gr.*	*m.*	*sig.*	*gr.*	*m.*
1608	10	20	35	7	22	20	1	26	13	7	3	13
1609	1	17	40	4	3	11	1	8	40	4	15	0
1610	4	14	45	0	14	2	0	19	37	1	26	47
1611	7	11	50	8	24	53	0	0	34	11	8	34
1612	5	2	20	8	17	1	1	1	45	9	11	50
1613	7	29	25	4	27	52	0	12	42	6	23	37
1614	10	26	30	1	8	43	11	23	38	4	5	24
1615	1	23	35	9	19	34	11	4	35	1	17	11
1616	11	14	5	9	11	42	0	5	47	11	20	27
1617	2	11	10	5	22	33	11	16	44	9	2	14
1618	5	8	15	2	3	24	10	27	41	6	14	1
1619	8	5	20	10	14	15	10	8	38	3	25	48
1620	5	25	50	10	6	23	11	9	50	1	29	4
1621	8	22	55	6	17	14	10	20	47	11	10	51
1622	11	20	0	2	28	5	10	1	44	8	22	38
1623	2	17	5	11	8	56	9	12	41	6	4	25
1624	0	7	35	11	1	4	10	13	53	4	7	41
1625	3	4	40	7	11	55	9	24	50	1	19	28
1626	6	1	45	3	22	46	9	5	47	11	1	47
1627	8	28	50	0	3	37	8	16	44	8	13	2
1628	6	19	20	11	25	45	9	17	56	6	16	18
1629	9	16	25	8	6	36	8	28	53	3	28	5
1630	0	13	30	4	17	27	8	9	50	1	9	52

Fig. 5.6 The table of the Jovian satellites, from the *Mundus Iovialis* (Jay and Naomi Pasachoff Collection)

Original Copies of *Mundus Iovialis*

Over 30 copies of *Mundus Iovialis* are known as of this writing; there were two editions, both in 1614; they are listed at the *Marius Portal.* Only the two copies of the 1612 almanac (*Prognosticon Astrologicum auf 1612* in the Staatsarchiv Nürnberg (State Archive) and in the Stadtbibliothek Nürnberg (City Library) are known.

Conclusion

Marius deserves much better recognition for his early seventeenth-century telescope work, including his independent discovery of the four major moons of Jupiter in addition to his work on the phases Venus, sunspots, and the Andromeda Galaxy.

Perhaps these proceedings and articles published in this quadricentennary year, including my own in the *Journal for the History of Astronomy* (Pasachoff 2015), will add to the general recognition of Marius's wonderful discoveries.

Acknowledgments I am grateful to Pierre Leich of the Nuremberg Astronomical Society for his collaboration and for various information that he has supplied. I thank Seth Fagen for convincing me to buy my first-edition Marius (1614). Prof. Andrew Ingersoll and the Planetary Sciences Department of Caltech have provided for hospitality and visitor status. Wayne Hammond of the Chapin Library of Williams College has continually provided assistance with my collection.

References

Bosscha, Johannes (1907): Simon Marius. Réhabilitation d'un astronome calomnié. *Archives Néerlandaises des Sciences Exactes et Naturelles*, Serie 2, vol. 12, pp. 258–307, 490–528

Gingerich, Owen; Van Helden, Albert (2003): From *Occhiale* to printed page: The making of Galileo's *Sidereus Nuncius. Journal for the History of Astronomy* 34, pp. 251–267

Gingerich, Owen; Van Helden, Albert (2011): How Galileo constructed the moons of Jupiter. *Journal for the History of Astronomy* 42, pp. 259–264

Marius, Simon (1614): *Mundus Iovialis Anno M.DC.IX. Detectus Ope Perspicilli Belgici.* Nürnberg: Johann Lauer

Meeus, Jean (1962): Galileo's first records of Jupiter's satellites. *Sky and Telescope*, vol. 24 (September), pp. 137–139

Meli, Domenico Bertoloni (1993): *Equivalence and Priority: Newton versus Leibniz*, Oxford: Clarendon Press

Pasachoff, Jay M. (2014): Messier, Copernicus, Flamsteed: The Société Astronomique de France (SAF) Rare-Book Collection in Paris. *Historical Astronomy Division, 223rd American Astronomical Society Meeting, National Harbor, MD*, 107.07

Pasachoff, Jay M. (2015): Simon Marius's *Mundus Iovialis*: 400th Anniversary in Galileo's Shadow. *Journal for the History of Astronomy*, vol. 46, no. 2 (May)

Pasachoff, Jay M.; Filippenko, Alex (2014, 2019): *The Cosmos: Astronomy in the New Millennium.* 4th ed., 5th ed., New York: Cambridge University Press (www.solarcorona.com)

Pasachoff, Jay M.; Leich, Pierre (2015): 400th Anniversary of Marius's Book with the First Image of an Astronomical Telescope and of Orbits of Jovian Moons. *Historical Astronomy Division, 225th American Astronomical Society Meeting*, Seattle, 215.05

Rosen, Edward (1947): *The Naming of the Telescope*. With a foreword by Harlow Shapley. New York: Schuhmann

Van Helden, Albert (1990): Mayr's *Mundus Iovialis* in German (book review). *Journal for the History of Astronomy* 21, p. 371f

Van Helden, Albert (1994): Naming the Satellites of Jupiter and Saturn. *HAD News*, The Newsletter of the Historical Astronomy Division of the American Astronomical Society, no. 32, August 1994 (http://had.aas.org/hadnews/HADN32.pdf)

Chapter 6
Simon Marius's Reports on the Comets of 1596 and 1618, in the Context of the Comet Research of His Times

Jürgen Hamel

Abstract In 1596, 1607, and 1618, several comets could be observed in Europe. They were widely commented upon in the contemporary literature, and numerous small tracts were published about these events. Marius himself published booklets about the comets of 1596 and 1618, the first as an "Alumnus" in Heilsbronn and the second as an experienced scholar. The differences in quality between the two publications are remarkable. In 1596, Marius had not yet developed an autonomous scientific view on the comets, and so his small publication fluctuates between two positions: astrological interpretation on the one hand and the increasingly accepted insights on the celestial nature of comets on the other. The comet of 1596 provided strong new arguments for the latter standpoint.

The difficulty concerning the nature of comets will first be briefly addressed. According to Aristotle's physics, comets were not heavenly objects but were assigned to meteorology as terrestrial atmospheric phenomena. This classification is mainly due to the normal nature of the heavens, which were characterized as immutable, eternal, and possessing mute circular motion—characteristics that gave the skies their divine character. Comets didn't fit this scheme at all. They appear suddenly, change their shape often daily, and then disappear again. Their tails appear miraculously in the sky. Such changes couldn't happen in the immutable celestial spheres but only in the terrestrial atmosphere (Fig. 6.1).

In accordance with the biblical creation myth, celestial objects were put in the sky by God to give "Zeichen und Zeiten"[1] to mankind; the sky is God's scripture. The observation of the sky and the signs that appeared there was a duty, an appreciation of God as the creator and sovereign (Marius 1596, sig. A2^{r}):

[1]"[...] signs and seasons"

J. Hamel (✉)
Archenhold-Observatory, Bartelshagen, Berlin, Germany

H. Gaab, P. Leich (eds.), *Simon Marius and His Research*, Historical & Cultural Astronomy, https://doi.org/10.1007/978-3-319-92621-6_6

Fig. 6.1 Comets are located in the upper layer of the atmosphere. Schöner 1531

> es redet Gott mit vns/nicht allein durch seine heilige Propheten/Apostel/vnd alle getreue Lehrer vnd Kirchendiener [. . .] Sondern auch durch andere natürliche Creaturen/als den Himmel/dahin er bißweilen/ehe grosse straffen herein fallen/Zornzeichen stellet/vnd zur Buß vermanet. Vnter welchen auch nicht für die geringsten gehalten werden die Cometen/deren niemals keiner ohne sonderliche grosse vnd schwere verenderung inn der Welt erschienen ist.[2]

Let us now turn to Marius's two booklets on comets that are dedicated to the comets of 1596 und 1618.[3] First it is notable that the text on the 1596 comet is his first ever publication: "dise mein geringe Arbeit vnd Erstlinge meines Studii,"[4] signed "Alumnus zu Hailsbronn" (Marius 1596, sig. A2^{v}). The second shows him as an experienced scholar, signed "Astronomus et Medicus." But first we turn to the 1596 text.

---

[2]God talks to us not only through His holy prophets, apostles, and all devoted teachers and ecclesiastical servants but also through other natural creations such as the sky on which He sometimes puts signs of His anger that remind us to do penance before the great punishment comes. Not the least among them are the comets. None of them ever appeared without great and severe changes in the world.

[3]Actually in 1618 three comets appeared, but they were often confused in temporary literature and counted as one.

[4]"[. . .] this my small work and first of my studies".

The Comet Script of 1596

Das erste Capitel/Von der Form vnd Gestallt dises Cometen[5] *(Marius 1596, sig. A3*^{v}*)*

Aristoteles teilt alle Kometen in comata, bei denen die Strahlen gleichermaßen in alle Richtungen geworfen werden, und barbata, die einen Bart oder Schwanz hinter sich werfen. Dieser ist ein barbata oder caudata, er wirft seine Strahlen nur nach einer Seite, "vnnd dieselbigen allezeit von der Sonnen abgewandt"[6] (Marius 1596, sig. A3^{v}).

Das ander Capitel/Von der Vrsach oder Constellatione, durch welche dieser Comet ist generirt vnd entzündt worden[7] *(Marius 1596, sig. A4*^{r}*–B2*^{r}*)*

Over four and a half pages, Marius discusses the cause for the emergence of the comet. This cause should be found in specific "schweren Aspecten oder Zusammenfügung"[8] of the planets. Marius thinks that the comet "inn den letzten 10. graden des Krebs entzündet worden"[9] (Marius 1596, sig. A4^{r}). "Daher hat ich einen argwohn geschöpfft/es würde die vrsach dises Cometen auch an disem ort verborgen ligen"[10] (Marius 1596, sig. A4^{v}) (Fig. 6.2).

Marius finds out now that the planets Jupiter, Mars, Venus, and Mercury met here in 1594. Furthermore there was also a solar eclipse on May 20, 1593, which exercised its influence on the position in question. In addition, Marius finds a similar constellation in 1564 that Cyprianus Leovitius recognized as "anzeigung künfftig erscheinender Cometen"[11] (Leovitius 1564). So Marius found a historic parallel for the conjunction and eclipse and assembles the terrible warlike events that can be attributed to both "Ich kann aber allhie nicht vmbgehen/die vergleichung der grossen vnd mercklichen effect, so sich auf diese beide Constellation haben zugetragen"[12] (Marius 1596, sig. B1^{r}). It should be added: "Es ist aber kunthbar auß der

[5]The first chapter/On the form and shape of this comet.

[6]Aristoteles classifies all comets as either *comate*, which spread their rays equally to all directions, or barbata, which throw a beard or tail behind them. This is a *barbata* or *caudata*, which throws its rays only to one side and always turn away from the sun.

[7]The other chapter/On the cause or constellation that generated and ignited this comet.

[8]Severe aspects or conjunctions.

[9]Was ignited in the last 10 degrees of Cancer.

[10]Though this raised my suspicion that the cause of this comet could be hidden in that place.

[11]A sign of future comets.

[12]I cannot avoid comparing the huge and remarkable effects that happened under these two constellations.

Fig. 6.2 Front page of script about the comet of 1596 by Simon Marius. Herzog August Bibliothek Wolfenbüttel: 42.7 Astron. (10)

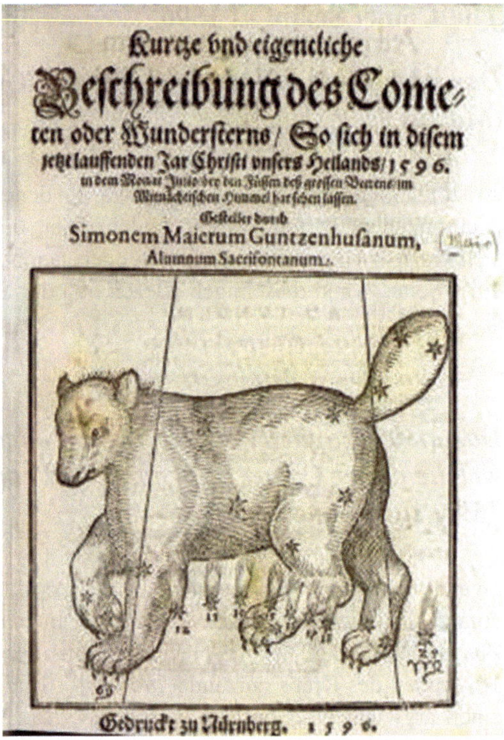

Astrologiae, daß die conjunctiones zwischen Merkur und Mars, unter andern Bubenstücken auch verrätherey bedeuten"[13] (Marius 1596, sig. B1^{v}). So the framework for this comet as a symbol of horror has been put initially into place, a subject to which Marius returns in detail in Chap. 7.

Das dritte Capitel/Von der Farb vnd Grösse des Cometen[14] *(Marius 1596, sig. B2^{r}–B2^{v})*

The comet's nature in comparison to the planets can be recognized through its color. According to Marius's observations, this comet appeared to be "doch etwas bleich vnd weißfarb geschienen/vnd hat sich mit Farb vnd Größ durchaus dem Mercurio verglichen"[15] (Marius 1596, sig. B2^{r}). Because of the smallness of the comet's head,

[13] It is known from astrology that the conjunctions between Mercury and Mars among other pranks also mean treachery.

[14] The third chapter/About color and size of the comet.

[15] Somewhat pale and white and could in color and size be compared to Mercury.

its size couldn't be measured, as the length of the tail varied considerably over time, in contrast to the usual habits of comets.

Das vierdte Capitel/Von der Natur des Cometen[16] *(Marius 1596, sig. B2*^{v}*–B3*^{v}*)*

In the fourth chapter, Marius justifies the mercurial nature of the comet, a subject that won't be discussed further here. However, this chapter offers other interesting information. First, it indicates that Marius had "seine distantiam von den nechsten Stern/durch ein langen radium Astronomicum, mit fleiß genommen"[17] (Marius 1596, sig. B3^{r}), but did not deem it necessary to reproduce the data here. Did he not consider the data worth publishing because he was insufficiently trained in using his equipment or because of its lack of quality? The observations were done with a Jacob's staff, a standard instrument to measure distances between celestial objects but by no means to be counted as one of the newer astronomical measuring instruments. Marius records his first observation on July 12, the last on July 25.

Also interesting is the information that Marius found Mercury's movement diverging from the advanced calculations in the Ephemerides of Stadius, which can be traced back to the fact "daß der Planeten lauf nicht allein in longitudinem, sondern auch in latitudinem, noch nicht gnugsam ergründet ist wie die tägliche erfahrung bezeuget."[18] This applies especially to Mercury, which is difficult to observe (Marius 1596, sig. B3^{r}). But also the position of Venus diverged on July 2 from the one in the Ephemerides by about 2 degrees to the south.[19]

---

[16]The fourth chapter/On the nature of the comet.

[17]The distance is measured to the next star with a long astronomic radium.

[18]That the movement of the planets, not only in longitude but also in latitude, are not yet sufficiently established as daily experience shows.

[19]For the questioned time, three editions of the Ephemerides by Johannes Stadius exist:

- *Ephemerides novae, auctae et expurgatae ab anno 1554. usque ad annum 1600.* Köln: Arnold Birckmanns Erben 1570
- *Ephemerides secundum Antvverpiae longitudinem, ab anno 1554. usque ad annum 1606.* Köln: Arnold Birckmann 1581
- *Ephemerides secundum Antvverpiae longitudinem ex tabulis Prutenicis supputatae ab anno 1583. usque ad annum 1606. & ad S.D.N. Gregorij XIII. anni reformationem accuratissime accommodatae.* Lyon: Philippus Tinghius 1585.

Das fünffte Capitel/Von dem Lauff vnd Bewegung/auff vnd nidergang des Cometen[20] *(Marius 1596, sig. B3^{v}–B4^{r})*

The movement of the comet was always clockwise. During the observation period, it traversed about 45° of the heavens.

Das sechste Capitel/Ob dieser Comet vnter oder ob der Sphaerae Lunae sey gewesen[21] *(Marius 1596, sig. B4^{r}–C1^{r})*

This topic is, of course, especially interesting, as the opinion of the Aristotelian physicists differed from those who trusted the new astronomical observational methods, such as Tycho Brahe, Wilhelm IV, Christoph Rothmann, or Michael Mästlin. Unfortunately Marius remains vague on this subject; his observations are not adequate to make a decision about the parallax of the comet, according to his own judgment.

Das sibende Capitel/Von der Bedeutung dises Cometen[22] *(Marius 1596, sig. C1^{r}–C4^{r})*

This subject has already been set up in the second chapter. With a total of 11½ printed pages, the 2 chapters set the major focus of the text and reveal the astrological orientation. The chapter, however, cannot be viewed as one of the many comet sermons in circulation. Marius starts the chapter with the words (Marius 1596, sig. C1$^{r\text{-}v}$.):

> Ob die Cometen künfftige grosse vnnd gefährliche zufäll in der Welt anzeigen vnd erregen/ bedarff keiner beweisung oder vil vergebens argumentieren/die allgemeine erfahrung bezeugt solches all zu klar vnd offenbarlich. Denn man lese nur inn den Historiis, so wirdt man finden/das niemals kein Comet geleuchtet/auff welche nit in concavo sphaerae Lunae, das ist/in Lufft/Meer/Erd/an Menschen vnd Viehe/vnd allen Gewächsen auff Erden/ sonderliche grosse vnd schädliche verenderung haben zugetragen. Es folgen gemeiniglich auff die Cometen/wie die gelehrten Meteorologi schreiben/vnd die erfahrung bekräfftiget/ grosse Hitz vnnd Dürre/Hungersnoth/Theurung/Pestilentz/Krieg/Blutvergiessen/Auffrhur/ Tödtliche abgang hoher Potentaten grosser Königen/Fürsten vnd Herrn/verheerung der Länder vnd Königreich/verenderung der Policey/Gesetz vnnd Statuten/grausame vngestümme Wind/Erdbidem/vnfruchtbarkeit der Erden/ergiessung der Wasser vnd ander vnglück. [...] So will ich jetzund kürtzlich vnd einfeltig/auß bißher erzehlten vmbständen/ die bedeutung dises vnserigen Cometen erklären/doch so gut ich es gelernt biß Gott der

[20] The fifth chapter/On the path, motion, and fall of the comet.

[21] The sixth chapter/Whether the comet was below or about the lunar sphere.

[22] The seventh chapter/On the meaning of this comet.

> Allmächtig durch ordentliche mittel mir andere vnd bessere Gelegenheit/dises herrliche Studium recht für die hand zunemen/verschaffet wirdt.[23]

While these comet-astrological derivations cannot be followed in detail here, a few examples from Marius's argumentation will suffice. For example, everything in the world can only create something according to its nature. So grapes cannot be found on thorns, figs cannot be found on thistles, and an eagle cannot breed a pigeon.

As the comet

- Is generated by the conjunction of planets in the constellation Leo
- Appeared in the constellation Cancer
- Is associated with a Mercury-ruled solar eclipse from 1593

It has to be associated with the heat of Leo (Marius 1596, sig. C2^{r}).

> So halte ich nun darfür/daß dieser Comet/ratione caussae efficientis, fürnemlich bedeute grosse Hitze und Dürre/das kein Regen erspriessen soll/vnd vnfruchtbarkeit der Erden/vnnd letztlich solche Kranckheiten/die von übriger hitz vnd truckenheit jhren vrsprung nehmen.[24]

Mercury would give the comet (Marius 1596, sig. C2^{v}).

> grosse trübseligkeit/angst vnd not villeicht gar den todt/allen Mercurialischen Menschen/als Gelehrten/die mit freyen Künsten vmbgehen/Kauffleuten/Schreibern/vnd dergleichen. [...] ein Comet von der Natur Mercurii bedeut grosse starcke vngestümme Wind/Auffrhur vnd Secten.[25]

Finally, Cancer signifies for the comet (Marius 1596, sig. C2^{v}).

> vil Krieg/Mörd/vnd Todtschläg/Rauberey/groß vngestümm/geschwind ergiessung der Wasser/hunger vnnd pestilenz/sterben der Fisch/vnnd vil gewürm/geschmeiß vnd vnzifer/ daß die frücht auff dem feld vnd bäumen verderbt.[26]

---

[23]Whether comets signify and cause future large and dangerous hazards in the world needs no proof or much unnecessary discussion, as general experience clearly manifests and testifies this. If one only reads the histories, one will find that a comet has never shone of which in the concave lunar sphere, in the air, sea, and earth has not caused massive and harmful changes affecting humans and animals and whatever grows on earth. Great heat and drought, famine, inflation, pestilence, war, bloodshed, turmoil; the deaths of potentates, great kings, princes and lords; devastation of lands and kingdoms; changes of policies, laws and statutes; fierce stormy winds, earthquakes, infertility of the soil, flooding, and other disasters usually follow comets, as the scholarly meteorologists write and as experience confirms. So I will now briefly and simply explain from heretofore-narrated circumstances the significance of our comet so well as I have learnt it, until God the Almighty through orderly means provides me with other and better opportunities to improve on these wonderful studies.

[24]So I consider that this comet's rational efficient cause in particular signifies great heat and drought, that no rain would come, leading to infertility of the soil as well as diseases that are caused by heat and drought.

[25]Depression, fear, and misery, maybe even death to all mercurial people, including scholars of the liberal arts, merchants, writers and such [...] a comet of mercurial nature means strong, fierce, stormy wind; turmoil; and sects.

[26][...] much war/murder/and manslaughter/robbery/massive impetuous/swift flooding/hunger and pestilence/death of fish/lots of worms/insects and vermin/spoilage of the fruit of the fields and the trees.

A further factor of influence was the astrological character of the constellations the comet passed through, “wenn ein Comet in constellatione navis erschiene, so bedeutet er Schiffbruch vnnd grossen vntergang zu Wasser.”[27] This comet passed the Great Bear, which “bedeut derwegen vntergang vnd sterben des grossen Viehes/ als Pferd/Ochsen/Kuh/Beern/Hirschen vnd dergleichen”[28] (Marius 1596, sig. C3^{r}).

Predictions are always problematic, as only God knows the future. Therefore, a theological conclusion necessarily belongs to a tract about comets. “Der allmächtige ewige Gott/wölle sich vnser aller erbarmen vns bekehren/so werden wir bekehrt werden”[29] (Marius 1596, sig. C4^{v}).

The Comet Booklet of 1619

We now turn our attention to the contrasts in Marius’s comet booklet of 1619. It is substantially different from the work of the Heilsbronn Alumnus; Marius acknowledges this in his writings very clearly. In 1596 he had been very restrained with his own opinion about the nature of comets and seemed not to have been able to produce reliable observations, but everything looks very different now. Already at the beginning of his report Marius dissociates himself from the physics of comets based on that of Aristotle (Marius 1619, sig. A4^{r}) (Fig. 6.3).

> Zuvor aber, will ich meine gedancken einfältig entdecken/von der vrsach der erscheinung deß Cometen/Andere Autores alle/so ich gelesen/vnnd jhre gedanken von diesem Cometen vnd seiner vrsach an tag geben haben/die folgen sämptlich der falschen opinion Aristotelis, auch anderer meinung die noch in nächsten seculis gelebt haben.[30]

Of course Tycho Brahe and Johannes Kepler would later be named as representatives of a new physics of comets. But a number of critics of Aristotelian physics already existed in 1618, who can also be found among the authors of small tracts about comets.

Let us look first at the words that characterize Marius’s own view based on the criticisms of the Aristotelian concept of comets, which he accepts and has absorbed into his own development (Marius 1619, sig. A4^{r}).

> wie ich denn selbsten der meinung vor 20. oder mehr Jahren gewesen bin. Nemblich das gewisse vnd vornehme Constellationes darzu vrsach geben. Aber damals hatte ich noch

---

[27] If a comet appears in a water constellation, it portends shipwreck and great destruction at sea.

[28] [. . .] that means the death of large animals such as horses, oxen, cows, bears, deer, and the like.

[29] The almighty God may have mercy on us all and convert us to Him so we will be converted to Him.

[30] First, I will show my thoughts about the cause and appearance of comets. Other authors I read who published their opinion about this comet and its cause all follow the wrong opinion of Aristotle, as well as other opinions of those who lived in the next age.

Astronomische vnd Astrologische beschreibung deß Cometen so im November vnd December vorigen 1618. Jahrs ist gesehen worden/ Genommen vnd Gestelt auß eygnen Observationibus dabey auch andere sachen kurtz eingemischet werden.

Durch Simon Marium Guntzenhusanum, Fürstlichen Brandenburgischen bestelten Mathematicum vnnd Medicum.

Gedruckt vnnd verlegt zu Nürnberg / bey Johann Lauern. Im Jahr Jesu Christi M. DC. XIX.

Fig. 6.3 Front page of the comet script of 1619 by Simon Marius. ZB Zürich: Rar 4261:1

> nicht erfahren/was ich Gott lob jetzo weiß/vnd mir glaubwürdig vorkompt/wiewol ich observationes veterum nicht allerdings verwerffe.[31]

This passage is very remarkable. Marius could perfectly well have skipped over his 25-year-old text silently—instead he refers self-critically to his earlier errors.

Marius observed the comet on November 24, 27, 28, and 30 and again on December 2, 8, 10, 11, 12, 13, 18, and 19. On the final day, the nucleus of the comet was already so weak that determination of its position was no longer possible. Contrary to 1596, he reports these observations in great detail in his comet tract.

[31] As I believed 20 or more years ago, namely, that certain constellations are the cause. But then I had not yet experienced what I know now, thanks to God, know and what seems credible to me. Though I do not reject earlier observations completely.

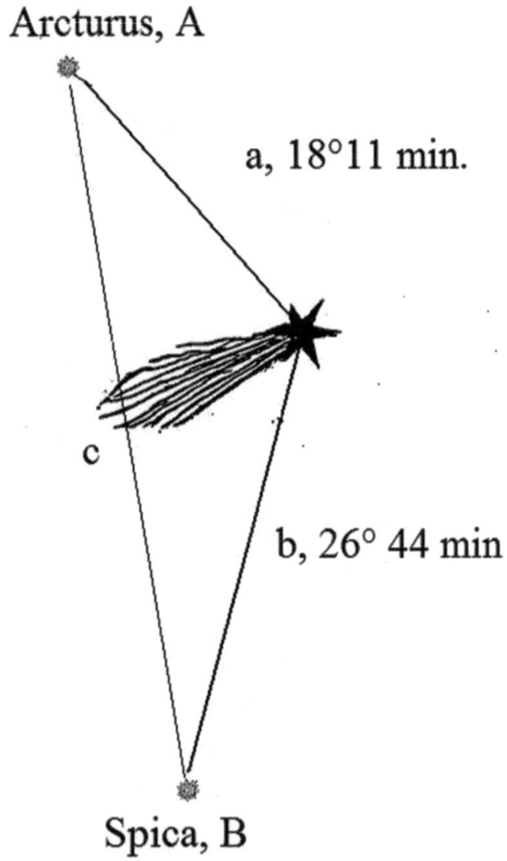

Fig. 6.4 Scheme of position determination of a comet by distance to neighboring stars

The first observation of the "Hauptsternlein,"[32]— that is the comet's head—Marius notes as follows: 6 hours, 5 minutes, 30 seconds in the morning; distance from the comet to Arcturus 18 degrees, 11 minutes, and to Spica 26 degrees, 44 minutes. The observations took place with Marius's measuring the comet's distance to neighboring bright stars with a Jacob's staff.

This is illustrated in Image 4: Marius creates a triangle ABC with the two stars whose positions are known and the comet. The positions of the stars A and B and the distances a and b (measured) and c (calculated) are known. So the comet's position is calculable. This approach corresponds to the usual, contemporary method of observation. Thus the precision of the determination of the position of a comet (or planet) depends not only on the accuracy of the measurement of the distance of the celestial object from the stars but also on the accuracy of the position of the stars (Fig. 6.4).

For the latter relationship, an interesting possibility is available to Marius. In his *Prognosticon auf 1618* he writes that he had brought "etlicher Fixstern veras

[32]"Small main star".

distantias von Tychone auß Prag"[33] (*Prog. 1618*, sig. 2^{v}), that means from his visit to Brahe in 1601. What sort of material would this have been? It is known that Brahe had made a handwritten register of fixed stars and had sent it to different scholars and sovereigns.[34] So it is possible that either Marius had received a copy or at least had seen these very carefully observed fixed star positions and was able to make a copy for himself.

Marius tracked the comet every night as far as the weather allowed "mit einem radio, dessen gebrauch ich wol weiß/auch selbsten sampt der außtheilung gemacht"[35] (Marius 1619, sig. A4^{r}), in other words, measuring the angular distance of the comet to neighboring stars. The instrument is, as in 1596, a Jacob's staff. In the meantime, better instruments had been designed, especially the astronomical sextant, which Tycho Brahe had demonstrated and which Christoph Rothmann together with Jost Bürgi had already utilized in Kassel. From the *Prognosticon auf 1618*, we know that Marius made test observations with a sextant together with his student Paulus Boyn in 1603 (*Prog. 1618*, sig. 2^{v}). It had been manufactured at the expense of his student, since Marius, despite his position as Court astronomer in Ansbach, had no financial resources to buy his own sextant. So he could only use the Jacob's staff to observe the comet of 1618.

Addressing the obvious question, one should not neglect the telescope here, as Marius was one of its earliest astronomical users. The early small instruments would have been ineffective for the fine structures of the comet's tail, should somebody have had the idea to examine a comet with this instrument. This realization was reserved for future scholars.

Marius is skeptical about his observations (Marius 1619, sig. B2^{v}):

> Diß seyn nun meine observationes, so ich an diesen Cometen hab verrichten können/wiewol ich particulariores hette/aber hierzu nit so hoch von nöthen/es werden ohne zweiffel andere Mathematici, die alicujus valoris seyn/sonderlich Herr Johann Kepler/Keyserlicher Mathematicus, als mein guter freund/auch jhr auffsehen auff diesen Cometen gehabt haben/vnd wo meine observationes nicht in minuto zutreffen/welches ich rationes instrumenti nicht versprechen kann/so will ich die correction oder mediation gern leiden.[36]

It was not possible for him to determine accurately the altitude of the comet (Marius 1619, sig. B2^{v}).

---

[33][...] the true distances of a number of fixed stars from Tycho in Prague.

[34]Brahe's presentation practice is not quite clear. It is, for example, known that Graf Simon IV. von der Lippe and Duke Ulrich von Mecklenburg received a handwritten register of fixed stars together with his *Astronomiae Instauratae Mechanica* (Wandsbek 1598), "Tychonis Brahe stellarum octavi orbis inerrantium accurata restitutio" (für Simon VI., Lippische Landesbibliothek Detmold, Mscr 21.1); cf. Bischoff 2014, pp. 109–141.

[35][...] with a Jacob staff, which I know how to use and I made the distributions myself.

[36]These are my observations of this comet that I was able to carry out, although I had the details they were not as accurate as necessary. Doubtless other mathematicians who are worthy, especially Herr Johann Kepler, Imperial Mathematicus, and my good friend, also observed this comet, and where my observations are not accurate enough, which I can't promise because of my instruments, I will accept correction or mediation.

> denn man der distants deß Monns oder Sonnen noch nicht vergwisset ist [. . .] Doch halte ich darfür er sey/seinem anfänglichen trieb nach nicht weit/weder höher oder niderer von der Sonnen gewesen/von da an höher gestigen/vnd wegen subtiler materia/daselben endlich ist dissipirt worden.[37]

Its brightness or "größ"[38] never reached the level of a first magnitude star. Its tail reached a length of about 45°. It was always directed away from the sun. The latter was described in the literature by Peter Apian in 1540 and became an important observation for the examination of the physical nature of comets and their tails.

Back to size, as the comet had been approximately at the altitude of the sun and had not even reached the size a first magnitude star, it must have had a quite small diameter, "aber der Schweiff/ist anfangs viel tausent Meilen lang gewesen"[39] (Marius 1619, sig. B3^r).

Now Marius examines the question how the comet arose and what its nature was. He doesn't try to conceal his rejection of Aristotle but starts Chap. 5, which is dedicated to this question, with the introduction (Marius 1619, sig. B3^v):

> Jetzo muß ich mich in das Feld vnnd wider die Feind begeben/dieweil ich ein besondere meynung von der materia dieses Cometen hab/damit ohn zweiffel die Aristotelici & Accademici jetziger zeit nicht allerdings werden zu friden sein.[40]

Of all the discussions and differing opinions he would "Allein Aristotelis meynung/ [. . .] in disem fall nicht hören/als wenn nicht was bessers könnte durch Gottes Gnad auff die Bahn gebracht werden."[41]

Initially no one could "läugnen/so diesen Cometen gesehen/daß es nicht warhafftig eine materia gewesen sey"[42] (Marius 1619, sig. B3^v). This sentence is important because at the end of the sixteenth century, a theory arose that considers the comet's head to be a transparent body that only simulated the appearance of a tail by collecting the sun's rays like a lens (Fig. 6.5). According to this theory, the tail wouldn't be a real object but only an optical phenomenon, the result of sunlight passing through the transparent body of the comet. This assertion can be found, for example, in the work of Helisaeus Röslin after the appearance of the 1596 comet. Röslin concluded that comets were not atmospheric phenomena—this was the new theory (Röslin 1597, sig. 7),

---

[37]As one is not yet sure about the distance of the moon or the sun [. . .] But I think the comet was at its beginning not far, neither higher or lower than the sun, then rose higher and because of subtle material was finally dissipated.

[38]Size.

[39][. . .] but at the beginning the tail was many thousand miles long.

[40]Now I have to enter the field and confront the enemy because I have a special opinion about this comet's material. Doubtless the current Aristotelians and Academics won't be satisfied.

[41][. . .] not hear the opinion of Aristotle in this case, as if something better couldn't be found with God's grace.

[42][. . .] no one who saw this could deny that it was of real material.

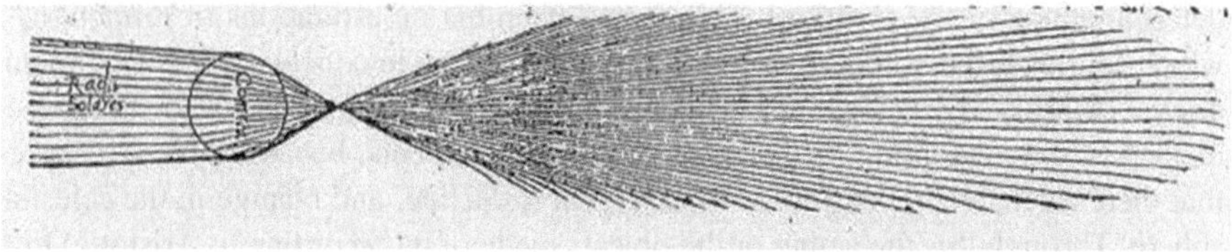

Fig. 6.5 The optical theory of comets' tails of Helisaeus Röslin (compare the text)

Sonder vil mehr ein aetherisch rundes dinnes durchleuchtendes Corpus, welches so es den schein von der Sonnen empfangen, die stralen davon durchgehn laßt, und also den schein von sich gibt mit langen strämen, als wenn es für sich selbs ein langes Corpus were.[43]

Initially, this was not a very unreasonable explanation, although there were difficulties with the linear propagation of light. A comet's tail is not directed in a straight line away from the sun but curved in multiple directions, even often combined with another adjacent or opposing tail. Christoph Rothmann in Kassel was not only the first to observe this phenomenon but also made it fruitful for the theory of comet tails. He stated that comet tails show no explicit relationship to the sun or the planets. So they can't be an optical phenomenon but must have an independent substantial existence.

Thus the question of where the material came from remained. Without going into the details of Marius's thought processes, it is clear that he was swayed by the biblical creation myth, where God created in His 7 days of labor two kinds of waters—the "schlecht Wasser"[44] below and the extremely fine, easy movable mercurial water above the firmament. There were two basic views, namely, that God created the waters completely from scratch or formed them from already created objects. This controversial issue seems, to Marius, hardly conclusively decidable. But he would not believe that the comets "von der Welt anfang gewesen"[45] (Marius 1619, sig. C1^{v}), and so they could only have been created later—by the order of God, without whom nothing can occur either on earth or in the heavens. Assuming that comets consisted of material risen up from the earth, the question emerges how such huge masses of material could rise up without being noticed on the earth. Secondly, indeed terrestrial vapors could not rise up from the lowly elementary regions to the celestial spheres because those were considered to be solid spheres. Under these preconditions, comets could only appear in the highest regions of the atmosphere (Marius 1619, sig. C2^{r}). It is remarkable that Marius mentions, at this point, the absolutely revolutionary idea of Christoph Rothmann that therefore there were no substantial differences between celestial and terrestrial air, so there was no material difference between the celestial and terrestrial regions, and so no celestial material was distinguished through its immutability. Rothmann derived this from exact

[43]Rather a round thin ethereal illuminated body, which receives its brightness from the sun and lets the sun's rays pass through it in long streams as if it were a long body.

[44][...] bad water.

[45][...] came from the beginning of the world.

determinations of the positions of stars and from his investigations of refraction,[46] which Marius refers to but does not draw any of the same conclusions as Rothmann does[47] (Marius 1619, sig. C2^{r}). Of course, these conclusions are also the consequence of the perception of the celestial nature of comets, because it demonstrates that there are indeed occurrences of emergence, decline, and change in the celestial sphere. Through this the aether or the objects made of it (according to Aristotle) lose their significant characteristics, namely, immutability and divinity. And if, according to Rothmann, the cosmos is considered to be heliocentric, the question arises of where the lower levels of the elements and the upper level of the ethereal objects should be considered to exist. This was completely clear in geocentric cosmology by the directional determinations of up and down but was dropped in a heliocentric cosmology.

Marius did not take the step to heliocentrism, I "glaube demnach/das die Erde [...] in centro universi stehe"[48] (Marius 1619, sig. B4^{v}), which for him leads to the Tychonic world system.[49]

After this digression Marius goes back to the question about the nature—or rather the origin—of cometary material. Marius is explicitly indecisive on this question. It seems to him "nicht unglaublich," "das Gott die exhalationes oder Dämpffe der Erden zu einem Cometen gebraucht."[50] Through the power of the sun, "exhalationes oder Dämpffe der Erden"[51] together with the "aetherischen Substantiâ durch Krafft der Sonnen/zu einem Cometen werden"[52] (Marius 1619, sig. C2^{v}).

> Also schliesse ich das Materia Cometica nicht allein in aethereâ regione, das ist/in Firmamento anzureffen/Sondern per communicationem aeris summi & Firmamenti, durch Gottes willen/vnnd der Sonnen anziehende Krafft/ein Cometische Materia entstehe. Man solle aber wissen/wie ich zuvor auch gedacht/das es nicht blosses Wasser sey/sondern ein solche feuchte/die jhren Subtilen Schweffel bey sich hat/vnnd derentwegen in der subtilheit der Aetherischen region von der Sonnen entzündet wirdt/vnnd eine formb doch vnvollkommen gibt/wie in Aethereâ regione die corpora sein/nemblich eines Sterns.[53]

[46]On Rothmann, see Hamel 1998, Revised 2nd ed. 2002 (*Acta Historica Astronomiae*; 2) and Granada, Hamel, Mackensen 2003, cap. 15–21.

[47]Marius refers here to *Prognosticon auf 1618*, where he in fact develops thoughts on refraction. However, they are only brief and, as already said, without consequences for cosmological inquiries (*Prog. 1618*, sig. A2^{v}–A3^{r}).

[48][...] believe that the earth is the center of the universe.

[49]See Pierre Leich's article in this volume: "In the Turmoil of the Early 17th-century Cosmology Debate – Simon Marius as a Supporter of the Tychonic System."

[50]Not unbelievable that God needs the exhalations or vapors of the earth for a comet.

[51]Terrestrial damps.

[52][...] the exhalations or vapors of the earth together with the ethereous materials through the power of the sun become a comet.

[53]So I conclude that the material of the comets comes not only from the ethereous region that is the firmament. Rather that a material comet comes into being through God's will and the attractive power of the sun by communication between highest atmosphere and the firmament. One should, however, know that I believed previously that it is not simple water but such moisture that contains a subtle brimstone, which therefore in the fineness of the ethereous region is ignited by the sun and becomes an imperfect body, unlike the perfect bodies in the ethereous region, namely, the stars.

Marius now draws a very interesting connection between comets and sunspots. He had seen, so he writes, during his observations of the sun since 1611, "etlichmal maculas caudatas in disco Solis"[54] that looked like comets. "Wie/wenn solche maculae ein refrigerium weren/summi caloris solis, vnnd hernacher per adunationem, vel potius conglobationem in einem Cometen würden"[55] (Marius 1619, sig. C3r).

As well as comets, Marius also observes the novae, whose material had to be of a different nature because, as opposed to the comets, they also appeared to be new but lasted longer and remained in one position in the heavens. Because of the "imperfectionis materiae,"[56] cometary material cannot last long, but it is "fortgetrieben"[57] by the power and the movement of the sun. So the material of the "Neuen Stern viel aus einer subtilern vnd perfectern materien"[58] is made by God's will "vnnd ihren stand in supremo aethere bey den Fixstern haben"[59] (Marius 1619, sig. C3r). He lists the appearances of 1572, 1601, and 1604, the Tychonic and Keplerian Novae, and the variable star in the constellation Swan.

Of course, Marius cannot totally ignore astrological prophecies. He had seen that many authors had given their predictions, all copied from old books. Because this "doch das aller vngewisest ist/ja nur auff einer muthmassung bestehet,"[60] Marius had wanted "in diesem Tractätlein mehr de Systemate mundano vnd materia Cometarum vnd Stellarvm novarum discutiren, als grosse Prophezeyhung an Tag geben wollen"[61] (Marius 1619, sig. C4r). In any case, he says, most of the horrible effects of this comet had already occurred; one needed think only of the conquest of Pilsen by Graf Ernst of Mansfeld. He explicitly dismisses all other astrological derivations: "Was nun vermuthlichen andere Bedeutungen anlangen thut [...] das lasse ich jetzund ganz vnnd gar fahren"[62] (Marius 1619, sig. D1v).

In the comet text of 1619, Marius mentions specifically the correspondence published in 1596 between Tycho Brahe, Wilhelm IV of Hessen, and Christoph Rothmann; he also mentions indirectly Rothmann's text *Scripta de cometa* from 1619 and Johannes Kepler's text about the comet of 1607 (Kepler 1608).

[54] Often sunspots on the sun's disc.

[55] What if such sunspots were cold areas in the high temperatures of the sun and later by combination or, rather, balling up, become a comet?

[56] Imperfect matter.

[57] Shied away.

[58] New stars many of a much more subtle and perfect material.

[59] [...] and have their place in the supreme aether by the fixed stars.

[60] [...] they are, all uncertain and consist only of speculations.

[61] [...] in this booklet to discuss more the system of the cosmos and the material of the comets and new stars rather than to pronounce great prophecies.

[62] [...] as far as other supposed meanings are concerned, I let them go completely, now and forever.

Summary

In summing up Marius's two comet booklets, we can conclude that they are of great interest for the author's development, as they originate from two very different periods of his life and work.

The text of 1619 considers without limitations the theory of the cosmic nature of comets. Marius is seeking exact determinations of the comet's position from his own observations, but it is striking that he doesn't touch either on the question of parallax or on the question of in which celestial sphere the comet must have been.

The theological dressing follows the necessary standards without setting textual priorities. In connection with this are the short thoughts about comets as signs of horror that Marius couldn't ignore, as they are derived from the Bible, but he does not explicate them.

Finally, it is worth emphasizing Marius's very remarkable idea that scientific progress would proceed slowly and only with the cooperation of many. And this he writes remembering, with certainty, earlier contrary experience, where it was not so (Marius 1599, sig. C3^{r}).

> Ich thue das meinige/andere thun auch das jhrige/nach deme jhnen Gott gnad verliehen hat/man muß der sachen ein anfang machen/vnnd einer dem anderen ohne verlesterung die Hand bieten/biß man endtlich was gewiesses schliessen kann.[63]

Credit I am grateful to Hans Gaab for inspiring discussions about and information on the topic.

References

Bischoff, Michael (2014): "… auch sonst von allerlei Sachen Bücher" – *Simon VI. zur Lippe und die Wissenschaften*. In: *Im Dienst des Kaisers*. Graf Simon VI. zur Lippe (1554–1613). Weserrenaissance-Museum Schloss Brake. Lemgo 2014

Granada, Miguel A.; Hamel, Jürgen; von Mackensen, Ludolf (eds.) (2003): Christoph Rothmanns *Handbuch der Astronomie* von 1589. Annotated edition of the manuscript Christoph Rothmanns "Observationum stellarum fixarum liber primus", Kassel 1589. Thun; Frankfurt a. M.: Harri Deutsch 2003 (*Acta Historica Astronomiae*; 19)

Hamel, Jürgen (1998): Die astronomischen Forschungen in Kassel unter Wilhelm IV. Mit einer Teiledition der deutschen Übersetzung des Hauptwerkes von Copernicus um 1586. Thun; Frankfurt a.M.: Harri Deutsch 1998, Revised 2nd ed. 2002 (*Acta Historica Astronomiae*; 2)

Kepler, Johannes (1608): *Ausführlicher Bericht von dem 1607 erschienenen Haarstern*. Halle 1608

Leovitius, Cyprianus (1564): *De coniunctionibus magnis insignioribus superiorum planetarum, Solis defectionibus, & cometis*. Lauingen: Emanuel Saltzer 1564

Marius, Simon (1596): *Kurtze und eigentliche Beschreibung des Cometen oder Wundersterns*. Nürnberg: Paul Kauffmann 1596

Marius, Simon (1599): *Tabulae Directionum Novæ*. Nürnberg: Christoph Lochner 1599

[63] I do my part; others do their parts; given the grace of God, one must start with it and should help the other without any hate, until one can conclude something with more certainty.

Marius, Simon (1619): *Astronomische vnd Astrologische beschreibung deß Cometen [...] 1618.* Nürnberg: Johann Lauer 1619

Röslin, Helisaeus (1597): *Tractatus meteorastrologiphysicus. Das ist / Auß richtigem lauff der Cometen / zusammenleuchtung der Planeten [...] eine Weissagung*. Straßburg: Bernh. Jobins Erben, 1597

Rothmann, Christoph (1619): *Scripta de cometa.* In: Willebrord Snellius, *Descriptio cometae qui anno 1618 mense Novembri primum effulsit*. Leiden: Elzevier 1619

Schöner, Johannes (1531): *Coniectur odder abnemliche außlegung vber den Cometen so jm Augstmonat / des M.D.XXXj. jars erschinen ist.* Nürnberg: Friedrich Peypus, o.J. [1531]

Chapter 7
Sunspot Observations by Simon Marius

Dagmar L. Neuhäuser and Ralph Neuhäuser

We present texts by Simon Marius, where he discusses his sunspot observations, ranging from 1611 to after 1618. While he had observed many sunspots since August 1611 (shown to him by Ahasverus Schmidtner), being one of the first telescopic sunspot observers, he wrote in early 1619 that sunspot numbers decreased in the last 1.5 years and that he had noticed spotless days in those 1.5 years, which had never happened before (i.e., since August 1611). In November 1611, he produced a drawing of sunspots, which is now lost. His maximum of sunspots was on 1612 May 30 (Julian), where he detected 14 spots, consistent with Galileo, Scheiner, and Harriot. We can conclude from these few statements that the starting minimum of the first telescopic sunspot Schwabe cycle was before August 1611, that the maximum was around 1612 (Marius) to 1615 (Tarde), and that the minimum at the end of this Schwabe cycle was after early 1619. All his information on spots is fully consistent with other observers. Marius seems to have observed spots through the Camera Helioscopica. We quote a number of additional important statements by Marius on spots: "I could see the sun and its spots clearly through the mentioned instrument during the bright day without harm for the face, including their daily motion," "spots would bring some kind of coolness to the extreme heat of the sun," and that "sunspots do not traverse the disk of the sun along the ecliptic, but that they build an angle with it." The observations by Marius were not fully considered in the telescopic sunspot compilations by Wolf (1857) and Hoyt and Schatten (1998).

D. L. Neuhäuser (✉) · R. Neuhäuser
Jena, Germany
e-mail: rne@astro.uni-jena.de

H. Gaab, P. Leich (eds.), *Simon Marius and His Research*, Historical & Cultural Astronomy, https://doi.org/10.1007/978-3-319-92621-6_7

Introduction

The reconstruction of past solar activity is essential to understand the internal physics of the sun and (sun-like) stars as well as to possibly predict future solar activity and space weather. Sunspots are used as proxy for solar activity and have been observed for millennia with the unaided eye and since 1610 also with the telescope.

R. Wolf studied solar activity and introduced the sunspot number: the daily Wolf or Zürich sunspot number R for an individual observer is defined as $R = k\,(10\,g + n)$ with the total number of individual sunspots n, the number of sunspot groups g, and the individual correction factor k of the respective observer. The international sunspot number is available for the time since 1700 at, e.g., www.sidc.be/silso/datafiles (see Clette et al. 2014 and also Hathaway 2010) and shows the Schwabe cycle with maxima and minima about every 10–11 years according to sunspot observations (Schwabe 1844; Hathaway 2010).

Hoyt and Schatten (1998, henceforth HS98) have then defined the daily group sunspot number. In HS98, dates of telescopic sunspot observations are listed together with the name of the observer, the place, and the number of sunspot groups observed by that observer on each day. If an observer detected only one spot, HS98 consider it as one sunspot group, but groups of course can also include several or even numerous (small and/or large) spots.

One can also quantify the spottedness with the so-called active-day fraction, namely, the fraction of all active days, where there was at least one sunspot, divided by the total number of observing days (with or without spots).

Here, we discuss the sunspot observations of Simon Marius from Ansbach, Germany (not far from Nuremberg), who observed sunspots from 1611 until at least 1619. We present a few clearly dated sunspot detections, which were not considered in HS98—with some impact to the daily, monthly, and yearly group sunspot numbers compared to HS98. The texts from Marius also deliver qualitative and quantitative input. The period 1611–1619 is of particular importance and relevance, because it is shortly after the invention of the telescope and shortly before the start of the Maunder Grand Minimum. The duration and depth of the Maunder Grand Minimum (first noted by Spörer 1887 and then amplified by Maunder 1890 and Eddy 1976) have received much attention since then (e.g., Ribes and Nesme-Ribes 1993; Usoskin et al. 2007, 2015; Vaquero et al. 2011; Vaquero 2012; Clette et al. 2014; Zolotova and Ponyavin 2015); it is usually dated from 1645 to 1715. Vaquero and Trigo (2015) argue that what they call the "extended Maunder Minimum" would have started in 1618 during or around a Schwabe cycle minimum at about that time.

The time period studied here is shortly after the calendar reform, when the new Gregorian calendar replaced the previous Julian calendar: 1582 Oct 4 was immediately followed by Oct 15, i.e., 10 days (Oct 5–14) were left out, while the sequence of weekdays was uninterrupted. Marius used the Julian calendar.

We will first consider the observations from 1617 to 1619, also mentioned in Wolf (1857) and HS98, and then also those from 1611 to 1615. More details and comparisons with other observers can be found in Neuhäuser and Neuhäuser (2016).

Reports from Simon Marius for 1617–1619

HS98 quote Wolf (1857) as source for their data for Marius. In Wolf (1857), we can read (our English translation of the partly German and partly Latin text can be found below):

> Simon Marius, astronomische und astrologische Beschreibung des Kometen von 1618, Nürnberg 1619. 4.Die Vorrede dieser Schrift ist *Anspach den 6. April 1619* datiert. Marius erzählt, dass er *nun über die anderthalb Jahr nicht mehr so viel maculas in disco solis habe finden können, ja gar offt kein einig maculam antroffen, das doch vorige Jahr niemals geschehen.* Dieser Fleckenarmuth stellt Marius das grosse Kometenjahr 1618 gegenüber, und fügt dann bei: *Ich erinnere es nur, und schliesse nichts.*

Wolf (1857) gave 6 April 1619, but the original manuscript clearly shows Apr 16; this is just a typo in Wolf (1857); furthermore, the text about the sunspots quoted here is in section V (5), not 4 as given by Wolf. Marius did not specify here, whether this date in the dedication is Julian or Gregorian; if Apr 16 is Julian, the Gregorian date is Apr 26. For the remainder of the paper, it does not matter much whether Marius meant here the Julian or Gregorian date, because his statement about "one and a half year prior to April 1619" may not be precise to better than about 1 or 2 months anyway. For a few dates given in Marius (1619), he adds explicitly that the date is given in the old style (Julian calendar), and for some other dates, he mentioned the weekday (being consistent with the given Julian calendar date), so that it is certain that he used the Julian calendar. He detected a comet on 1618 Nov 11 and 21 (Julian) and then obtained positional measurements of a comet 1618 Nov 24 to Dec 19 (Julian), i.e., Dec 4–29 Gregorian (his section III), while others have observed comet 1618 W1 from 1618 Nov 23 or 25 (Gregorian) until 1619 Jan 22 (Gregorian), the latter by Cysat with a telescope (according to Kronk 1999, who does not mention Marius for the comets in 1618). When Marius mentioned that he saw a comet on 1618 Nov 11 early in the morning with tail but without the comet head (his section II) or since "Martini Alten Calendars" (i.e., Nov 11 Julian) (his section IV), he is referring to the other comet (1618 V1 in Kronk 1999), which was observed by others from 1618 Nov 10 or 11 until Nov 29 (Gregorian) or even until Dec 9 (also seen by others with tail but without head); note that some dates given in Kronk (1999) are incorrectly shifted from a presumable Julian date (but truly being Gregorian) by 10 more days, namely, the last Jesuit observation of comet 1618 V1 obviously incorrectly shifted from Nov 29 to Dec 9. It is quite clear that comet 1618 V1 was discovered earlier than W1; V1 was observed in China since Nov 14 and W1 since Nov 26; their dates should not be affected by a wrong Julian-Gregorian calendar conversion. It is then clear that Marius first observed comet V1

(1618 Nov 11/21) and then comet W1 from 1618 Nov 21/Dec 1 until Dec 19/29—claiming that all observations pertained to only one comet.

We translate Wolf's citation from Marius (1619) to English as follows:

> Simon Marius, astronomical and astrological description of the comet of 1618, Nuremberg 1619, 4. The foreword of this text is dated and given as foreword 1619 Apr 6 [correct: Apr 16, see above], Marius reports that he now, for one and a half years, could not find as many spots on the solar disk, indeed rather often not even a single spot, as was never the case in the years before. Marius compares this dearth of spots with the comet year 1618 and adds I just recall it, but I do not conclude anything.

The text in the book by Marius (1619) on the comet of 1618, dated 1619 Apr 16 (Julian), gives even more details, in our English translation (square brackets our additions):

> [. . .] while I now, for one and a half year, could not find as many spots [maculas] on the solar disk, yet rather often not even a single spot [maculam], as was never the case in the years before, I have therefore written this in my observational log books, [the remaining part of this sentence is in Latin in the original] this appears strange to me, that rather few [raras] or more often [saepius] no spots could be detected on the disk of the sun, which was never observed before. As if something would be covered at this location. I just recall it, but I do not conclude anything, I let other high, healthy, and sharp-thinking genius (people) think further on those things, I do my part, others do their parts, given the grace of God, one must start with it, and should help the other without any hate, until one can conclude something with more certainty. I have thought about it a lot since the year 1611, what those spots could be, and how they would form, but have not come to a conclusion yet, which I could rest on. But this I say: that I several times have clearly seen tail-like longish spots on the disk of the sun, indeed somewhat similar to a comet, so that I was often surprised. Like, if those spots would bring some kind of coolness to the extreme heat of the sun, and later would become a comet by merging or rather combining, I do not conclude anything, I cannot do it, but just indicate my thoughts.

The observational logbooks (observationibus) mentioned are unfortunately considered lost. After mentioning his observationibus, he probably had checked them before continuing the writing (and then fell into Latin). Then, he specified some information with more precision:

1. Few [spots] in addition to "could not find as many spots on the solar disk"; the word "raras" could mean "few or here and there or isolated/single" regarding the number and spacing of spots or "rare" regarding the frequency of spots; "few" might be the best compromise; "raras" is connected grammatically to "maculas."
2. "More often no spots" in addition to "yet rather often not even a single spot"; the Latin "saepius" is a comparative degree and translates to more often or more frequently, so that on more than 50% of the observing days the sun was spotless. Hence, the active-day fraction was below 0.5 but larger than 0 from fall 1617 to spring 1619.
3. "Never observed before" in addition to "never the case in the years before"; with "never observed before," Marius obviously means since the start of his own sunspot observations in August 1611; see also below; since this (spotless days) was "never the case in the years before," the active-day fraction was 1.0 from August 1611 to fall 1617 (at least on his observing days).

If the statement by Marius above, that sunspots may provide "some kind of coolness to the extreme heat of the sun", should mean that they are cooler than the surrounding photosphere, then this would be an acceptable consideration. His hypothesis regarding the formation of comets from sunspots was not confirmed, but indeed, cometary tails are blown by the solar wind; Marius states the observed fact that tails point away from the sun. He supports his connection of comets with sunspots by the observational fact of having observed a large comet (of 1618) but very few spots (in 1.5 year prior to spring 1619), i.e., at the same time; in the time before fall 1617, he observed many spots but no comets.

Marius mentioned a period of (roughly) 1.5 years until sometime in the first 3.5 months in 1619. The comet reported by Marius (1619) was detected by him until 1618 Dec 19/29 (and he continued to try to observe it until 1618 Dec 25 (Julian), i.e., 1619 Jan 4 (Gregorian), and the dedication of his book is dated 1619 Apr 16 (Julian), so that the book was written sometime in the first 3.5 months of 1619 (the very earliest possible date for the end of those 1.5 year would be 1619 Jan 4 (Gregorian), when his comet observations ended, because he connected the reduction of spots with the appearance of the comet). In the remainder of this paper, we assume that the period ended in April 1619, given that the generic statement about spots is located toward the end of the book in section V (of six sections); the length of the period (1.5 year) is probably meant to be "roughly 1.5 year," maybe to within 1 or 2 months. (The period of 1.5 year therefore started roughly between July and October 1617.) Marius does not mention any evolution of spottedness within those 1.5 years; it is likely that there was a significant change at the beginning of those 1.5 years (fall 1617) regarding spottedness, the appearance of spotless days, and/or (much) less spots than before.

From Marius's text, one cannot conclude that he never saw a spot (as incorrectly done by HS98). Also, the observational period 1617 Jun 7 to 1618 Dec 31 as given in HS98 cannot be deduced from that text. (Even if Marius had reported something like that he would have observed all of 1618, one should consider this as the Julian calendar year of 1618, while HS98 let him end his monitoring on 31 Dec 1618 on the Gregorian calendar.) That HS98 lets Marius start his successless monitoring on 1617 June 7 is not justified, neither by some 1.5 years before the end of 1618 nor before early 1619 nor before April 1619 nor by any statement at all from Marius.

Marius's text as quoted above clearly shows that he did detect spots, both in those roughly 1.5 years before 1619 Apr and even more in the years before those roughly 1.5 years. In the period before fall 1617, he noticed "several times" spots being lengthy like a comet, obviously describing unresolved groups or double spots. Marius noticed the decrease from high(er) spot numbers in the years before about fall 1617, explicitly without any spotless days on observational days, to much smaller numbers in those roughly 1.5 years before 1619 Apr—in the latter period of roughly 1.5 years, most of the observational days were spotless for him (as specified in his Latin sentence, i.e., active-day fraction below 0.5, but not zero), contrary to previous years; he noticed a decrease in solar activity from the previous Schwabe cycle maximum to a minimum.

HS98 furthermore cite Zinner (1952), where it is just briefly (one line) mentioned that Marius would have observed spots from 1611 to 1619. Zinner (1952) gives as reference Zinner (1942), which relates to previous years and which we will investigate below.

Observations from 1611 to 1615

We will now present additional texts by Marius about his sunspot observations for the time before fall 1617, namely, 1611–1615.

1. *Marius and Schmidnerus on 1611 August 3/13*

In Zinner (1942), a letter from Marius to Maestlin, the teacher of Kepler, is cited, which is dated to 1611 Dec 29 (Julian). We translate this into English as follows:

> I praise You most for those things about which I write to you, Your Excellency, namely the irradiation of Venus and Mercury from the Sun in the same way as the moon, and about the spots on the sun, which I have observed in very large numbers and always in different form since August.

Hence, Marius has observed sunspots "in very large numbers ... since August" 1611. Unfortunately, except the fact that the spots were "always in different form," he does not give exact dates here but just adds that he has to hurry to finish the letter, because the courier is waiting and pressing.

Marius mentions the start of his sunspot observations in his work *Prognosticon auf 1613* (Marius 1612), i.e., the yearly forecast for 1613, finished and dated 1612 June 30 (Julian). We translate this into English as follows:

> Regarding the spots in the sun, which were first observed by Johannn Fabricius and his father, David Fabricius, which I had seen for the first time last year 1611 in August, as they were shown to me by Ahasverus Schmidnerus from the Preussian Königsberg, who had visited me at that time.

The person mentioned above, Ahasverus Schmidnerus (called David Schnidner in Klug 1904), had shown sunspots to Marius. Schmidnerus may well have known about sunspots from Johann Fabricius, who studied at the same time in Wittenberg and had observed spots in early 1611; the first publication about telescopic sunspots (Fabricius 1611) also was printed in Wittenberg. Marius knew about those early observations; he was in contact with David Fabricius, whom he got to know during a visit to Brahe years earlier.

In his Latin book on the moons of Jupiter, *Mundus Iovialis* (Marius 1614)—in which Marius uses the Julian calendar—dedication dated to 1614 Feb 18/28, Marius gives the exact date in the foreword (Praefatio ad Candidum Lectorem); he discusses sunspots as follows[1]:

> It had been my intention, according to my former proposal, to deal now with the spots on the Sun, setting out all my observations upon them from August 3, 1611, to the present time. However, I do not wish—and, indeed, am unable—to make any definite statement about them at present, not only from the causes originally pointed out, but for the further reason that I find the greatest authorities in disagreement, and am unable to satisfy myself. I therefore pass these matters by, and will take up here four other points not yet mentioned by me in the dedications of my yearly forecasts.

Marius started his sunspot observations on 1611 Aug 3/13, i.e., Aug 3 on the Julian calendar, but Aug 13 on the Gregorian calendar. We can reasonably assume that he started the observations with his first positive detection of at least one spot or group. HS98 do not list any observer for 1611 Aug 3/13 (and only one other telescopic detection before that date: Harriot on 1610 Dec 8/18).

Regarding the question, what Marius has mentioned before about sunspots (*Mundus Iovialis*, Marius 1614), we can read a few pages earlier at the beginning of the same foreword (Praefatio), as follows[2]:

> It had been my intention, Candid Reader, to deal with you at some length in this preface, and to give a lengthened statement of all the objects which I have observed up to the present time through the Belgian instrument commonly called a spy-glass, in the Sun, the Moon, the other stars, and in the heavens generally, as you may see in various passages of this little book. But, as bad health and interruptions caused by other business have kept me back, and also the Frankfurt fair was close at hand, and the book was already going through the press, I have been unable to keep my promise, and find myself unwillingly compelled to reserve for another time the publication of my observations.

To summarize, from all the texts cited above, we can clearly conclude that Marius had been detecting spots since 1611 Aug 3/13, namely, "in very large numbers and always in different form" until at least 1611 Dec 29 Julian (his letter to Maestlin), even until at least 1614 (Marius 1614). Multiple spot detections in 1611 are well possible considering the data in HS98: Scheiner and Harriot had seen 1–6 spots or groups on each of 42 different days from October to December 1611. Furthermore, David Fabricius wrote on 1611 Dec 11 in a letter to Maestlin (citing from Reeves and Van Helden 2010):

> Indeed, this summer [1611] I often observed ten or eleven spots scattered on the Sun's disk at one time.

This is also fully consistent with "very large numbers" reported for that time by Marius.

2. *Marius October 1611*

---

[1]pp. 42ff in the Latin-German edition translated by Schlör (Marius 1614/1988); preface in the English edition translated by Prickard (Marius 1614/1916) and Prickard/Van Helden (Marius 1614/1916/2019).

[2]Marius 1614/1988, Marius 1614/1916, Marius 1614/1916/2019.

In his work *Prognosticon auf 1613*, cited above, Marius (1612) continues (our translation into English):

> When that [original] way of observing them [spots] was not sufficient any more for me, namely through the light ray in a dark room [Camera Helioscopica] by using the Belgian instrument, I had thought and implemented on Oct 11 [1611] a different way, so that I could see the sun and its spots clearly through the mentioned instrument during the bright day without harm for the face, including their daily motion. But later more about this.

Here, we see that Marius observed regularly and on subsequent days ("daily motion"). The text says that Marius already used the Camera Helioscopica before 1611 Oct 11/21; the improvement on Oct 11/21 was probably a better way to see "spots clearly" and their "daily motion"; he explains the projection method with the telescope onto a white screen in a dark room in more detail in the foreword of *Mundus Iovialis* (Marius 1614); the improvement may have been regarding the placement of the white screen perpendicular to the telescope ("spots clearly") and then by drawing the spots day to day ("daily motion") onto some paper.

Since Marius states that he changed his observing technique on 1611 Oct 11/21, he detected at least one spot group on that day, too.

After the appearance of *Mundus Iovialis*, whose main part was finished and published in 1614 (cited here as Marius 1614), Marius wrote an appendix or addendum "Ad candidum lectorem," dated 1615, in Latin and translated to German recently by Gaab and Leich (2014), which we translate into English as follows (considering the German translation by Gaab and Leich 2014):

> I add at least this and confirm it with holy emphasis that I do not possess anything else from Galilei than *Sidereus Nuncius* and that I also did not read anything else. Also, I could not yet get hold of the book by Apelles [by Scheiner as pseudonymous author on spots]; I do not know why, even though I have searched for it carefully in Nuremberg. The first discoverers and observers of sunspots are the two Fabricius, father and son, but because they are considered heretics, their names are not cited. [sig. G4^{v}]
>
> In the year 1611, I have found a method to observe the colors of the stars. Also in the same year on October 3/13, I have invented a method to observe sunspots on the sun itself through a tube, without any harm for the eye; in addition, [I add] that sunspots do not traverse the disk of the sun along the ecliptic, but that they build an angle with it. [sig. H1^{r}]

With the "tube" Marius may have meant what he described in the foreword to *Mundus Iovialis* (Marius 1614), namely, that he observed the sun and its spots with the naked eye at low altitude (hence, "without any harm for the eye") through some sighting "tube: if the sun stood low, I used a black paper arranged as narrow tube, whose narrow opening hole was put to the eye, but its wider opening hole towards the Sun" (Marius 1614).

Marius mentioned two slightly different dates for implementing new observing techniques, namely, 1611 Oct 11/21 in Marius (1612) and 1611 Oct 3/13 in the 1615 appendix to *Mundus Iovialis* (Marius 1614). Either the two different dates indicate two steps in the implementation of the new observing technique or one of them is given by mistake. Hence, Marius has observed sunspots on 1611 Oct 3/13 and/or 11/21. In Oct 1611, also Scheiner observed regularly but not daily: on 1611 Oct 21 (Gregorian), he detected four groups (HS98), but there were no observations on

1611 Oct 13 (Gregorian) according to HS98. Scheiner's drawing of the spots of 1611 Oct 21 appeared in *Tres epistolae de maculis solaribus* under the pseudonym "Apelles," where it is the drawing with the earliest date; Marius mentioned above that he did not have available a copy of Apelles at the time of writing the *Mundus Iovialis* appendix—it may well be that Marius did not know at this time that Scheiner was the author of those letters of *Apelles*.

In the 1615 appendix to *Mundus Iovialis*, Marius gives an important result from his observations: "sunspots do not traverse the disk of the sun along the ecliptic, but ... they build an angle with it." Both for spots moving on the solar surface and especially for small solar system bodies transiting the sun, it might have been expected at that time that they would traverse the disk of the sun on the ecliptic. To notice the inclined path of the spots, it may have been necessary to draw a spot from day to day into the same drawing with, e.g., the Camera Helioscopica. The statement that the spots (observed almost daily) form an angle with the ecliptic includes the notion that the solar equator is inclined to the ecliptic. The amount of this effect, or whether it changes with time, e.g., within a year, is not reported by Marius.

Marius gives some more details about his solar observations in the foreword of *Mundus Iovialis* (Marius 1614), continuing the citation from above. We translate this into English as follows:

> The fourth observation is a very special one on the sun in connection with the spots; I and Mr. David Fabricius, a theology scholar from eastern Frisia, a very excellent astronomer and my dearest friend, have written a few times about them. ... Sometimes, the ray [from the sun] seems to stand almost still in its motion, which is otherwise the usual daily motion. But sometimes it [the ray] seems to move further like jumping. The same uneven motion also applied to sunspots. ... This motion either originates from the sun or from the Earth or from the air. I think it is not due to the air.

Hence, Marius seems to have noticed effects of the sun and its spots, which we now call "seeing"; Marius (1614) also mentioned that David Fabricius considered the air to cause this effect, which indeed is correct. This effect was also reported by the Chinese naked-eye observers, e.g., "several black spots rocking to and fro" for 1617 Jan 11 (e.g., Xu et al. 2000).

3. *Marius November 1611*

Directly after the above quotation from Marius in his 1615 appendix to *Mundus Iovialis*, where he listed various observations and inventions, Marius continues (our translation into English):

> ... and that I have shown a figure, which I had drawn on the 17th/27th day of November of the year 1611, to the previously mentioned Holsteinian, who looked at it with admiration and added that this had been shared with him in secret by Scheiner. [sig. H1^r]

It is known that the person from Holstein (Saxonius) had visited Marius on 1615 Jul 4/14 (Gaab and Leich 2014). We will discuss his possible observation on 1615 Jul 4/14 with the visitor from Holstein below.

In the 1615 appendix to *Mundus Iovialis*, Marius clearly states that he has drawn sunspots for 1611 Nov 17/27 and that he has shown the drawing to the visitor from Holstein (Saxonius), who was in contact with Scheiner and who told Marius that he (Saxonius) had seen such (a) drawing(s) from Scheiner. It seems that Saxonius did not mention the *Apelles*, where such drawings were published pseudonymously by Scheiner. As mentioned above, Marius did not have available a copy of *Apelles* with the drawings from Nov 17/27 by now (1615). Unfortunately, the drawing from Marius has not yet been found. We can judge what Marius has seen and drawn that day, namely, from the drawings by Scheiner on this very date.

Why did Marius show a figure with sunspot(s) drawn by him to Saxonius from Holstein? One could imagine that Marius and his visitor would have at least tried to observe spots together during the visit. If the weather had been too overcast on that or those day(s), then he would instead have shown a drawing to him. On the other hand, it is also well possible that the drawing shows a particularly large number of spots and/or spots of unusual form and/or on unusual location(s), so that Marius did show it to his visitor anyway (possibly in addition to the collaborative observation that day). Two drawings of sunspots by Scheiner for that day, 1611 Nov 17/27, indicate a special situation around that day: Many spots are distributed over a large range of heliographic latitudes, about half of them near the equator and all others only on one hemisphere. This particular sunspot distribution may have been discussed in connection to the nature of spots, as such a large range of heliographic latitudes may not be consistent with one of the theories discussed (namely, that they are transits of small solar system bodies).

It is well possible that Marius often produced drawings (given that the technique was available to him, namely, the Camera Helioscopica—see Marius 1614). He observed the "daily motion" of spots (Marius 1612) and that their paths are inclined to the ecliptic (in the 1615 appendix to Marius 1614).

4. *Marius May 1612*

In his *Prognosticon auf 1613*, finished and dated to 1612 June 30 (Julian), Marius wrote (again our translation):

> On May 30 [Julian] of this year [1612], I have seen 14 such [spots] at once. They were [would be], however, not on the solar body themself, but they were [would be] bodies orbiting the sun.

Given that the *Prognosticon auf 1613* was dated to 1612 June 30 [Julian], this text about those 14 sunspots observed on May 30 (Julian), i.e., 1612 Jun 9 (Gregorian), means they were observed only shortly before the text was written.

On that very same day, also Galileo and Jungius observed: Galileo had seen 7–10 spot groups (HS98), his largest daily number, and Jungius in Gießen had seen five groups (HS98). The drawing by Jungius shows five spot pairs or groups with a total of ten spots, the one by Galileo shows 7–9 groups, partly resolved into smaller structures with a total of some 25–30 spots. In the drawing by Jungius, only the smallest spots seen by Galileo are missing. Those large numbers appear to be consistent with Marius giving 14 spots. It is of course somewhat subjective how many groups there are and how many individual spots are present inside or outside of groups. Marius saw those spots in this way (same day) or slightly differently (different instrumentation).

According to HS98, Harriot would also have observed spots on that day; HS98 gave five groups on Harriot's drawing. However, the catalogue of the Harriot's drawings (digilib.mpiwg-berlin.mpg.de) does not contain a drawing for 1612 May 30 (Julian), June 9 (Gregorian). Harriot did observe on both the day before and after that date and had detected basically the same five groups as Jungius on May 30 and 31 (Julian), but a few more spots (13 or 14 for Harriot, 10 for Jungius). Galilei saw additional smaller spots, both inside the groups detected by the others but also a few more weak groups with two weak spots each. Marius reported 14 spots for May 30/June 9.

It is well possible and understandable that several European observers monitored the sun closely on those days because of the solar eclipse visible in Europe on 1612 May 30 (Gregorian). According to HS98, Galileo, Harriot, and Jungius reported sunspots for that day.

It is quite obvious that Marius gave this particular day (1612 May 30/June 9) as an example, because he never had seen so many spots on any others days until the date of this statement (1612 June 30/July 10). We can then assume 13 spots/groups as an upper limit for Marius for the time before 1612 June 30/July 10. For 1612 May 30/June 9, also Galileo reported his largest spot/group number for this period.

There are no naked-eye sunspots known for 1612 May/June (e.g., Vaquero 2012), but from 1612 Aug 19–21, Galileo saw sunspots with both the telescope and the unaided eye, and he had drawn the telescopic sunspots for 1612 Aug 19 (Vaquero 2004). Adam Tanner reported that he has detected spots *almost daily* in 1612 from Ingolstadt (see Neuhäuser and Neuhäuser 2016).

5. *Marius and Saxonius on 1615 Jul 4/14*

We read above from Marius written 1615 in the appendix to *Mundus Iovialis*:

> and that I have shown a figure, which I had drawn on the 17th/27th day of November of the year 1611, to the previously mentioned Holsteinian, who looked at it with admiration and added that this had been shared with him in secret by Scheiner.

Regarding the person from Holstein, Marius had mentioned before (our translation to English):

> Namely on [1615] July 4/14 there was a highly educated man here, Mr. Petrus Saxo from Holstein, student of mathematics, who undertook a travel from Ingolstadt [southern Germany] from Scheiner directly to me.

Petrus Saxonius (1591–1625) was from Husum in northern Germany; he was travelling in southern Germany in 1614, also visiting Scheiner in Ingolstadt. According to HS98, he had observed sunspots in Feb and Mar 1616; since September 1617, he had been professor for mathematics in Altdorf (Gaab 2011). Petrus Saxonius visited Marius on (or after) 1615 Jul 4/14. It is quite likely that Marius and Saxonius observed sunspots together that day, but we have no firm statement about it.

Tarde (1620) drew some 30 spots for 1615 Aug 25 (image reprinted in Vaquero and Vázquez 2009). Since they (also) are spread over a large range of heliographic latitudes, a similar pattern 1 month earlier during the visit of Saxonius to Marius may have motivated their discussion of the nature of spots, given their large range of heliographic latitudes, so that Marius has shown him another example (1611 Nov 17/27). There are no naked-eye sunspots known for 1615 (e.g., Vaquero 2012).

6. *Spotless days*

From the text cited above from Marius (1619) that there were more often spotless days in those roughly 1.5 years before 1619 Apr, but that there were no spotless days before those roughly 1.5 years, i.e., until fall 1617, we can conclude that Marius either did not observe on those days in the period 1611 Aug 3/13 to fall 1616 when the sun was spotless—or that he detected spots when others did not detect any. There were 16 days in that period, when other observers noticed a spotless sun, namely, as follows (according to HS98, all dates Gregorian—with reservations, because we noticed some shortcomings in HS98, and we did not check the sources of these observations):

1611 Dec 29 (Harriot: no spots)
1612 Mar 2, 4, 5, 6, Apr 13–17, 23 (Harriot: no spots)
1612 Mar 2, 4 (Cigoli: no spots)
1616 Nov 13–15, 22, 23 (Scheiner: no spots)

On these dates, there are no naked-eye sunspots known (e.g., Vaquero 2012).

The fact that Harriot reported spotless days for 10 days in 1612 does not need to be in contradiction to the statement by Marius that there were no spotless days before fall 1617. Even though Marius and Harriot saw about the same number of spots around 1612 June 9, it is possible that Marius saw one (or a few) more spots on other days, which were not spotted by Harriot—or, Marius may not have observed on those days. Also, from Tanner's record, there is no evidence for spotless days in 1612, despite "almost daily" observations (see Neuhäuser and Neuhäuser 2016).

We can conclude that all essential elements in the statements by Marius can be confirmed, while no parts were falsified.

Summary

Hoyt and Schatten (1998) list Marius only for 1617 and 1618, but without detecting any spots. They cite, but misinterpret, Wolf (1857) and Zinner (1942, 1952). Zinner (1942, 1952) in fact wrote that Marius observed sunspots from August 1611 until at least 1619, and Wolf (1857) gave explicitly the essential quotation from Marius regarding the time 1617–1619.

With the original texts written and published by Marius, we could find the following information on his sunspot observations:

- Simon Marius and Ahasverus Schmidnerus together saw at least one spot or group on 1611 Aug 3/13.
- Marius observed spots "in very large numbers" from 1611 Aug 3 to Dec 29 (Julian), consistent with David Fabricius, Scheiner, and Harriot; Marius saw spots "always in different form, with their daily motion, and that they do not cross the disk of the sun on the ecliptic, but build an angle with it." This indicates regular observations.
- Marius improved his observing technique on 1611 Oct 3/13 and/or Oct 11/21 (Camera Helioscopica), and he detected at least one spot on that (or those) day(s).
- Marius drew sunspots at least once, namely, on 1611 Nov 17/27 (also drawn by Scheiner for that day).
- Marius reported 14 spots for 1612 May 30 (Julian), which is probably his largest daily number until 1612 June 30 (same for Galileo).
- Simon Marius may have observed spot(s) together with Petrus Saxonius on 1615 Jul 4/14.
- Marius had observed not only many spots since 1611 Aug 3/13 but had no spotless days before fall 1617, i.e., before the period of 1.5 years ending 1619 Apr; this implies an active-day fraction of 1.0 before fall 1617.
- Marius observed spots in a period of (roughly) 1.5 years before 1619 Apr (Sect. 3.3) but much less than before.
- With his statement that there were more often spotless days in those (roughly) 1.5 years before 1619 Apr, together with limits on the active-day fractions from Malapert et al. and Scheiner (HS98), we could constrain the active-day fraction to larger than 0.08 and smaller than 0.5 for 1618 (see Neuhäuser and Neuhäuser 2016).
- The generic statement by Marius also constrains the Schwabe cycle minima to lie before August 1611 (no spotless days seen Aug 1611 to fall 1617) and in or after 1619, consistent with other observers and a typical cycle length; the maximum was somewhere 1612 (14 spots by Marius and others) to 1615 (30 spots by Tarde).

The observations made by Marius since 1611 Aug 3/13 were among the very first telescopic sunspot records. Harriot had detected his first three telescopic spots in England on 1610 Dec 8/18 (HS98), and Johann and David Fabricius in northern Germany had detected their first spot on 1611 Feb 27 and 28 (Julian), hence 1611 Mar 9 and 10 (Gregorian); also, Scheiner and Cysat observed spots in March 1611.

Given that Marius reported in 1619 that there were no spotless days before about fall 1617, we could conclude that he did not observe on those days from 1611 Aug 3/13 until at least end of 1616, when others noticed a spotless sun, namely, on 16 days—or that he could detect spot(s) when others missed them.

Marius clearly said that there were (significantly) less spots in those roughly 1.5 years before April 1619, i.e., from about fall 1617 to spring 1619. Group sunspot numbers cannot be calculated with trends or limits or undated observations, as reported by Marius (and others), which demonstrates a major problem in the group sunspot number system by HS98.

All essential elements in the statements on spots by Marius for 1611–1619 can be confirmed, while no parts were falsified, so that his texts are highly credible. We could date the first telescopic Schwabe cycle with the information from Marius to run from shortly before Aug 1611 to shortly after early 1619.

In addition, we found evidence for the following facts that could also be relevant for the history of telescopic sunspot observations (found mostly in texts by Marius):

- Marius and Schmidnerus with their observation on 1611 Aug 3/13 were among the first known telescopic sunspot observers.
- Marius changed the observing technique on 1611 Oct 3/13 and/or 11/21: he used the Camera Helioscopica (Marius 1612), apparently earlier than Castelli, Galilei, and Scheiner, and improved it by directing the telescope onto a (perpendicular) white screen, explained in the foreword to *Mundus Iovialis* (Marius 1614); he could then detect and draw the daily motion of spots. His observing techniques would be worth another investigation.
- A theory of connecting spots with comets may have been invented by Marius, partly based on writings by Cardanus about a connection between comets and the sun (discussed later by Riccioli).
- Marius noticed that the path of spots form an angle with the ecliptic. In contrast, Tarde wrote that the path of spots is on the ecliptic or parallel to it, and he did not draw the path with curvature, while Malapert has drawn the path with correct curvature.
- In the year 1611, Marius had "found a method to observe the colors of the stars." He added in the *Prognosticon auf 1616*: "the significant change in the colors of several fixed stars ... was first noticed with naked eye by Mr. M. Mastlino and Mr. Keplero, and it was rather clearly seen by myself through the Perspicillum" [telescope], cited after Zinner (1942).

The HS98 and Wolf databases with their large bibliographies are of great value. For further improvement of sunspot (group) numbers and our understanding of solar activity, in particular for the time before and during the Maunder Minimum, it is absolutely essential to check carefully the material from all observers in that time—and also to take into account lower limits, monthly or yearly averages, and trends mentioned by the observers.

Acknowledgments In particular, we thank Hans Gaab for sharing with us his knowledge about Marius's sunspot observations. We are also grateful to Klaus-Dieter Herbst, Pierre Leich, and Klaus Matthäus for valuable advice on the work of Marius. We would especially like to acknowledge Pierre Leich, who motivated us for this study and who organized a Marius conference in September 2014, where we could present some first results. He also built up the Marius portal www.simon-marius.net. We are grateful to the Simon-Marius high school in Gunzenhausen for translating the full *Mundus Iovialis* from Latin to German as part of a special school project with eight school students under the supervision of their teacher, Joachim Schlör (Marius 1614/1988). We thank Chris Graney for discussing the astronomical debates in which Marius was involved, in talks in Nuremberg and Bamberg in 2014 on "Simon Marius—an astronomer too good." We consulted the database of Hoyt and Schatten (1998) at ftp://ftp.ngdc.noaa.gov/STP/space-weather/solar-data/solar-indices/sunspot-numbers/group. We thank Regina von Berlepsch (AIP Potsdam) for providing the publication by Zinner (1952) on the work of Marius—and Rainer Arlt (AIP Potsdam) for providing some works by Wolf and Tarde as well as other helpful advice. We also thank T. Friedli for advice on R. Wolf.

References

Clette, Frederic; Svalgaard, Leif; Vaquero, Jose M.; Cliver, Ed W. (2014): *Space Science Review* 186, pp. 35–103

Eddy, J.A. (1976): The Maunder Minimum, *Science*, 192, 1189

Fabricius, Johannes (1611): *De Maculis in Sole observatis*, Wittenberg

Gaab, Hans (2011): Astronomie in Altdorf, *Altnürnberger Landschaft e.V.*, Sonderheft 2011. Neuhaus, pp. 16–22

Gaab, Hans; Leich, Pierre (2014): Marius' Replik auf Scheiner – Der Anhang zum Mundus Iovialis von Simon Marius [Nachwort], *Globulus – Beiträge der Natur- und kulturwissenschaftlichen Gesellschaft e.V.*, vol. 18, pp. 11–14

Hathaway, David H. (2010): The Solar Cycle, *Living Reviews in Solar Physics* 7, pp. 1–65

Hoyt, Douglas V.; Schatten, Kenneth H. (1998): Group sunspot numbers: A new solar activity reconstruction, *Solar Physics* 189, pp. 491–512 (HS98)

Klug, Josef (1904): Simon Marius aus Gunzenhausen und Galileo Galilei, in: *Abhandlungen der mathematisch-physikalischen Klasse der königlich Bayerischen Akademie der Wissenschaften*. vol. 21, pp. 385–526

Kronk, Gary W. (1999): *Cometography*, Cambridge University Press

Marius, Simon (1612): Prognosticon for 1613, Johann Lauer Nürnberg

Marius, Simon (1614/1916): Arthur Octavius Prickard, The 'Mundus Jovialis' of Simon Marius, *Observatory* 39 (1916), pp. 367–381, 403–412, 443–452, 498–503

Marius, Simon (1614/1988): Joachim Schlör, *Mundus Iovialis – Die Welt des Jupiter. Die Entdeckung der Jupitermonde durch den fränkischen Hofmathematiker und Astronomen Simon Marius im Jahr 1609 – lateinisch und deutsch* (= *Fränkische Geschichte*, Vol. 4). Gunzenhausen: Johann Schrenk 1988

Marius, Simon (1619): Astronomische und Astrologische Beschreibung des Kometen von 1618, Johann Lauer Nürnberg

Maunder, E.W. (1890): Prof. Spoerer's researches on Sun-spots, *Monthly Notices of the Royal Astronomical Society*, 50, 251

Neuhäuser, Ralph; Neuhäuser, Dagmar L. (2016): Sunspot numbers based on historic records in the 1610s – early telescopic observations by Simon Marius and others, *Astronomische Nachrichten* 337, pp. 581–620

Reeves, Eileen; Van Helden, Albert (2010): *Galileo Galilei & Christoph Scheiner on sunspots*, Chicago/London: University of Chicago Press

Ribes, J.C.; Nesme-Ribes, E. (1993): The solar sunspot cycle in the Maunder minimum AD1645 to AD1715, *Astronomy & Astrophysics*, 276, 549

Schwabe, Heinrich (1844): Sonnen-Beobachtungen im Jahre 1843, *Astronomische Nachrichten* 21, pp. 234–235

Spörer, F.W.G. (1887): Vierteljahresschrift Astronomische Gesellschaft, 22, 323

Tarde, Jean (1620): Borbonia Sidera, Paris

Usoskin, I.G.; Solanki, S.K.; Kovaltsov, G.A. (2007): Grand minima and maxima of solar activity: new observational constraints, *Astronomy & Astrophysics*, 471, 301

Usoskin, I.G.; Arlt, R.; Asvestari, E., et al. (2015): The Maunder minimum (1645-1715) was indeed a grand minimum: A reassessment of multiple datasets, *Astronomy & Astrophysics*, 581, 95

Vaquero, José M. (2004): A forgotten naked-eye sunspot recorded by Galileo, *Solar Physics* 223, pp. 283–286

Vaquero, José M.; Vázquez, Manuel (2009): *The Sun Recorded Through History*, Springer

Vaquero, José M.; Gallego, M.C.; Usoskin, Ilya G.; Kovaltsov, Gennady A. (2011): Revisited sunspot data: A new scenario for the onset of the Maunder minimum, *Astrophysical Journal* 731, no. 2, L24–L27

Vaquero, José M. (2012): Historical records of solar grand minima: a review, in: Characterizing quiet times in the Sun and stars, in Mandrini, C.H. et al. (Eds.) *Proc. IAU Symp.* 286, pp. 383–392

Vaquero, José M.; Trigo, R.M. (2015): Redefining the limit dates for the Maunder Minimum, *New Astronomy* 34, pp. 120–122

Wolf, Rudolf (1857): *Mitteilungen über die Sonnenflecken*, vol. 1, 44 (Ref. 1)

Zinner, Ernst (1942): Zur Ehrenrettung des Simon Marius, *Vierteljahrsschrift der Astronomischen Gesellschaft* 77, pp. 23–75

Zinner, Ernst (1952): Ältere Beobachtungen von Sonnenflecken, *Naturforschende Gesellschaft Bamberg,* Sonderdruck aus dem XXXIII. Bericht 1952, pp. 36–37. In: Zinner, Ernst: Zur Erklärung des Lichtwechsels der vermissten Sterne. Bamberg: Remeis-Sternwarte 1952

Xu, Zhenoao; Pankenier, David W.; Jiang, Yaotiao (2000): *East Asian archaeoastronomy*, Gordon and Breach

Zolotova, Nadezhda V.; Ponyanin, Dmitri I. (2015): The Maunder Minimum is not as grand as it seemed to be, *Astrophysical Journal* 800, pp. 1–14

Chapter 8
An Astronomer Too Excellent: Simon Marius, the Telescope, and the Problem of the Stars During the Copernican Revolution

Christopher M. Graney

Abstract Simon Marius argued in his 1614 *Mundus Iovialis* that telescopic observations of stars supported Tycho Brahe over Copernicus. Prior to the advent of the telescope, Brahe's was a powerful voice against the Copernican theory. Brahe used observations and calculations regarding the sizes of stars to produce what, at the time, appeared to be a formidable scientific case against Copernicus. The advent of the telescope raised questions about the true sizes of stars. Marius appears to have been the first astronomer to argue that the telescope supported Tycho. Today Marius's support for Tycho might seem to have been an error. Yet it in fact illustrates Marius's skill as an astronomer. It also contrasts Marius favorably with Galileo, who also made telescopic studies of stars but did not share all his results. The tricky nature of telescopic observations of stars in the early seventeenth century, and why Marius was right, even though he was wrong (while Galileo was wrong, even though he was right), will be the focus of this paper.

> Although Marius was in the possession of the most important astronomical discoveries of the early 17th century, he opposed the heliocentric view of the cosmos and favoured the Tychonic one. The latter, he claimed to have arrived at independently of Brahe after reading Copernicus during the winter of 1595–1596. (Gaab/Leich, 2014)

Indeed, Simon Marius *should* have favored the Tychonic world system. Only the Tychonic system was the logical choice for Marius, granted the telescopic observations of the stars that he made. Those observations were excellent, owing to Marius's skills as an astronomer.

Astronomers in our time look at the stars in the sky and think themselves to observe merely points of light without size. However, over a period of many centuries, astronomers used the language of size in their descriptions of the stars.

C. M. Graney (✉)
Jefferson Community & Technical College, Louisville, KY, USA
e-mail: christopher.graney@kctcs.edu

H. Gaab, P. Leich (eds.), *Simon Marius and His Research*, Historical & Cultural Astronomy, https://doi.org/10.1007/978-3-319-92621-6_8

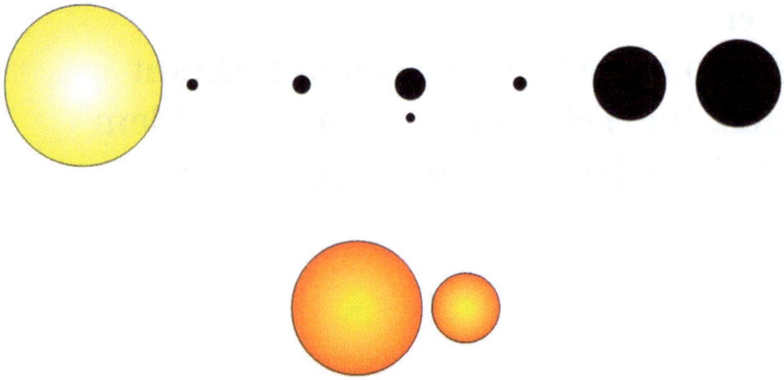

Fig. 8.1 Relative sizes of celestial bodies in the case of a geocentric universe (where the stars lie just beyond Saturn) as calculated by Tycho Brahe, based on his observations and measurements. Bodies are (from left to right, 1st row) the sun, Mercury, Venus, the Earth and moon, Mars, Jupiter, Saturn, and (2nd row) a first and third magnitude star. Sun, stars, and planets all fall into a fairly consistent range of sizes. From Graney (2013), p. 167

This is because, in fact, the keen eye perceives the stars as dots of varying size. *Magnitudo* was the Latin word used for this purpose, a word that means *size*. Stars were divided into six classes of *magnitudo*—six classes of size. The largest visible stars were assigned to the first class; the smallest were assigned to the sixth class. These size classes developed into the modern concept of *brightness*, so that today astronomers say the star Regulus is a rather bright star of apparent magnitude 1.38, not a rather large star of the first magnitude.

For Tycho Brahe, stellar magnitudes or sizes formed a powerful argument against the Copernican system. Brahe sought to precisely determine the apparent diameters of stars, and determined that the apparent diameter of a typical star of the first magnitude was approximately 2 minutes of arc. The disk-like appearance of such a star is spurious—an artifact of optics—but this was unknown to Brahe. Moreover, the diameters Brahe measured were comparable to those obtained by earlier astronomers such as Ptolemy (Van Helden, 1985, p. 27, 30, 32, 50). Brahe combined his measured apparent diameters with the distances to the stars under geocentric and heliocentric cosmologies to determine the true physical sizes of the stars.

Under a geocentric cosmology, stars would be located just beyond Saturn. Hence a star of magnitude comparable to Saturn would be similar to Saturn in physical size. Thus Brahe calculated the stars to be comparable in size to the other celestial bodies in the case of a geocentric universe (Fig. 8.1).

But under a heliocentric cosmology, stars were located far beyond Saturn—indeed, they had to be at least hundreds of times more distant than Saturn, or else they would show a detectable annual parallax. Therefore, a star of magnitude comparable to Saturn would be hundreds of times larger than Saturn in physical size. Brahe calculated the stars to be enormous—absurdly so, dwarfing the sun—in the case of a heliocentric universe (Fig. 8.2). Brahe proposed a hybrid geocentric cosmos, in which the sun, moon, and stars circled the Earth, while the planets circled the sun (Fig. 8.3). This Tychonic cosmology would turn out to be compatible with

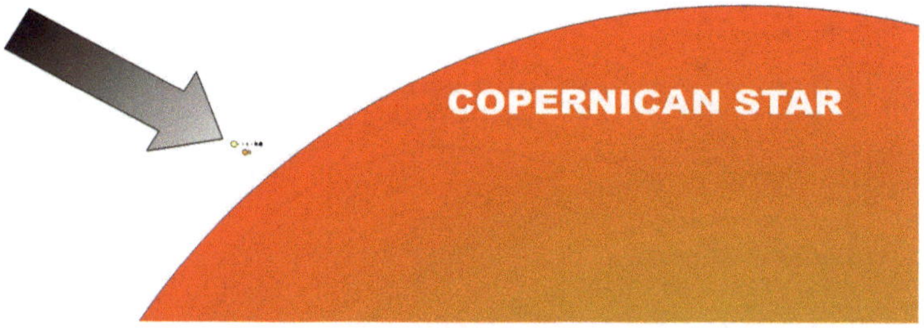

Fig. 8.2 The arrowed dots are Fig. 8.1, reproduced to scale compared to Brahe's calculated relative size for a third magnitude star in the case of a heliocentric universe (where the stars lie at vast distances and thus must be enormous to explain their apparent sizes as seen from the Earth). From Graney (2013), p. 167

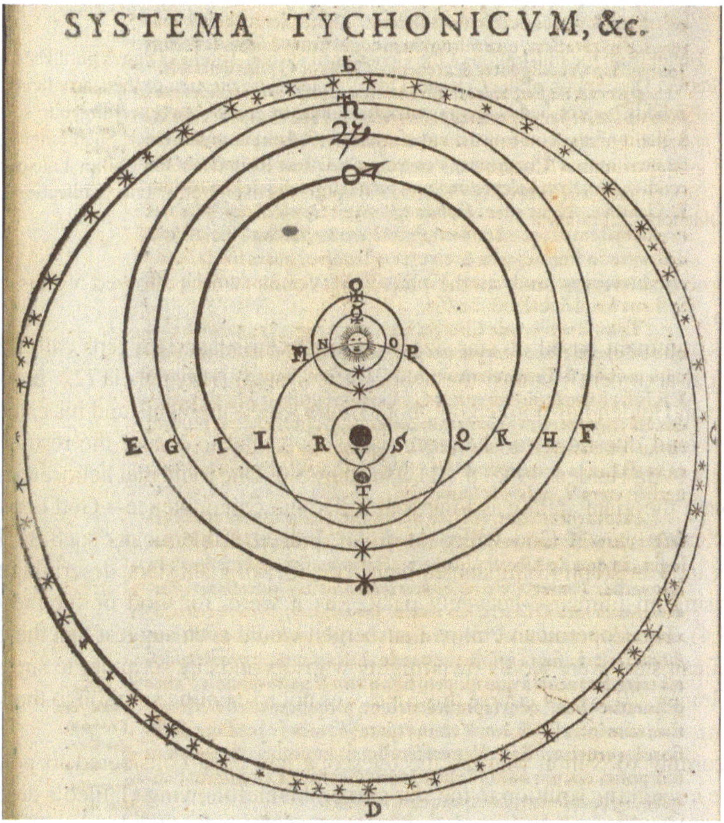

Fig. 8.3 The hybrid geocentric system of Tycho Brahe. The planets circle the sun as in the Copernican system, while the sun, moon, and stars circle the Earth. From Locher/Scheiner 1614, p. 52. ETH-Bibliothek Zürich, Alte und Seltene Drucke

Fig. 8.4 Detail from the illustration of the Copernican system Thomas Digges published under the title "A Perfect Description of the Celestial Orbs." Note his description of the starry heavens as the "palace of felicity" garnished with innumerable glorious lights "far excelling our sun both in quantity and quality," the court of the angels, and the dwelling of the elect. Such language is also reflected in the writings of Copernicans such as Christoph Rothmann and Philips Lansbergen. See Graney 2013, Graney 2015, pp. 77–79. Image(s) courtesy History of Science Collections, University of Oklahoma Libraries

telescopic discoveries such as the phases of Venus (which showed Venus to circle the sun).[1]

This argument based on star sizes was what Christiaan Huygens called Brahe's "principal argument" against the heliocentric system (Huygens, 1722, p. 145). As Albert van Helden has pointed out, both Brahe's measurements and his calculations were beyond question, and heliocentrists simply had to accept the results of this argument (Van Helden, 1985, p. 51). This they did. One route that heliocentrists took to explain the giant stars was to appeal to Divine Omnipotence—God could make sun-dwarfing stars if God willed to do so. Indeed, Copernicans such as Thomas Digges and Christoph Rothmann embraced the idea of giant stars, describing them as comprising an immense heavenly palace, as it were, for God or for the blessed (Fig. 8.4). The Copernican Philips Lansbergen would even suggest that the gigantic Copernican stars were God's own mighty warriors and cite scripture to support this idea (Graney, 2013). Of course this route was not a scientific solution to the star size problem.

According to Simon Marius, the telescope had been considered to potentially provide a scientific solution to the star size problem, following Galileo's description in his 1610 *Sidereus Nuncius* of stars as seen through a telescope. In his 1614 *Mundus Iovialis*, Marius says,

[1]For a detailed discussion of Brahe and the sizes of stars, see Graney (2015), Chapter 3.

> Indeed [Galileo] writes in his Sidereus Nuncius, the fixed stars to appear in no way restricted by a circular periphery—something which certain persons since have considered grounds of the greatest of arguments. In truth, by this statement itself they confirm the Copernican world system: it is on account of the immense Copernican distance of the fixed stars from Earth that their globe shape cannot be perceived from Earth by any method at all. (own translation; original Marius 1614, sig.)( )( )(1^{r})[2]

But Marius declared this idea to be erroneous, because indeed the globe shape of stars could be perceived.

Marius was a highly skilled observer. One illustration of his skill is his description of the Andromeda galaxy, also published in the *Mundus*. He describes it as—

> a fixed star or kind of star of remarkable form which I came upon and saw by means of a telescope the night of 15 December of the Year 1612. In the whole heaven I am not able to discover another such star. But it is near the third and northernmost star in the belt of Andromeda. Without the instrument it is discerned as a kind of little quasi-cloud in that spot; with the instrument no distinct stars are seen (like in the nebula of Cancer, and other nebulous stars), but whitish rays, which where they are nearer the center there grow brighter. The light is dull and pale in the center. It occupies almost the quarter part of a degree in diameter. The luster appears almost like if a candle shining through translucent horn were to be discerned from far off. It appears not unlike to that Comet in the Year 1586 ... (Marius 1614, sig.)( )(4^{r})[3]

Persons familiar with the view of this object through a telescope (who know that, even with modern telescopes far superior to what was available to Marius, many observers cannot describe the Andromeda galaxy so clearly) will fully appreciate what Marius was able to accomplish with such a small instrument. A further illustration of Marius's outstanding skill as an astronomer is his observations of Jupiter's moons—he derived better values than did Galileo for their periods of revolution and other orbital elements, something immediately noted by Georg Johann Locher in his 1614 *Disquisitiones Mathematicae* (even though Locher speaks of Marius disparagingly versus Galileo) (Locher/Scheiner, 1614 p. 78, 80). And, as J. L. E. Dreyer has noted, Marius observed the disks of stars telescopically (Dreyer, 1909, p. 191).

Regarding the disks of stars, Marius writes:

> I obtained an instrument, through which not only the planets, but also all the more conspicuous fixed stars I observed, are seen round (especially the great dog, the small dog, and the brighter stars in Orion, Leo, Ursa Major, etc.). Before that time I had never happened to see this. I am truly surprised Galileo did not see this with his most excellent instrument. (Marius 1614, sig.)( )(4^{v}–)( )( )(1^{r})[4]

It should be noted that these disks Marius saw were also spurious, again a product of optics. However, to a skilled observer using a telescope of very small aperture, such as was used by early telescopic astronomers, they are clearly visible (Grayson and Graney, 2011). They can be viewed in modern telescopes if the aperture is

---

[2]In "Praefatio ad candidem lectorem." English translation from Graney (2015), p. 51.

[3]In "Praefatio ad candidem lectorem." English translation from Graney (2015), p. 50.

[4]In "Praefatio ad candidem lectorem." English translation from Graney (2015), p. 51.

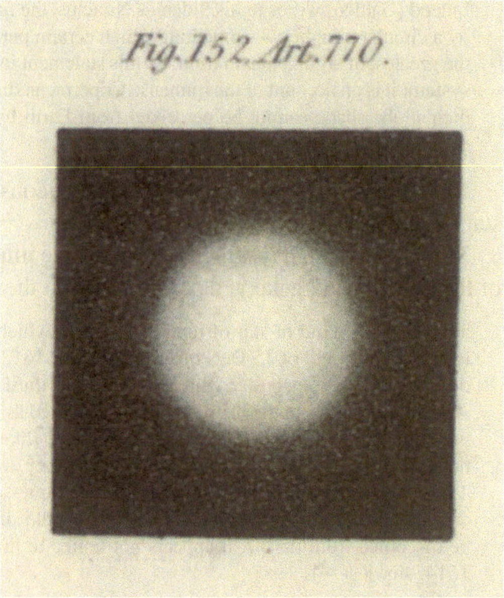

Fig. 8.5 A star as seen through a small aperture telescope (see Herschel 1828, 491 & Plate 9). This appearance of a sphere of measurable size is entirely spurious—an artifact of optics known as an *Airy disk* (after George Biddell Airy, who worked out the mathematical theory of the spurious disk). However, early telescopic astronomers took such telescopic images to be the physical bodies of stars (Grayson and Graney 2011). ETH-Bibliothek Zürich, Alte und Seltene Drucke

masked down to approximately 1–2 cm (Fig. 8.5). Marius thought he was seeing the physical bodies of the stars and that this then showed that they could not be at the distances required by the Copernican cosmos:

> Since truly now it may be most certainly established, that by this telescope on the Earth even the fixed stars to be seen to be circular, this line of argument [that on account of the immense Copernican distance of the fixed stars from Earth their globe shape cannot be perceived from Earth] surely falls, and the contrary is plainly built up: specifically, that the sphere of the fixed stars is by no means removed from the Earth by such an incredible distance as the speculation of Copernicus produces. Rather, such is the segregation of the fixed stars from the Earth, by the harmonious Tychonic ordering of the spheres of the heavens and those of mine, as the structure of those bodies may nevertheless be distinctly seen the shape of a circle by this instrument. (Marius 1614, sig.)( )( )(1^r)[5]

In brief, Marius's observations of the disks of stars (disks that he did not understand to be spurious) showed that the telescope did not provide a solution to the star size problem. Thus he endorsed the Tychonic system, and not the Copernican one.

It is worth noting another skilled observer who observed the (spurious) telescopic disks of stars: Galileo. Indeed, writing to Marc Wesler in 1612 concerning sunspots, Galileo sought to refute Christoph Scheiner's suggestion that stars and sunspots may have in common differing shapes. To that effect he states that:

[5]Marius (1614), sixth-seventh page of "Praefatio ad candidem lectorem." English translation from Graney (2015), p. 51.

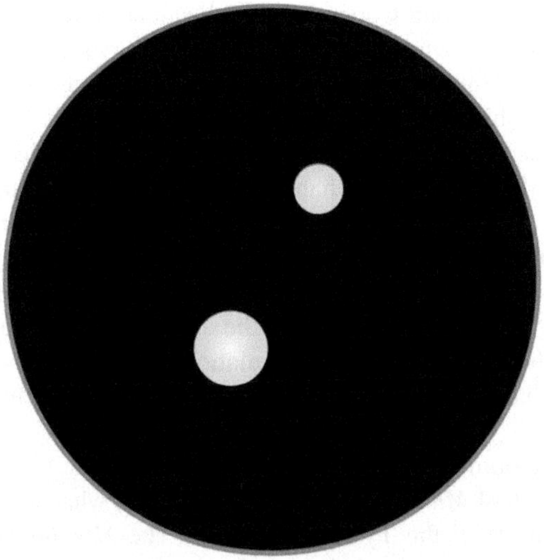

Fig. 8.6 Diagram showing the telescopic appearance of the double star Mizar, according to observing notes by Galileo from 1617. Galileo recorded the components of Mizar as having diameters of 6 and 4 seconds of arc and a separation of 15 seconds of arc. He assumed that Mizar was two stars at differing distances along a line of sight. On the assumption that stars were of the same physical size as the sun, he calculated that since the larger component was 1/300th the apparent diameter of the sun's 1800 seconds, it must be 300 times more distant than the sun. At such a distance, these two stars must reveal prominent differential parallax if the Earth is in motion. They do not, which would have suggested to Galileo that either the Earth is not in motion or the stars are at vast distances—and thus enormous—by virtue of their 6- and 4-second apparent diameters (the star size problem of Tycho Brahe). See Ondra (2004), Graney (2007), Graney (2008), Graney and Sipes (2009)

> stars, whether they are fixed or wandering, are always observed to maintain their shape, and that shape is circular. (Reeves and Van Helden, 2010, p. 101)

and also:

> this assertion [of stars having differing shapes, like sunspots] could not be condemned as a manifest falsehood if the telescope had not resolved this uncertainty by showing us that all stars, fixed and wandering alike, are absolutely round in shape. (Reeves and Van Helden, 2010, pp. 286–287)

Indeed, Galileo would make numerous references to the telescopic disks of stars in his writings—disks which he too, like Marius, took to be the true bodies of stars. He would even observe the double star Mizar and measure both the separation of the two component stars and their apparent diameters (Fig. 8.6), assuming them to be two stars at differing distances along a line of sight. Logically Galileo should have arrived at a conclusion similar to that of Marius—especially as regards Mizar, whose two components, according to Galileo's assumptions, should have showed a dramatic differential parallax were the Earth in motion about the sun. However, Galileo made no mention of the anti-Copernican implications of his telescopic star observations (and no mention whatsoever of his Mizar observations).[6]

Thus we reach an interesting question. Marius supported a Tychonic universe, based in part upon his telescopic observations of stars. His support was logical, granted the knowledge of the time, but ultimately the Tychonic system was shown to be invalid. Galileo, on the other hand, supported a Copernican universe, despite his observations of stars which argued to the contrary. Arguably, granted the knowledge of the time, this was not so logical, but ultimately the Copernican system was shown to be valid. Is the better astronomer the one who follows the observations and calculations logically or the one who ends up supporting the valid system, despite the observations and calculations? Was Marius an astronomer too excellent? Might he have been viewed by history more favorably had he ignored certain results?

There is much room for further study of Simon Marius, especially in regard to the impact of his 1614 *Mundus Iovialis*. For example, what was the impact of the *Mundus* endorsement of the Tychonic system? The *Mundus* is cited by Locher (who worked under the Jesuit Scheiner) in Locher's *Disquisitiones*, also published in 1614. Although the *Disquisitiones* adopts a tone that is favorable to Galileo and unfavorable to Marius, it both notes the telescopic appearance of stars and cites the star size problem as a primary argument against Copernicus.[7] Moreover, the star size problem is cited by Francesco Ingoli in the essay he wrote to Galileo just prior to the 1616 condemnation of the Copernican theory by agents of the Roman Inquisition[8] [Galileo considered Ingoli to have been of influence in regards to that condemnation (Finocchiaro, 1989, p. 155)]. How central is Marius's *Mundus*—perhaps the first work to cite telescopic observations in support of a geocentric cosmos—to all the events that followed in the 2 years after the *Mundus*? Indeed, is the *Mundus* the first work to cite such telescopic observations, or did Marius make earlier mention of them? Or did another astronomer precede Marius in this regard? There is much yet to study on the subject of the most excellent astronomer Simon Marius.

---

[6]Ondra (2004), Graney (2007), Graney (2008). Galileo's Mizar observations lay overlooked in his notes until recently discovered by Ondra.

[7]See Locher/Scheiner (1614), pp. 25–28, 53–54, 78, 80. On Locher and the star size problem, see also Graney (2015), pp. 64–67.

[8]Graney (2015), pp. 66–76. Interestingly, Ingoli suggests to Galileo that the solution to the star size problem is that the stars might behave differently than other celestial bodies—as indeed they did, for the disks the telescope revealed when turned to the stars were spurious, whereas the disks the telescope revealed when turned to planets were genuine. Appendix A of Graney (2015) contains a complete translation of Ingoli's essay.

References

Dreyer, John Louis Emil (1909): The Tercentenary of the Telescope, *Nature* 82 (December 16), pp. 190–191

Finocchiaro, Maurice A. (1989): *The Galileo Affair: A Documentary History*. Berkeley, California: University of California Press

Gaab, Hans; Leich, Pierre (2014): Life and his Research, *Marius-Portal* (http://www.simon-marius.net/index.php?lang=en&menu=2)

Graney, Christopher M. (2007): On the accuracy of Galileo's observations, *Baltic Astronomy* 15, pp. 443–449

Graney, Christopher M. (2008): But Still, It Moves: Tides, Stellar Parallax, and Galileo's Commitment to the Copernican Theory, *Physics in Perspective* 10, pp. 258–268

Graney, Christopher M. (2013): Stars as the Armies of God: Lansbergen's Incorporation of Tycho Brahe's Star-size Argument into the Copernican Theory, *Journal for the History of Astronomy* 44, pp. 165–172

Graney, Christopher M. (2015): *Setting Aside All Authority: Giovanni Battista Riccioli and the Science against Copernicus in the Age of Galileo*. Notre Dame, Indiana: University of Notre Dame Press

Graney, Christopher M.; Sipes, Henry (2009): Regarding the Potential Impact of Double Star Observations on Conceptions of the Universe of Stars in the Early 17th Century, *Baltic Astronomy* 18, pp. 93–108

Grayson, Timothy P.; Graney, Christopher M. (2011): On the telescopic disks of stars – a review and analysis of stellar observations from the early 17th through the middle 19th centuries, *Annals of Science* 68, pp. 351–373

Herschel, John F. W. 1828, *Treatises on Physical Astronomy, Light and Sound Contributed to the Encyclopædia Metropolitana*. London and Glasgow: Griffin and Co. (http://www.archive.org/details/treatisesonphysi00hersrich)

Huygens, Christiaan (1722): *The Celestial Worlds Discover'd: or, Conjectures Concerning the Inhabitants, Plants and Productions of the Worlds in the Planets, 2nd ed.* London: James Knapton

Locher, Johann Georg; Scheiner, Christoph (1614): *Disquisitiones Mathematicae, De Controversiis et Novitatibus Astronomicis*. Ingolstadt: Angermaria 1614

Marius, Simon (1614): *Mundus Iovialis Anno M.DC.IX. Detectus Ope Perspicilli Belgici*. Nürnberg: Johann Lauer 1614

Ondra, Leos (2004): A New View of Mizar, *Sky & Telescope* 108 (July), pp. 72–75

Reeves, Eileen; Van Helden, Albert (2010): *On Sunspots: Galileo Galilei and Christoph Scheiner*. Chicago: University of Chicago Press

Van Helden, Albert (1985): *Measuring the Universe*. Chicago: University of Chicago Press

Chapter 9
In the Turmoil of the Early Seventeenth-Century Cosmology Debate: Simon Marius as a Supporter of the Tychonic System

Pierre Leich

Abstract The early seventeenth century had a great diversity of astronomical world system models hotly debated by the astronomers of the period. With hindsight, in comparison with Galilei, one cannot avoid noticing that the margravial astronomer Simon Marius held the wrong position. This article will show firstly that a reliable scientific proof for heliocentrism had not yet been accomplished and secondly, that Marius's arguments were in good accordance with the empirical observations. In some areas, one would even like to call on him, not to take individual observations too "literally." An examination of his research is not only a serious reconsideration of an underappreciated astronomer but also promotes an understanding of how the world systems were negotiated and how we progress toward certain knowledge.

Following the introduction of the telescope[1] at the end of September 1608 at the court of Prince Moritz of Nassau (Prins Maurits van Oranje, 1567–1625) during a peace conference in the Netherlands, the announcement of the new invention spread rapidly throughout the whole of Europe via diplomatic correspondence, nunciatures, and merchants. By April 1609 the French King Henry IV, his prime minister the Duke of Sully, Archduke Albert VII of Austria, and Pope Paul V had all apparently received one (Sluiter 1997).

Born in 1573 in Gunzenhausen, Simon Mayr, who called himself Marius and who had been court mathematician in Ansbach since 1606, was very likely the first professional astronomer outside of the Netherlands, who knew about the telescope.

[1]The leaflet "Ambassades du Roy de Siam envoyé à l'Excellence du Prince Maurice, arrivé à La Haye le 10 Septemb. 1608" reports about the presentation, presumably Paris 1608, reprinted in November in Lyon, sig. $B1^r$–$B2^r$ (Facsimile-reprint in Stillman Drake 1976 and Huib J. Zuidervaart/Henk Zoomers 2008). An astronomical usage had already been mentioned, since stars, usually not visible, could be seen with this instrument.

P. Leich (✉)
Simon Marius Society, Nuremberg, Germany
e-mail: pierre.leich@simag-ev.de

H. Gaab, P. Leich (eds.), *Simon Marius and His Research*, Historical & Cultural Astronomy, https://doi.org/10.1007/978-3-319-92621-6_9

Shortly before the presentation of the telescope in Den Haag his patron, Hans Philipp Fuchs von Bimbach (circa 1567–1626), visited the Michaelmas Fair in Frankfurt, which started its main trading week on Monday, September 12/22, 1608.[2] Fuchs von Bimbach was the most influential official at the Ansbach court, before he switched to the imperial services in 1616 and died as colonel general (Generaloberst) in the army of the Danish King Christian IV during the Battle of Lutter am Barenberge in 1626. Obviously he believed that aspects of Marius's mathematical work could assist the military, for example, in surveying, and therefore he was very interested in inventions of that nature. In Frankfurt he was offered a telescope through the mediation of a known merchant. However one lens was broken and the requested price was too high, so he did not purchase. Nevertheless the news of the instrument now reached Ansbach, where Fuchs von Bimbach and Marius tried a replicate the telescope.

This report is due to Marius, who refers to the encounter in Frankfurt and the following efforts in Franconia in the preface of his main work *Mundus Iovialis* from 1614 (Marius 1614, sig.)( )(2^{r-v}). Since this corresponds with the alleged reason as to why the States General of the Netherlands refused to grant patents to the opticians Hans Lipperhey (circa 1570–1619) and Jacob Adriaansz Metius (after 1571–circa 1630), there is no doubt about the accuracy of this report. By October 1608 there was no more exclusivity, and the knowledge of manufacture was obviously widespread. Even a third, unknown person claimed to be able to manufacture telescopes.

The vendor's identity at the Frankfurt autumn fair is not recorded, but there is a plausible connection to Jacob Metius.[3] In 1608 his brother Adriaan (1571–1635) had prepared a version of his previously compiled book *Institutionum astronomicarum* for international distribution. The Frankfurt fair was the most important transshipment point for the German market and Adriaan got his brother's help with his publications. Therefore it is possible that Jacob Metius was at the autumn fair for the promotion of Adriaan's book. This argumentation is backed up by the fact that Marius tells us in *Mundus Iovialis* that "Belgian" (here Belgian refers to inhabitants of the United Provinces and not those of modern Belgium) had not only offered the device as merchandise; he had also "developed" it.[4] Despite his excellent Europe-wide contacts, Galileo (1564–1642) claimed, he had only heard about these spy-glasses by the middle of May 1609. This would be more than half a year after Marius. After a merchant in Padua appeared with the instrument by the end of July 1609, Galileo quickly managed to build a replica, and he was already able to demonstrate an instrument with eight times magnification to the nobles and senators

[2]For further details on dating and on Fuchs von Bimbach, see the contribution by Dick in this book.

[3]I owe this and the following references to Leiden, Delft, and Franeker Huib J. Zuidervaart. For the supplier's identity, see Arjen F.B. Dijkstra (2012, p. 137).

[4]"Qui excogitarit instrumentum quoddam," sig.)( )(2. In contrast to this, Dick argues in this edition that the "Belgian" probably stayed in Frankfurt until the end of the fair and could have hardly returned to the Netherlands in the middle of October for the patent application. Since the fair last week usually was used for residual sales to the mostly rural customers, in my opinion, this argument doesn't seem to be mandatory for suppliers of astronomical literature.

of Venice at St Mark's Tower in August 21 and present it to Leonardo Donato, the Doge of Venice, on August 24, 1609.

By contrast the efforts in Franconia couldn't initially make any progress. Marius reports how Fuchs von Bimbach drew a concave and a convex lens on the table with chalk, and by using regular glasses, they both figured out that they had "to a certain extent ascertained the truth of the matter" (Marius 1614/1916/2019, preface; Marius 1614, sig.)( )(2^{v}). With the help of a gypsum cast, Fuchs von Bimbach assigned lens makers in Nuremberg to produce exact lenses, but according to Marius, they lacked the appropriate tools and the true knowledge of the manufacturing method.[5] And so it took until summer 1609 before Marius could hold a Belgian telescope in his hands.

In the meantime the Leiden professor of mathematics Rudolph Snel (1546–1613)—father of Willebrord van Roijen Snel, who discovered Snell's law in 1621—demonstrated a telescope in one of his lectures, and his students could even acquire devices too (Zuidervaart/Rijks 2015, pp. 58–60). One of this students could have been the Frankfurt-born Adam Valentin Fuchs von Bimbach, who had matriculated at the University of Leiden on June 20, 1609, before he transferred to Franeker, where he is verifiable until 1610 (Dijkstra 2012, p. 138), thereby establishing a plausible connection, as to how Hans Philipp Fuchs von Bimbach could have had access to a telescope and from which source it could have come. Corresponding to the memories of Snel's student Théodore Deschamps, the telescope shown by Snel during his lectures came from a "Lunetier de Delft," suggesting that additional instruments were also coming from Deft. In turn the only optician in Deft in this time was Evert Harmansz, who later adopted the surname Steenwijck (Zuidervaart 2012). Therefore Harmansz was very probably the manufacturer of Johannes Fabricius's instrument, who had matriculated at the Universiteit Leiden in 1609 and was a student of Willebrord Snel (Zuidervaart/Rijks 2015, p. 60).

If Fuchs von Bimbach's relative had relations to Franeker before his departure from Leiden, however, a connection via Adriaan Metius would also be plausible. He was a professor of mathematics at the Universiteit van Franeker, the second oldest university of the Netherlands, after Leiden. With his brother Jacob Metius from Alkmaar, we arrive again at one of the telescope's inventors.

In the autumn of 1609, Fuchs von Bimbach increasingly entrusted his telescope to the court astronomer from Ansbach:

> Interdum dabatur mihi potestas port domum, præsertim circa finem Novembris, ubipro more in meo observatorio confiderabam astra.[6] (Marius 1614, sig.)( )(2^{r})

Since the duties of the margravial court mathematicus, besides astronomy and astrology, also included work as a medical doctor and the creation of calendars, Marius knew about Jupiter's good visibility during the turn of the year 1609/1610

[5]Marius 1614, sig.)( )(2^{v}: "instrumentis idoneis & veram conficiendi ratione."

[6]"Sometimes I was allowed to take it home, especially around the end of November; there I usually watched the stars in my observatory."

and announced the Jupiter opposition in his *Schreibkalender auf 1609*.[7] With the closest approach between the Earth and Jupiter on the 6th and the opposition on the 8th of December, the planet was particularly bright. Around the winter solstice, it was visible for a particularly long period, and following the beginning of its retrograde motion on the 9th of October, it stood out especially well against the fixed stars by seeming to run against the direction of the stars.

Marius probably started his observations of the Jupiter system 1 month before Galileo, but this can't be verified by independent sources and is pointless anyway, since Galileo unequivocally secured his priority by publishing first in March 1610.

Sensational Observations

In *Sidereus Nuncius* Galileo describes having seen three little points around Jupiter on January 7, 1610, which he initially considered to be fixed stars. Nevertheless he was amazed, "because they were perfectly aligned parallel to the ecliptic and seemed to be brighter than the other stars of the same size."[8] In the following night, when Jupiter stood east of the little stars, he suspected them to be common stars and Jupiter to be prograde, although the astronomical tables already indicated retrograde motion. He soon realized that he had made a sensational discovery, possibly making him world-famous, if only he could be the first one to publish it. Still writing in his diary, he immediately began to note instructions for the wood engraver and switched from Italian to Latin—the international language of science (Galilei 1613) (Fig. 9.1).

Marius started his record of the Jupiter's moons, by his own statement, on December 29, 1609, according to the Julian calendar (os), 1 day after Galileo, who was already making his records in the new style, since Pope Gregory XIII had implemented a calendar reform in the year 1582, which wasn't recognized in protestant countries for many years. The differing dating systems have sometimes led to some confusion in the modern literature, but in the seventeenth century, they were quite common, and Marius's calendars always have columns for both the old and the new calendars.

Unfortunately Marius failed to publish his findings quickly, and he mentions his telescopic observations for the first time in print in the *Prognosticon Astrologicum auf 1612* (dedication from March 1, 1611). He disclosed that by the use of his telescope, he identified the Milky Way and some nebulae as accumulations of a myriad of stars. Observations of the Earth's moon and the Jupiter's moons are only hinted at. In the dedication from June 30, 1612, to the *Prognosticon auf 1613*,

[7]Marius miscalculated only by a few hours. ☍♂ see 27.11/07.12.1609 in *Alter vnd Newer SchreibCalender*, sig. B3r; in *Newer vnd Alter SchreibCalender*, sig. B4v. December, 8th 1609, greg., 01:35 UT would've been correct.

[8]Galilei (1610), Er: "eo quod secundum exactam lineam rectam atque Eclipticæ parallelam dispositæ videbantur, ac cæteris magnitudine paribus splendidiores."

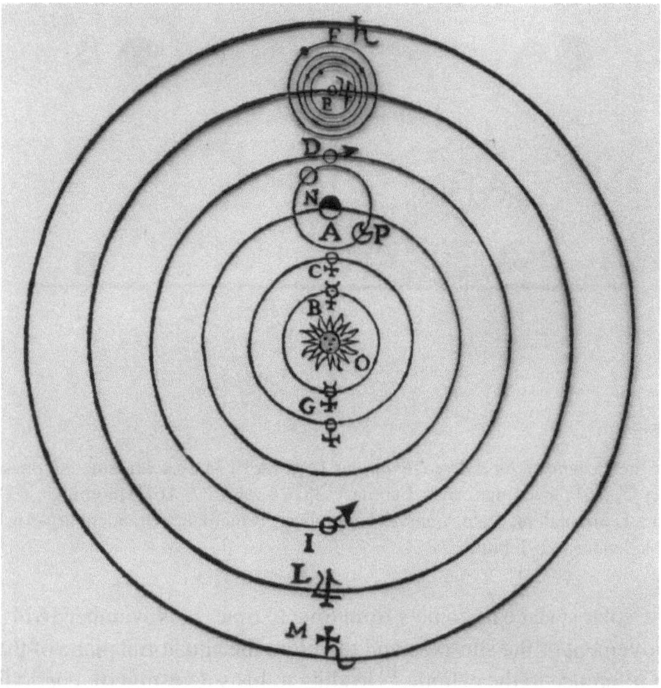

Fig. 9.1 The Copernican world system with the Earth's moon and the Jupiter's four moons; in Galileo Galilei, *Dialogo sopra i due massimi sistemi del mondo, Tolemaico e Copernicano,* Firenze 1632, p. 320; State- and Municipal Library Augsburg, signature 4 Math 207

Marius mentions sunspots and informs about distances and orbital period times of the Jupiter's moons. In the *Prognosticon auf 1614* (dedication May 16, 1613), he presents his observations more precisely, which he then presented completely in 1614's *Mundus Iovialis.*

Marius also observed the Sun with his telescope, although he wasn't one of the first to do so. In December 1610, the Englishmen Thomas Harriot (1560–1621) had already started with 199 drawings of sunspots and Johann Fabricius (1587–1616) saw in East Frisia dark spots on the Sun of which he and his father David Fabricius (1564–1617) determined from March 1611 onward the rotation periods.[9] In the same month, Christoph Scheiner (1573–1650) and his student Johann Baptist Cysat (1586–1657) started observing the Sun in Ingolstadt. For Galileo records are known as of February 1612.

According to his own account (Prog. 1613, sig. A4^{v}), Simon Marius was shown the techniques by Ahasver Schmidner, improved them and had observed sunspots since August 1611, which he thought to be slag, resulting from the Sun's fire, which

[9]In June 1611 Johann Fabricius issued *De Maculis in Sole observatis et apparente earum cum Sole conversione narratio* in Wittenberg, as the first publication about sunspots.

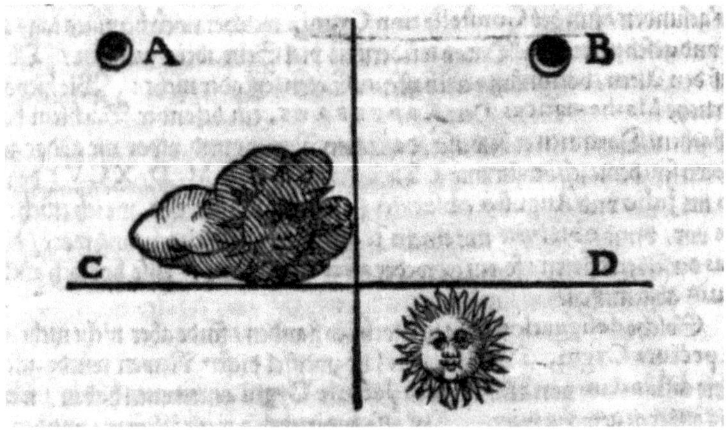

Fig. 9.2 In the *Prognosticon Astrologicum auf 1612* (A3^r) Marius explains the phases of Venus: left February 5, 1611 evenings, right February 25, 26, and 27, 1611 mornings, CD horizon, C occidentalem, D orientalem; State Archives Nuremberg, Principality Brandenburg-Ansbach, Staats- und Schreibkalender (129), number 274

falls off the solar surface as comets from time to time. In November 1611, he figured that the movement of the sunspots and therefore the equatorial plane of the Sun were tilted with reference to the ecliptic,[10] leading to his suggestion of a periodicity of the sunspots.[11]

The phases of Venus, which were very important for determining which astronomical system was the correct one, had been observed by Marius since winter 1610–1611, a quarter of a year after Galileo, who started his Venus observations shortly after arriving in Florence in October 1610. Galileo informed his friend and correspondent Giuliano de Medici (1574–1636) in Prague about this phenomenon on December 11, 1610, by the use of an anagram, for which he gave the solution a few weeks later: "Cynthiae figuras aemulatur mater amorum" (Galilei 11, 1966, p. 12 in document 451): The mother of love emulates the shapes of Cynthia. The "mother of love" stands for Venus and Cynthia is the epithet of the Greek moon goddess Artemis, whereby the effect is clearly addressed. In 1623 in *Il Saggiatore* Galileo published illustrations of the phases of Venus (Fig. 9.2).

Marius reports in a lost letter to Nikolaus von Vicke, who later informs Johannes Kepler on July 6, 1611 (os), verbatim:

> Tertiò demonstrabo, Venerem non secus illuminari à Sole ac Lunam, eamque Corniculatam, διχότομον, etc. reddi, prout à fine anni superioris, usque in Aprilem praesentis, à me ope perspicilli Belgici multoties et diligentissimè observata et visa est.[12]

[10] Cf. the article by Neuhäuser/Neuhäuser in this book.

[11] For full details of Marius's sunspot observations, see Neuhäuser and Neuhäuser in this book.

[12] Johannes Kepler (1611), p. 383, in document 618. "Thirdly I will prove, that Venus is illuminated by the sun in the same way as the Moon and that she is horned and becomes half, as I have thoroughly observed many times with the Belgian perspicillum from the end of the last year to April of the present year."

Fig. 9.3 The planets perform loops with stationary periods and retrograde motion as observed from the Earth. From: Giovanni V. Schiaparelli, Die homocentrischen Sphären des Eudoxus, des Kallippus und des Aristoteles, *Abhandlung zur Geschichte der Mathematik* 1 (1877)

In the dedication from March 1, 1611, of the *Prognosticon auf 1612* (sig. A3^r), Marius is convinced:

> Daß also gar kein zweiffel mehr ist/denn das Venus von der Sonnen erleuchtet wird/wie der Mond/Welcher Meinung wol etliche auß den Alten gewesen/aber nie von keinem mit Augen gesehen worden.[13]

Kepler first published the Venus observations of both Galileo and Marius in the preface of the *Dioptrice* (1611), whereby the priority is acknowledged to Galileo. Galileo published his findings in the introduction of the *Discorso al Serenissimo D. Cosimo II. Gran Duca di Toscana intorno alle cose, che stanno in sù I'aqua, ò che in quella si muovano*, Florence 1612. Marius gives first graphic presentation in the *Prognosticon auf 1612* (sig. A3^r). The first graphic display of the moons of Jupiter and their courses can also be found there (sig. C3^r). In 1620 Kepler published an illustration of the Venus phases in *Epitome Astronomiae Copernicanae* (Kepler 1620, p. 536) (Fig. 9.3).

---

[13] "That there is no doubt, that Venus is enlightened by the sun like the moon. An opinion shared by the old ones, but never was seen by anyone."

In January 1611 Galileo drew the conclusion:

che Venere necessariissimamente si volge intorno al sole[14]

While Galileo doesn't mention the shining light of Mercury, Marius noted in the dedication from June 30, 1612, in the *Prognosticon auf 1613* "that Mercury is illuminated by the Sun in the same way as the Venus and the Moon" and reports his observations of the brightness. Johannes Bosscha (1831–1911) therefore calls him the discoverer of the Mercury phases (Bosscha 1907, p. 518f.).[15] But he probably deduced this only from the brightness,[16] and the Mercury phases were first observed by Giovanni Battista Zupi (Zupus, around 1590–1650) in Naples.[17]

His whole life Marius was also interested in phenomena, which strictly speaking fell in the category of meteorology. In the pre-telescopic era, he was already a well-versed observer and had kept meteorological recordings since 1594. He published on the comet of 1596 as well as on the big comet of 1618[18] and determined the position of the supernova in the constellation of the Serpent Bearer of 1604. In 1599 he published an extensive tabular spreadsheet, and his annual calendar was issued for the years 1601 to 1629.[19]

During his lifetime his telescopic observation of the Andromeda nebula in December 1612 was already acknowledged. As the first European, he described the pale radiance, which couldn't be attributed to single stars and instead compared it to the glow of a burning candle through a translucent horn (Marius 1614, sig.)( )(4^{r}).[20] Of course he couldn't fathom its true nature as a galaxy, so he speculated a faraway comet as a possible explanation. The earliest description of the Andromeda nebula can be found in a parchment manuscript by Persian astronomer Abd ar-Rahman al-Sufi (Al Sufi) about fixed stars from circa 964 (cf. Strohmaier 1984, text p. 50, fig. p. 52).

Not only was Marius active in every important fields of observation, he also had significant mathematical expertise and published as a translation from the Greek *Die*

[14]Letter to Giuliano de' Medici in Prague; in Galilei 11 (1901), p. 12 (Lett. 451); "that necessarily Venus as well as Mercury orbit around the sun."

[15]Kepler honors Marius's arguments in *Epitome astronomiae Copernicanae*, Linz 1620, p. 536.

[16]In the *Dialogo* Galileo makes Salviati say on the third day: "come anco le mutazioni di figure, che assolutamente bisogna, che seguano, come in Venere," p. 331; "It is the same with his [Mercury's] changes of its phase, that are necessarily taking place as with Venus."

[17]Francesco Fontana (around 1580–1656) reports about whose observations from May 23, 1639, and January 26, 1646, with two illustrations by Zupi in *Novae coelestium terrestriumque rerum observationes, et fortasse hactenus non vulgatae*, Napoli 1646, p. 89f. I thank Fabio Ferrario for this source, who confirmed the accordance of the modern recalculation with the description of the phases by Zupi.

[18]For further details, see the chapter by Hamel in this book.

[19]For further details, see the contributions of Matthäus and Kremer in this book.

[20]"Similis fere splendor apparet, si à tonginquo candela ardens per cornu pellucidum de nocte cernatur".

Ersten Sechs Bücher Elementorum Euclidis in Ansbach in 1610. Therefore it is historically significant upon which astronomical world system he settled.

Arguments for Heliocentrism

A mathematical formulation of heliocentrism had been in existence since the publication of Nicolaus Copernicus's *De revolutionibus orbium coelestium* in Nuremberg in 1543. With the epicycle system, the circular motion remained unchanged, but now the system, first made known by Aristarchus, could be discussed by professional astronomers and entered public awareness. Of course Copernicus couldn't offer any new physics with a theorem of inertia or laws of motion, but a number of arguments seemed much clearer.

His aversion to the Ptolemaic equant led Copernicus to a change of his reference point, which allowed an elegant explanation for the apparent loop motion of the planets (Fig. 9.4). While the movements of the moon, the Sun, and the sphere of the fixed stars could be described to a certain degree with spherical geometry since antiquity, the planets' periodic stagnation and retrograde motion proved difficult to explain, with the best solution being Apollonius's model, that the planets were rolling on circles, whose centers in turn were circling the Earth.

With his change in perspective, Copernicus could show that the planetary loops could also be explained, if we assume that we are watching the celestial spectacle

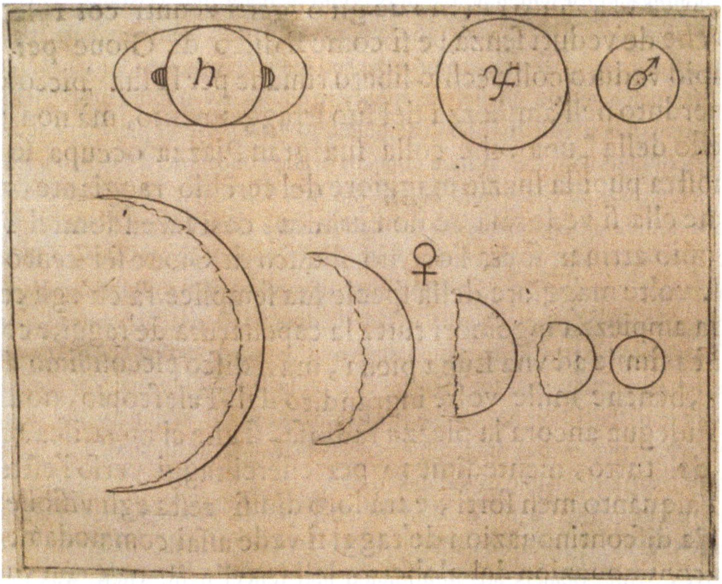

Fig. 9.4 The Venus has phases and has "handles," from: Galileo Galileo, *Il Saggiatore*, Rom 1623; Istituto e Museo di Storia della Scienza, Firenze

from a moving Earth. Every time the Earth passes one of the outer planets or is passed by one of the inner planets, those loops occur, which therefore prove not to be real but only a trick of perspective. This clarifies why the Sun and the moon show no loop movement, which you would expect in a consistent geocentric system. This also explains why Mercury and Venus, as seen from the Earth, are always near the Sun.

All of this is conclusive and can be reconciled with the observed phenomena, even without any additional presumptions that are necessary in the Ptolemaic system. Looking at the outer planets, for example, the straight lines joining the Earth and the (middle) Sun and the epicycle center and the true planet have to be parallel and have to have the same direction of rotation. Therefore a planet could have arbitrary angle distances to the Sun, but retrograde motion would only occur during opposition, and the planet would be nearer to the Earth and, respectively, brighter—both as observed.

Furthermore Copernicus could point out that the fixed stars no longer had to rotate through a vast distance, so that—anachronistically speaking—enormous centrifugal forces could be dispensed with. Yet, whereas materially undetermined stars at the time belonged to the nonmaterialistic aether, the argument taken seriously could also be turned against the Copernicans, and an answer could be demanded as to why the rotating Earth did not break apart.

More discrepancies with the Ptolemaic-Aristotelian worldview came to light, when Danish astronomer Tycho Brahe (1546–1601) observed a new star in the northwest of the constellation Cassiopeia in November 1572, which stayed visible for 18 months and for which no parallax could be determined (Brahe 1573). By now we know, that it was a supernova type Ia of a white dwarf approximately 10,500 light years away. For the sixteenth century, the sudden emergence and disappearance of a star in the unchangeable sphere was highly irritating. Even more problematic was a comet 5 years later, which Tycho calculated to cross several planetary spheres and therefore collided with the antique concept of the concentric crystal spheres.

All of this was already valid before the new telescopic observations. With the telescope additional arguments for the heliocentric system were found.

The rough surface of the moon with lofty mountains and deep valleys, "not smooth, regular and of perfect curvature, like most of the philosophers believed the moon and other celestial bodies to be" (Galilei 1610, p. 12), points to a similarity of the Earth and other planets, which disavows the dualism that has been valid since ancient times, namely, that for the heavens, other laws apply than for the Earth. On the other hand, the moon lies on the frontier between the sublunary and supralunary spheres, which makes the comparable aspects plausible, as illustrated by the spots on the moon's surface, visible to the naked eye.

Marius and Galileo agree on a lot of phenomena, although they often differ in their interpretation. Both consider it to be proven that the fixed stars shine by themselves, while planets are illuminated and reflect light. There was increasing evidence that the stars are not nailed to the firmament, thereby making the issue of the planet's motion controversial.

Also there are more stars than one could see with the naked eye. Some nebulae can be resolved into many stars. Marius recognized the Milky Way as a multitude of individual stars and mentioned the "cloud of the crab" as an example. Galileo demonstrated the multitude of the stars in the constellation of Belt and Sword as well as in the Pleiades.

The sunspots contradict the classical opinion, which stated the Sun is a flawless, unchanging crystal sphere a concept, unfortunately used by the church as a metaphor for the "immaculate" Virgin Mary. For an upright Christian, it was hard to abandon this doctrine carelessly and possibly prematurely, as can be seen in the work of the Jesuit Christoph Scheiner of Ingolstadt. Once the sunspots were accepted as real, the rotation of the Sun can be deduced, fitting perfectly into Johannes Kepler's idea of a vortex power, carrying the planets along and giving a hint to a modern concept of force.

The occurrence of supernovae and comets was a strong indicator that the Aristotelian-Ptolemaic worldview regarding its arrangement and "mechanism" is fundamentally flawed.

From this discomfort one can hardly construct a proof for heliocentrism, but obviously the traditional astronomy had substantial explanatory gaps. With the discovery of the Jupiter's moons, suspense builds up, because now it is obvious that there are celestial bodies not orbiting the Earth. The moons of Jupiter don't circulate on an epicycle in front of or behind Jupiter, but rather around it. However, the movement of the whole Jupiter system with its moons can't initially be deduced from the lunar orbits.

Copernicanism therefore wins no compelling arguments. Nevertheless a new structural model is indicated, consisting of a central star, planets, and moons. The existence of the Jupiter's moons also terminates the special position of the Earth's moon, which circulates the Earth in the heliocentric system. It is recognized that planets can have moons, the number seven is inconsequential for the planets, and obviously there is a kind of "ligation" in a pre-gravitational sense. This facilitates the idea of Earth being accompanied by the moon during its orbits around the Sun.

Refutation of the Ptolemaic System

The fatal blow for the classical geocentric system came with the discovery of the phases of Venus. This involves not only the simple observation that there are phases. Ptolemy even demanded those for the "bottom" planets—meaning our inner planets. These just remained unobservable for the astronomers of ancient times. The sequence of shining lights revealed in the telescope proved that Venus orbits the Sun. Simultaneously it was confirmed that Venus and probably all other planets were not self-illuminating. However, even this strong argument does not determine if the Sun-Venus-system is moving and if so around what. With this once again heliocentrism had not been proven, however, the Ptolemaic-Aristotelian system of the world had been falsified, with respect to its statements about the inner planets.

Systematically this insight requires at least the so-called Egyptian system, in which the Earth is still orbited by the moon, Sun, Jupiter, and Saturn, but Mercury and Venus orbit the Sun.

This state of affairs, which we can consider to have been reached by Galileo in December 1610 and Marius in spring 1611, can be expanded by repeated and precise observations of the Jupiter satellites. Here Simon Marius comes into focus.

In the *Prognosticon auf 1613* (sig. A4^r), he already described his difficulties in measuring the periods especially of the fourth moon, until he found a solution:

> Nemlich/daß solche Newe Planeten mit jhrer *æqualitate*, sampt jhrem *Centro* ♂ nicht *terram*, sondern *solem respicirn* [. . .][21]

In *Mundus Iovialis* he also writes about an astonishing discovery he made while observing the Jupiter's moons (Marius 1614, sig. B3^v, V):

> Post plurimas observationes factas, atq; post deprehensas cuiuslibet quam proximè periodos revolutionum, animadverti etiam aliud phenomenon. Nimirum quod æqualitate motus sui principaliter quidem Iovem: cum Iove autem non terram sed Solem respiciant.[22]

In his theory he determined (Marius 1614, sig. E2^v):

> Verum observationes meæ [...] restantur aliam adhuc inæqualitatem subesse, & Iovem non terram sed Solem pro centro habere [...][23]

The moons only orbit uniformly, with reference to the Sun. This remarkable conclusion is indicated by the precision of both Galileo's and Marius's observations, but neither Galileo nor Kepler remarks upon this. Why is that? There is a strong suspicion that for a convinced Copernican, there is nothing much to gain, since the mere existence of the moons already sheds light on heliocentricity. Primarily this question is interesting for those astronomers who believe in a geocentric system. These astronomers are forced by the phases of Venus to allow that the inner planets orbit the Sun. The precise observation of the Jupiter system shows that Jupiter also orbits the Sun. Since Marius suggests that Mars and Saturn also orbit the Sun, he arrives at the model known as the "Tychonic system." Therein all other planets orbit the Sun, which together with the moon orbit the Earth. There are some so-called semi-Tychonic models in between, with varying numbers of the planets orbiting the Sun, but here only the full Tychonic system is relevant, since it is kinematical equivalent to the Copernican model. Both systems equally allow the reproduction of all effects of illumination, size, and location in the solar system.

---

[21] "That is, that these new planets move uniformly around their center Jupiter, not with respect to Earth but to the Sun [. . .]".

[22] "After I made a great deal of observations and collected the orbital periods of every satellite as exactly as possible, I noticed another phenomenon: in the regularity of their movement they were aligned with Jupiter as their center; but together with Jupiter they are not aligned with Earth, but instead with the Sun as their center."

[23] "My observations [...] proof, that there is another disparity left and that Jupiter does not have the earth as its center, but the Sun."

Within the Tychonic system, the apparent "planet loops" can be explained with the relative movement of the involved bodies, and it is logical that the Sun and the moon don't show any loops. The phases of Venus as well as the size and brightness of the planetary discs are also explained, and Jupiter revolves in both systems around the Sun. Thereby the Tychonic system was by all means an advanced model at the beginning of the seventeenth century.

Marius describes in *Mundus Iovialis* (Marius 1614, sig. C3^{r}) how he came to this conclusion:

> Occasionem autem hujus inventi mihi præbuit mea de systemate mundano opinio, quæ in genere cum Tychonis consentit, in quam incidi hyeme, quæ erat inter annum 1595. & 1596. quando primum legi Copernicum [...][24]

Marius first encountered the Tychonic system in the following autumn in a draft version. Shortly before this, he is said to have handed over a manuscript about his system with an explanation to the consistory in Ansbach. This is mentioned for the first time as *Hypotheses de systemate mundi* in Vocke's *Geburts- und Todten-Almanach Ansbachischer Gelehrten* of 1797[25] and from there was adopted in a lot of reference books. This manuscript is thought to be lost and was probably never printed. Marius only notes in *Mundus Iovialis* that this conception "was discovered by myself'[26] (Marius 1614, sig. C3^{v}), and after Easter 1596, he "offered his assumptions with an explanation" to the consistory[27] in Ansbach. Marius calls on all former members of the consistory—two persons by name—as witnesses. Most of his teachers and members of the consistory councils were already deceased by 1614. But he also mentions George Friedrich I of Brandenburg-Ansbach-Kulmbach, who had allegedly shown his appreciation of it. To claim something like this without any legitimacy would have been an atrocity, excluding Marius from any following employment at the margravial court.

In the absence of any sources, nothing more could be said about this point, but it may be stated that Marius not only was at the top level internationally in terms of his observational skills but also his theoretical knowledge predestined his attempts to find an answer for the world system problem.

Although Marius not only knew about the findings falsifying the Ptolemaic system but also discovered them independently, he remained a proponent of the Tychonic system. Why didn't he decide to "change sides?"

[24]"The possibility to find this [the fact, that the moons move uniform in relation to the sun], formed my opinion about the world system, which corresponds with Tycho's. I encountered it in the winter between the years 1595 and 1596, when I read Copernicus for the first time."

[25]"Marius, oder Mair, Simon". In: Vocke 2 (1797/2001), p. 415. From there *Hypotheses de systemate mundi*, Norimbergae 1596 got sorted under number 2832 in. Houzeau, Lancaster 1964, p. 611.

[26]"hujus mei invéti"

[27]For further details about the consistory of Ansbach, see the contribution by Gaab in this book.

Arguments Against the Copernican System

To answer this question, we have to reconstruct his line of reasoning. Copernicus had provided a lot of good arguments for heliocentrism, but a compelling argument was missing. The new telescope observations elucidated above partially contradicted the Ptolemaic system but supported both heliocentrism and the Tychonic system. Even worse, apart from the dynamical arguments—and you have to exclude even these without a new physics—every observation within the solar system which supports Copernicus also supports Tycho and vice versa.

Differences or even an "Experimentum crucis" are only imaginable for two groups of phenomena: quasi-physical phenomena and optical observations of objects beyond the planetary system.

The first group is examined for expected consequences and their conformity or contradiction—depending on the standpoint—to reality. Supporters of geocentricity orientate themselves on Aristotle and Ptolemy, amounting to "common sense," which just doesn't perceive the double movement of the Earth. If the Earth orbits the Sun while rotating, three phenomena would be expected, or it had to be explained why they would not be expected:

Why isn't the Earth's movement noticeable in the clouds? Shouldn't they be blown away against the direction of motion? Airstream can be experienced on the Earth and even comets and shooting stars have a visible tail.

Why do falling objects not lag behind the Earth's rotation? Since the times of Eratosthenes, the Earth's size had been known roughly. With modern data the speed of the ground at middle European latitudes is 1000 km/h. This should lead to observable effects.[28]

Shouldn't this huge Earth break apart during its daily rotation like a cream cake in a centrifuge? Of course there was no theory of centrifugal forces, but watching water drift outward, when you whirl a bucket of water tied to a rope, is not really a high-tech experiment. How strong would the effect be, if the rope were 6000 km long?

To all these questions, the Copernicans couldn't give any satisfactory answers. Only Newton's physics would deliver arguments to clarify the issue in favor of heliocentrism. Galileo—and this should be emphasized—had the right intuition with his much criticized theory of the tides[29] to look for effects of the Earth's movement (Coriolis-group) on the terrestrial processes of motion. However, with the means available in his times, this was a futile endeavor.

At the beginning of the seventeenth century, a lack of explanations spoke against the Copernicans, because the burden of proof normally is the duty of the proponents of a new theory and not of those who can call on a one and a half thousand-year-old theory.

---

[28]The argument is basically correct. However, an opposing forward motion results when falling bodies take their impulse downward so that they are now "too fast".

[29]On January 8, 1616, "Discorso sopra il flusso e reflusso del mare" was directed to Alessandro Orsini for the first time; Galilei 2 (1843), pp. 387–406.

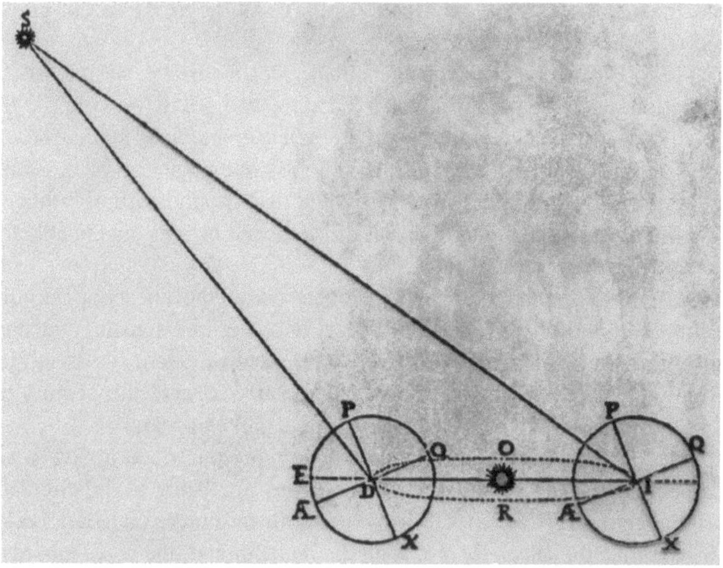

Fig. 9.5 During the circulation around a stationary Sun, nearby stars should appear under a different angle in spring than in autumn. From: John Wallis, *Opera mathematica*, vol. 3, Oxford 1699, p. 706

Worse still, a serious optical objection can be proposed. When the Earth orbits the Sun, shouldn't nearby fixed stars during their annual course be observable under different angles? This fixed star parallax should be noticeable through a shift against the heavenly background, which wasn't observed until the nineteenth century (Fig. 9.5).[30]

This argument had already been brought up by Aristotle (Aristoteles, De caelo, 2. book 296 b 3ff.) and Ptolemy (Ptol. Almagest 1,6; 7,1), and the correct answer had been given by Aristarchus, who referred to the enormous distances involved. This can be found in the metaphorical claims that the Earth's orbit corresponds to the fixed star sphere like the center of a sphere to its radius, found in the dedication to King Gelon in Archimedes' *Sand Reckoner*. Knowing about Aristarchus, at the end of the 10th chapter of the first book in his main work, Copernicus writes:

> Quod enim à supremo errantium Saturno ad fixarum sphæram adhuc plurimum intersit, scintillantia illorum lumina demonstrant.[31] (Copernicus 1543, Lib. I, Cap. X, p. 10)

[30]Friedrich Wilhelm Bessel, Bestimmung der Entfernung des 61sten Sterns des Schwans, *Astronomische Nachrichten* 16 (1838), col. 65–96, at the Königsberg observatory; further early observations were made by the Scot Thomas Henderson, who observed the southern parts of the sky in South Africa during 1831–1833, and Friedrich Georg Wilhelm von Struve 1835–1837 with a refractor at the Dorpat observatory; cf. Hans Strassl, Die erste Bestimmung einer Fixsternentfernung, *Die Naturwissenschaften* 33 (1946), Heft 3 (15. August), pp. 65–71.

[31]"From Saturn, the highest of the planets, to the sphere of the fixed stars, there is an additional gap of the largest size. This is shown by the twinkling lights of the stars."

Hence the recognition of fixed star parallax, although its extent couldn't be measured with the available instruments.

With Tycho Brahe this problem intensified, because Brahe, despite his superior instrumentation, described the brighter fixed stars as discs. If we concede the huge distances of space to the Copernicans—thereby letting the fixed star parallax be less than observational accuracy—you also have to assume huge star sizes. Stars would not only be immeasurably faraway; they would also be immeasurably huge. Brahe calculated diameters reaching to the orbit of Mars, and he was thus unable to accept the heliocentric view.

How was this issue affected by the invention of the telescope? At first, the assumption, that planets are reflecting objects of a certain extension and stars shine by themselves, was confirmed. However, while even modern observatories can't show bright stars as discs, the telescopes of the early seventeenth century not only showed planets but also fixed stars as having a disclike appearance.

Of course this appearance is an optical illusion, but our modern view is based on the understanding of the so-called airy disc (rings appearing at the circular screen when a light beam is diffracted). Glowing, punctiform sources on a dark background appear bigger than they actually are. With the realization of the wave nature of light, the limits of the resolution capacity become comprehensible—especially with reference to small openings. Therefore the effect has to be accepted as real at the beginning of the seventeenth century.

In *Mundus Iovialis* Marius announces against Galileo, but also against the tradition, that "all stars twinkle [. . .], some more, some less,"[32] except the moon. He speaks of a "flashing or overflowing of star matter" and is aware of his outsider role: "There will be a lot of know-it-alls, screaming and accusing me of insanity and the biggest misconception. They can do what they want: Regardless, I will inform the reasonable reader about everything I have seen with my eyes and have observed with the utmost care" (Marius 1614, sig.)( )(4^{r-v}). With the instrument that he had possessed since November 1613, he concentrated on the greater stars and realized that "also the fixed stars are of round shape." This confirmed his opinion that "that the sphere of fixed stars is in no way as incredibly faraway from the Earth as Copernicus assumes"[33] (Marius 1614, sig.)( )( )(1^{r}).

Appropriately Marius strives to keep the celestial bodies small and near; by doing so, their (geocentric) high velocities then appear plausible.[34] For Marius Saturn is only 3 times as large as the Earth's diameter, and Jupiter fits 5 times into the Earth's

[32]"Omnes enim stellæ in cælo scintillant, etiam ipse Sol, Luna solâ exceptâ, at quædam plus quædam minus."

[33]"Cum vero nunc certissime constet, etiam fixas orbiculari in terris hoc perspicillo videri, cadit profectò hæc argumentatio, & plane contrarium astruitur, nimirum sphæram stellarum fixarum nequaquam adeo incredibili distantiâ à terris removeri, uti fert speculatio Copernici."

[34]Already at Neptune's orbit the sky would spin with faster than light speed. A model calculation by Kepler yields the circumference of the sphere of the fixed stars 12.566.370 Sun diameters, which would yield 300 times the speed of light on the celestial equator with the modern data, cf. Kepler 1620, p. 500.

diameter, Mars 145, Venus 91, and Mercury 506 times. Marius estimate for the size of Jupiter is particularly inaccurate. Therefore he assumes that Regulus (Cor Leonis) is hardly a quarter of the diameter of Jupiter and is thus four times smaller than the Earth (Prog. 1613).[35]

This size specifications are intended to demonstrate that the stars are only insignificantly further away than the planets, whose closeness is clearly shown by the parallax phenomenon in the form of "loops." Despite the inappropriate size specifications, one has to acknowledge that the disclike perception of the stars was initially an empirical and legitimate basis for this conclusion. It explains why Marius couldn't force himself to accept Copernicus, while Galileo ignored the evidence and assumed the disc-shaped appearance of the stars could somehow be explained away. The American historian of astronomy, Christopher M. Graney, found a fitting formulation for the fact that Galileo had no justification but was right in the end: how Marius was right and Galileo was wrong even though Galileo was right and Marius was wrong.[36]

Galileo's strategy was consequently diametrical to Marius's, and he disposed of the parallax argument with the huge distances between the stars. Of course Galileo tried in silence to verify the existence of stellar parallax, because the argumentative situation is asymmetrical: By not finding the fixed star parallax, the Copernicans wouldn't lose more they had already lost but could win with a single hit. In contrast the supporters of the Tychonic system had to show that fixed star parallax doesn't occur, and all the stars are so close that the parallax should have already been found.

Galileo was certainly aware of his chances of making a revolutionary observation, when in July 1611, Lodovico Ramponi asked him in a letter, including a sketch and the description of a method, to determine the parallax shift of stars close to each other (cf. Siebert 2006, p. 171ff.). Visual binary stars in particular are very promising for this endeavor—meaning pairs of stars that are in a distinctively different distance from the Earth, so that their different parallaxes should be visible as differences in shift. In November 1616, Galileo received a letter from his friend and student Benedetto Castelli (1578–1643) with this suggestion. In January 1617 Castelli urged him to focus his telescope on Mizar, where a binary star was visible (cf. Ondra 2004). Under the assumption all stars are equal in size; Galileo classified the apparent size of the star discs, as an indication of their distance and searched for the parallax effect. Unfortunately, of all things, the Mizar system is a real binary star[37] with an estimated 2000-year period, making it impossible to monitor any movement. The "Experimentum crucis" failed, but the first observation of a binary star was successful.

[35]Apparently Marius confused the numbers here, which of course, is criticized by Klug (1906, p. 477); an excerpt of this prognostic is given by Klug on pp. 521–524.

[36]Graney (2009) as well as the introduction of Chris Graney's contribution in this book.

[37]In fact Zeta Ursae Maioris in the Great Bear constellation is a double system consisting of two binary stars.

Until 1632 every effort was obviously fruitless, but on the third day of the *Dialogo*, Galileo deals extensively with the issue. In print he de facto admits the disclike appearance of the stars and blames the irradiation[38] for it. In *Dialogo* he explains: "Glowing and far away objects are not reflected easily and sharply by it [the eye], but instead it delivers images, surrounded with a corona. These additional and unfamiliar beams are so long and dense, that the actual core appears to be ten, twenty, a hundred and a thousand times bigger, than it would without the unrelated corona" (Galilei, 1632, p. 327). This could be verified by stretching a cord in front of the star (Galilei, 1632, p. 354).

Whether Galileo, who barely appreciated the optical works of Kepler, possessed the means of Hermann von Helmholtz (Helmholtz 1867, pp. 90–103; as presumed by: Plateau 1838) (1821–1894), who held the spherical and chromatic aberration responsible, may be left open. From Galileo's point of view, he had eliminated the alleged huge stars from heliocentrism and mentioned the possibility of verifying a fixed star parallax (Galilei 1632, p. 375):

> talchè quando si trouasse co'l Telescopio qualche piccolissima stella, vicinissima ad alcuna delle maggiori, e che però quella fusse altissima, potrebbe accadere, che qualche sensibil mutazione succedeße tra di loro, rispondente a quella de I pianeti superiori.[39]

However, Galileo couldn't deliver a result and instead let Salviati announce the construction of a splendid instrument. He already "has chosen a location, that is excellently suitable for such an observation" (Galilei 1632, p. 381). He blamed the nonappearance of the expected shifts on their "imperceptible" extension and insufficient precision of the instruments.

As appropriate as this explanation may be, during Marius's lifetime the missing stellar parallax spoke against the heliocentric system. If one abstains from the unfounded allegation a scientifically correct proof for heliocentrism had already been achieved, the telescopic observations cannot deliver a decision between the Copernican system and the one of Brahe and Marius. Moreover, the geo-heliocentric model avoids the contradiction to the Aristotelian-Ptolemaic system, to common sense and of course to the literal reading of the Bible.

The best evidence that heliocentrism could claim for itself during the first half of the seventeenth century were Kepler's laws, even though they lacked the terms "mass" and "force," necessary for dynamics. Especially the third law, describing the connection between distance and period time of bodies in a gravitational center, would have been suited to show that the elaborate Copernicanism—meaning Kepler's—has a lot of explanation power (Fig. 9.6).

At the same time, Marius certainly recognizes the differential motion of Jupiter's satellites in the *Mundus Iovialis* and reports it as a fourth phenomenon:

[38]In *Il Saggiatore*, published in 1623, where Galileo already accuses Marius of plagiarism in the second paragraph of the Prefazione, he also discusses the irradiation of the planets.

[39]"If it is proven by means of the telescope, that a small star is really close to a bigger one and the small one is of higher standing, then it could probably happen, that a noticeable change of their mutual position occurs, analogously to the apparitions on the upper planets."

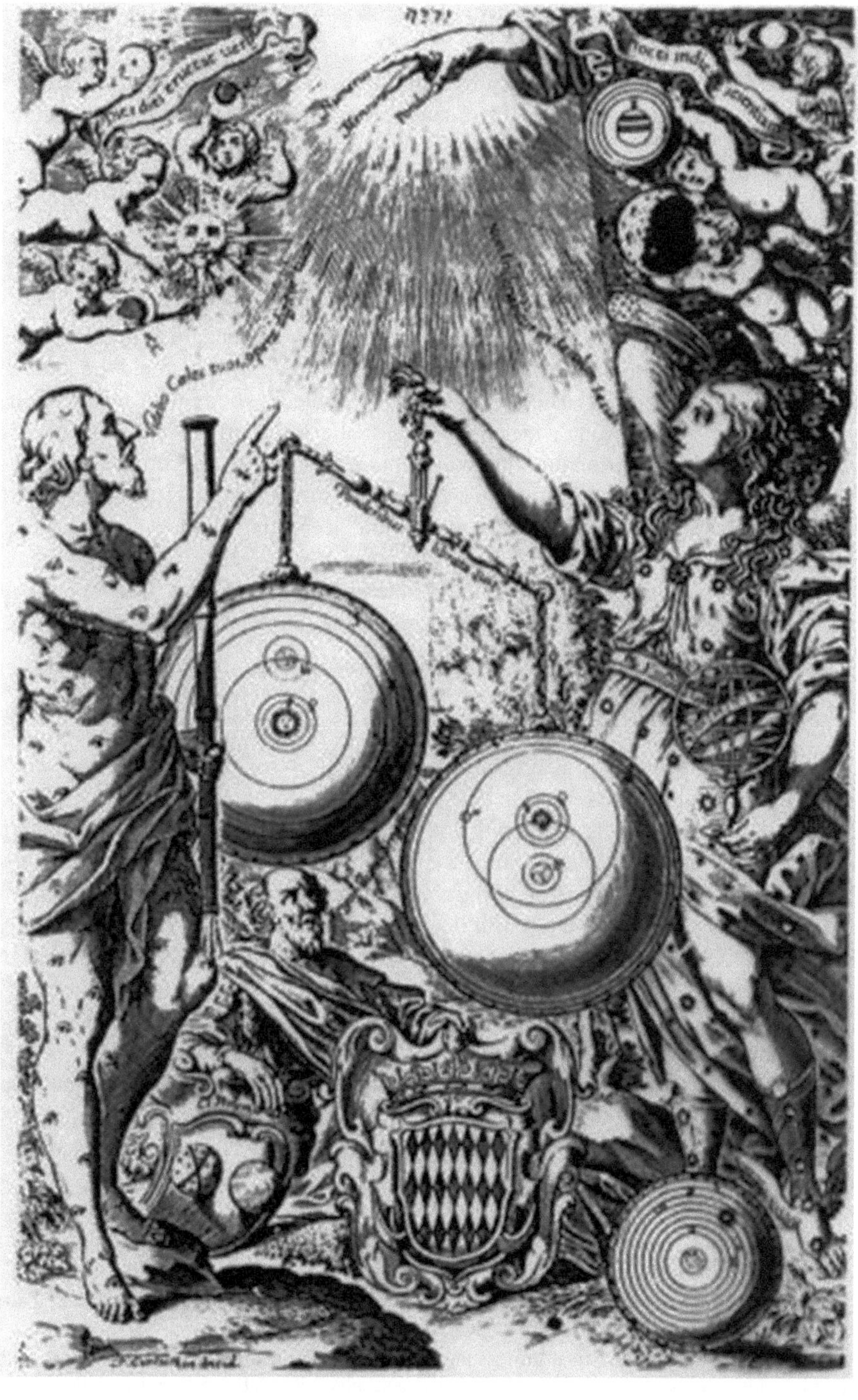

Fig. 9.6 When weighing the world systems in the middle of the seventeenth century, the arguments for the Copernican system appeared to be too light for Riccioli. The arguments for his semi-Tychonic system seemed to be heavier. Giovanni Battista Riccioli, *Almagestum Novum*, Bologna 1651; edition 1653 from State- and Municipal Library Augsburg, signature 2 Math 81-1, frontispiece

> Periodicas restitutiones circa Iovem inæquales deprehendi, propioris celeriorem, remotioris tardiorem.[40] (Marius 1614], sig. B3^{v}, IV)

Already considering the dimensions of the spheres of the four Jovian planets, he conjectured: "Whether, however, this increase or decrease of speed depends on the revolution of the real Jupiter or not, as Kepler, the Imperial Astronomer, has argued with some probability about the Sun and his planets, Mercury, Venus, Mars, Jupiter, and Saturn, is so far unascertained by me, and unobserved." Although he doesn't want to voice his opinion about this issue he adds: "But, to speak the honest truth, I wholly disapprove of this method of reckoning speed or its opposite" (Marius 1614/1916/2019; Marius 1614, sig. A3^{r}).

What Rheticus and William Gilbert could only guess with regard to the increasing orbital periods of the planets in the heliocentric system and could be proven mathematically by Kepler. The orbital period "doubled"—once with regard to the circumference of the orbit, which requires for the same orbital velocity linearly increasing times proportional to the radius, and additionally in accordance with the greater distance between the planet and the Sun in the center. This leads to the conclusion that the Sun is the cause of the motion, decreasing in effect as the distance increases (Fig. 9.7).

The third planetary law can be found for the first time in 1619 in the fifth book of the *Harmonices Mundi*, Chapter 3, the eighth of the thirteen main propositions:

> Sed res est certissima exactissimaque, quod proportio quæ est inter binorum quorumcunque Planetarum tempora periodica, sit præcisè sesquialtera proportionis mediarum distantiarum, id est Orbium ipsorum[41] (Kepler 1619, p. 189f.).

In the summer of 1620, Kepler's fourth book of *Epitome Astronomiae Copernicanae* was printed (Caspar 1948, p. 348). Those seven books were published between 1617 and 1621, delivering the first systematic presentation of modern heliocentrism. In the first three books, Kepler rebuts the pertinent objections against the Earth's movement, which in popular presentations are normally first associated with Galileo's *Dialogo*. In the fourth book, he tried to justify his third law causally (Kepler 1620, Liber IV, Pars II, Section VI, pp. 549–569) and realizes with regard to Marius's figures that it also proves to be approximately accurate for the moons of Jupiter (Kepler 1620, p. 554).[42]

Unfortunately Galileo and Marius never deduced a relationship between orbital period and orbital radius, and they ignored all three of Kepler laws throughout their

[40] "The periodic orbits are uneven, as I have determined: Those of a nearby satellite are faster, those of a more distant one are slower."

[41] "But it is absolutely certain and perfectly correct, that the proportion existing between the periods of two planets is exactly one and a half times the proportion of their mean distances, that is of the orbits. Whereby one has to consider, that the arithmetic mean between both diameters of the elliptical orbits is a little bit smaller than the longer diameter."

[42] 1643 Godefroy Wendelin noted this assumption in a letter to Giovanni Battista Riccioli, who returned to it in his *Almagestum novum* of 1651 in a scholia, Liber VII, Sectio I, Caput III, p. 492, r. col.

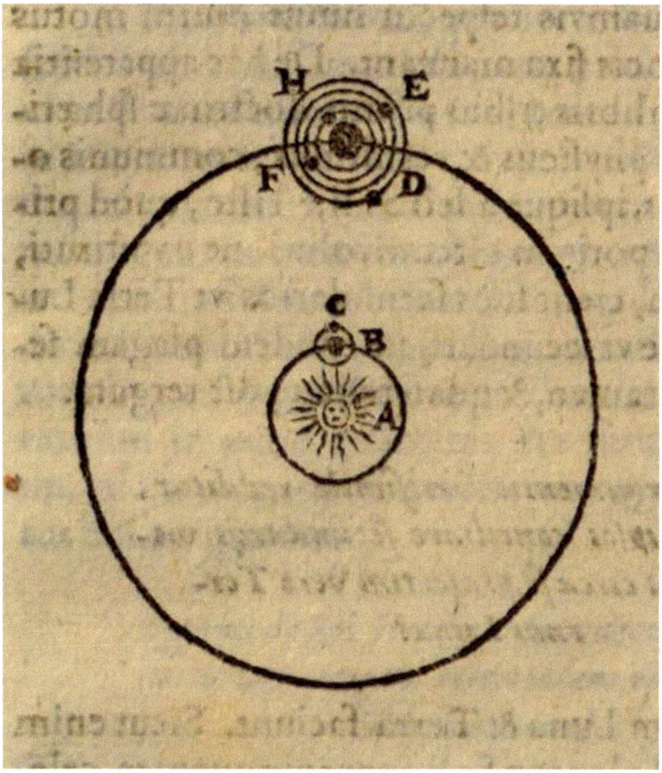

Fig. 9.7 In the movement of Jupiter's moons, Johannes Kepler finds his third law, too; *Epitome Astronomiae Copernicanae*, [printed] Linz [Lentiis ad Danubium] 1620, Liber IV, Pars II, Paragraph 6, p. 554, State Library Regensburg, signature 990/Philos. 1655. The illustration is also used on page 450

lives. While Galileo would have had time until 1642, it is fascinating that in 1614 Marius speaks of a conjecture of Kepler's in connection with the differential motion of the Jupiter's moons, which he, according to Kepler himself, didn't think of until March 8,1618.

An interesting question for further research seems to be whether there are any references in letters or if only verbally at the meeting in Regensburg in October 1613, where Kepler encouraged Marius to name the moons after the affairs of the mythological Jupiter (Marius 1614, sig. B2^{r-v}).

Be that as it may, it shows that Marius was among those at the forefront of contemporary astronomical research. But from a modern perspective, he gave his findings regarding the appearance of the stars too much credit. Although he already opposed the Ptolemaic system at the end of the sixteenth century, he didn't want, because of his impressive observational data, to lend his support to heliocentrism. Neglecting him in the history of astronomy is to misjudge the astronomical debate at the beginning of the seventeenth century and doesn't do justice to Simon Marius's research contributions.

References

Aristoteles, De caelo: *Περὶ οὐρανοῦ* (Peri uranú)

Bessel, Friedrich Wilhelm (1838): Bestimmung der Entfernung des 61sten Sterns des Schwans. *Astronomische Nachrichten*, vol. 16, issue 5, no. 365/366 (13. Dezember 1838); Altona 1839, col. 65–96

Bosscha, Johannes (1907): Simon Marius. Réhabilitation d'un astronome calomnié. *Archives Néerlandaises des Sciences Exactes et Naturelles*, vol. 12 (1907), serie 2a, La Haye, pp. 258–307, 490–528

Brahe, Tycho (1573): *De nova et nullius aevi memoria prius visa stella*. Copenhagen: Laurentius Benedictus 1573

Caspar, Max (1948): *Johannes Kepler*. Stuttgart/Zürich/Salzburg: Kohlhammer 1948

Copernicus, Nicolaus (1543): *De revolutionibus orbium coelestium, Libri VI*. Nürnberg: Petreius 1543

Dijkstra, Arjen F.B. (2012): Between Academics and Idiots. A Cultural History of Mathematics in the Dutch Province of Friesland (1600–1700), Dissertation Universiteit van Twente, Leeuwarden 2012

Drake, Stillman (1976): *The Unsung Journalist and the Origin of the Telescope*. Los Angeles: Zeitlin & Verbrugge 1976

Fontana, Francesco (1646): *Novae coelestium terrestriumq[ue] rerum observationes, et fortasse hactenus non vulgatae*. Napoli: Apud Gaffarum 1646

Galilei, Galileo (1612): *Discorso al Serenissimo D. Cosimo II. Gran Duca di Toscana intorno alle cose, che stanno in sù l'acqua, ò che in quella si muovano*. Florenz: Cosimo Giunta 1612; translated by Albert Van Helden. *Galileo Galilei, Sidereus Nuncius, or The Sidereal Messenger*. Chicago/London: The University of Chicago Press 1989

Galilei, Galileo (1613): I Pianeti Medicei, Osservazioni 7 Gennaio 1610 – 29 Maggio 1613. In: Segni, Antonio: *Le Opere di Galileo Galilei*, Nuova Ristampa della Edizione Nazionale, vol. III, part 2. Firenze 1964, p. 427f.

Galilei, Galileo (1632): *Dialogo sopra i due massimi sistemi del mondo, Tolemaico e Copernicano*. Firenze: Batista Landini 1632

Galilei, Galileo Le *Opere di Galileo Galilei*. Edizione Nazionale, ed. Antonio Favaro, 20 vol. Firenze: Barbèra 1890–1909. Reprint 1929–1939, 1964–1966

Graney, Christopher M. (2009): How Marius Was Right and Galileo Was Wrong Even Though Galileo Was Right and Marius Was Wrong. eprint arXiv:0903.3429, 03/2009

Helmholtz, Hermann Ludwig Ferdinand von (1867): *Handbuch der physiologischen Optik*. Leipzig: Leopold Voss 1867

Hozeau, Jean-Charles; Lancaster, Albert (1880): *Bibliographie générale de l'Astronomie jusqu'en* 1880. Bruxelles: Académie Royale de Belgique 1887–1889, 2 vols. Reprint: General Bibliography of Astronomy to the Year 1880, ed. David Dewhirst. London: The Holland Press 1964

Kepler, Johannes: *Gesammelte Werke*, vol. 16: Briefe 1607–1611, ed. by Max Caspar. München: Beck 1954

Kepler, Johannes (1611): *Dioptrice sev Demonstratio eorum quae visui & visibilibus propter conspicilla non ita pridem inventa accidunt*. Augsburg: Francus 1611; *Gesammelte Werke*, vol. 4 (1941): Kleinere Schriften. Ed. Max Caspar and Franz Hammer. München: C.H. Beck'sche Verlagsbuchhandlung

Kepler, Johannes (1619): *Harmonice mundi Libri V*. Linz: Godofred Tampachius 1619; English translation as *The Harmony of the World by Johannes Kepler*, E.J. Aiton, A.M. Duncan, and J.V. Field, Philadelphia, Pennsylvania: American Philosophical Society 1997

Kepler, Johannes (1620): *Epitome Astronomiae Copernicanae*, book 4. Linz: Johannes Plancus 1620

Klug, Josef (1906): Simon Marius aus Gunzenhausen und Galileo Galilei: ein Versuch zur Entscheidung der Frage über den wahren Entdecker der Jupitertrabanten und ihrer Perioden. *Abhandlungen der mathematisch-physikalischen Classe der Königlich Bayerischen Akademie der Wissenschaften* 22 (1906), pp. 385–526

Leich, Pierre (2012): Die Copernicanische Wende bei Galilei und Kepler und welche Rolle Simon Marius dazu einnimmt. In: Wolfschmidt, Gudrun (ed.): *Simon Marius, der fränkische Galilei, und die Entwicklung des astronomischen Weltbildes* (= *Nuncius Hamburgensis – Beiträge zur Geschichte der Naturwissenschaften*, vol. 16), Hamburg: tredition 2012, pp. 163–193

Leich, Pierre (2014): Die Marius-Renaissance. Die Beobachtung des Jupitersystems durch den markgräflichen Hofastronomen Simon Marius, *Sterne und Weltraum*, Nov. 2014, Heidelberg 2014, pp. 44–53

Leich, Pierre (2015): Sensationen am Himmel – Galilei und Marius entdecken die Jupitermonde und der Streit ums richtige Weltsystem. *Astronomie + Raumfahrt im Unterricht* 52 (2015) 3–4, pp. 15–20

Marius, Simon (1610): *Die Ersten Sechs Bücher Elementorum Evclidis*. Ansbach: Paul Böhem 1610

Marius, Simon (1614/1916/2019): The World of Jupiter, translated by A.O. Prickard and Albert van Helden, in *Simon Marius and his Research*, Hans Gaab and Pierre Leich (eds.), Cham: Springer 2019

Ondra, Leon (2004): A New View of Mizar. *Sky & Telescope* 108 (July 2004), pp. 72–75

Plateau, Joseph Antoine Ferdinand (1838): *Mémoire sur l'irradiation*. Brüssel 1838

Ptolemäus, Almagest: *μαθηματική σύνταξις* (Mathematike Syntaxis)

Riccioli, Giovanni Battista (1651): *Almagestum novum astronomiam veterem novamque complectens observationibus aliorum et propriis novisque theorematibus, problematibus ac tabulis promotam in tres tomos distributam*. Bologna: Victor Benatius 1651

Siebert, Harald (2006): Die große kosmologische Kontroverse. Rekonstruktionsversuche anhand des Itinerarium exstaticum von Athanasius Kircher SJ (1602–1680) (= *Texte und Abhandlungen zur Geschichte der Mathematik und der Naturwissenschaften*, vol. 55). Stuttgart: Franz Steiner 2006

Sluiter, Engel (1997): The Telescope Before Galileo, *Journal for the History of Astronomy* 28 (1997), part 3, pp. 223–234

Strohmaier, Gotthard (1984): *Die Sterne des Abd ar-Rahman as-Sufi*. Hanau/Main: Müller & Kiepenheuer 1984

Vocke, Johann August (1796): *Geburts- und Todten-Almanach Ansbachischer Gelehrten, Schriftsteller, und Künstler*; oder: Anzeige jeden Jahrs, Monats und Tags, an welchem Jeder derselben gebohren wurde, und starb, nebst ihrer kurz zusammengedrängten Lebens-Geschichte und dem Verzeichnis ihrer Schriften und Kunstwerke. Augsburg: Georg Wilhelm Friedrich Späth. part 1 1796, 2. part 1797. Reprint Neustadt a.d. Aisch: Verlag für Kunstreproduktionen 2001

Wohlwill, Emil (1926): Der Betrug des Simon Marius von Gunzenhausen; In: Wohlwill, Emil: *Galilei und sein Kampf für die Copernicanische Lehre*, 2. vol., appendix III. Leipzig: Leopold Voss 1926; reprint Wiesbaden. Sändig 1669, pp. 343–426

Zuidervaart, Huib J. (2012): The 'invisible technician' made visible. Telescope making in the seventeenth and early eighteenth-century Dutch Republic, in: Alison D. Morrison-Low, Sven Dupré, Stephen Johnston, Giorgio Strano (eds.), *From Earth-Bound to Satellite. Telescopes, Skills and Networks*. Leiden/Boston: Brill 2012, pp. 41–102

Zuidervaart, Huib J. (ed.), Zoomers, Henk (transcribed, translated and introduced), *Embassies of the King of Siam sent to His Excellency Prince Maurits, arrived in The Hague on 10 September 1608*. An early 17th century newsletter, reporting both the visit of the first Siamese diplomatic mission to Europe and the first documented demonstration of a telescope worldwide. Wassenaar: Louwman Collection of Historic Telescopes 2008

Zuidervaart, Huib J.; Rijks, Marlise (2015): Most rare workmen: optical practitioners in early seventeenth-century Delft, *The British Journal for the History of Science*, vol. 48/01 (March 2015), pp. 53–85.

Chapter 10
Simon Marius as a Calendar Writer

Klaus Matthäus

The Calendar Writer Simon Marius

The reason for printing yearly calendars, the type of which we are still using today, was to provide a sound health guide.[1] Printed annual calendars were intended primarily to provide dependable information on when, on which days, and on where, on which body part, bloodletting might most conveniently be performed. Since antiquity bleeding had been one of the most common treatments for health care and hygiene, the practice continued through the Middle Ages and into the nineteenth century. Correct application had to take the position of the moon into consideration in accordance with the instructions of *astrologia medica*, *iatromathematics*. Favorable days for phlebotomy were first to be found in a so-called Lasstafeln (bloodletting tables), also called an almanac (Fig. 10.1).

The new art of book printing immediately recognized a lucrative opportunity in publishing the information essential for phlebotomy every year, single-sheet prints showing the status of the lunar phases throughout the year. These dates, changing year by year, as the lunar year does not coincide with the solar year, could hardly be interpreted by a layman without assistance. The first printed broadsides were gradually expanded to their present appearance as annual calendars, presenting the days of the month and the festive days of the church year.

From the middle of the sixteenth century onward, the single-page calendars segued into a multi-leaved calendar that came to be called the "Schreibkalender." One resourceful printer[2] came up with a momentous idea, viz., to break the typesetting of the annual wall calendar in such a way that it could be published as a

[1]For the following explanations, see Matthäus (1969), col. 981–1006.

[2]The resourceful printer was probably Hans Guldenmund in Nuremberg. See footnote 6.

K. Matthäus (✉)
Simon Marius Society, Erlangen, Germany

H. Gaab, P. Leich (eds.), *Simon Marius and His Research*, Historical & Cultural Astronomy, https://doi.org/10.1007/978-3-319-92621-6_10

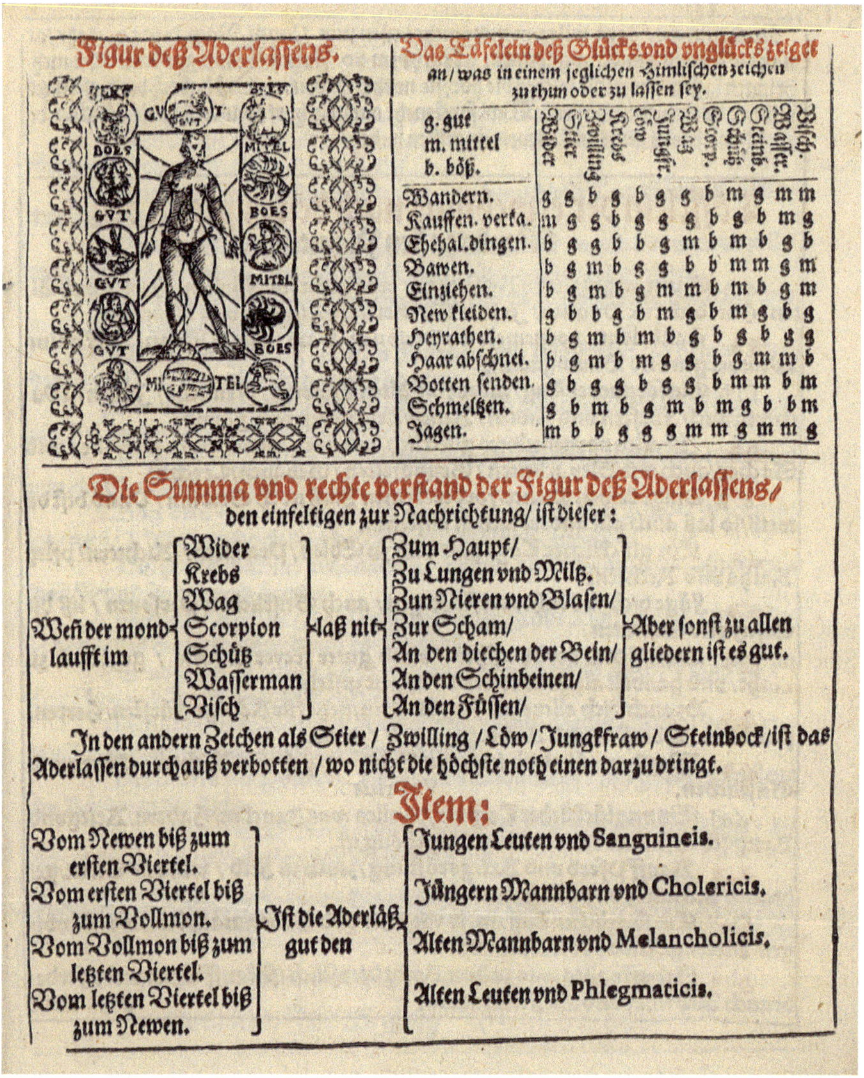

Figur deß Aderlassens.

Das Täfelein deß Glücks vnd vnglücks zeiget an / was in einem jeglichen Himlischen zeichen zu thun oder zu lassen sey.

g. gut m. mittel b. böß.	Wider	Stier	Zwilling	Krebs	Löw	Jungfr.	Wag	Scorp.	Schütz	Steinb.	Wasser.	Visch.
Wandern.	g	g	b	g	b	g	g	b	m	g	m	m
Kauffen. verka.	m	g	g	b	g	g	g	b	g	b	m	g
Ehehal. dingen.	b	g	g	b	b	g	m	b	m	b	g	b
Bawen.	b	g	m	b	g	g	b	b	m	m	g	m
Einziehen.	b	g	m	b	g	m	m	b	m	b	g	m
New kleiden.	g	b	b	g	b	m	g	b	m	g	b	g
Heyrathen.	b	g	m	b	m	b	g	b	g	b	g	g
Haar abschnei.	b	g	m	b	m	g	g	b	g	m	m	b
Botten senden.	g	b	g	g	b	g	g	b	m	m	b	m
Schmeltzen.	g	b	m	b	g	m	b	m	g	b	b	m
Jagen.	m	b	b	g	g	g	m	m	g	m	m	g

Die Summa vnd rechte verstand der Figur deß Aderlassens / den einfeltigen zur Nachrichtung / ist dieser:

Weñ der mond laufft im [Wider / Krebs / Wag / Scorpion / Schütz / Wasserman / Visch] laß nit [Zum Haupt / Zu Lungen vnd Miltz. / Zun Nieren vnd Blasen / Zur Scham / An den diechen der Bein / An den Schinbeinen / An den Füssen /] Aber sonst zu allen gliedern ist es gut.

In den andern Zeichen als Stier / Zwilling / Löw / Jungkfraw / Steinbock / ist das Aderlassen durchauß verbotten / wo nicht die höchste noth einen darzu dringt.

Item:

[Vom Newen biß zum ersten Viertel. / Vom ersten Viertel biß zum Vollmon. / Vom Vollmon biß zum letzten Viertel. / Vom letzten Viertel biß zum Newen.] Ist die Aderlåß gut den [Jungen Leuten vnd Sanguineis. / Jüngern Mannbarn vnd Cholericis. / Alten Mannbarn vnd Melancholicis. / Alten Leuten vnd Phlegmaticis.]

Fig. 10.1 Rules for phlebotomy in the calendar by Marius for 1610 (WLB Stuttgart: HBF 3708)

quarto, in a layout that left the recto page largely empty while the 12 monthly calendars were printed on the verso pages.

The owner of an almanac could then fill these unprinted pages with notes that corresponded to the calendar. This particular design is still in use today. These new quarto calendars were also called writing calendars, the name deriving from the unprinted pages for the owner's notes. The possibility offered here of a chronologically ordered diary met a silent demand, which subsequently easily stimulated large-scale editions of such writing calendars. To quote Kepler[3]:

> [there] is no book under the sun, which sells so many copies and is renewed every year, as just those calendars and prognostica of an infamous astrologer.

The foundation and starting point of the genre of the annual calendar and its annual attachment, the "practica" or "prognostica," were the doctrines of astrology. In their application, the calendar provided the diagnosis and forecasted the cure. The prognostica, in the same format as the writing calendar and on a similar scale, supplemented the calendar by way of further astrological prognostications on the expected weather, on the threats and plagues to be feared, as well as on the threat of war. Since the sixteenth century, it was common for a calendar maker to publish both almanacs and prognostica.

Calendar makers were generally well-educated, skilled mathematici versed in the noble art of *astrologia*, i.e., astronomers and physicians. The proper handling of phlebotomy required a calendar calculated for the locations in which it would be used. For Nuremberg, a series of official mathematici, approved by the city council, were approbated, and the barbers of the imperial city had to execute phlebotomy adhering to their calendars.[4]

Calendars had strong regional links, despite the fact that a notorious astrologer was capable of reaching a wider audience. This in turn demanded the presence of printers and publishers, a demand Nuremberg had been fulfilling since the sixteenth century. The imperial city established itself as the single most important publishing place for calendars within the German Empire.

This presented both opportunity and a place for a creator of calendars, who supplied the Brandenburger Margraves of Kulmbach and Ansbach with his works. The court physician of Margrave Georg der Fromme (the Pious), Dr. Georg Seyfridt, born in Sulzfeld near Kissingen, should be mentioned first. In 1540, he produced a single-sheet calendar printed in Ansbach, the first known calendar for this town.[5] Seyfried published two more calendars covering the years 1544 and 1545, naming him as a physician now serving the court of Margrave Albrecht Alcibiades (1522–1557). The calendars were printed in Nuremberg.[6] Seyfried even left an English single-sheet calendar that had probably been printed in Antwerp.[7] His reputation was, by all accounts, not insignificant. However, it took another

[3]Kein Buch unter der Sonnen ist, dessen soviel Exemplaria verkaufft und alle Jahre wieder erneuert werden, als eben die Calendaria und Prognostica eines beschreyeten Astrologi. Kepler (2004), p. 45.

[4]See Matthäus (1969), col. 1007–1069 and Matthäus (2010).

[5]See Matthäus (1969), col. 1086.

[6]The calendars, like their single-sheet counterparts, bore the title "almanach." Both were printed by Hans Guldenmund in Nuremberg. The masthead for 1545 lists Hans Guldenmund (the Elder) as the printer. For Guldenmund, see Reske (2007), p. 670 ff. The calendars/almanacs for the years 1544/1545 are both preserved in the Ratsschulblibliothek Zwickau.

[7]See Capp (1979), p. 379, for Seyfried, and also see Pültz (1973), p. 179 (B 318).

30 years before M. Georg Caesius (1543–1604),[8] a priest, appeared in print with calendars covering the Ansbach area.

Caesius, born in Rothenburg ob der Tauber, had acquired a working knowledge of astronomy and astrology alongside his studies in theology at the University of Wittenberg.[9] In 1565 he had been ordained as a priest in his hometown, a position he had to give up in 1574 due to a theological dispute. He subsequently made his way to Ansbach where he took up the post of town chaplain. Caesius had already tried calendar publishing as early as 1566, and he had successfully established himself in the market after finding a publisher in Nuremberg. His new sovereign, Margrave Georg Friedrich (1539–1606), reigning from 1556, favored Caesius.[10]

He promoted him to lucrative parishes and granted him the additional position of appointed astronomus.[11]

Next to Caesius another priest, M. Johannes Schulin (1561–1606), hailing from Crailsheim, tried to gain a foothold in Ansbach.[12] His course did not run as smoothly since the consistory took to regarding his calendrical activities with suspicion and demanded censorship over his output, demands that obviously Caesius did not have to meet. At least Schulin was able to dedicate his last calendar to the margravine. These two calendar makers, Caesius and Schulin, active in the Ansbach area of Brandenburg-controlled Franconia, were joined in 1600 by another, Simon Marius (1573–1624) from Gunzenhausen.[13] He had no academic qualifications as a mathematician. The year before he had described himself as a "stipendarius" and "alumnus" at the Margravial College in Heilsbronn[14] to which he had been admitted in 1586. Marius was, as a mathematician and astronomer, as he later stressed, an

[8]Compare Matthäus (1969), col. 1087–1092, and also Dieter Kempkins' chapter (Chap. 4).

[9]Matthäus (1969), col. 1087 and 1091 footnote 756.

[10]Compare Schuhmann (1980), pp. 101–118.

[11]The additional salary of 25 fl. can be viewed as a generous honorarium for the yearly dedicated calendars. Caesius did not call himself appointed astronomus, which by such a post would not normally be omitted. The title woodblocks of his calendars display a certain level of officialdom with the Brandenburger coat of arms. Karl Heinrich Lang noted 1811 (p. 354): "was the margrave's personal astrologer." The vita on p. 351 is however inaccurate.

[12]On Schulin compare Matthäus (1969), col. 1093–1096. Schulin also studied in Wittenberg. See also Heischmann (1974), col. 1684–1688 and 1833f. Schulin owned Marius' *Tabulae Directionum Novae* (1599) as well as Kepler's earlier text *Prodromus Dissertationum cosmographicarum* (1596). Compare Hamel (1998), p. 4f., no. 6.

[13]See Matthäus (1969), col. 1096–1099. Regarding Marius I was able to use the substantial preliminary efforts by Ernst Zinner (1942), whose intentions are now being resumed on the *Marius-Portal*, edited by Pierre Leich. For biographical notes on Simon Marius please consult the chapter by Hans Gaab (Chap. 3). I am very grateful to Hans Gaab for the many amicable conversations regarding Marius. I would also like to express my gratitude to Klaus-Dieter Herbst and Richard L. Kremer for their notes on astronomical matters. Richard Gugel assisted in translating passages from Latin.

[14]Marius (1599), title page

autodidact. The college had little to offer to someone like him[15] with a special interest in the celestial from childhood onward. It did, however, afford him the necessary freedom to acquire a profound knowledge of astronomy for himself. Without ever having been taught by a teacher, he obtained the necessary knowledge in studio astronomy and astrology by himself, subjects he had desired and loved since his early youth, and the study of which cost him, according to his own account, both time and health.[16]

In 1593 Simon Marius first mentions making his own astronomical observations.[17] The "ingenious juvensis"[18] received ample opportunity to conduct his own studies in the field of astronomy through the benevolence of his sovereign, Margrave Georg Friedrich.[19] After the publication of his first work, a treatise on the comet of 1596 that he dedicated to the margrave, he hoped for a grant that would enable him to further his education at the university; but despite several recommendations, the grant was not forthcoming. Despite a grant toward the publication of his *Tabulae Directionum Novae* in 1599 from his sovereign, his personal circumstances still remained uncertain.[20] Marius himself talks about wicked, ill-fated times. Nefarious elements were hindering his ascent. In the end, however, society at court yielded him some benefactors as well. Maria von Eyb, born von Crailsheim,[21] was the widow of the former steward of Margravine Sophie (1564–1639),[22] who supported him during those years with "vielfeltiger [. . .] Ehr und gutthaten" (diverse honors and good deeds). At the beginning of 1599, he gratefully promised to dedicate a calendar to her, since she shared his interest in the celestial arts.[23]

This duly occurred in 1600 when Marius first dedicated a calendar with prognosticon for 1601 to his benefactress. It was an attempt to put his knowledge of the celestial arts to practical use. He found his publisher in the Nuremberger bookseller Johannes Lauer.[24] Lauer was already Schulin's publisher but apparently felt this was not a hindrance. Lauer's publishing house initially employed two printers in the city, Abraham Wagenmann (until 1609) and after that Christoph Lochner, until Lauer was able to open his own printing workshop in 1613 against

---

[15]See Lang (1811), p. 349, who noted to the curriculum for IV highest grade "the sudden break in the classical studies [. . .] the complete disregard of mathematics and history."

[16]*Prog. 1607*, sig. A2^{r-v}, *Prog. 1619*, sig. A3^{r}, [4.2].

[17]Marius (1596), sig. A4^{v}.

[18]Marius (1596), sig. A1^{v}, dedication (in verse) Ad ingeniosum juvenem Simonem Marium.

[19]*Prog. 1602*, sig. A4^{r}; [4.2], p. 101.

[20]Büttner (1813), pp. 70–82.

[21]Eyb (1984), p. 216f.

[22]Schumann (1980), pp. 105–124.

[23]*Prog. 1601*, dedication preface dated Heilsbronn June 29 1600.

[24]See Reske (2007), p. 711ff. In 1584 Lauer married one of the daughters of Michael Endter, bookseller.

the opposition of his former printers.[25] In 1606 Lauer even became Marius' father-in-law, when Marius married his daughter Felicitas.

The new star began to shine brightly. Noble gents and good friends had pleaded with him to continue, even though this seemingly had not been his intention. In his calendar work, he should not let "other strange astrologers" drive him crazy. Marius complied, but the next calendar created something of a problem due to lack of time. The year 1601 finally opened fresh perspectives. Margrave Georg Friedrich offered him a grant, affording him a stay in Prague with Tycho Brahe, spending the summer as part of his team of observers and calculators.[26] It was only for 3 months, but Marius gained deeper insight into the practice of astronomy while being fully accepted by Brahe's knowledgeable staff. His friendly collegiality with David Fabricius dates back to his stay in Prague.[27] This, however, cuts into the time needed to prepare the next calendar, and he was not able to access the works he usually consulted for its compilation.[28] He was allowed, like Caesius, to dedicate the slightly rushed calendar for 1602 to the margrave. At the end of the year, the margrave finally came forward with a grant, enabling him to study medicine in Padua, where he stayed from December 1601 until July 1605.

The circumstances, which led him to study medicine, are not known. This age placed great importance on the teachings of *iatromathematics*, delineating medical practice along the guidelines of astrology. Medicine as a further subject of study was thus a plausible choice, and the higher social status of a physician is indisputable. Marius also continued to send his calendar writings from Padua. Until Georg Friedrich's death in 1603, they were dedicated to the margrave and after that from 1605[29] onward to his successors from the Berlin line of Electors, Margrave Joachim Ernst von Ansbach (1583–1625)[30] and Christian von Kulmbach-Bayreuth (1581–1655).[31] Marius did not complete his academic studies in Padua with a title. His abrupt return home in August 1605 was due to important circumstances "auß wichtigen ursachen [...] vornehmen müssen"; however no reason is given. He humbly appeals to the benevolence of his sovereign.[32]

Once again he offered himself up for service. The following year he was formally appointed by both margraves to the position of principal mathematic with a salary of

---

[25]See Matthäus (1969), col. 1132ff. For Wagenmann, see Reske (2007), p. 700f.; for Lochner, see p. 698ff. Paul Böheim of Ansbach printed the *Prognosticon auf 1613*, probably due to the conflict with the Nuremberg printers (Reske 2007, p. 217ff.). The statement that he had been printing the Ansbach calendars by Marius since 1606 remains to be corrected.

[26]See Büttner (1813), p. 81f. Letter of recommendation to Tycho Brahe by the Margrave, 12 May 1601.

[27]See below, Sect. 2.2.1.

[28]*Prog. 1602*, Dedicatory preface to the margraves, dated 21st September 1601 without location.

[29]The calendar for the year 1604 has not survived.

[30]See Schuhmann (1980), pp. 127–142. Schuman writes that Marius ranked highly in the margrave favors.

[31]Margrave Christian moved his residence from Kulmbach to Bayreuth.

[32]*Prog. auf 1606*, dedication, dated 12 September 1605, Gunzenhausen (sig. 4^{v}).

150 gulden per annum[33] with the proviso not only to continue his medical but also his mathematical—i.e., astronomical—studies.[34]

Johannes Kepler received the same sum as a district mathematician in Graz in Styria.[35] Marius described himself as "Princely Brandenburger Appointed Mathematicus and Medicus" on the title pages of the prognostica from 1607 onward. From 1609 onward, the meridian he based his astronomical computations on ran no longer through Heilsbronn, where the Fürstenschule was, but Ansbach. He continued to dedicate the calendars to both rulers, in Ansbach and Bayreuth, signifying that he continued to hold the position of mathematicus for the districts of both courts.

Calendars compiled by Marius, who died in 1624, continued to be published by Lauer up until the year 1629, since he had amply provided for future editions during his last years.[36]

Fees from his publisher supplemented Marius' insubstantial income. The fees were of great importance to him, and because of his positions, they would have exceeded those of Johannes Schulin who received 12 fl. for a volume. Further dedications to the Nuremberg council also yielded additional income.[37] There is uncertainty about the number of copies printed. In 1610 Lauer was requested by the Nuremberg city council to replace the first sheet of the *prognosticon*; as the calendar maker, Georg Halbmayer felt injuriously attacked.[38] At the beginning of the dispute, there was talk of 11,000 copies, but Lauer claimed that the initial print run for calendars and prognostica had not exceeded 6000 copies; it was in his interest to keep numbers deliberately low.

Since very few comparative accounts are available, Marius' position in the marketplace is difficult to define. He sold more of his writings than Kepler, for whom a print run of 400–600 copies is cited.[39] Marius had seemingly established

---

[33]Compare the chapter by Gaab (Chap. 3).

[34]*Prog. 1607.* Dedication dated Ansbach July 17 1606 (sig. 4^{v}).The assumption that he must have become a successor to either Caesisus or Schulin cannot be sustained in this context. At least there is no contemporary statement affording proof. Likewise there is no historical statement naming Marius as a consulting astrologer to the margraves. According to a statement in Schumann (Annotation 30), this seems to have been the case with Margrave Joachim Ernst von Ansbach.

[35]See Volker Bialas, follow-up report, in Kepler 11,2 (1993), p. 448.

[36]Editions from 1608 onward were usually compiled 2 years in advance. The dedications were usually dated from the first half of the preceding year because the printing had to be finished for the autumn fairs. The years, 1626–1628, were compiled in 1623; the year 1629 was compiled in 1624. See Zinner (1942), p. 34. Perhaps Marius, already in poor health, wanted to provide for his family. The calendars for the years 1626–1629, published after his death, lack the usual dedicatory preamble. The dedication on the title page of the prognostica now cites Sophia, neé Countess of Solms-Laubach, widow of Margrave Joachim Ernst and her brother Count Friedrich of Solms who served as regents for Prince Friedrich, the heir.

[37]See Diefenbacher for a list for 1603–1611, Fischer-Pache (2003), p. 437f.

[38]Compare Matthäus (1969), col. 1099–1102.

[39]See Bialas for follow-up report in Kepler 11,2 (1993), p. 448 (calendar and prognostica for 1599). The print run for 1618 was about 800 copies. See Caspar (1968), p. 63. Polymath Leonhardt Thurneysser (1513–1596), who also acted as a calendar maker; the number of copies is cited at around

himself in the calendar market. Otherwise his publisher, Johannes Lauer, would not have applied for imperial protection against reprints in 1613, which was granted to him the following year. The expenditure for such privileges was not insubstantial.[40]

It should be noted that there are print variations in Marius' prognostica. For the year 1608, the copy held at the ETH Library in Zurich differs significantly from the one held in the State Archives in Nuremberg. It remains to be determined whether this is a later or a pirated edition. Efforts by the publisher Lauer suggest the latter. For the year 1609, copies with different signatures on the single sheet exist, which indicates a hasty reprint.

Who read Marius' calendars? They were not like the popular calendars of later years, intended for the ordinary citizen and farmer, which were also read aloud to the illiterate. Marius' texts were more demanding, often containing longer passages in Latin. Thus the four pages describing the forthcoming total eclipse of the sun of 1605 are completely in Latin. Occasionally he used Greek words. Marius' audience were educated layman and academics; the nobility also took notice of him. Marius expected that his fellow authors would be watching him. The writing calendars of this period were not intended for an illiterate milieu.

Writing Calendars and Prognostica

The Calendars

Marius' calendars correspond to the style in which calendars of his times generally appeared (Figs. 10.2, 10.3, 10.4 and 10.5).[41] The title page cites him by name, holding him responsible for the quality of the work that bears his name. The title *Old and New Writing Calendar* indicates that the calendar was intended for those regions within the empire that had not subscribed to the Gregorian calendar reform of 1582, and therefore the first column displayed the old Julian calendrical calculations.

The *Marius-Portal* (www.simon-marius.net) displays a copy for the year 1615 held in Graz, which starts the first column with the Gregorian calendar and therefore must have been published under the title *New and Old Writing Calendar*.[42]

2000, without further evidence. As he occasionally published two calendars simultaneously, further evidence will be forthcoming once his correspondence, of more than 4000 letters, has been evaluated. See Herbst (1750), *Biographical Handbook* (I am indebted to Klaus-Dieter Herbst for this information).

[40]See Koppitz (2008), p. 312, no. 58.

[41]See Matthäus (1969), col. 1175–1178.

[42]This copy is missing its title page. This edition—the single surviving copy of its kind—was apparently intended for distribution in the Austrian counties. The second half of the calendar, starting with the month of July, however, was compiled by Caesius's pupil, Georg Halbmeyer (see Matthäus 1969, col. 1099–1102).

Fig. 10.2 Title page of the calendar for 1611 (Staatsarchiv Nürnberg: Fürstentum Brandenburg-Ansbach, Staats- und Schreibkalender (129), Nr. 274, 11. Stück). See illustration in Lerch (2015, p. 167): Josephus Scala, Ephemerides. Venedig 1589

Changing woodcut borders, similar styles of which were used by various printers for the editions of other calendar makers, frame the title page. The first surviving writing calendar for the year 1602 displays a border showing the famous

Fig. 10.3 Title page of the calendar for 1606 (UB Augsburg: 02/IV.5.4.15)

Fig. 10.4 Title for the calendar for 1616 [StB Nürnberg: Amb. 4. 261 (3)]

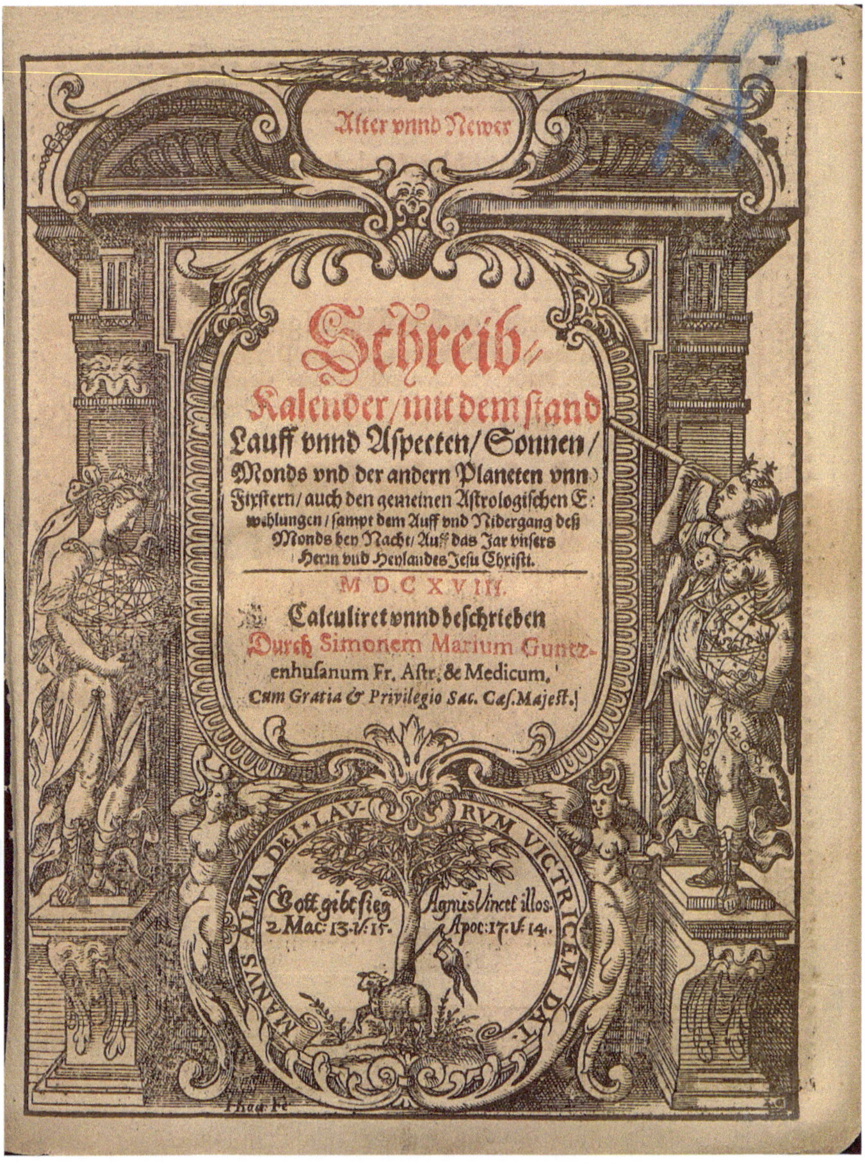

Fig. 10.5 Title page of the calendar for 1618 (Staatsarchiv Nürnberg: Fürstentum Brandenburg-Ansbach, Staats- und Schreibkalender (129), Nr. 274. 18. Stück)

astronomers from Pythagoras to King Alfonso of Castile,[43] another the planetary gods of antiquity.[44]

From 1607—Marius was now appointed mathematicus—the border contains the coat of arms of the princedoms and dominions of the House of Brandenburg.[45] They first appeared in the calendars of Caesius and were still used until the middle of the seventeenth century. Usually these borders appear stenciled in with color. A new version is introduced first for the year 1618: a portal, flanked by personifications of "geometria" and "astronomia," a depiction still in use for calendars of the nineteenth century.[46]

On the reverse of the title page was room to insert a dedication to one of the rulers of the country with a woodcut of the Brandenburg coat of arms. In the editions with a Brandenburg coat-of-arms border, this was already displayed on the title page.

The almanac, the calendar, starts on the next page,[47] with the standard page of explanations, whose first rubric lists the specific dates of the relevant year. To begin with the epoch-making years, the birth of Christ and the creation are named and then the pertinent chronological basic data for the year such as the golden number and the dominical letters. A second rubric explains the abbreviations used in the almanac and calendar for the different phases of the moon and the planetary aspects with red and black print signifying good and bad, respectively. Furthermore ephemeral elections are mentioned, symbols for opportune days for bleeding up to ones for successful child weaning. The symbols remained the same from calendar to calendar, varying only in detail.

Marius attempted to prove his competence with these schematics. The year he gave for the creation of the world was not the usual 3952 BC as calculated by the Venerable Bede; instead he initially used the more recent assessment of Leonhard

[43]The central position is held by Ptolemy, flanked by Timocharis (c. 320–260 BC) and Callippus (fourth century BC). To the left are portraits of Menton and Almeon, meant are obviously the Greek astronomer Meton (fifth century BC) and the Arabic astronomer Almansur (eighth century). To the right stand Hipparch (c. 190–125 BC) and presumably Eratosthenes (thirrd century BC). Below are the Greek mathematicians Pythagoras and Euclid, the Arabic astronomers Messahala (ca. 800 AD), and Albategnius (around 900 AD) as well as later astronomers like King Alfonso X. (1221–1284) and Georg von Peuerbach (1423–1461). Bringing up the rear is Archimedes, in the center, bottom row.

[44]By the gods of the planets, the sun god Apollo is placed in the center, with his corresponding star sign Leo. To is right are Venus, Mercury, and Luna and to the left Mars, Jupiter, and Saturn.

[45]On the upper left is the coat of arms of Prussia. Progressing clockwise is the coats of arms of the Margraves of Brandenburg and the Burggraves of Nuremberg and Rügen. The "Blutfann" (blood banner) signifies the Blutgerichtsbarkeit the highest penal authority in the Holy Roman Empire. Furthermore the coats of arms of Zollern, Jägerndorf, Silesia, Pomerania, Stettin, Wenden, and Cassuben.

[46]Matthäus (1969), n. 1168–1175. The portal first appears on a calendar by David Herlicius (1557–1636) for 1613. It is signed by the Nuremberg artist Johann Hauer (1586–1660). Similarly, a title copperplate designed by Hauer: Caspar Uttenhofer: *Pes mechanicus*. Nuremberg, 1615. See Matthäus (1969), n. 1172, and note 1260 for Hauer; see Grieb 2 (2007), p. 588.

[47]Marius here uses explicitly the term Almanach for the Kalendarium, as it was already common practice in the bloodletting tables.

Krentzheim (1532–1598),[48] the year 3970 BC. From 1618 onward he quoted Joseph Justus Scaliger (1540–1609), who dated it to 3949 BC.[49] After 1612 he quoted three new planetary aspects, defined by Kepler in his *De Stella Nova* from 1606, introduced to provide more accurate weather forecasts.[50]

After this preamble the actual calendar begins on the next (verso) page with the month of January (Fig. 10.6); the following (recto) page is the corresponding writing page for notes, which gives this quarto calendar its name.

The monthly calendars, always on the verso side, usually contain, after the introduction of the Gregorian calendar, four columns. It starts with the "old calendar" and the days of the months, dominical letters, and the day's Christian patron saint. On Sundays, the corresponding scripture passages are printed in red. Next is the column containing the lunar phases and its daily position in the zodiac. The third column was the most extensive. It contains ephemeral elections for the individual days, the aspects and constellations of the planets with timelines, as well as brief weather forecasts according to the course of the stars. A more detailed interpretation of this data was provided by the prognosticon or practica/praktik in the second part of the calendar.

On one occasion, the advice printed in the third column "Besihe hievon mit fleiß die Practica von dieser Constellation" (view diligently the practica of this constellation)[51] can be viewed as self-evident (Fig. 10.7). The fourth and last column contains the "new (Gregorian) calendar." Calendars edited for the Catholic counties had the "new calendar" in the first and the old calendar in the fourth column, as the Graz edition for 1615 demonstrates.[52]

The right-hand side of the calendar, which initially was left completely empty for notes, was soon reduced by another text column (Fig. 10.8a, b). The histories, short extracts of history, were included into the calendar, thereby possibly increasing its appeal.[53] More usual to a calendar were the columns indicating the lengths of day and night or sunrise and sunset. Marius initially wanted to dispense with such common material and opted for a more sophisticated table displaying the daily

[48]Krentzheim (1577), p. 1: "We set the creation of the world 3970 years before Christ our Saviour's birth." Krentzheim, who was born in Iphofen, was a clergyman in Liegnitz for many years; he lost his office as an alleged Calvinist, in 1593.

[49]Grafton (1993), p. 262: "Like most of his predecessors." A later calendar writer, Johann Meyer (1607–1665), in his Quedlinburg writing calendar for 1644, notes a list with 24 different calculations (sig. A2^{r}—friendly note from Klaus-Dieter Herbst). However, Meyer confuses the dates Scaliger puts the creation of the world with the date he defined for the beginning of the Julian period. For Johann Meyer, see Herbst, *Bibliobiographical Handbook*.

[50]See Kepler (1606). Marius gives in the *Prognosticon auf 1612* a detailed explanation for "diligent but privati Mathematici" (sig C6^{v}–C7^{v}); see furthermore, Bialas (2013), pp. 138–143: 4. Aspektenlehre.

[51]*Cal. 1615*, November.

[52]Compare footnote 42.

[53]Hieronymus Lauterbach first expanded his *Neuen Historien und Schreibkalender* for 1572 with a history column. Compare Matthäus 1969, col. 1192 footnote 1338.

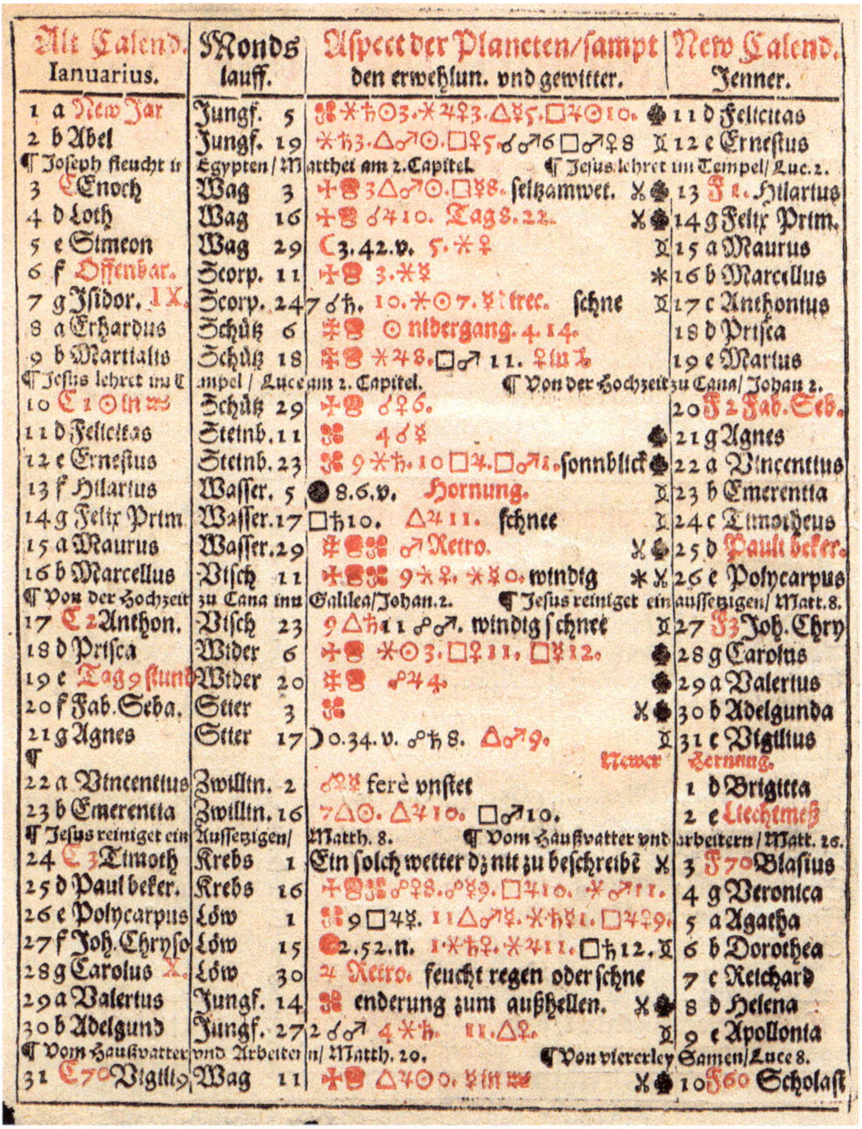

Fig. 10.6 Calendar page for January 1602, verso (Staatsarchiv Nürnberg: Fürstentum Brandenburg-Ansbach, Staats- und Schreibkalender (129), Nr. 283)

position of the planets and the dragon, the ascending point of intersection of the paths of the sun and the moon.[54] It was intended for readers, "so dieser kunst zimlich verstand," who had a working knowledge of the matter without the opportunity to

[54]12.00 o'clock midday civil time calculation,corresponds to 0.00 o'clock astronomical time calculation. The column moon trajectory in the calendar also references midday.

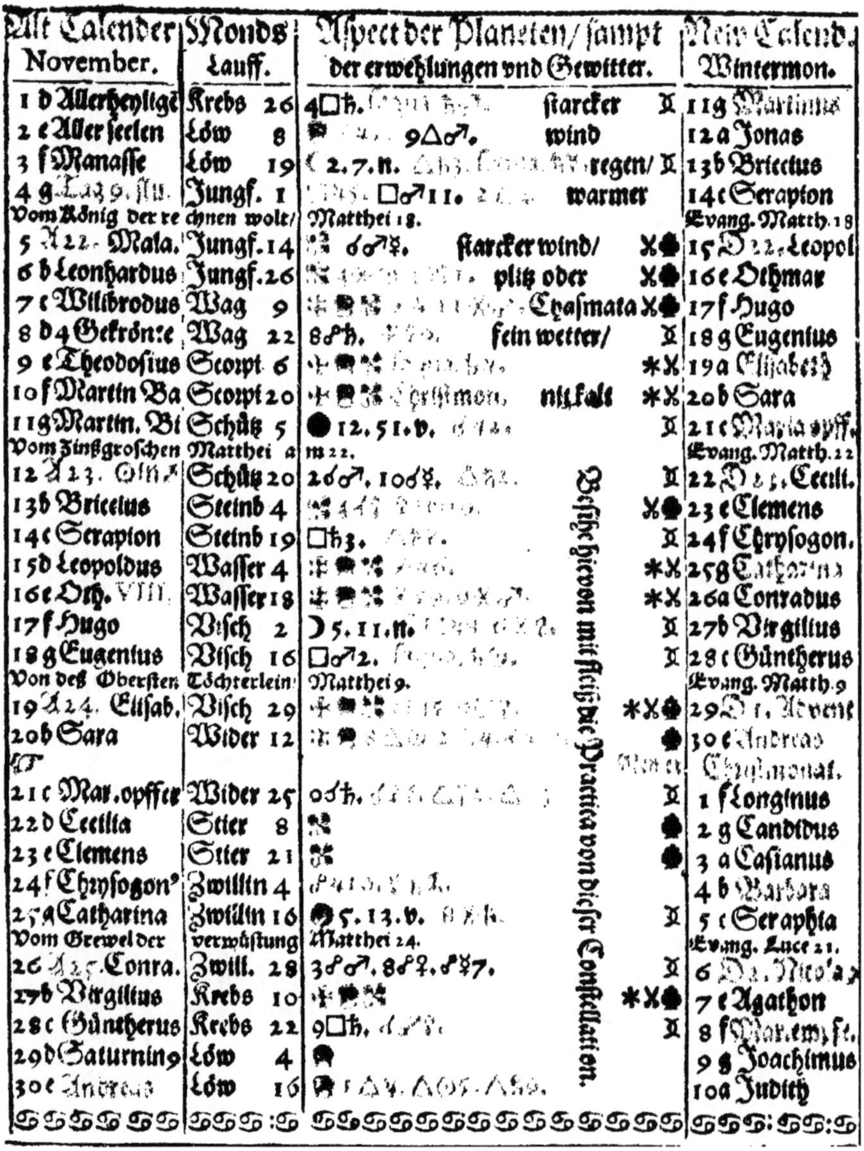

Alt Calender November.	Monds Lauff.	Aspect der Planeten/ sampt der erwehlungen vnd Gewitter.	New Calend. Wintermon.
1 d Allerheylige	Krebs 26	4□♄. … starcker	11 g Martinus
2 e Aller seelen	Löw 8	… 9△♂. wind	12 a Jonas
3 f Manasse	Löw 19	☾ 2.7.n. … regen/	13 b Briccius
4 g [illegible]	Jungf. 1	… □♂ 11. … warmer	14 c Serapion
Vom König der re	chnen wolt/	Matthei 18.	Evang. Matth. 18
5 [illegible] Mala.	Jungf. 14	… ☌♂☿. starcker wind/	15 [illegible] Leopol
6 b Leonhardus	Jungf. 26	… plitz oder	16 c Othmar
7 c Willibrodus	Wag 9	… Chasmata	17 f Hugo
8 d 4 Gekrönte	Wag 22	8☍♄. … fein wetter/	18 g Eugenius
9 e Theodosius	Scorpi 6	…	19 a Elisabeth
10 f Martin Ba	Scorpi 20	… nit kalt	20 b Sara
11 g Martin. Bi	Schütz 5	● 12.51.v. …	21 c Maria opff.
Vom Zinßgroschen	Matthei a	m 22.	Evang. Matth. 22
12 [illegible]	Schütz 20	2☌♂. 10☌☿. …	22 [illegible] Cecili.
13 b Briccius	Steinb 4	…	23 e Clemens
14 c Serapion	Steinb 19	□♄ 3. …	24 f Chrysogon.
15 d Leopoldus	Wasser 4	…	25 g Catharina
16 e Oth. [illegible]	Wasser 18	…	26 a Conradus
17 f Hugo	Visch 2	☽ 5.11.n. …	27 b Virgilius
18 g Eugenius	Visch 16	□♂ 2. …	28 c Güntherus
Von deß Obersten	Töchterlein	Matthei 9.	Evang. Matth. 9
19 [illegible] Elisab.	Visch 29	…	29 [illegible] Advent
20 b Sara	Wider 12	…	30 c Andreas
			Christmonat.
21 c Mar. opffer	Wider 25	☌♄. …	1 f Longinus
22 d Cecilia	Stier 8	…	2 g Candidus
23 e Clemens	Stier 21	…	3 a Casianus
24 f Chrysogon	Zwillin 4	…	4 b Barbara
25 g Catharina	Zwillin 16	● 5.13.v. …	5 c Seraphia
Vom Grewel der	verwüstung	Matthei 24.	Evang. Luce 21.
26 [illegible] Conra.	Zwill. 28	3☍♂. 8☍♀. ☌☿♂.	6 [illegible] Nicola
27 b Virgilius	Krebs 10	…	7 e Agathon
28 c Güntherus	Krebs 22	9□♄. …	8 f Mar. empf.
29 d Saturnin9	Löw 4	…	9 g Joachimus
30 e Andreas	Löw 16	…	10 a Judith

Besihe hievon mit fleiß die Practica von dieser Constellation.

Fig. 10.7 Calendar and practica belong together: Marius had that information printed horizontally on the calendar page for November 1615: besihe hievon mit fleiß die Practica von dieser Constellation (WLB Stuttgart: HBF 3713)

access the ephemerides. It would then be possible to use an astrolabe or a "tabula domorum" of the astrological houses to determine a celestial constellation (i.e., to cast a horoscope).[55]

[55] Cal. 1602, sig. B6^v, End.

a

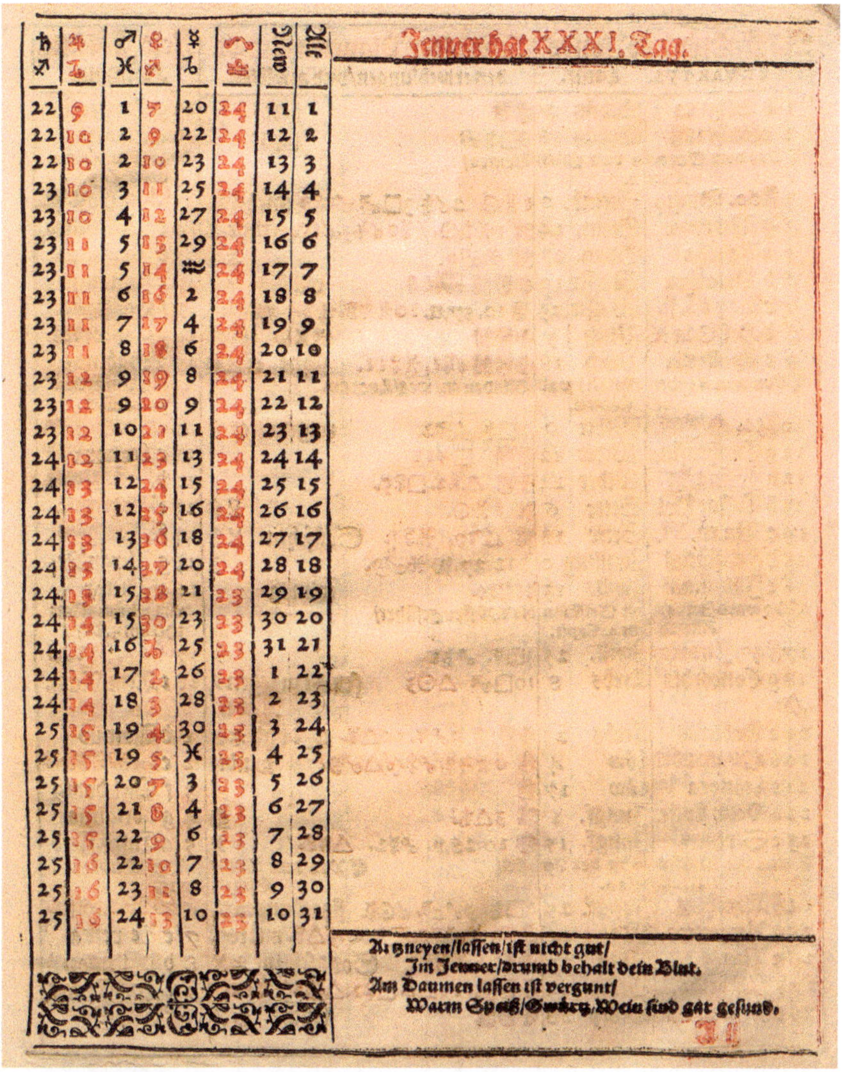

Fig. 10.8 The recto of January in the calendars of (**a**) 1605 and (**b**) 1618 (following page). Staatsarchiv Nürnberg: Fürstentum Brandenburg-Ansbach, Staats- und Schreibkalender (129), Nr. 285 und Nr. 274, 18. Stück

From 1607 onward Marius also includes tables for the lengths of day and night and sunrise and sunset and, as Brandenburger mathematicus from 1609 onward, a column detailing the history of the house. Each month lists the respective births, deaths, and marriages.[56]

[56]This could be the manuscript calendar, cited on the *Marius-Portal* (*Brandenburgischer Historischer Calender im Concept beym Brandenburgischen Archiv*), the full title cited by Zinner (1942), p. 27.

b

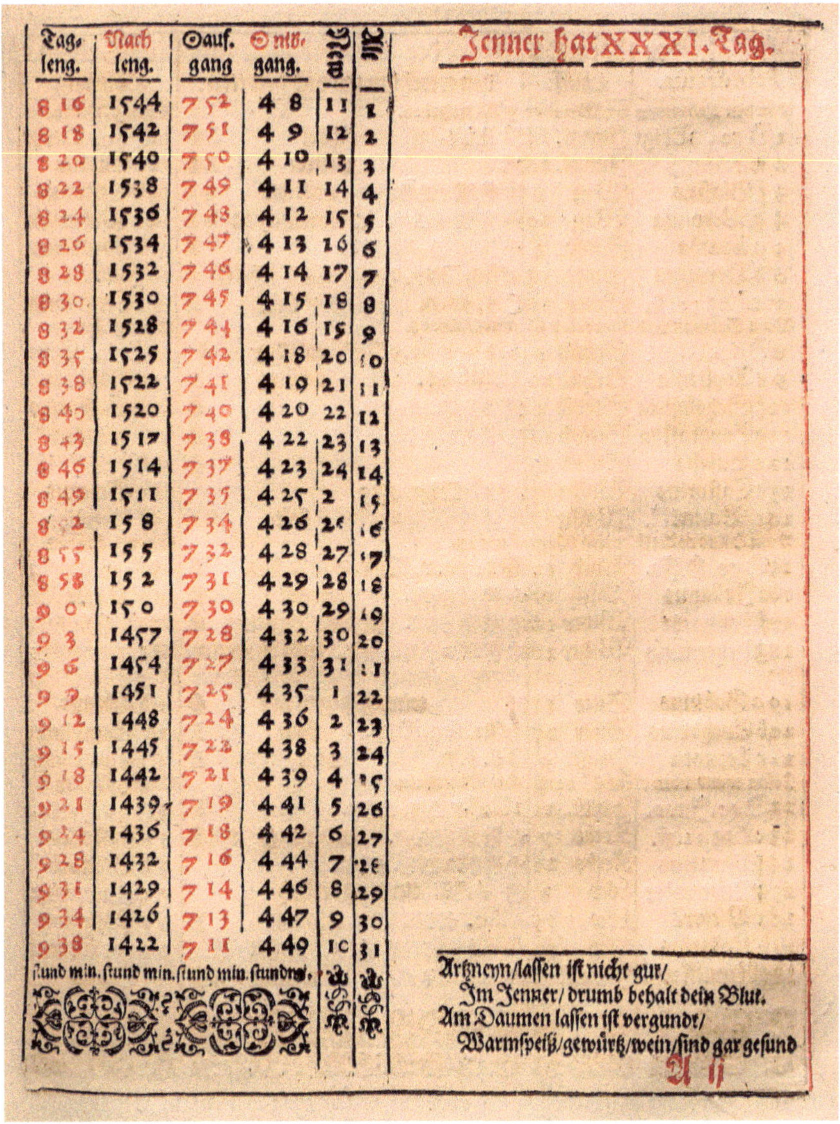

Jenner hat XXXI. Tag.

Tag. leng.	Nacht leng.	☉ auf. gang	☉ nidg. gang.	New	Alt
8 16	1544	7 52	4 8	11	1
8 18	1542	7 51	4 9	12	2
8 20	1540	7 50	4 10	13	3
8 22	1538	7 49	4 11	14	4
8 24	1536	7 48	4 12	15	5
8 26	1534	7 47	4 13	16	6
8 28	1532	7 46	4 14	17	7
8 30	1530	7 45	4 15	18	8
8 32	1528	7 44	4 16	19	9
8 35	1525	7 42	4 18	20	10
8 38	1522	7 41	4 19	21	11
8 40	1520	7 40	4 20	22	12
8 43	1517	7 38	4 22	23	13
8 46	1514	7 37	4 23	24	14
8 49	1511	7 35	4 25	2	15
8 52	158	7 34	4 26	25	16
8 55	155	7 32	4 28	27	17
8 58	152	7 31	4 29	28	18
9 0	150	7 30	4 30	29	19
9 3	1457	7 28	4 32	30	20
9 6	1454	7 27	4 33	31	21
9 9	1451	7 25	4 35	1	22
9 12	1448	7 24	4 36	2	23
9 15	1445	7 22	4 38	3	24
9 18	1442	7 21	4 39	4	25
9 21	1439	7 19	4 41	5	26
9 24	1436	7 18	4 42	6	27
9 28	1432	7 16	4 44	7	28
9 31	1429	7 14	4 46	8	29
9 34	1426	7 13	4 47	9	30
9 38	1422	7 11	4 49	10	31
stund min.	stund min.	stund min.	stund min.		

Artzneyn/lassen ist nicht gut/
Im Jenner/ drumb behalt dein Blut.
Am Daumen lassen ist vergundt/
Warmspeiß/gewürtz/wein/sind gar gesund

A ij

Fig. 10.8 (continued)

The text remained unchanged in subsequent years. It was not until 1619 that Marius replaced it with an up-to-date article on the wars with the Ottoman Empire.[57] From 1623 onward the history column included histories from the prodigy

[57]Under the header *Fragmentum. Ex Chronologie Astronomica Authoris(!) Magnoque ejus Prognostico de Periodo fatali Regni Turcici*. Marius frequently used the prognostica to pose the question whether the constellations pointed toward the decline of Turkish power (see below, Sect. 2.2.2).

a

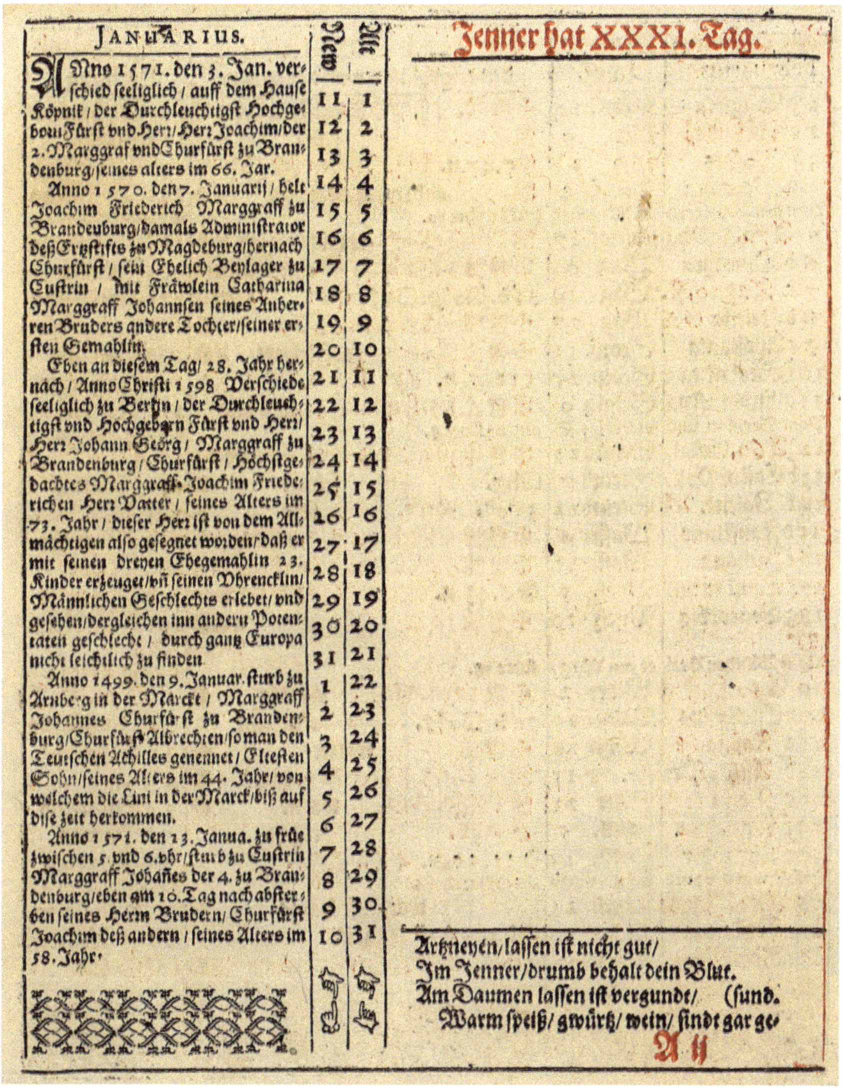
JANUARIUS.

ANno 1571. den 3. Jan. ver-
schied seeliglich / auff dem Hause
Köpnik / der Durchleuchtigst Hochge-
born Fürst vnd Herr / Herr Joachim / der
2. Marggraf vnd Churfürst zu Bran-
denburg / seines alters im 66. Jar.
Anno 1570. den 7. Januarij / helt
Joachim Friderich Marggraff zu
Brandenburg / damals Administrator
deß Ertzstifts zu Magdeburg / hernach
Churfürst / sein Ehelich Beylager zu
Custrin / mit Fräwlein Catharina
Marggraff Johannsen seines Anher-
ren Bruders andere Tochter / seiner er-
sten Gemahlin.
Eben an diesem Tag / 28. Jahr her-
nach / Anno Christi 1598 Verschiede
seeliglich zu Berlin / der Durchleuch-
tigst vnd Hochgeborn Fürst vnd Herr /
Herr Johann Georg / Marggraff zu
Brandenburg / Churfürst / Höchstge-
dachtes Marggraff Joachim Friede-
richen Herr Vatter / seines Alters im
73. Jahr / dieser Herr ist von dem All-
mächtigen also gesegnet worden / daß er
mit seinen dreyen Ehegemahlin 23.
Kinder erzeuget / vnd seinen Vhrencklin /
Männlichen Geschlechts erlebet / vnd
gesehen / dergleichen inn andern Poten-
taten geschlecht / durch gantz Europa
nicht leichtlich zu finden.
Anno 1499. den 9. Januar. starb zu
Arnberg in der Marckt / Marggraff
Johannes Churfürst zu Branden-
burg / Churfürst Albrechten / so man den
Teutschen Achilles genennet / Eltesten
Sohn / seines Alters im 44. Jahr / von
welchem die Lini in der Marck / biß auf
dise zeit herkommen.
Anno 1571. den 13. Janua. zu frue
zwischen 5. vnd 6. vhr / starb zu Custrin
Marggraff Johañes der 4. zu Bran-
denburg / eben am 10. Tag nach abster-
ben seines Herrn Brudern / Churfürst
Joachim deß andern / seines Alters im
58. Jahr.

Neu	Alt
11	1
12	2
13	3
14	4
15	5
16	6
17	7
18	8
19	9
20	10
21	11
22	12
23	13
24	14
25	15
26	16
27	17
28	18
29	19
30	20
31	21
1	22
2	23
3	24
4	25
5	26
6	27
7	28
8	29
9	30
10	31

Jenner hat XXXI. Tag.

Artzneyen / lassen ist nicht gut /
Im Jenner / drumb behalt dein Blut.
Am Daumen lassen ist vergundt / (sund.
Warm speiß / gwürtz / wein / sindt gar ge-

A ij

Fig. 10.9 Column with the history of the (**a**) House of Brandenburg (1615), (**b**) the Ottoman Wars (1619) and (**c**) Histories (1623, following pages) (WLB Stuttgart: HBF 3713, 3717; Staatsarchiv Nürnberg: Fürstentum Brandenburg-Ansbach, Staats- und Schreibkalender (129), Nr. 275, 4. Stück)

collections of Conrad Lycosthenes (1518–1561), whose compilation of divine miraculous omens was often referred to by calendar makers to confirm astrological statements (Fig. 10.9a–c).[58]

[58]See also Matthäus (1969), col. 1191–1195: The inclusion of histories in the writing calendars.

b

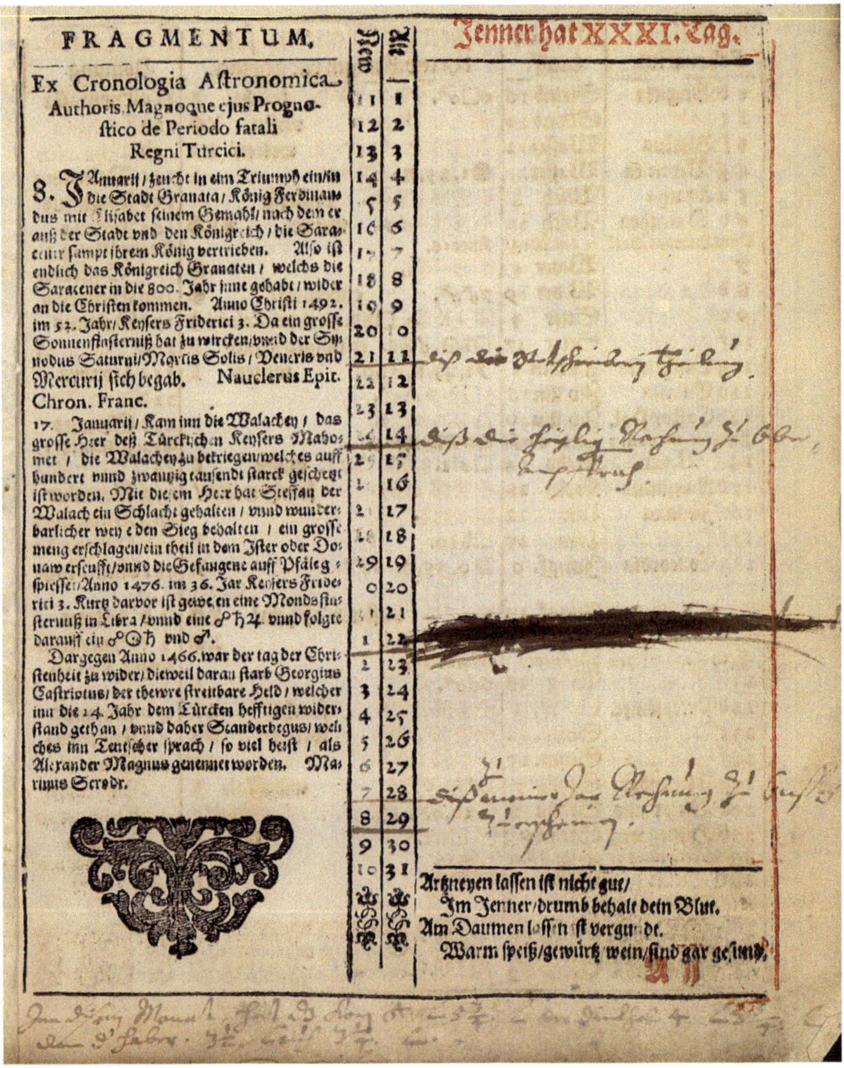
FRAGMENTUM.

Ex Cronologia Astronomica Authoris Magnoque ejus Prognostico de Periodo fatali Regni Turcici.

Jenner hat XXXI. Tag.

Artzneyen lassen ist nicht gut/
Im Jenner/drumb behalt dein Blut.
Am Daumen lassen ist vergut dt.
Warm speiß/gewürtz wein/sind gar gesund.

Fig. 10.9 (continued)

The diary pages consequently dispense with the usual wood vignettes of monthly images.[59] The usual monthly verse with health advice in traditional unchanging form is printed on the lower edge.[60]

The epilogues of the writing calendars—the last pages—are, as was usual, filled with stereotypical rubrics of the rules of phlebotomy, diagrams depicting the blood-letting points as well as horary charts, which are usually contributions of the printer.

[59]Matthäus (1969), col. 1179–1185.

[60]Matthäus (1969), col. 1185–1191.

c

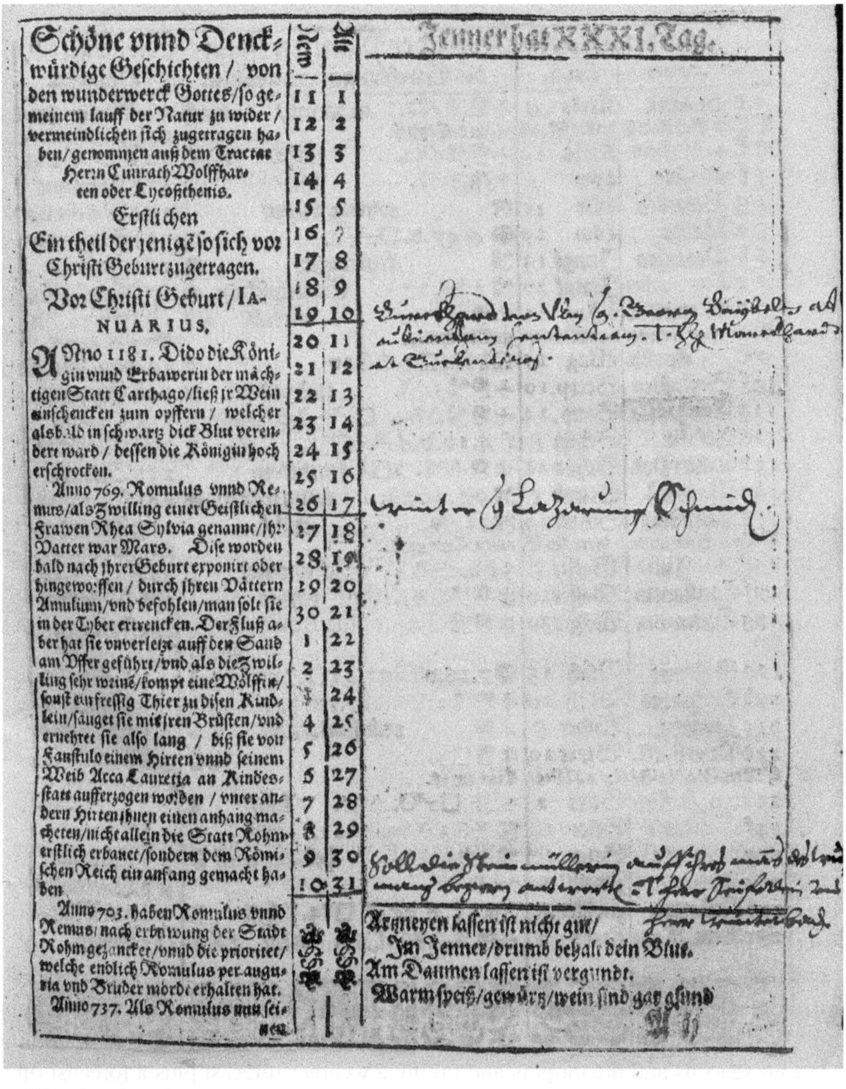

Fig. 10.9 (continued)

Marius displays, however, a more personal note. He provides information on all the eclipses throughout the year, always referring to the more detailed accounts provided by the prognosticon. The rules for phlebotomy he cites are impressively brief, and he concludes with some stern advice to the "Einfältigen" (simpletons) "Was sonst mehr darzu gehört, das wissen die Medici, denen ich hierinnen nicht will zu weit eingriffen haben" (What also belongs to this is known by the Medici, which I don't want to go too far into here.).

The vulgar astrological “Täfelein der Erwehlung nach den Aspecten des Monds zu den Planeten” (Table for elections according to the aspects of the moon relative the planets) with recommendations and warnings for all circumstances is at first conspicuously absent from his accounts. Here Marius, like other like-minded colleagues, had to give in to popular demand.[61] Occasionally at this stage, he inserted astronomical miscellanies, which were not part of the usual contents of a calendar. In 1610 he included a new table of the 28 mansions of the moon he had created using Tycho Brahe’s catalogue of fixed stars. It was intended to be valid for the next 60 years (Fig. 10.10a, b). The volume of 1618 offered the reader a table of moonrise and moonset times during the night for the year 1618, a useful tool for travellers, instead of the usual rules for bloodletting. Marius had composed the table at the suggestion of Prince Christian von Anhalt (1568–1630),[62] uncle to Joachim Ernst, Margrave of Ansbach (Fig. 10.11a, b). He certainly had his reasons for accommodating the special wishes of a person of rank. The majority of his readers, however, continued to demand their usual contents, and the matter was quietly discontinued.

The writing calendars of Marius appeared from 1609 onward in two editions, judging from the copies that survive with differing title pages and writing pages. One of them confined itself to the specification of the length of day and night and celestial information; the other had a column with historical data. The edition with histories displayed the woodcut with the coat of arms of the Brandenburg principalities on the title page. This division probably increased sales potential.

The Prognosticon Astrologicum

Ultimately the writing calendar displays a very schematic structure. The annual changing astronomical and astrological data are indicated throughout the calendar only as abbreviations. An account of what is to be expected during the respective year is presented in the second part of the calendar, called “prognosticon” or “practica.”[63] It gives the annual astronomical data in greater detail, and more detailed astrological forecasts are provided. Marius devotes the largest space to the chapter on “Von den vier unterschiedlichen zeiten” (Of the four different times), detailing the four seasons and the months and giving a weather forecast plus a forecast on the prospective harvest. Following that the chapter on the season also contains conjectures on diseases and the threat of wars. The following chapter discusses the eclipses and their meanings; Marius discusses the dire consequences of the upcoming solar and lunar eclipses. The final chapter again summarizes briefly the fertility of the year.

[61] Matthäus (1969), col. 1196–1198.

[62] Prince Christian was the leader of the Protestant Union of 1608 and became commander of the troops of Friedrich V. of the Palatine after he had been elected King of Bohemia.

[63] Marius uses both terms synonymously. The prognosticon was printed on plain paper, since it was not intended to be written on.

a

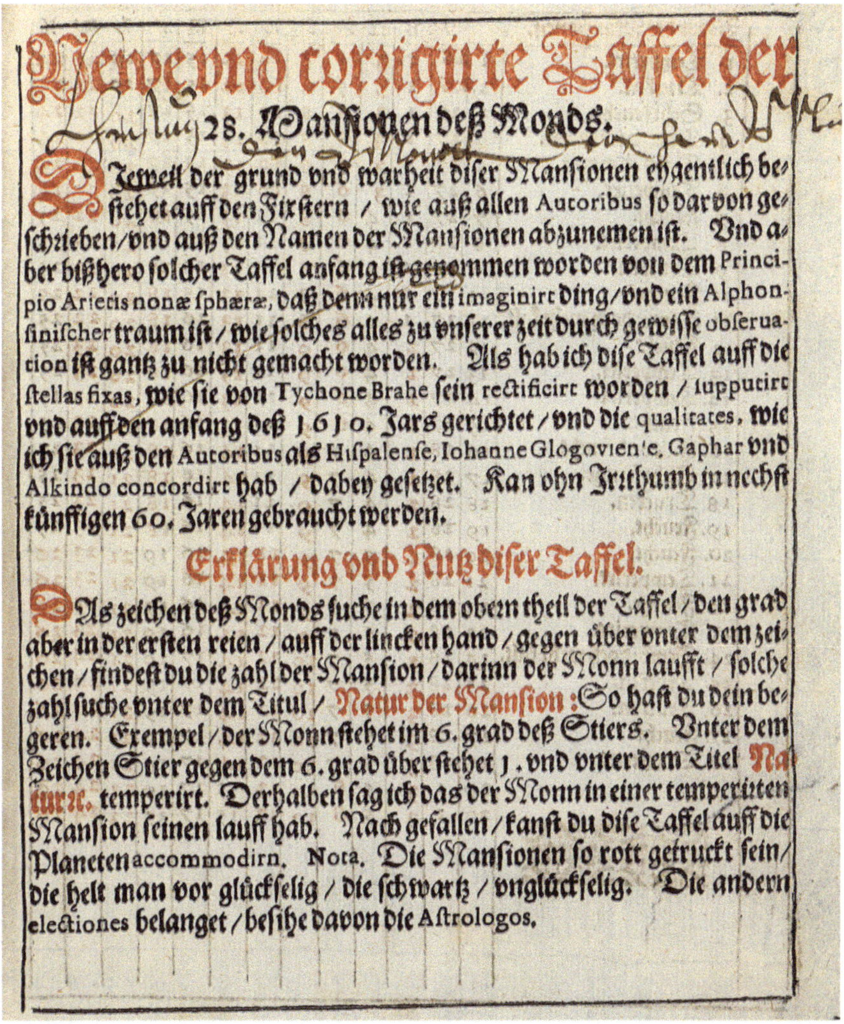

Newe vnd corrigirte Taffel der 28. Mansionen deß Monds.

Dieweil der grund vnd warheit diser Mansionen eygentlich bestehet auff den Fixstern / wie auß allen Autoribus so darvon geschrieben/vnd auß den Namen der Mansionen abzunemen ist. Vnd aber bißhero solcher Taffel anfang ist genommen worden von dem Principio Arietis nonæ sphæræ, daß denn nur ein imaginirt ding/vnd ein Alphonsinischer traum ist / wie solches alles zu vnserer zeit durch gewisse obseruation ist gantz zu nicht gemacht worden. Als hab ich dise Taffel auff die stellas fixas, wie sie von Tychone Brahe sein rectificirt worden / supputirt vnd auff den anfang deß 1610. Jars gerichtet/vnd die qualitates, wie ich sie auß den Autoribus als Hispalense, Iohanne Glogoviense, Gaphar vnd Alkindo concordirt hab / dabey gesetzet. Kan ohn Irrthumb in nechst künfftigen 60. Jaren gebraucht werden.

Erklärung vnd Nutz diser Taffel.

Das zeichen deß Monds suche in dem obern theil der Taffel / den grad aber in der ersten reien / auff der lincken hand / gegen über vnter dem zeichen / findest du die zahl der Mansion / darinn der Monn laufft / solche zahl suche vnter dem Titul / Natur der Mansion: So hast du dein begeren. Exempel / der Monn stehet im 6. grad deß Stiers. Vnter dem Zeichen Stier gegen dem 6. grad über stehet 1. vnd vnter dem Titel Natur rc. temperirt. Derhalben sag ich das der Monn in einer temperirten Mansion seinen lauff hab. Nach gefallen / kanst du dise Taffel auff die Planeten accommodirn. Nota. Die Mansionen so rott getruckt sein/ die helt man vor glückselig / die schwartz / vnglückselig. Die andern electiones belanget / besihe davon die Astrologos.

Fig. 10.10 (**a**) and (**b**) (following page): The mansions of the moon from the calendar for 1610 (WLB Stuttgart: HBF 3708)

In the last rubric, Marius also gives a "Register der Städt, Länder und Königreich, so unter den zwölf Himmlischen Zeichen" ("Register of the towns, countries and kingdom, so under the twelve celestial signs") as is to be found in most prognostica of that period. By common consensus of the astrologers, beginning with Aries, towns and countries are listed under their sign in the zodiac. This affords the reader the opportunity to adapt general prognostications more locally. In contrast to the table of the elections in the calendar part this compilation and its relevance is generally accepted.

b

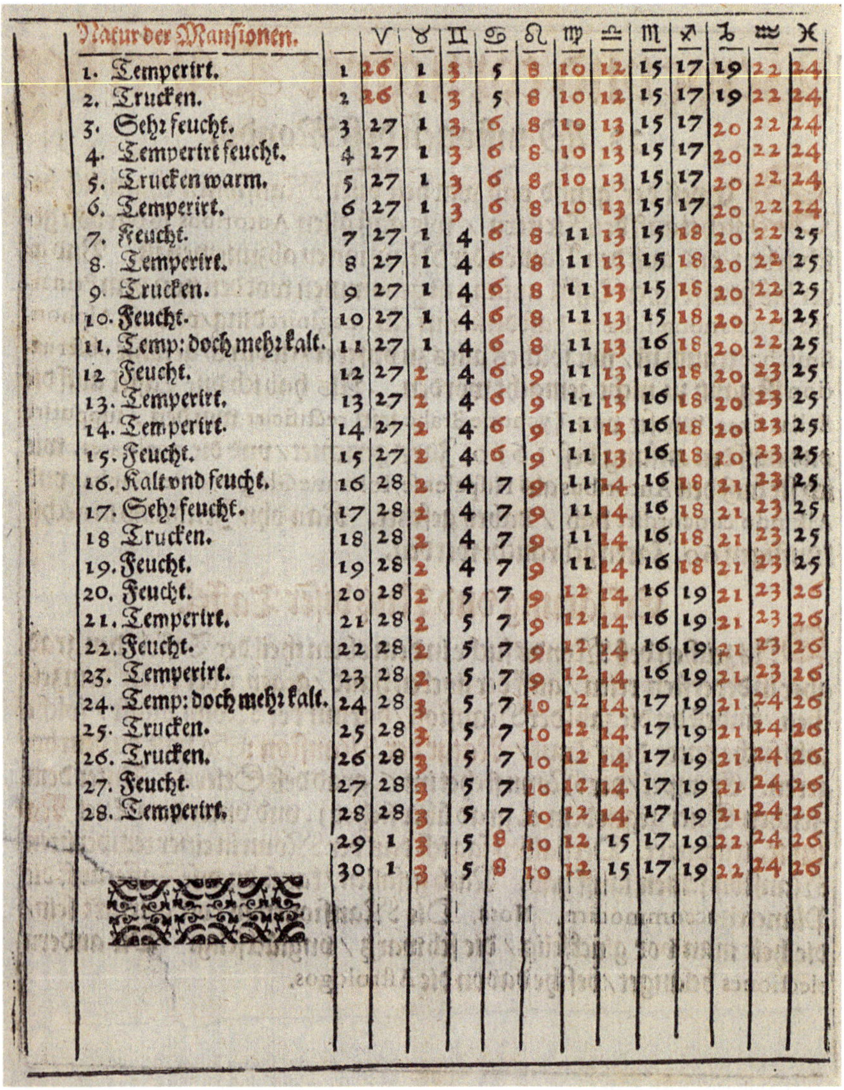

Natur der Mansionen.		♈	♉	♊	♋	♌	♍	♎	♏	♐	♑	♒	♓
1. Temperirt.	1	26	1	3	5	8	10	12	15	17	19	22	24
2. Trucken.	2	26	1	3	5	8	10	12	15	17	19	22	24
3. Sehr feucht.	3	27	1	3	6	8	10	13	15	17	20	22	24
4. Temperirt feucht.	4	27	1	3	6	8	10	13	15	17	20	22	24
5. Trucken warm.	5	27	1	3	6	8	10	13	15	17	20	22	24
6. Temperirt.	6	27	1	3	6	8	10	13	15	17	20	22	24
7. Feucht.	7	27	1	4	6	8	11	13	15	18	20	22	25
8. Temperirt.	8	27	1	4	6	8	11	13	15	18	20	22	25
9. Trucken.	9	27	1	4	6	8	11	13	15	18	20	22	25
10. Feucht.	10	27	1	4	6	8	11	13	15	18	20	22	25
11. Temp: doch mehr kalt.	11	27	1	4	6	8	11	13	16	18	20	22	25
12 Feucht.	12	27	2	4	6	9	11	13	16	18	20	23	25
13. Temperirt.	13	27	2	4	6	9	11	13	16	18	20	23	25
14. Temperirt.	14	27	2	4	6	9	11	13	16	18	20	23	25
15. Feucht.	15	27	2	4	6	9	11	13	16	18	20	23	25
16. Kalt vnd feucht.	16	28	2	4	7	9	11	14	16	18	21	23	25
17. Sehr feucht.	17	28	2	4	7	9	11	14	16	18	21	23	25
18 Trucken.	18	28	2	4	7	9	11	14	16	18	21	23	25
19. Feucht.	19	28	2	4	7	9	11	14	16	18	21	23	25
20. Feucht.	20	28	2	5	7	9	12	14	16	19	21	23	26
21. Temperirt.	21	28	2	5	7	9	12	14	16	19	21	23	26
22. Feucht.	22	28	2	5	7	9	12	14	16	19	21	23	26
23. Temperirt.	23	28	2	5	7	9	12	14	16	19	21	23	26
24. Temp: doch mehr kalt.	24	28	3	5	7	10	12	14	17	19	21	24	26
25. Trucken.	25	28	3	5	7	10	12	14	17	19	21	24	26
26. Trucken.	26	28	3	5	7	10	12	14	17	19	21	24	26
27. Feucht.	27	28	3	5	7	10	12	14	17	19	21	24	26
28. Temperirt.	28	28	3	5	7	10	12	14	17	19	21	24	26
	29	1	3	5	8	10	12	15	17	19	22	24	26
	30	1	3	5	8	10	12	15	17	19	22	24	26

Fig. 10.10 (continued)

Annual prognostica had been published in a slim quarto format since the days of the incunabulum, a format, which had been adapted for writing calendars since the middle of the sixteenth century. It had also become customary for a mathematicus to compile both annual calendars and annual prognostica.[64] It is in the prognostica that

[64]For practica and prognostica, see Matthäus (1969), col. 1001–1006 and col. 1199–1234 also B. Bauer (1994), pp. 167–173: Die Normalform einer Jahresprognostik.—From the first Brandenburg calendar writer, Georg Seyfridt, no Praktik has survived.

a

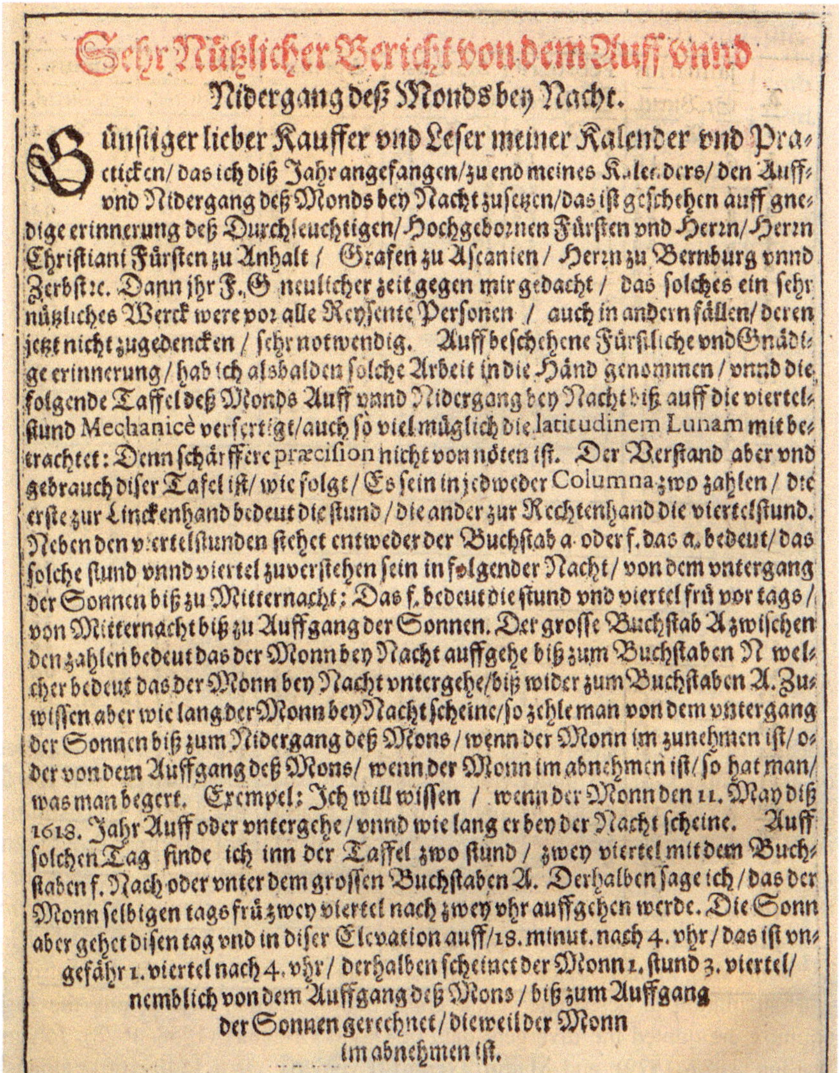

Sehr Nützlicher Bericht von dem Auff vnnd
Nidergang deß Monds bey Nacht.

Günstiger lieber Kauffer vnd Leser meiner Kalender vnd Practicken/ das ich diß Jahr angefangen/ zu end meines Kalenders/ den Auff- vnd Nidergang deß Monds bey Nacht zusetzen/ das ist geschehen auff genädige erinnerung deß Durchleuchtigen/ Hochgebornen Fürsten vnd Herzn/ Herzn Christiani Fürsten zu Anhalt/ Grafen zu Ascanien/ Herzn zu Bernburg vnnd Zerbst/ rc. Dann jhr F. G neulicher zeit gegen mir gedacht/ das solches ein sehr nützliches Werck were vor alle Reysende Personen/ auch in andern fällen/ deren jetzt nicht zugedencken/ sehr notwendig. Auff beschehene Fürstliche vnd Gnädige erinnerung/ hab ich alsbalden solche Arbeit in die Händ genommen/ vnnd die folgende Taffel deß Monds Auff vnnd Nidergang bey Nacht biß auff die viertelstund Mechanicè verfertigt/ auch so viel müglich die latitudinem Lunam mit betrachtet: Denn schärffere præcision nicht von nöten ist. Der Verstand aber vnd gebrauch diser Tafel ist/ wie folgt/ Es sein in jedweder Columna zwo zahlen/ die erste zur Linckenhand bedeut die stund/ die ander zur Rechtenhand die viertelstund. Neben den viertelstunden stehet entweder der Buchstab a. oder f. das a. bedeut/ das solche stund vnnd viertel zuverstehen sein in folgender Nacht/ von dem vntergang der Sonnen biß zu Mitternacht; Das f. bedeut die stund vnd viertel frü vor tags/ von Mitternacht biß zu Auffgang der Sonnen. Der grosse Buchstab A zwischen den zahlen bedeut das der Monn bey Nacht auffgehe biß zum Buchstaben N welcher bedeut das der Monn bey Nacht vntergehe/ biß wider zum Buchstaben A. Zu wissen aber wie lang der Monn bey Nacht scheine/ so zehle man von dem vntergang der Sonnen biß zum Nidergang deß Mons/ wenn der Monn im zunehmen ist/ oder von dem Auffgang deß Mons/ wenn der Monn im abnehmen ist/ so hat man/ was man begert. Exempel: Ich will wissen/ wenn der Monn den 11. May diß 1618. Jahr Auff oder vntergehe/ vnnd wie lang er bey der Nacht scheine. Auff solchen Tag finde ich inn der Taffel zwo stund/ zwey viertel mit dem Buchstaben f. Nach oder vnter dem grossen Buchstaben A. Derhalben sage ich/ das der Monn selbigen tags frü zwey viertel nach zwey vhr auffgehen werde. Die Sonn aber gehet disen tag vnd in diser Elevation auff/ 18. minut. nach 4. vhr/ das ist vngefähr 1. viertel nach 4. vhr/ derhalben scheinet der Monn 1. stund 3. viertel/ nemblich von dem Auffgang deß Mons/ biß zum Auffgang der Sonnen gerechnet/ dieweil der Monn im abnehmen ist.

Fig. 10.11 (**a**) and (**b**) (following page): Table with moonrise and moonset from the calendar for 1618 (January–June) with an introduction. Staatsarchiv Nürnberg: Fürstentum Brandenburg-Ansbach, Staats- und Schreibkalender (129), Nr. 274, 18. Stück

Marius reveals more of his persona than in the sparse, pared-down calendar. Here he gives the clearest indication of his convictions, his attitude toward his chosen métier, and occasional glimpses into his private life. Without the prognostica we would know very little about him (Fig. 10.12).

b

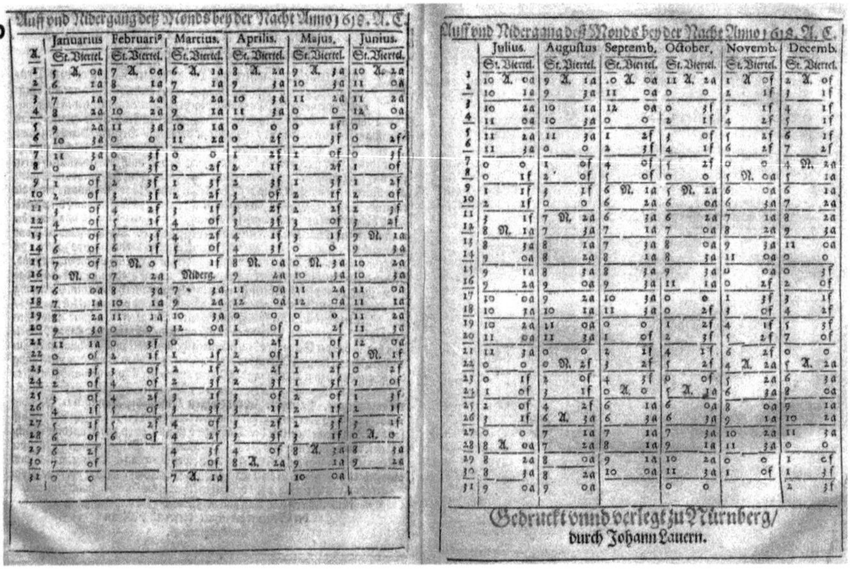

Fig. 10.11 (continued)

Astronomical Data as a Prerequisite for the *Prognosticon*

For the astrological believer, the reliability of the prognosticon is highly dependent on the accuracy of the underlying astronomical calculations. Even with his first calendar for the year 1601, compiled in 1600, Marius, the novice, places himself apart from the great majority of commonplace calendars of his time. Programmatically the second page held the yearly horoscope, the constellation of the stars for the astronomical beginning of spring starts with the note: *Constitutio Coeli* [...] *iuxta Calculum geonerosi and magnifici Domini Tychonis Brahe*. To Marius the *Calculus* Tychonis was, in his conviction, the most modern and reliable method of astronomical calculation to determine stellar constellations. When determining the spring equinox, he quoted the divergent data of David Origanus (1558–1628), Johannes Stadius (1527–1579), and Martin Everard (Everaerts, circa 1540–1601) alongside those of Brahe and explained that "according to daily observations," those of Tycho were the best and most reliable.[65]

By 1600 Marius knew of Brahe's latest method of calculation, on which he had not yet drawn in his *Tabulae Directionum* of 1599. It dealt with the parameters of astronomical calculations as modified by Brahe; astronomers by that time might have gained the knowledge via correspondence.[66] In an extensive chapter on solar

[65] *Prog. 1601*, sig. A6$^{r–v}$.

[66] Brahe had issued a first private print run of letters, Uraniborg, 1596. A reprint was published in 1601 in Nuremberg. See. also: Kepler 14 (1949), pp. 16–21: Letter from Herwart von Hohenburg to

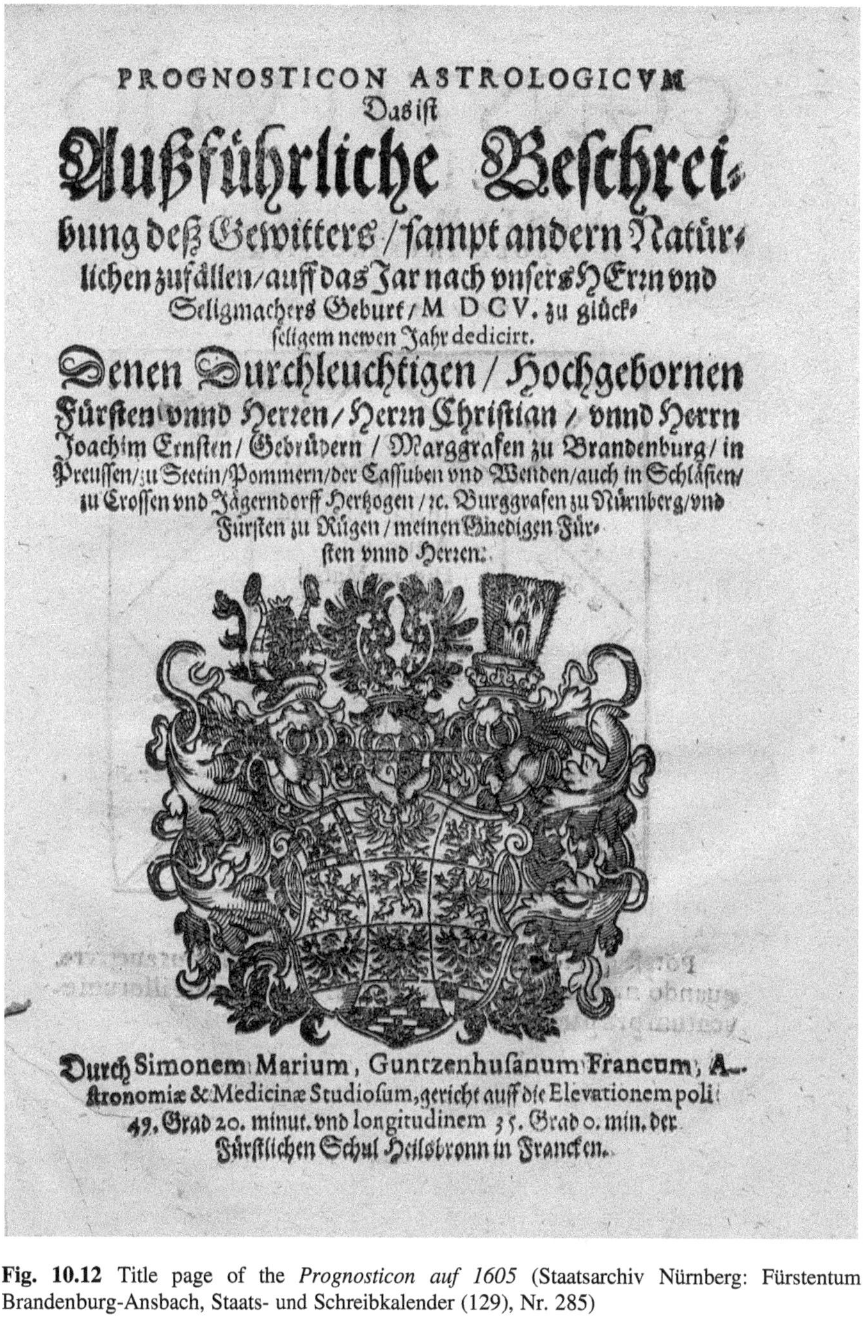
PROGNOSTICON ASTROLOGICVM
Das ist
Außführliche Beschrei-
bung deß Gewitters/sampt andern Natür-
lichen zufällen/auff das Jar nach vnsers HErrn vnd
Seligmachers Geburt/M DCV. zu glück-
seligem newen Jahr dedicirt.
Denen Durchleuchtigen/Hochgebornen
Fürsten vnnd Herren/Herrn Christian/vnnd Herrn
Joachim Ernsten/Gebrüdern/Marggrafen zu Brandenburg/in
Preussen/zu Stetin/Pommern/der Cassuben vnd Wenden/auch in Schlesien/
zu Crossen vnd Jägerndorff Hertzogen/rc. Burggrafen zu Nürnberg/vnd
Fürsten zu Rügen/meinen Gnedigen Für-
sten vnnd Herren.
Durch Simonem Marium, Guntzenhusanum Francum, A-
stronomiæ & Medicinæ Studiosum, gericht auff die Elevationem poli
49. Grad 20. minut. vnd longitudinem 35. Grad 0. min. der
Fürstlichen Schul Heilsbronn in Francken.

Fig. 10.12 Title page of the *Prognosticon auf 1605* (Staatsarchiv Nürnberg: Fürstentum Brandenburg-Ansbach, Staats- und Schreibkalender (129), Nr. 285)

Kepler 20 July 1599; here p. 20f.: Herwart refers to the type of calculation Brahe applied (Klaus-Dieter Herbst alerted me to this letter).

and lunar eclipses, Marius determined the lunar eclipse of 1601 using calculations by Melchior Jostel (Jostelius 1559–1611) based on Brahe's calculatory method, published in 1599.[67] For a humble scholarship student from Heilsbronn to gain access to these latest methods of astronomical observations must have come courtesy of Margrave Georg Friedrich and his court. Beyond doubt Marius had earned at this stage a reputation as a competent mathematicus in "Astronomia instrumentalis et numeralis" in the methods of observation and calculation. As already mentioned, Marius was able, after his debut as an author of calendars in 1601, to join the staff of Tycho Brahe in Prague for a short time as a servant and observer, which only helped to increase his admiration for Brahe.[68] "Alle Mathematici wie auch tota futura posteritas" seien diesem "nechst Gott" ewiges Lob und Dank schuldig. (All mathematici as well as all of the future following owe him, next to God, praise and thanks).[69]

The *Calculus Tychonis* remained in all his future prognostica the point of reference for his own astronomical calculations (see Kremer, this volume). Marius continuously emphasized that he calculated his own data for the beginning of the seasons and eclipses and did not simply rely on those given the ephemerides of Origanus or others. In Padua, however, it seems that the relevant papers had not been available to him while editing the prognostica for 1602 and 1603.[70] The volume he compiled in 1604 for 1605 (the one for the year 1604 is missing) again refers emphatically to Brahe's methods, specifically mentioning the terrifying solar eclipse. A three-page Latin explanation gives an exact procedure so that common astronomers can practice the same methods of calculations and become acquainted with them. It would be a disgrace, wrote Marius, for practitioners to simply rely on the ephemerides.[71]

Meanwhile, Brahe's *Progymnasmata* (1602) had been published, and Kepler, too, had declared himself in his prognostica for 1604 against calendar writers, who were pandering to the same old style when at the same time, Tycho Brahe's improved method of calculation had already been available for purchase in all

---

[67] See *Prog. 1602*, sig. D2^{v}: Marius stated the following year (1601) that Jöstels 'restitution in motu lunae' did not quite apply but had been corrected by Brahe and his staff. For Jöstels publication, see Zinner (1964), p. 320, no. 3817. In the *Prognosticon auf 1603* (sig. D1^{r}), Marius again points to "neue restitutio curriculi Solaris & Lunaris," helping him to correctly place the lunar eclipse of December14/24 1601 and proved by sighting in Padua. See Zinner (1942), p. 48; see Thoren (1972), Swerdlow (2004).

[68] *Prog. 1607*, sig. B4: "Ich bleibe bey der restitution Tychonis, dessen instrumenta ich nit allein gesehen, sondern selbst gebrauchet." See also *Prog. 1606*, sig. A3^{r}: 1604 in Padua Marius had Instrumenta Astronomica constructed like those he had seen in Prague in 1601 "nach der Form und weiß [. . .] wie ich sie anno 1601 zu Prag gesehen."

[69] *Prog. 1601*, sig. A4^{r}. *Prog. 1610*, sig. E2^{v}: "der Edle und vere Magnus Astronomus Tycho Brahe, cuius celebre nomen merito cum mundo coaevum erit."

[70] Marius had brought several of Brahe's true distances for fixed stars with him to Italy "etlicher Fixstern veras distantias von Tychone auß Prag mit in Italiam gebracht" (*Prog. 1618*, sig. A2^{v}).

[71] *Prog. 1605*, sig. B1^{r} und D3^{r-v} and *Supputatio* D4^{v}–E1^{r} and *Prog.* 1606, sig. E1^{r}. However, Kremer (this volume) has found that Marius himself often relied on ephemerides.

bookshops for a year ("sintemahl Hern Tychonis Brahe verpesserte Rechnung jnen schon ein Jahr lang für der Nasen und in allen Buechläden zu finden ist").[72] Marius found his procedure confirmed when making his annual calculations according to Brahe's specifications since they later corresponded most accurately with appearances.[73] He deemed Brahe's calculations the best of the period despite the fact that, after nearly 20 years of compiling calendars, they had yet to reach their fullest perfection.[74] The frequent references to his own calculations do not signify, however, that Marius computed all his astronomical data himself. Kepler did this, in his own calendars, emphasizing correctly that other calendar writers had hardly 100th of his labors.[75] Aspects and position of the planets, as well as the course of the moon, had been adapted by Marius from the ephemerides of the "excellent" Origanus.[76] A slight modification was announced for the *Prognosticon for 1611*. The data for the new and the full moon are now calculated according to the new tables of Brahe and not taken from Origanus. Therefore it should be of no surprise if they differed from those of other writers.[77] Marius always acknowledged that he owed his skills to Brahe.[78] He felt justified in distancing himself in his prognostica from lazy calendar writers,[79] who either from a lack of skills or pure ignorance failed to compile their own calculations.[80] He displayed some understanding for such colleagues who due to old age were no longer able to follow such "subtleties."[81] This seems to have included M. Georg Caesius, a priest from Burgbernheim, who had established himself a generation earlier as Brandenburg's from Margrave Georg Friedrich

[72]Kepler 11, 2 (1993), p. 93.—Marius owned a copy of the *Progymnasmata*; see *Prog. 1621*, sig. B3^{r}.

[73]*Prog. 1612*, sig. C4^{v}.

[74]*Prog. 1619*, sig. A4^{r}.

[75]Kepler 11, 2 (1993), p. 192 (*Prognosticum auf 1620*). Kepler obviously sees his calendar work in connection with the production of the *Tabulae Rudolphinae*. This demonstrates a special function, for writing calendars by especially qualified writers, which lasted well into the mid eighteenth century, See Herbst (2012), p. 17 for additional literature.

[76]*Prog. 1616*, sig. E3^{v}. They were based on the *Tabulae Prutenicae* of Erasmus Reinhold (1511–1553). Ephemerides were, as Kepler (11, 2 (1993), p. 192) puts it, the mother of the calendar. It has to be mentioned that Origanus calculated his calendars emphatically according to the latest discoveries in astronomy, i.e., incorporating the works of Brahe. (Writing calendar for 1604 Stadtarchiv Frankfurt (Oder)—friendly communication from Klaus-Dieter Herbst).

[77]*Prog. 1611*, sig. B2^{r}.

[78]*Prog. 1616*, sig. B1^{r}: "dessen ich denn billich ehrlich und danckbarlich wegen grosser auffgewendten Unkosten, mühe und arbeit gedencke."

[79]*Prog. 1612*, sig. B1^{v}, B4^{r} and C4^{v}.

[80]*Prog. 1607*, sig. B1^{r} and *Prog. 1608*, sig. D4^{v}: "Die Practicanten oder Calenderschreiber werden diß 1608 Jahrs in beschreibung der Finsternußen nit allerding ubereintreffen" (because they only used the ephemerides of Everard or Origanus or Erasmus Reinhold's *Tabulae Prutenicae* (1511–1553).

[81]*Prog. 1605*, sig. D3^{v}.

more or less appointed calendar maker. Marius seems to have respected the elder colleague, but he wanted to provide a well-founded, more modern prognostica.[82]

He was a master of his art. Ernst Zinner argued that the observations of Marius could be compared with the best of his times.[83] He used exact data, which he was always anxious to improve. He generously shared them with the guild and requested, where appropriate, additions and corrections. His intent had always been:

> die Astronomiam, soviel mir müglich, helffen illustrirn, welches dann auch mein beruff erfordert, und andern mit meiner mühe und arbeit die Hand [zu] bieten (to help to illustrate the astronomy as much as I am possible, which my profession requires and to offer a hand to others with my efforts and work).

He hoped that his opinions and observations were "of true sincere astronomers [. . .] to more perfection of this art diligently taken into consideration."[84] With his observations recorded in the prognostica, he liked to give other sharp minds cause to diligently think things through [. . .] Observata enim non obstant observaturis.[85]

Marius includes his astronomical discussions and notes into the relevant passages of the prognostica. In his struggle for astronomical certainty,[86] he used common calendar techniques as a medium for scholarly correspondence.[87] Like Johann Krabbe (1553–1616),[88] a calendar maker from Wolfenbüttel, Marius was one of the first who in their prognostica opened such a calendar discussion forum, of which intensive use was made continually in the seventeenth century—especially in the second half—then the now established journals took over this function.[89] It has to be noted here that Marius already commented critically on Krabbes's calculations and observations in his *Prognosticon auf 1608*.[90]

---

[82]Caesius's prognostica in the hands of the theologians, who were carried by the conviction that the Day of Judgment was due, became more or less astrological penitential sermons.

[83]See Zinner (1942), p. 40.

[84]*Prog. 1618*, sig. A3.

[85]*Prog. 1610*, sig. E2^{v}.

[86]As formulated by Herbst (2010a), p. 45 (here in conjunction with the solar eclipse of 1654).

[87]This aspect of calendar writing has been emphatically pointed out by Klaus-Dieter Herbst. See Herbst (2009a). However, Kepler's teacher Michael Mästlin (1550–1631) had reservations about making the discourse between scholars public in calendars. See Bialas in Kepler 11, 2 (1993), p. 449f.

[88]Herbst (2009a), pp. 199–204: Gelehrte Kommunikation bei Johannes Krabbe. For Krabbe, see Herbst, *Biobibliographisches Handbuch*.

[89]See Herbst (2010b); pp. 160–162 early examples of calendars as discussion forum with examples from Marius and Krabbe quoted here. See also Herbst (2008, 2009c), here pp. 116–122 examples from the 2nd half of the seventeenth century. For the first half of the seventeenth century, calendar makers Peter Crüger (1580–1639), David Frölich (1595–1648), and Lorenz Eichstädt (1596–1660) are named.

[90]*Prog. 1608*, sig. B4^{r} (ETH-Bibliothek Zürich: Rar. 1379: 2). The edition does not conform to the one quoted by Zinner (1942), pp. 53–55, held in the Staatsarchivs Nürnberg.

His observations of the moons of Jupiter entered the *Prognosticon auf 1612*, which he compiled in the years 1610 and 1611, this way.[91] Unlike Galileo, Marius did not press his discovery into a rushed separate publication, a move which brought him those well-known accusations of plagiarism. Marius makes no further mention of this in his prognostica apart from a passing brief rebuff of the accusations of Christoph Scheiner (1573–1650), then teaching at Ingolstadt University.[92] Even before Galileo denounced him as a plagiarizer, in his *Il Saggiatore* of 1623, Marius noted, without mentioning a name, in his prognosticon for that year, that there had already been differences of opinion about astronomical observation methods with him in Padua.[93]

Besides his constant references to Brahe, Marius led a lively discourse on calendars in his prognostica with two well-known astronomers. With David Fabricius (1564–1617),[94] pastor in East Friesland, and with Johannes Kepler (1571–1630), leading to the assumption that both were familiar with his calendars.[95] Marius had already formed an acquaintance with Fabricius in Prague when they both were visiting Brahe there in 1601. He reestablished the contact and friendship[96] with the most excellent astronomer, his esteemed master and good friend [97] taking him up in his prognostica and correspondence, following his erudite and excellent prognosticon[98] for 1609.[99]

Fabricius had engaged in an intense correspondence with Kepler between 1601 and 1609. A final communication took place when Kepler responded publicly in 1616 to questions put forward by Fabricius in his prognostica for 1615, 1616, and

[91]Compare Zinner (1942), p. 34; Herbst (2009b), p. 538ff: 5. Simon Marius' discovery in a calendar of 1612; Pellengahr (2012). In the preamble of the *Prognosticon auf 1612*, Marius reports his observations on the phases of Venus and in the one for 1613, he concluded that Mercury and Venus were orbiting the sun. See Leich (2012), p. 180, pp. 186–188.

[92]*Prog. 1618*, sig. A2^r. Compare also Gaab and Leich (2014).

[93]*Prog. 1623*, sig. A2^v: The distinguished "professor philosophiae" was probably Galilei. Galilei spoke of Marius in 1607 as "mio antico avversario, invido inimico non sol de me, ma di tutto 'l genero umano" (Galileo 1607, p. 519).

[94]*NDB* 4 (1959), p. 731f. Kepler awarded Fabricius first rank among observing astronomers after Brahe's death.

[95]Kepler was by all means a critical reader of Marius. He commented on the calendar (not surviving) of 1617 in his prognosticon (Kepler 11, 1 (1983), p. 21, 560).

[96]*Prog. 1610*, sig. C4^v.

[97]*Prog. 1616*, sig. C4^v.

[98]*Prog. 1610*, sig. C4^v and C6^r; *Prog. 1612*, sig. C7^r; *Prog. 1616*, sig. B3^v and C4^v; *Prog. 1619*, sig. B4^v.

[99]Fabricius corresponded with Marius about the reappearance in 1609 of the variable star in the constellation Cetus, which Fabricius had already observed in 1596; sig. B1: Marius on the discovery of the moons of Jupiter; see also *Prog. 1616*, sig. C2; *Prog. 1618*, sig. A2: reference to a letter from Fabricius as well as sig. A3; *Prog. 1620*, sig. B5.

1617.[100] Marius valued his efforts in perfecting astronomy no less than those of Brahe.[101]

On his visit to Prague, Marius did not meet Kepler in person.[102] The latter had already benevolently taken notice of the *Tabulae* in 1599,[103] but further contact did not arise for the time being. From 1608 onward, Marius also referenced Kepler's observations in his prognostica.[104] In 1612, as already stated, he adopted with a detailed explanation Kepler's introduced expansion of the aspects.[105] It was evident that Marius was anxious to align his calendar calculations with those of the Imperial Mathematicus.

The *Dioptrice* of 1611, however, temporarily casts a shadow over relations as Marius was unpleasantly affected by the publication with comments of one of his letters to another addressee.[106] Kepler relented somewhat after an intervention from Ansbach at the Imperial Court. Marius appeared conciliatory and hoped for a meeting in person.[107] The meeting took place in Regensburg in October 1613.[108] Marius continued to regard Kepler, the excellent astronomer, as like-minded,[109] and from 1619 onward repeatedly called him a good friend.[110]

Kepler, in turn, kept a greater distance. In 1611/1612 he had already shown no understanding for Marius who continued to support the geocentric system. In a letter from the year 1619, Kepler expressed himself in unfriendly terms, and in his *Prognosticon auf 1620*, he criticized Marius' detailed predictions very strongly. A subsequent criticism of Marius' observations on a presumed natural light in the

[100]Kepler 17 (1955, Letters 1612–1620), p. 192, no. 746 und p. 481. The calendars and prognostica were published by Johann Lauer, Nuremberg and not, as previously, in Hamburg, a move surely orchestrated by Marius, Lauer's son-in-law. Compare Matthäus (1969), colp. 1131.

[101]*Prog. 1621*, sig. C1. On the tragic death of Fabricius—he was bludgeoned to death by a member of the congregation—Marius noted: "Also pflegt Gott bißweilen solche Ingenia sampt jhren Inventis und laboribus der Welt widerumb zunehmen, dieweil sie von der Welt nicht geacht, sondern nur verlacht unnd veracht werden." On the evaluation of Fabricius's calculations see also: *Prog. 1610*, sig. C4^{v}; *Prog. 1616*, sig. C4^{v}; *Prog. 1618*, sig. A3^{v}; *Prog. 1619*, sig. B4^{v}; *Prog. 1620*, sig. B5^{v}; *Prog. 1621*, sig. C1^{r}.

[102]See chapter by Gaab (Chap. 3).

[103]Kepler 14 (1949), p. 131 (Letter to Herwart von Hohenburg 12th July 1600).

[104]*Prog. 1610*, sig. E2^{v}, compiled 1608. Kepler's "optica" mentioned here: his *Ad Vitellionem Paralipomena, quibus Astronomiae Pars Optica traditur*. Frankfurt a.M. 1604). See also *Prog. 1611*, sig. B4^{v}: On Brahe's und Kepler's observations of Mars. He was waiting for *Astronomia Nova*, which was being printed in Heidelberg (1609); see also *Prog. 1620*, sig. B5^{v}.

[105]*Prog. 1612*, sig. C6^{v}–C7^{v}.

[106]Compare Kepler 4 (1941, Kleinere Schriften 1602/1611, Dioptrice), p. 353; Annotation p. 516. The estimation that Marius had shown himself to be pathologically sensitive did not appear again after Zinner's essay in the letters edition from 1942.

[107]Kepler 17 (1955, Briefe 1612–1620), pp. 33–37, no. 604 (Kepler); pp. 72–74 (Marius). Marius pointed out that this had been a private letter not intended for a wider audience, at least in this form.

[108]Marius (1614), sig. B2^{v} and *Prog. 1615*, sig. C1^{v}.

[109]*Prog. 1618*, sig. A3^{v}; *Prog. 1619*, sig. B2^{r}; *Prog. 1620*, sig. B5^{v}.

[110]First in Marius 1619, sig. B2^{v}.

moon remains rather more factual.[111] Both were keeping up appearances, as Kepler continued to correspond with Marius[112] and Marius found his celestial observations justified in Kepler's prognostica.[113]

In his last prognostica for 1627 and 1628, published after his death, Marius once again wished his very good friend Kepler after all his effort, diligence, and expense might finally succeed in producing a complete edition of his ephemerides.[114] As early as 1602, Marius had already emphasized that the usual ephemerides left much to be desired. Better tables were a prerequisite for better practice.[115] In his opinion Kepler and David Fabricius were following Tycho Brahe as excellent astronomers and guarantors for exact astronomical observations and calculations.[116]

Did other calendars makers[117] take note of Marius? Calendars preserved from that time are fragmentary, and for a more precise appraisal, it would probably be more prudent to wait for a complete digital survey. At least two references are, however, known: Firstly, with the Royal Hungarian Astronomus, David Frölich (1595–1648), active in Upper Hungary, whose calendar was widely distributed.[118] In his prognosticon for 1634, published in Breslau, he introduced those two "diligent Mathematicii," Galileo and Marius, who with the aid of a telescope had discovered incredible, new things in the heavens from which "in Astronomicis viel unerhörte und selzame [...] theils unglaubliche Paradoxa drauß schliessen (könne)" (one (could) conclude in astronomy many outrageous and strange [...] partially unbelievable paradoxes).[119] Marius appears on an equal footing with Galileo. Secondly, Martin Horky's prognosticon for 1657,[120] published in Nuremberg, recalled Marius' nova observation of 1607. Now it appears that the Horky calendar, published by Wolfgang Endter, was compiled by the Franconian priest and prolific calendar writer Marcus Freund (1603–1662) at that time.[121] Marius was, at least in Franconia, not forgotten.

---

[111]Kepler, Calendaria (see Ann. 35), p. 198, 202 and 205 (Prognosticum f. 1620). Marius claimed that the moon was the light emitting in his *Prognosticon auf 1620* (sig. C6^{v}). See also the remarks below.

[112]*Prog. 1622*, sig. A2^{v}–A3 ^{v}. Kepler regretted that inclement weather had prevented beautiful astronomical discourse about the latest solar eclipse. Letters do not survive.

[113]*Prog. 1625*, sig. C2^{v}; as well as *Prog. 1626*, sig. C1^{v} and D2^{r}.

[114]*Prog. 1627*, sig. B4^{r} and *Prog. 1628*, sig. A4^{v}.

[115]*Prog. 1602*, sig. A3^{v}.

[116]Marius was hoping for new ephemerides from either of them (*Prog. 1613*, sig. C1^{r}).

[117]The essay volumes listed on the *Marius-Portal* on prognostications regarding the war taken from the prognostications made by known calendar writers for the years 1627 and 1628 name the usual suspects. They roughly correspond to Ernst Zinner's bibliography (1964).

[118]Compare Herbst, *Biobibliographisches Handbuch*, Entry: David Frölich.

[119]*Prognosticon astrologo-physicum für 1634*, sig. A2^{v}–A3^{v}, A4^{v}–B1^{r}. Copy in the Nationalbibliothek Széchényi Budapest, RMK III 1499. Frölich list all presently visible celestial phenomena. I owe the reference mentioning Marius to Ilona Pavercsik, Budapest.

[120]For Horky compare Matthäus (1969), col. 1146f.

[121]Matthäus (1969), col. 1155; for Marcus Freund see Herbst: *Biobibliographisches Handbuch*.

The Astrological Interpretation of Astronomic Data

The astronomical data offered Marius the prerequisite for the other part of *astronomia*, viz., to handle the *astrologia* and her assertions with due diligence.[122] Astrology remained for him knowledge of the influence of celestial bodies, by divine commandment, upon this earth.[123] It was for him a valid empirical science, whose secrets could only be unlocked through ceaseless, scrupulous observation. He was convinced that[124]:

> [. . .] in den astris grosse geheimnuß sein, unnd sie sterckere kräfft haben in dieser unteren Welt, als jemand glauben kan oder mag, unnd das kein grossere unnd vornemere ursach fortunae mundanae, nechst Gott dem Allmechtigen, könne erdacht oder vorgebracht werden. ([. . .] in the stars there are great secrets and they have a stronger influence in our lower world than anybody could believe or want to and that no greater or distinguished cause of mundane fortune, after God the Almighty, could be imagined or produced)

Marius admitted that he initially failed to grasp the teachings of astrology until he was quasi-forced to insight by diverse experiences.[125] Critics and those who despised the noble art of reading the stars, which is an especial great gift of God, could not make him waver.[126] He was adamant, however, to distinguish between noble "astrologia" and false or deceptive superstitious handling. In his prognostica Marius repeatedly emphasized that these were compiled diligently with huge efforts completely without superstition.[127] A proper astrological interpretation was guaranteed through observing the doctrines of Ptolemy, which he had lain down in his *opus quadripartitum* or *Tetrabiblos*. The *Prognosticon auf 1601* opened with a statement that they were complied "nach der lehr deß nun mehr über die 1400 Jar Hochberümbten Astrologi Ptolo(!)maei" (according to the guidelines of the now for more than 1400 years highly renowned astrologer Ptolemy) and not according to the superstitions guidelines and fantasies of the Arab, Chaldean, and Jewish astrologers. Unfortunately, a critical edition of Ptolemy's works was not available. If one were available, then astrologia [. . .] would regain its previous, if not higher, dignity.[128] Later Marius even planned to publish his own version of the *opus quadripartitum*

---

[122] *Prog. 1607*, sig. A3^{v}.

[123] *Prog. 1615*, sig. A2^{r}.

[124] *Prog. 1605*, sig. A3^{v}.

[125] *Prog. 1607*, sig. A3^{v}, A4^{r}.

[126] *Prog. 1605*, sig. A3^{v}.

[127] *Prog. 1603*, sig. A4^{v} and later repeatedly. *Prog. 1612*, sig. C5^{r}: He prognosticates "Astrologice doch Christlich wohlmeynent."

[128] *Prog. 1601*, sig. B6^{r}. Marius hardly references the Arabic authorities. He quotes only Messhalla occasionly (*Prog. 1620*, sig. C7^{r}, *Prog. 1621*, sig. C5^{r}). When he refers to other astrologers "anderer Astrologen lehr," no names are given (*Prog. 1611*, sig. A4^{v}). He admits there are good things to be found among the Arabian, Indian, and Chaldean authors (*Prog. 1610*, sig. A4^{r}).

with commentary.[129] While discussing astrology he had to admit that the high art of forecasting "auß deme bißhero gebräuchlichen Astrologische Fundament [. . .] zur Zeit noch nit gnugsam ist ergündet worden"[130] (by the up till now usually astrological foundation [. . .] at this time not sufficiently well founded). His constant concern was, therefore, to refine the interpretation of astrological theorems through even more accurate astronomical observations of the actual stellar constellations.

Both Marius' approach and his reasoning thereby document the simultaneousness of the non-simultaneous,[131] the linking of the novel, and the new dimensions of astronomy with the ancient doctrines of astrology. Adhering to Ptolemy, Marius focuses in his prognostica on the "universalia," the principal tendencies suggested by the course of the stars to which he refers.[132] There are possibilities, which can changed by the intervention of God at any time and can also be corrected by rational and agreeable-to-God actions of man.

This does not diminish the intrinsic value of his astrological speculations.[133] "Astra inclinant, non necessitant"[134]—the stars merely display a tendency. He does not want to give a "particular" statement, a direct prediction, or an "astrologia iudiciaria" about whatever. This is the restriction he applies to himself in contrast to other writers. While looking at the forecasts by 18 different astrologers, he could find no justifications for their "particularia" and awaited an explanation.[135] Marius claimed to know what could be attributed to the stars' aim and measure and declined any advice from those who were not versed in his craft.[136]

The main objective of prognostication is, as indicated on the title page, the description of the thunderstorm, the weather, and, if applicable, the description of other natural happenstances. Most calendar writers usually dealt with the latter in a separate chapter on the feared plagues and dangers of war. Marius employs this pattern only for the first year, which perhaps can be attributed to his marked detachedness to statements on particulars. He did not want to drop the subject altogether, though, if only to please his circle of readers. The topic is discussed in a subsequent chapter on lunar eclipses for each year, to which those knowledgeable of the stars attributed lasting influence on earthly occurrences.

The weather forecasts are dealt with in detail on a quarterly basis, by which the discussion of the stellar constellations at the beginning of the season takes up much

---

[129] *Prog. 1612*, sig. C2^{r}. As he did not expect a lot of gratitude, he was inclined to postpone affairs. See also *Prog. 1603*, sig. D3^{v}: false Latin Version; and *Prog. 1624*, sig. E2^{v} (Cardanus also comments erroneously); *Prog. 1625*, sig. B2^{v}: Critical thoughts on the Greek Ptolemaic text.

[130] *Prog. 1610*, sig. A3^{v}.

[131] Statement by Christoph Meinel, Regensburg, on the state of the natural sciences in 17. Jahrhundert. In: Gaab (2011), p. 14.

[132] *Prog. 1601*, sig. A3^{r}.

[133] *Prog. 1611*, sig. E1^{v}.

[134] Die Origin of this 'dictum' is unknown. See Bauer (1994), p. 173, 179.

[135] *Prog. 1602*, sig. A2^{r-v}.

[136] *Prog. 1607*, sig. A4r and *Prog. 1616*, sig. E1^{r}: "will von einem in dieser Kunst unerfahrnen ungereformirt seyn."

space. In contrast, the monthly rubrics are quite brief. Notes on the respective visibility of the planets are given. On one occasion he mentions a favorable opportunity to witness a conjunction of Mars and Venus with the new perspicillum, the telescope.[137]

Already in his first prognosticon, Marius distances himself clearly from incompetent calendar writers, who offer wildly fantastic prophecies but were unable to give a proper prediction of the weather.[138]

> [...] das sie billich sollten fleissig in acht nehmen, als dessen Augenscheinliche und unlaugbare enderung von dem stand unnd lauff der Planeten herkommet. ([...] that they should fairly and diligently take into account, as its evident and undeniable changes come from the position and course of the planets)

This was also Kepler's conviction,[139] which is the reason why he extended the range of the aspects. The weather forecasts, however, were not solely based on an astrological interpretation of accurate astronomical data.[140] Calendar makers willingly substantiated the conclusiveness of the forecasts in those days with historic weather reports or their own weather records as empirical proof that similar celestial constellations suggest similar weather.[141] For this Marius used the entries in his old ephemerides[142] published by Johann Stöffler and a weather diary[143] he had kept since 1594. He had the weather records of his old teacher Vogtherr[144] at hand, and during his Italian sojourn, his brother Jakob had kept records of the weather at home.[145] He would have liked to have had access to the weather observations of the late Caesius.[146] Like other calendar writers, he often referred to the extensive weather records in Cyriacus Spangenberg's (1528–1604) *Mansfeld Chronicle* (Mansfeldische Chronik), which first appeared in 1572.

---

[137] *Prog. 1620*, sig. B4^r.

[138] *Prog. 1602*, sig. A2^v.

[139] Kepler 11, 2 (1993), pp. 103–113 (*Prog. 1605*, Dedication). Anhand der Witterung des vergangenen Jahres der versuchte Nachweis, "das die verenderung des gewitters gewißlichen sich von tag zu tag nach den aspecten und stationibus Planetarum richte" (On the basis of the actual weather of the previous year the attempt to establish that the change in the weather had day by day conscientiously complied with the aspects and the stations of the planets).

[140] Compare here also Bialas in Kepler 11, 2 (1993), p. 458.

[141] See: Herbst (2010a), pp. 214–232: Neues Material – Berichte zu Naturbeobachtungen (p. 215f. for Marius). Calendar writers worked quasi on a weather statistics on an astrological basis. Compare also Kepler 11, 2 (1993), p. 104 (*Prog. 1605*): Kepler gratefully registered that Caesius recounted his weather prognosis with the actual weather and pleaded with other practitioners to follow suit.

[142] *Prog. 1607*, sig. B4^r (Johann Stöffler, 1452–1531. His ephemerides were published from 1499 till 1551 in 13 editions).

[143] *Prog. 1616*, sig. C3^v.

[144] *Prog. 1618*, sig. A4^v; for Vogtherr see Vogtherr (1908), pp. 52–55.

[145] *Prog. 1612*, sig. B4^v.

[146] *Prog. 1612*, sig. B2^r. Kepler, too (1993), p. 104, 11,2), appreciated these weather observations (*Prog. 1605*).

Great uncertainties remained, however. Marius admitted frankly that the "doctrina meteorologica" still left a lot to be desired, "for I in my industrious observationibus Meteororum have different findings than the common doctrina meteorologica provides." On this topic he is also prepared to offer a helping hand to others.[147] The interpretation of multiple aspects is not easy,[148] as is the consideration of retrospective effects, the long-term influences of previous constellations.[149] Like Kepler, Marius viewed the absence of accurate ephemerides[150] as the biggest deficiency in forecasting the weather. He shared his conviction that the weather resulted from influence of the aspects on the particular configuration of the Earth.[151]

Shortly before his death, Marius gave a summary of the subject of weather forecasting after many years of discussion that leaves many questions unanswered[152]:

> Es ist warlich den Aspecten im gewitter allein nicht zu trauen, man muss die qualitates signorum, unnd stellarum fixarum, bey welchen solche Aspect geschehen, auch in acht nehmen [. . .] Das Fundamentum stehet penes observationem et experimentiam, wer hierinnen nur schlecht mit rationibus will umbgehen, wiewol solche, wo es sein kann, nit zu verwerfen seyn der wird warlich ein schlechte Astrologiam endlich schmiden. (One should not trust the aspects alone in storms, one has to also take into account the qualities of the star signs and the fixed stars by which such aspects occur, the foundations are always precise observation and experiment, who within this only handles with bad calculations albeit such where it can be are not to be rejected, he will truly forge a bad astrology)

He did not wish to instruct anyone.

From the weather forecast sprang the data on fertility and harvest expectations. Marius was obviously particularly keen on the wine harvest. He gave his prognostication for each year, although he would have preferred to stay silent on the subject. He was afraid that his prognostication would only serve the usury manipulations of the rich wine and corn Jews, baptized and unbaptized, to the detriment of the common men and paupers.[153] However God was so omnipotent that he could give a gloriously good potion in abundance to the poor, the elderly, and the infirm against the "influence of the stars."[154] A forecast for a particularly cold September in 1620 concludes with the advice[155]:

---

[147] *Prog. 1610*, sig. A3^v.

[148] *Prog. 1616*, sig. E2^v.

[149] *Prog. 1626*, sig. C1^v. Here Marius refers to Keplers *Discurs von der Grossen Conjunction [. . .] uber dass 1623. Jahr.* In: Kepler 11, 2 (1993), pp. 230–244.

[150] Kepler 11, 2 (1993), p. 157 (*Prog. 1618*). For Marius compare footnote 76.

[151] Kepler 11, 2 (1993), p. 158; Marius: *Prog. 1611*, sig. A4^v–B1^r; *Prog. 1616*, sig. E2^v–E3^r; *Prog. 1621*, sig. B3^r: "Non omnes tempestates simpliciter a stellis excitantur [. . .]." From the Kepler's prognostica he can gather that Kepler shares this opinion.

[152] *Prog. 1628*, sig. B4^r.

[153] *Prog. 1606*, sig. E2^v.

[154] *Prog. 1602*, sig. E2^r.

[155] *Prog. 1620*, sig. C3^r. The predictions for the wine harvest were noted. In the *Prognosticon auf 1623*, the announcement of a good vintage has been underlined by the reader (copy of the Staatsarchiv Nürnberg, sig. C3^r).

O jhr Francken, und alle die jhr gern Wein zur notturft trinckt, betet fleissig, so ich leb, will ich auch mit euch beten. (Oh you Franconians, and all those who like to drink wine to quench their thirst, pray diligently, as I live, I will also pray with you.)

Weightier were the remarks on the eclipses for a year, those of the sun and the moon and their effects. Primarily their exact time and astronomical location within the framework of the other celestial constellations had to be determined, which Marius did explicitly "ex hypothesi and tabulis"of the "praestantissimi Astronomi Tychonis Brahe."[156] By the expected effects of the eclipses on global events, Marius again followed the instructions of Ptolemy,[157] which he had presented in detail for the years 1601 and 1603. According to this (1) the locations affected by the eclipse are determined; (2) the time, when this is likely to happen; (3) the "genus," the way it effects humans or other objects; and (4) the quality, whether they are to be assessed as good, evil, or something else. These instructions of the new calendar maker were also attached to an admonition to his reader that he should preferably keep the prognosticon as the effects could stretch over longer periods of time.[158] Thus he began the *Prognosticon auf 1603* with the discussion of the eclipses of the year 1601.

The details on the "natural fortunes" that dominate the chapter on eclipses are, to the modern reader, mostly vague hints or general references on occurrences, wars and catastrophes, plagues, and misfortunes poised to happen anywhere at any given time. More than the weather forecast, however, the prognostications came with the proviso of being speculations about possible tendencies "astra inclinant, non necessitant," a caveat of all calendar makers who did not want to transgress the boundaries of "astrologia christiana." A benevolent God could always change everything for the better if he was moved by prayer and a pious way of life. A sensible man could, if he took celestial influence on natural earthly matters into account, make the necessary provisions—"vir sapiens dominabitur astris."[159] On the other hand, it could hardly be helpful if he, Marius, promised mountains of gold or the land of milk and honey.[160] To the modern reader, for the most part, the texts appear as prolonged astrological litanies. A recommendation to the reader, drawn from a celestial constellation, recommending to be careful when dealing with the authorities to avoid falling into disgrace,[161] certainly rang as true as a prediction for the autumn of 1608 with Saturn in the seventh house[162]:

[...] will unrath zwischen Eheleuten anrichten. Ist aber nichts seltzames, geschicht wol, wenn gleich Saturnus nit an diesen ort stehet. ([...] will cause discord between married

[156] *Prog. 1606*, sig. E1^r.

[157] *Prog. 1601*, sig. B3^r.

[158] *Prog. 1602*, sig. D2^r; a table with useful explanations was included (sig. D4^r–E1^v).

[159] B. Bauer (1994), p. 173. This "dictum" is not explicitly quoted by Marius, as with the one noted in footnote 134; it is however analogously always present.

[160] *Prog. 1609*, sig. D4^r (StAN 287).

[161] *Prog. 1603*, sig. D1^v.

[162] *Prog. 1608*, sig. D6^r.

couple, which, however, is not strange and happens even when Saturn is not located in this place)

It would not be appropriate, however, to reduce the calendar forecasts to such trivia. One can sense that Marius is seriously attempting to give the astronomical data a careful astrological interpretation. Without going into "particularia," he wants in all his astrologies to only present that which he could with best knowledge and conscience say and put it forward with the prognostic for the discussion by knowledgeable astrologer.[163] His often cryptic modes of expression were, he explained thus, that he as an astrologer spoke parabolically. Bright people, "id est prudentes politici," would easily recognize how and who, even where he meant it.[164] The codes also point in the direction of *particularia*, but Marius did not want to be more precise.[165] This was, however, not always so perceived. In his prognostic for 1619, he wrote about the Kingdom of Bohemia, that, if the astrologers were correct, without any doubt, something special would happen among great potentates, yes, the greatest—but he may not write more.[166]

Allein man wird wunder hören, ehe diß prognosticon in gemeiner Leut Händ wird kommen, wer lebt und zu Prag ist, [. . .] der wird wunder hören oder wol gar sehen [. . .] etwas besonders, vor gute wird es gut sein, vor böse aber bös. (Solely one would hear wonders before this prognosticon reaches the common people, who live and are in Prague, [. . .] he will hear wonder and even see [. . .] something special for the good it will be good, for the bad however bad)

The foreword bears the date of March 5, 1618, 3 months before the Defenestrations of Prague, but the text must have, given Marius' mode of working, already been written in 1617. The tone of the prognostication gives us a clear indication of the mood at the Ansbach Court and in the Protestant camp. Margrave Joachim Ernst, co-founder and general of the Protestant Union of 1608, stood at the side of Frederick V of the Palatine and supported the election of the Elector as King of Bohemia in 1619. Marius's sympathies, as a confirmed supporter of the Reformation, would have matched the politics of his chosen sovereign, who probably read

[163]*Prog. 1602*, sig. D3^{v}, sig. A4^{v}: "[. . .] damit hohen ingenijs solchen sachen besser und eygentlicher nach zudencken, hab ursach geben wollen." "[. . .] so that higher intelligences consider such things better and more realistically, wanting to give them a cause."

[164]*Prog. 1620*, sig. B4^{v}–B5^{r}; In the *Prognosticon auf 1621* (sig. C4^{r}) he gives the clue: "Wer wissen will, welche land und Personen ich hiermit meine, der schlag dz Täffelein der 12 Himlischen Häuser auff," das den Prognostiken angehängte "Register der Städt" (Those who wish to know, which country and which people I mean here should open the table of the 12 celestial houses, and the register of the town appended to the Prognostica). See also: *Prog. 1625*, sig. A2^{v}: "meine einfältige und keineswegs ärgerliche oder ehrnrührige meinung [. . .] meistentheils Parabolischer weiß angezeigt" (my simple and by no means annoying or dishonourable opinion [. . .] mostly expressed parabolically).

[165]*Prog. 1615*, sig. A4^{r} [= B4^{r}]: He did not want (in published writings) to make predictions about (recognisable) persons, as this only brings great danger.

[166]*Prog. 1619*, sig. A4^{v}–B1^{r}.

those lines with the proper amount of satisfaction.[167] The Imperial Mathematicus, Kepler, would have been less pleased to read them. His unfriendly words from 1619, with which he described Marius as "vates invisus et audax" (an unpopular and brash prognosticator) who should refrain from getting on the nerves of his acquaintances,[168] could be linked to these passages as they contained particulars which also touched Kepler. The manner in which Marius described planets and their constellation with quasi-human characteristics and their subsequent interpretation evidently went too far for Kepler.[169] Kepler criticized Marius for his prognostication about Bohemia once again in the *Prognosticon für 1620*, without naming him directly.[170] Subsequent global events made Marius more cautious again. In his Prognosticon for 1622, compiled in 1619, he explained that he would offer little or nothing at all about the high and mighty and announce that no longer so clearly[171]

> [...] dieweil ich augenscheinlich vermercke, dass man mir mein reden und schreiben zum ärgsten ausslegt, und ich damit nit mehr aussgericht, als mir dadurch freund zu feinden worden unnd mit meiner sauren arbeit [...] nur feindschafft verdient hab. ([...] because I notice currently that my talk and my writing are interpreted direly and I achieve nothing more than my friends have become enemies and with my hard work [...] have only earned animosity)

What he naturally knows from the stars and shows in his prognostica does not refer to any person in particular, because one could not find anybody's name among the stars. He had never attempted to intervene in secular matters, as this kind of knowledge is not his profession.[172]

With the obvious expectations of the readers of his prognostica of news about the Great War in Germany, Marius grew ever more reserved. He could not possibly write about it as earthly constellations proved to be more powerful than celestial ones,[173] but the aspects led him to have hope until the end.[174] The *Prognosticon auf 1629*, compiled in 1624, the year of his death, listed constellations under which peace treaties had previously been signed.[175] The convoluted present, however, made it more and more difficult to provide conclusive forecasts for the future.[176]

[167]See footnote 34.

[168]Kepler 17 (1955), p. 376 (no. 850, from Linz 31.08.1619).

[169]Compare B. Bauer (1989), p. 103f., pp. 110–112.

[170]Kepler 11, 2 (1993), p. 198, 202 (*Prognosticum auf 1620*): It had never been his opinion, that "irdische Handlangen," (earthly acts), statement about people "nach ihren Umbständen im Gestirn praedestinirt" (were predestined through celestial constellations). The preamble is dated November 10th 1619. He had had ample time to digest in depth Marius' *Prognostik* for 1620.

[171]*Prog. 1622*, sig. B5^r. He had already explained in the *Prognosticon auf 1615*, he did not want "öffentlichen schrifften" (in public and in writing) to make predictions about (recognisable) persons, as this brought great dangers. (sig. A4^r [= B4^r]).

[172]*Prog. 1622*, sig. A3^r.

[173]*Prog. 1622*, sig. A4^r; *Prog. 1623*, sig. C4^r.

[174]*Prog. 1624*, sig. B1^v, D2^r.

[175]*Prog. 1629*, sig. B2^r.

[176]See Seethaler (2000), p. 244; Matthäus (1969), col. 1223–1225.

In his foreword for 1610, Marius had already acknowledged that the foundations of astrology had not yet been sufficiently investigated.[177] More and more frequently occurring unusual celestial phenomena of his time, the unusual constellations, the suddenly appearing novae, as well as the plethora of comets appeared to him[178]

> [. . .] die ordenliche influentiam Coeli vel siderum mercklichen Turbirn, und die fundamenta Astrologiae, so auff langer unnd fleissiger oberservation bestehen, gleichsam durch ein Erdbidem auß seinem ort [zu] bewegen und zweiffelhafftig [zu] machen. ([. . .] the regular influence of the heavens or the stars are perceptibly turbulent and the fundamental astrology based on long and diligent observations quasi through an earthquake moved out of position and made doubtful)

He could, however, also gain a positive view of this assessment: God has[179]

> [. . .] die neuen Stern vnnd Cometen darzu verordnet, die Menschen, sonderlich die die Mathematicos, auffzumundern, und den sachen scherpfer nachzudencken. ([. . .] decreed the new stars and comets for the purpose of encouraging the people and especially the mathematicians to reflect more astutely about the situation)

To think more astutely had always been Marius' ambition. This concern is taken up in the dedications to the sovereign that also introduce his prognostica. Preambles to prognostica were quite common among calendar makers. They present varied justifications of the noble art of astrology—sometimes there were also detailed explanations of the basics of calendar computations. As far as I can see, Marius is one of the first to venture beyond this narrow topic, which is naturally also included.[180] In his dedications to the sovereigns, he wanted to "alle zeit von einer wichtigen frage oder sonsten nützlichen sachen [. . .] handeln"[181]—"philosophice unnd mathematice[182] (always to treat an important question or other useful subject [. . .] "philosophice and mathematice").

In his first preface for 1601, Marius had initially planned to address the intransigent dispute and altercation with our adversaries, the papists, about the introduction of the Gregorian calendar. He dropped the subject, however, as he feared it would be deemed imprudent for an author whose work was being published for the

---

[177] *Prog. 1610*, sig. A3^v.

[178] *Prog. 1616*, sig. A4^v. Already more reserved in the *Prognosticon auf 1611*, sig. E2^r: "Darzu so lesset es jetzt etlich Jahr am gewitter und anderm ansehen, das entweder die frequentia stellarum novarum und Cometarum eine Jrrung in die Astrologiam machen, oder welches denn glaublicher, das Gott selbsten einmal zum Regiment recht greifft, und besihet wie alle und jede ständ bißhero haußgehalten haben." (In addition now several years of storms and other let it appear that either the frequent new stars and comets cause an error in the astrology or with believers that God himself has really grasped command and inundates everything and everybody who has till now kept house.)

[179] *Prog. 1620*, sig. A3^v–A4^r.

[180] Names to be mentioned here are Peter Crüger (1580–1639), mathematicus of the city of Danzig, whose essays excerpted from his writing calendars were republished under the title *Cupediae Astrosophicae Crügerianae* (Breslau 1631). David Frölich (compare footnote 115) included profound essays on diverse topics to his prognostica. For Crüger compare also Kremer (2014).

[181] *Prog. 1606*, sig. A2^r.

[182] *Prog. 1622*, sig. A2$^{r–v}$.

first time. Fresh from his appointment as a princely mathematicus, he could now strike out again. His dedication for 1607 defined the geometria and astronomia containing scientia mathematica that he practiced, in which the astrologia as applied astronomia brought manifold and great benefit. Both here and in the preface for 1619, he clearly denounces all superstitious practices, indicating respect for God's omnipotence. Marius asserts that his reflections on astrology, which he bases on natural or astral influence, should not be taken as resorting to Christian theology.[183] Therefore, he also opposed making a superstition of "ex naturali influentia Coeli"[184] (to turn the natural influences of the heavens into superstitions). In the preface of 1608, Marius tries to sketch the beautiful order of the heavens and the Earth created by God as the interplay of microcosm and macrocosm and the "aurea catena." The sequel from 1609 contains the proof that the devil must be a creature contrary to this natural order and the cause of all disorder.[185]

Topics from the "Astronomia instrumentalis et numeralis" were also dealt with: the observation of the New Star of 1604 in Padua (1606), whether the Earth is immovable, as well as the size of the planets (1613); whether the impact of the stars was currently growing weaker (1616); refraction while observing the lunar eclipse (1618); and the assumption of an own light of the moon that is observable during a lunar eclipse (1621). Marius wanted to display his opinions on these subjects not least to his colleagues and "andern gelehrten und fridfertigen Mathematicis dz urtheil befehlen"[186] (and to demand a judgment from other scholars and peaceable mathematicians).

Slightly more out of the ordinary were discussions whether the liberal arts and sciences published in foreign languages should be translated into German. Reasons for and against were explained in the two editions for 1610 and 1611. Seventy years before Christian Thomasius (1655–1728), Marius saw no reasons against a scientific German language.[187] For the edition for 1615, Marius the physician investigated the question whether a "panacea" a "Medicamentum universale," a cure for all diseases, existed and had to concede that up till now no such panacea had been found. The text found its recipient, as a person of the highest rank admonished him during a private conversation not to openly discuss the secrets of nature he had alluded to.[188] Later Marius contradicted a severe Biblicism, which claimed that according to Preacher

[183] *Prog. 1619*, sig. A2^{r-v}. Das seien auch die Grenzen der von ihm auf Wunsch erstellten Nativitäten (limits of the nativities).

[184] *Prog. 1621*, sig. A2^{r}.

[185] Matthäus (1969), col. 1098 must accordingly be corrected. Marius wanted to treat the topic in a separate publication "weittleufftiger von der Dispostion der gantzen Natur [. . .] handeln" (extensively according to disposition of the whole of nature). But ill health and the complexity of the material prevented him from addressing the subject satisfactorily (*Prog. 1609*, sig. A3^{r}).

[186] *Prog. 1621*, sig. A3^{r}.

[187] This discourse was probably initiated not least by Marius' translation of Euclid, which the highest official of the Ansbach government, Johann Philipp Fuchs von Bimbach (ca. 1568–1626), had requested from him. Marius referred to Fuchs's preface in his preamble for 1611 (sig. A3^{r}).

[188] *Prog. 1616*, Preamble.

Solomon, there was nothing new under the sun, and therefore book printing and the telescope had already been known in the ancient world.[189]

With the preamble for 1622, he initially wanted to present a very personal topic. It is "nicht allein in privatis Colloquiis, sondern auch in offentlichen Schrifften und declamationibus" (not only in private discourse but in public writings and declarations) that his credibility had been questioned. He had, however, been strongly discouraged from expressing personal statement of rejection. The large print of the dedication had to mask this fact. Presumably it had been insinuated that he had, like Caesius, crypto-Calvinistic tendencies.[190] Marius did not reply in print but he talked about it. "I am a Medicus, Mathematicus, Musicus and in my heart a simple theologian"[191]—that was his credo in public at that time.

One recurring theme Marius had been dealing with since 1603 was the discernible, fantastical change or reformation gripping the world "bald in Occident und Orient [. . .] der gleichen bey Manns gedencken nicht geschehen" (soon in occident and the orient [. . .] nothing similar having ever occurred in the thoughts of humanity) clearly indicating that Judgment Day was imminent.[192] The great conjunction of Saturn and Jupiter of 1603, the first since 800 years, at the beginning of the fiery triangle,[193] tremendous heavenly phenomena, not seen since the world began,[194] as well as the four new stars since 1572 were, in his opinion, clear signs of a global change starting in Germany, which would encompass the whole world. The astrologers agreed on this, as there was more than enough astrological evidence.[195] Already in the preamble, which he had written in Padua for the 1603 edition, Marius presented at length the significant circumstances that had led him to this assessment.

With the transition into the age of the fiery trigon, one could assume that the power of the Turkish Empire would be weakened, but it was obvious "daß nun mehr die Welt zu jhrem ende eyhlet" (that the world was hurrying toward its end).[196] In the end his conviction grew that in the present a direct intervention by God was apparent that could not be explained by any sensible astrologer alone through the influence of the stars.[197] However, his forecasts were neither drawn "ex solo enthusiasmo"[198] nor did he adhere to chiliasts, the superstitions of the Kabbalah or prophecies, rather he

---

[189] *Prog. 1614*, Preamble.

[190] In a brief justification, Marius refers to the Augsburg Confession but not to the Formula of Concord of 1577, to which the Ansbach scholars and clergy had to pledge. As an astrologer Marius probably had more affinity to Melanchthon than to Lutheran orthodoxy.

[191] *Prog. 1622*, sig. A3$^{r–v}$.

[192] *Prog. 1603*, Preamble, sig. A3^{r}–A4^{v}, sig. D3^{v}.

[193] Compare Bauer (1994), p. 183; Ernst (1986); Hamel (2012), pp. 396–398 (Konjunktion, große) and 641f. (Triangel, Trigon).

[194] *Prog. 1612*, sig. A2^{v}; *Prog. 1620*, sig. A3^{v}–A4^{r}.

[195] *Prog. 1612*, sig. A2^{r}.

[196] *Prog. 1603*, sig. A2^{r}–A4^{v}, here sig. A3$^{r–v}$.

[197] *Prog. 1625*, sig. A2^{v}.

[198] *Prog. 1612*, sig. A2^{r}.

remained for them by his regular celestial constellations.[199] Slightly more unusual is therefore a quote by the elevated Philosophus Paracelsus in the prognosis for 1609 on the subject of global change, which should be preceded by a "pacifico mundi seculo"[200]:

> Auch müssen die Ständt [– das Ständewesen –] untergehen und gar auß der Welt gereuth werden ... Als dann wird der Mensch in sein rechten verstand kommen, und Menschlich leben, nicht Viehisch. (Also the estates must perish and even be uprooted out of the world ... and then humanity will come to its senses and live like humans and not like animals)

However, in a postscript the princely court mathematicus dissociated himself—in Latin—from Paracelsus's political demands,[201] and he never returns to such conjectures. His statement that God had revealed the Day of Judgment remains predominant. In his prognosticon for the year 1623—compiled in 1621—he speculates in depth about his assumption, first brought forward 19 years previously, on the imminent end of the world, "wenn auch diese sachen mehr Theologica als Astrologica sein" (even though these matters are more theological than astrological).[202]

In the penultimate *Prognostica auf 1628*, which had already been composed in 1623 and was published posthumously, Marius summarized the ancient discipline of *astrologia*: In the course of 28 years, I have learned what I should maintain about this art.[203]

I do not believe everything; therefore I do not reject everything.

As in the past, he saw no reason to relinquish astrology and its teachings.[204]

Like other "saniores astrologeri," he wanted[205]

> [...] bey den Constellationibus oder causis physicis ohne superstition also verbleiben, das man der providentiae divinae nicht zuweit eingreiffe. ([...] to remain without superstition by the constellation or the physical causes, so that one does not interfere to deeply with divine providence)

[199] *Prog. 1623*, sig. A3^{v}; *Prog. 1625*, sig. A2^{v}–A3^{r}.

[200] *Prog. 1609*, sig. C4^{r}. Paracelsus in Chap. 8 of his book de rebus naturalibus.

[201] *Prog. 1609*, sig. C4^{r}: "Haec Theophrastus; quae de eversione ordinum & statuum politicorum scribit, mihi non probantur, anabaptisticum enim quoddam sapiunt; reliqua vero maxime." Brahe at the end of part 1 of his *Progymnasmata* had also written extensively about such a great change. See *Prog. 1612*, sig. A2^{v}–A4^{r}.

[202] *Prog. 1623*, Preamble sig. A2^{r}–A4^{r}. This topic had already been announced in the prognostic for 1622 (sig. A3$^{r–v}$).

[203] *Prog. 1628*, sig. B4^{v}.

[204] *Prog. 1611* sig. E1^{v}–E2^{r}: "[...] mein bißhero gethanes progosticirn, welches ich noch also als eine Astrologische mutmassung nicht aller dings will auffgehoben, sondern an seinem werth bleiben lassen." – falls Gott anders entscheidet. (the predictions that I have made up till now, which I still want not as an astrological conjecture cancelled by the way but rather left standing on their merits).

[205] *Prog. 1626*, sig. D4^{r}. *Prog. 1611*, sig. E1^{v}–E2^{r}.

In his final prognostic for 1629, which he had composed in 1624, the year of his death, Marius ended the chapter on eclipses with the words[206]:

> Der Astronomorum Bedencken sein natürliche Mutmassung auß langer Erfahrung genommen, welche Gott enden und wenden kann, nach seinem [. . .] Wolgefallen. (The astronomer conceives from long experience his natural assumptions, which God can end or change according to his [. . .] pleasure)

He had hoped that the advances in astronomical science, to which he had contributed with all his skills, could lay the foundations for a more certain astrology. He shared this expectation with Kepler, who also tried to claim the renewal of astrology in science. In order to improve the precision and certainty of astrological predictions, committed astrologers pushed for a continuous improvement in astronomical observations and calculations. Kepler justified these efforts self-critically in the *Tertius intervienens*, with the admonition not to throw the baby out with the bathwater.[207]

> Wann zuvor nie niemandt so thöricht gewest were, daß er auß dem Himmel künfftige Dinge zu erlernen Hoffnung geschöpfft hette, so werest auch du Astronome so witzig nie worden, daß du deß Himmels Lauff von Gottes Ehr wegen, zu erkündigen seyn, gedacht hettest: Ja du hettest von deß Himmels Lauff gar nichts gewust. (When never in the past anybody had been so misguided that he had hoped to learn future occurrences out of the heavens then you astronomers would never have been so clever that you would have thought to investigate the motion of the heavens by God's grace: Yes, you would have known nothing about the motion of the heavens)

The efforts to construct astrology definitely as a legitimate science had failed.[208] The further founding of astronomy finally led to astrology's demise".[209] The impetus that the astrological vision had given to astronomy led from the middle of the seventeenth century visibly to a gradual but unstoppable diminishing of astrological interpretations. Astrology had lost the mood[210] that had carried her. The new discoveries in astronomy were one of the forces that pushed the gates into the modern period open. One of the driving forces had been astrology, ultimately an unscientific explanation of the world. This is the constellation in which we see Simon Marius and his calendars.

---

[206] *Prog. 1629*, sig. D3. Until 1612 Marius had ended his prognostica with verses from the Bible "Thus saith the Lord, Learn not the way of the heathen, and be not dismayed at the signs of heaven; for the heathen are dismayed at them." (Jeremiah 10,2) "Commit thy way unto the Lord; trust also in him; and he shall bring it to pass." (Psalm 37,5).

[207] Kepler (2004), p. 56ff.: VII. Der Fürwitz in Astrologia lehret und ernehret die Astronomiam. Marius did not want to throw out the baby with the bathwater either. Already in the *Prog. 1611* (sig. A4^r) he likewise used this particular phrasing. He seems to have received a copy of the *Tertius Interveniens*, published in 1610, quite quickly.

[208] See Grössing (2005), p. 182.

[209] Graubard (1958), see also Herbst (2010a), pp. 140–144: Die Erosion des astrologischen Glaubens; Gaab (2011), pp. 339–341: Thesen zum Niedergang der Astrologie im 17. Jahrhundert.

[210] I have adopted the term "mood" from Kürnberger (1874), p. 339 "The belief in witchcraft did not submit to evidence, but to the mood. This would have to be the briefest formula."

References

Bauer, Barbara (1994): Sprüche in Prognostiken des 16. Jahrhunderts. In: Haug, Walter and Wachinger, Burghart (eds.): *Kleinstformen der Literatur* (= *Fortuna vitrea*; 14). Tübingen: Niemeyer 1994, pp. 164–204

Bauer, Barbara (1989): Die Rolle des Hofastrologen und Hofmathematicus als fürstlicher Berater. In: Buck, August (ed.): *Höfischer Humanismus* (= *Mitteilung XVI der Kommission für Humanismusforschung*). Weinheim: VCH, Acta Humaniora 1989, pp. 93–117

Bialas, Volker (2013): Johannes Kepler, *Astronom und Naturphilosoph* (= *Schriftenreihe Geschichte der Naturwissenschaften und der Technik*; 25). Linz: Trauner 2013

Büttner, Heinrich-Christoph (1813): Aktenstücke den Simon Marius, seinen Aufenthalt zu Kloster Hailsbronn und seine Empfelung(!) an Tycho Brahe betr. In: *Franconia. Beiträge zur Geschichte, Topographie und Litteratur von Franken*. Vol. 2, Ansbach: Gassert 1813, pp. 70–82

Capp, Bernard (1979): *Astrology & Popular Press. English Almanacs 1500–1800*. London: Faber & Faber 1979

Caspar, Max (ed.) (1968): *Bibliographia Kepleriana*. 2. Edition produced by Martha List. München: Beck 1968

Diefenbacher, Michael; Fischer-Pache, Wiltrud (eds.) (2003): Das Nürnberger Buchgewerbe. Buch- und Zeitungsdrucker, Verleger und Druckhändler vom 16. bis zum 18. Jahrhundert. Edited by Manfred H. Grieb. With a contribution by Peter Fleischmann. Compiled form archives by Lore Sporhan-Krempel and Theodor Wohnhaas (= *Quellen und Forschungen zur Geschichte und Kultur der Stadt Nürnberg*; 31). Nürnberg: Selbstverlag des Stadtarchivs Nürnberg 2003

Ernst, Germana (1986): From the watery Trigon to the fiery Trigon: Celestial Signs, Prophecies and History. In: Zambelli, Paola (ed.): *„Astrologi hallucinati". Stars and the End of the World in Luther's Time*. Berlin: De Gruyter 1986, pp. 265–280

Eyb, Eberhard Freiherr von (1984): *Das reichsritterschaftliche Geschlecht der Freiherrn von Eyb* (= *Veröffentlichungen der Gesellschaft für Fränkische Geschichte*; IX, 29). Neustadt a.d. Aisch: Degener 1984

Gaab, Hans (2011): *Der Altdorfer Mathematik- und Physikdozent Abdias Trew (1597–1669). Astronom, Astrologe, Kalendermacher und Theologe* (= *Acta Historica Astronomiae*; 42). Frankfurt a. M.: Harri Deutsch 2011

Gaab, Hans; Leich, Pierre (2014): Marius' Replik auf Scheiner. Der Anhang zum Mundus Iovialis von Simon Marius. In: *Globulus. Beiträge der Natur- und kulturwissenschaftlichen Gesellschaft für Ingolstadt*; 18 (2014), pp. 11–14

Galilei, Galileo (1607): Difesa di Galileo Galilei [...] contro alle callunnie & imposture di Baldessar Capra. Venedig 1607. In: Galilei, Galileo: *Le Opere*. Vol. II, Florenz: Barbèra 1891

Grafton, Anthony (1993): *Joseph Scaliger. A Study in the History of Classical Scholarship*. Vol. 2: Historical Chronology. Oxford: Claredon Press 1993

Graubard, Mark (1958): Astrology's Demise and its Bearing on the Decline and Death of Beliefs. In: *Osiris* 13 (1958), pp. 210–261

Grieb, Manfred (ed.) (2007): *Nürnberger Künstlerlexikon. Bildende Künstler, Kunsthandwerker, Gelehrte, Sammler, Kulturschaffende und Mäzene vom 12. bis zur Mitte des 20. Jahrhunderts*, 4 Vol. München: Saur 2007

Grössing, Helmuth (2005): Gedanken zu Keplers Astrologie. In *Miscellanea Kepleriana. Festschrift für Volker Bialas* (= *Algorismus*; 47). Augsburg: Rauner 2005, pp. 175–182

Hamel, Jürgen; Caspar, Max (eds.) (1998): *Bibliographia Kepleriana*. Supplementary Volume München: Beck 1998

Hamel, Jürgen; Caspar, Max (eds.) (2012): Begriffe der Astrologie. *Von Abendstern bis Zwillingsproblem*. Frankfurt a.M.: Harri Deutsch 2012

Heischmann, Günther (1974): Die Bibliotheken der Freien Reichsstadt Rothenburg ob der Tauber. *Archiv für Geschichte des Buchwesens* 14 (1974), col. 1589–1878

Herbst, Klaus-Dieter (2008): Das Neueste im Jahresrhythmus. Zur Professionalisierung des Kalenderwesens im 17. Jahrhundert. In: Blome, Astrid; Böning, Holger (eds.): *Presse und Geschichte. Leistungen und Perspektiven der historischen Presseforschung* (= *Presse und Geschichte – Neue Beiträge*; 36). Bremen: edition lumière 2008, pp. 97–124

Herbst, Klaus-Dieter (2009a): Die Jahreskalender – Ein Medium für gelehrte Kommunikation. In: Herbst, Klaus-Dieter; Kratochwil, Stefan (ed.): *Kommunikation in der Frühen Neuzeit*. Frankfurt a. M.: Peter Lang 2009, pp. 189–223

Herbst, Klaus-Dieter (2009b): Galilei's astronomical discoveries using the telescope and their evaluation found in a writing-calendar from 1611. In: *Astronomical Notes/Astronomische Nachrichten* 330 (2009), pp. 536–539

Herbst, Klaus-Dieter (2009c): Der Schreibkalender der Frühen Neuzeit – eine noch wenig genutzte Quelle für die Astronomiegeschichtsschreibung. In: Hamel, Jürgen (ed.): 400 Jahre Kepler, Galilei, das Fernrohr und die neue Astronomie (= *Sitzungsberichte der Leibniz-Sozietät der Wissenschaften*; 103). Berlin: Trafo-Verlag 2009, pp. 31–48

Herbst, Klaus-Dieter (2010a): Die Schreibkalender im Kontext der Frühaufklärung (= *Acta Calendariographica – Forschungsberichte*; 2). Jena: Verlag Historische Kalenderdrucke 2010

Herbst, Klaus-Dieter (2010b): Die erstmalige Benutzung von Keplers Rudolphinischen Tafeln für die Herstellung eines Schreibkalenders. In: Gaulke, Karsten; Hamel, Jürgen (eds.), *Kepler, Galilei, das Fernrohr und die Folgen* (= *Acta Historica Astronomiae*; 40). Frankfurt a. M.: Harri Deutsch 2010, pp. 160–169

Herbst, Klaus-Dieter (2012): Von Astronomie bis Volksaufklärung. Neue Forschungen und Perspektiven. In: Herbst, Klaus-Dieter (ed.): *Astronomie – Literatur – Volksaufklärung. Der Schreibkalender der Frühen Neuzeit mit seinen Text- und Bildbeigaben*. Bremen, Jena: Verlag Historische Kalenderdruck 2012, pp. 15–44

Herbst, Klaus-Dieter: Biobibliographisches Handbuch der Kalendermacher von 1550 bis 1750. Noch im Erscheinen. Online einsehbar unter: http://www.presseforschung.uni-bremen.de/dokuwiki/doku.php?id=Startseite

Kepler, Johannes (1606): *De Stella Nova In Pede Serpentarii, Et Qui Sub Ejus Exortum De Novo Iniit, Trigono Igneo: Libellus Astronomicis, Physicis, Metaphysicis, Meteorologicis & Astrologicis Disputationibus endoxois & paradoxois plenus. Accesserunt I. De Stella Incognita Cygni: Narratio Astronomica. II. De Jesv Christi Servatoris Vero Anno Natalitio, consideratio novissimae sententiae Lavrentii Svslygae Poloni, quatuor annos in usitata Epocha desiderantis*. Prag: Sessius 1606

Kepler, Johannes (1938–2009): *Gesammelte Werke*. 21 vol. München: Beck 1938–2009

Kepler, Johannes (2004): *Tertius Interveniens*. Ed. by Jürgen Hamel (= *Ostwalds Klassiker der Exakten Wissenschaften*; 295). Frankfurt a. M.: Harri Deutsch 2004

Koppitz, Hans Joachim (2008): Die kaiserlichen Druckprivilegien im Haus-, Hof- und Staatsarchiv Wien (= *Buchwissenschaftliche Beiträge aus dem Deutschen Bucharchiv München*; 75). Wiesbaden: Harrassowitz 2008

Kremer, Richard L. (2014): Galileo in Danzig, as Portrayed in Peter Crüger's Schreibcalender. In: Albrecht, Andrea; Cordibella, Giovanna; Remmert, Volker R. (ed.): *Tintenfass und Teleskop. Galileo Galilei im Schnittpunkt wissenschaftlicher, literarischer und visueller Kulturen im europäischen 17. Jahrhundert* (= *Spectrum Literaturwissenschaft*; 46). Berlin: De Gruyter 2014, pp. 101–124

Krentzheim, Leonhard (1577): *Chronologia, Das ist, Gründliche vnd Fleissige JahrRechnung / Sammpt verzeichnung der fürnemsten Geschichten / Verenderungen vnd Zufell / so sich beyde in Kirchen vnd WeltRegimenten zugetragen haben / zu jeder Zeit / Von anfang der Welt / biß auff vnsere*, Teil 1. Görlitz: Ambrosius Fritsch 1577

Kürnberger, Ferdinand (1874): *Siegelringe. Eine ausgewählte Sammlung politischer und kirchlicher Feuilletons*. Hamburg: Meißner 1874

Lang, Karl Heinrich von [Ritter von Lang] (1811): Neuere Geschichte des Fürstenthums Baireuth. Band 3, welcher die Regierungszeit des Markgrafen Georg Friedrich von 1557 bis 1603 begreift. Göttingen: Schneider 1811

Leich, Pierre (2012): Die Copernicanische Wende bei Galilei und Kepler und welche Rolle Marius dazu einnimmt. In: Wolfschmidt 2012, pp. 162–193

Lerch, Andreas (2015): Scientia astrologiae. Der Diskurs über die Wissenschaftlichkeit der Astrologie und die lateinischen Lehrbücher 1470–1610 (= *Acta Historica Astronomiae*; 56). Leipzig: AVA – Akademische Verlagsanstalt 2015

Marius, Simon (1596): *Kurtze und eigentliche Beschreibung des Cometen oder Wundersterns / So sich in disem jetzt lauffenden Jar Christi unsers Heilands / 1596. in dem Monat Julio / bey den Füssen deß grossen Beerens / im Mitnächtischen Himmel hat sehen lassen.* Nürnberg: Paul Kauffmann 1596

Marius, Simon (1614): *Mundus Iovialis*. Nürnberg: Johann Lauer 1614

Marius, Simon (1614/1916/2019): *The World of Jupiter*, translated by A.O. Prickard and Albert van Helden, in *Simon Marius and his Research*, Hans Gaab and Pierre Leich (eds.), Cham: Springer 2019

Marius, Simon (1619): *Astronomische vnd Astrologische beschreibung deß Cometen so im November vnd December vorigen 1618. Jahrs ist gesehen worden.* Nürnberg: Lauer 1619

Matthäus, Klaus (1969): Zur Geschichte des Nürnberger Kalenderwesens. Die Entwicklung der in Nürnberg gedruckten Jahreskalender in Buchform. In: *Archiv für Geschichte des Buchwesens* 9 (1969), col. 965–1396

Matthäus, Klaus (2010): Die offiziellen Nürnberger Kalenderschreiber. In: Wolfschmidt, Gudrun (ed.): *Astronomie in Nürnberg* (= *Nuncius Hamburgensis*; 3). Hamburg: tredition 2010, pp. 184–195.

NDB: *Neue Deutsche Biographie*. Vol. 1–25. Berlin: Duncker & Humblot 1953–2013

Pellengahr, Hans-Georg (2012): Simon Marius – die Erforschung der Welt des Jupiter mit dem Perspicillum 1609–1614. In: Wolfschmidt 2012, pp. 72–141

Pültz, Otto (1973): *Die deutschen Handschriften der Universitätsbibliothek Erlangen*. Ed. by Armin Dietzel and Günther Bauer (= *Katalog der Handschriften der Universitätsbibliothek Erlangen*, Neubearbeitung, vol. 4). Wiesbaden: Harrassowitz 1973

Reske, Christoph (2007): *Die Buchdrucker des 16. und 17. Jahrhunderts im deutschen Sprachgebiet* (= *Beiträge zum Buch- und Bibliothekswesen* 51). Wiesbaden: Harrassowitz 2007

Schuhmann, Günther (1980): Die Markgrafen von Brandenburg-Ansbach. Eine Bilddokumentation zur Geschichte der Hohenzollern in Franken (= *Jahrbuch des Historischen Vereins für Mittelfranken*; 90). Ansbach: Selbstverlag des Historischen Vereins 1980

Seethaler, Josef (2000): Kalender und Prophezeiungen. In: Petschar, Hans (ed.): *Alpha & Omega. Geschichten vom Ende und Anfang der Welt.* Wien: Springer 2000, pp. 242–255

Swerdlow, Noel (2004): Tycho Brahe's early lunar theory and the lunar eclipse of 31 January 1599. *Centaurus* 46 (2004), pp. 1–40

Thoren, Victor (1972): An 'unpublished' version of Tycho Brahe's lunar theory. *Centaurus* 16 (1972), pp. 203–230

Vogtherr, Friedrich (1908): *Geschichte der Familie Vogtherr im Lichte des Kulturlebens*. 2. ed. Ansbach: Seybold 1908

Wolfschmidt, Gudrun (ed.) (2012): *Simon Marius, der fränkische Galilei, und die Entwicklung des astronomischen Weltbildes* (= *Nuncius Hamburgensis. Beiträge zur Geschichte der Naturwissenschaften* 16). Hamburg: tredition 2012

Zinner, Ernst (1942): Zur Ehrenrettung des Simon Marius. *Vierteljahrsschrift der Astronomischen Gesellschaft* 77/1 (1942), pp. 23–75

Zinner, Ernst (1964): *Geschichte und Bibliographie der astronomischen Literatur in Deutschland zur Zeit der Renaissance*. 2nd ed. Stuttgart: Anton Hiersemann 1964

Chapter 11
Simon Marius as a Tychonic Calendar Maker

Richard L. Kremer

Abstract This chapter analyzes the mathematical astronomy in the printed annual *Schreibkalender* and prognostications authored by Simon Marius for the years 1601–1629. It considers how Marius determined the times of the new and full moons, eclipses, and Sun's entry into the four cardinal points of the year and finds frequent discrepancies between his actual procedures (copying from published sources) and his descriptions of those procedures (independent computation). This chapter suggests that the highly competitive world of calendar production, especially in Nuremberg, may have prompted Marius to deploy combative rhetoric against other calendar makers and to exaggerate his own originality. And the chapter briefly examines Marius's description, in his calendars, of his relationships with two contemporary astronomers, David Fabricius and Kepler. The goal of this chapter is to explore how Marius represented himself in the world of print calendars.

In 1611, Kepler accused Simon Marius (1573–1624) of plagiarizing Galileo's telescopic discoveries. In 1614, the Jesuit professor in Ingolstadt Christoph Scheiner would repeat the charge. In 1623, Galileo would expand and elaborate the accusation with considerable heat in *Il Saggiatore*. Since then, and especially since 1899 when the Société Hollandaise des Sciences à Harlem sponsored a prize contest on the plagiarism question, most scholarship on Marius has focused on matters telescopic and Galilean (Programme de la Société Hollandaise 1899).[1]

This paper, however, examines Marius's relationship with another contemporary astronomer, Tycho Brahe, and his *familia* of young assistants. Following Kepler, I shall refer to Tycho and his defenders as the "Tychonics" (Kepler to Longomontanus, early 1605, Kepler, *GW*, Vol. XV, 1951, p. 140. Cf. Voelkel 2001, Chap. 7; Christianson 2000). My interest is not in Marius as a telescopic

[1]For a survey of some of the scholarship, see Folkerts 1990; for an early account, see Weidler 1741, pp. 430–432; most recently see Pasachoff 2015.

R. L. Kremer (✉)
Dartmouth College, Hanover, NH, USA
e-mail: Richard.L.Kremer@dartmouth.edu

H. Gaab, P. Leich (eds.), *Simon Marius and His Research*, Historical & Cultural Astronomy, https://doi.org/10.1007/978-3-319-92621-6_11

observer or as an author making claims about astronomical discoveries. Rather I want to consider Marius as the annual calendar maker for the small, Franconian principalities of Brandenburg-Ansbach and Kulmbach-Bayreuth with its residence located in Ansbach, about 45 km west of Nuremberg. In particular, I shall examine the mathematical, astronomical foundations of Marius's calendars, i.e., his computation of predicted planetary positions, and shall try to determine the sources he used for those computations. An analysis of how Marius did (or did not) reveal those sources to his readers can inform us about his authorial etiquette, about how he chose, in the world of early seventeenth-century calendrical printing, to characterize his relations with other contemporary astronomers. Marius as a calendar maker, I shall suggest, might help us understand Marius as a telescopic observer since both activities involve an author self-reflexively reporting his own working procedures; both take the working astronomer into the world of print. Studying Marius as calendar maker will also reveal how knowledge of Tycho's new solar and lunar theories spread among Europe's calendar makers and astronomers.

Our essay consists of five sections. First, we consider how Marius learned mathematical astronomy from the available printed sources. We then examine his early calendars for the years 1601 to 1603. Our third section, treating the calendars for 1605 to 1612, finds Marius learning to use Tycho's newly published lunar theory but misleading his readers about his computations. The fourth section examines his calendars for the years 1614 to 1629, when Marius began to explore Tychonic planetary calculations in collaboration with David Fabricius, another astronomical observer and calendar maker living in East Frisia. Our final section considers Marius's relations with Kepler, as depicted in his calendars. For this study, I have consulted copies of all existing editions of Marius's annual *Schreibkalender* (writing calendar) and *Prognosticon astrologicum*, spanning the years 1601 to 1629. No copies have survived for the years 1604 and 1617; for 1601, only the Prognosticon is extant.[2]

Becoming a Calendar Maker

As is well known, Marius, born in Gunzenhausen, ca. 20 km south of Ansbach, would be largely self-taught in astronomy.[3] In 1586 he entered the Fürstenschule (founded in 1581 by the Margrave Georg Friedrich for sons of poor families) in nearby Heilsbronn, where he remained until 1601. During these years his attention turned to astronomy, a topic not formally taught at the school although the third class

[2]No new editions have come to light over the past 70 years. See Zinner 1942, pp. 27–32. Digital copies of many of the editions are conveniently available at the *Marius-Portal*, http://www.simon-marius.net (accessed 1 July 2015). For an overview of Marius's calendars, see Matthäus in this volume.

[3]For biographical details, see Gaab in this volume.

did include "Questions on the sphere, church computus, arithmetic with simple proofs and useful examples"[4] (Lang 1811, p. 348). The earliest results from Marius's study of astronomy appear in a short tract on the comet of 1596 that he published later that year. Although the 23-year-old claimed to have measured the comet's angular distance from nearby stars with a "langen *radium astronomicum*" (Jacob's staff), he did not publish any measured data and refused to take a position on whether the comet was moving above or below the Moon (Marius 1596, sig. B2^{v}).[5] In 1618, Marius would again measure a comet's distance from nearby stars, now reported in arc minutes, with a Jacob's staff of his own construction (Marius 1619, sig. A4^{v}–B2^{v}).[6]

More important for our purposes, the comet treatise reveals the young Marius's skepticism toward published ephemerides with their daily predictions of planetary positions. The comet had moved through the sign of Leo with the planet Mercury, both remaining north of the ecliptic. Yet, Marius noted, the ephemerides of Johann Stadius, in predicting when Mercury would reach its northernmost extent of latitude (he offered no specific dates), was "perhaps false and erroneous," a rather vague accusation suggesting that Marius lacked the apparatus to measure accurately the planet's latitude. Nonetheless, Marius widened his criticism (Marius 1596, sig. B3^{r})[7]:

> For it is undeniable that the planet's path not only in longitude but also in latitude is not adequately ascertained, as daily experience proves.

Offering a quantitative, historical example, Marius noted that Bernhard Walther's measured latitude of Venus varied by 3 degrees from the position predicted in ephemerides of Stöffler. Perhaps assuming that his readers were well acquainted with sixteenth-century astronomy, Marius did not indicate that Walther had made that observation in Nuremberg back in 1504 and that Stöffler's ephemerides, based on the medieval Alfonsine Tables, had been printed in 1499. He also did not indicate that Stadius had based his ephemerides, first published in 1556, extended in 1585 to cover the years through 1606, on the Copernican Prutenic Tables. Marius also reported a quantitative example of his own. On 2 July 1596, he had observed Venus in conjunction with Saturn, with the former's latitude 2 degrees "south" of that listed in Stadius's ephemerides (Marius 1596, sig. B3^{r-v}; Schöner 1544, sig. 54^{v}; Stöffler, Pflaum 1499; Stadius 1585).[8] Our young astronomer, already by 1596, had

---

[4] "Questiones Sphaericae, Calendarium Ecclesiasticum [computus], Arithmetica, mit kindischen Demonstrationibus und dienlichen Exempeln."

[5] Cf. Hamel in this volume.

[6] For a useful summary of Marius's astronomical observations, see Zinner 1942, pp. 36–40.

[7] "Denn vnlaugbar, daß der Planeten lauff nicht allein in longitudinem, sondern auch in latitudinem, noch nicht gnugsam ergründet ist wie die täglich erfahrung bezeuget."

[8] For 2 July 1596, Stadius listed the latitude of Venus as 2;26 "south," a typographical error as the edition failed to mark when the latitude had shifted to north. Marius silently corrected this error in reporting his observation "south" of the ephemerides. Modern computation (JPL Horizon) gives Venus's latitude for this date at 0;44 north. Marius here showed himself to be a careful critic of printed texts but a beginning observer, at least when measured against Tycho's standards.

announced an astronomical program, viz., testing received mathematical astronomy by means of quantitative measurement of planetary motions.

Marius later claimed that during the winter of 1595–1596, he had read Copernicus's *De revolutionibus* and had arrived at a "system of the universe in general identical with that of Tycho." Yet at that time, "Tycho was not known to me even by name, much less his hypothesis, which I only saw in the following year, in outline," in a sketch that had been sent to a local pastor by a student at the university in Wittenberg (Marius 1614/1916/2019).[9] The dedication of the comet tract is dated September 1596, presumably after Marius had come upon a Tychonic arrangement for the heavens; but he did not discuss such cosmological considerations in the comet publication.

Marius's first publication in mathematical astronomy would be his *Tabula directionum novae*, printed in Nuremberg in 1599. Containing 8 numerical tables and an introductory canon of 18 rules, Marius's *Tabula* offered astrologers a set of tabular procedures to divide any given horoscope into 12 houses and to find its "direction" (a technical procedure for advancing a chart to a later time), ostensibly according to methods presented in Ptolemy's *Tetrabiblos*. A century earlier, Regiomontanus had prepared a similar set of tables for finding directions and had offered a new, "rational" method for house divisions. The young Marius, however, accused Regiomontanus of misunderstanding Ptolemy.[10] His new tables, he proclaimed, reject the "repugnant method of Regiomontanus" and not only restore but discover anew the "true foundation of Ptolemy and other ancients" (Marius 1599, sig. A1^{r}–A2^{r}).[11] As is well known, Regiomontanus's "rational" method of house divisions had, by 1600, become widely accepted. The young Marius, who signed his dedication "stipendiarius et alumnus Heilsbrunensis," is revealing another side of his emerging authorial persona; he would not shy away from provocative, even pugnacious rhetoric as he challenged existing astrological practice.

Elsewhere I shall analyze Marius's claim for novelty in these tables. Here we need note only that his account of their creation might have confused his readers about who did what. His preface correctly states that three of the tables were copied verbatim from Regiomontanus (Marius 1599, sig. E2^{v}–I2^{r}; Regiomontanus 1490, sig. D1^{v}–D7^{r}, D8^{v}–F4^{r}, G2^{v}–G3^{r}).[12] The tables for the Ptolemaic house divisions, he added, were computed by his friend August Lanius of Ansbach, following a method Marius had developed using an astrolabe *before* he had read a description of

[9]"Ne nomen quidem Tychonis, multo minus hypothesis ipsius mihi cognita erat; quam tandem sequenti anno in Autumno delineatam vidi" (Marius 1614, sig. C3^{r}; Marius 2019, 2. part, of the fifth). For a later catalog of books, including the sixteenth-century astronomical imprints, held in the Heilsbronn monastery library, see Hocker 1731, pp. 268–76. Nothing is known about Marius's personal library.

[10]For house systems used in the sixteenth century, see North 1986; Kennedy 1994.

[11]Marius's title page reads: "Verissimus antiquorum astrologorum ipsiusque Ptolemaei duodecim coeli domicilia distribuendi modus non tam restitutus, quam de nouo inuentus."

[12]Marius also (silently) used Regiomontanus's value for the obliquity of the ecliptic (23;30) although by 1600 most astronomers were using the more precise value of 23;31,30.

Ptolemy's method published in 1548 (Marius 1599, sig. B1$^{r–v}$; Heller 1548, sig. D2^{v} –D3^{r}). Contemporary astronomers confessed difficulty in understanding the tables, asking an ever more exasperated Kepler for help. Indeed Marius would later admit to their impenetrability even as he emphasized their novelty (Marius to Nikolaus Vicke, quoted in Vicke to Kepler, July 1611, Kepler, *GW*, Vol. XVI, 1954, p. 382)[13]:

> When writing these tables as an autodidact in this skill and little accustomed to geometrical demonstrations, I did what I could. For I had been seriously studying astronomy hardly for two years, without any teacher of mathematics [. . .] Who before me ever published tables of erections and directions to be constructed in the Ptolemaic matter?

We might wonder how a young astronomer could claim priority for a computational method published 50 years earlier in a book he admitted reading. Yet Marius's tabular algorithms for computing Ptolemaic houses do contain some novel elements (e.g., one "conjectures" the ascendent in what we might now call an approximation technique). And they surely show that he had mastered the basic spherical trigonometry required for astronomical and astrological computation as well as for understanding contemporary astronomical theory.

Marius's connections to Tycho Brahe would expand over the next several years. In the summer of 1601, the Danish astronomer was struggling to establish his family, his astronomical program, and his financial base at the court of the Emperor Rudolph II in Prague. He also struggled to find assistants, as many of his long-time subordinates at Hven had refused to move to Bohemia. Among those who did join Tycho that summer was Simon Marius, bearing a letter of introduction apparently from the Margrave Georg Friedrich who had funded his many years of schooling in Heilsbronn (NN to Tycho, 12 May 1601, transcribed in Büttner, Vol. 2, 1813, pp. 81–82).[14] His arrival attracted some attention. On 27 May 1601, one of Tycho's young assistants, Johannes Eriksen, wrote Johannes Kepler, then back in Graz, that the "mathematician of the Margrave of Ansbach, Simon Marius, will in several days increase the number of our *familia* and from what I can tell from conversations here will, I trust, free me from observations, not being condemned to other astronomical heresies." Four days later, Barbara Kepler, who had remained in Prague, wrote her husband that Tycho "hat ein Matematiguß aufgenumen von annspach es ist ein

[13]"Ego tunc temporis velut αυτοδιδαχτοζ in hac facultate et geometricis demonstrationibus minus assuefactus, feci quod potui. Vix enim per biennium serio tunc astronomica tractaueram, omni carens praeceptore mathematico . . . Quis enim ante me tabulas erectionis et directionis Ptolomaico modo instituendae unquam publicavit?" See Gaab, this volume, at note 113; Klug 1906, pp. 404–405. Kremer a, forthcoming.

[14]Scholars have long assumed that Georg Friedrich drafted this letter. See Christianson 2000, pp. 319–321. The letter refers to unnamed "acquaintances" of Marius's who had, on his behalf, previously approached Tycho. Concerning Marius's abilities, it merely asserts that "he has already made a relatively good beginning in his studies and now especially can experience more of the art with you in front of others" ("er in solchem seinem studio allbereit einen ziemlichen guten Anfang habe vnd er jezo sonderlich der Art bei euch vor andern weisen mehrers erfahren kan").

lötiger gesöll" (Tycho has taken in a mathematician from Ansbach, who is a capable assistant) (Kepler, *GW*, Vol. XIV, 1949, p. 168, 170).[15] Apparently, the 28-year-old from Ansbach had made a good, initial impression. Although Marius had never matriculated at a university, Frau Kepler saw him as a "mathematician" and a "capable fellow"; Eriksen expected him to be able to contribute to Tycho's program of systematic astronomical observation.

Unfortunately, only scattered reports of Marius's activities in Prague during the summer of 1601 are known. In 1608, he claimed that he had not only seen but used Tycho's instruments in Prague. In 1610, Marius recalled that his "Kundt- und Freundschaft" with David Fabricius, another of Tycho's assistants, had begun in Prague (see below). In 1611 he remembered that "students of Tycho," in 1601 in Prague, had told him about Tycho's conclusion that the solar eccentricity was incorrectly known. In 1619 Marius claimed that he had been Tycho's "Diener vnd Observator" (servant and observer) (*Prog. 1608*, sig. B4^{r}, *Prog. 1610*, sig. C4^{v}; 1611, sig. B4^{v}, Marius 1619, sig. A4^{v}).[16] However, no letters by Tycho are extant after the spring of 1601. Kepler returned to Prague in early September and Tycho would die on 24 October 1601. It is not clear, therefore, whether Marius and Tycho ever worked together directly. But as we shall see below, it is clear that Marius during that summer established working relationships with some members of Tycho's *familia*, relationships that would continue for many years.

Marius would complete his formal education by studying medicine at the university in Padua from December of 1601 through the summer of 1605 (he did not complete a degree) (see Gaab, this volume, Sect. 6 "Simon Marius in Padua"). His three elder brothers each had attended the university in Wittenberg, with support from the Margrave. Why Simon did not follow in their footsteps is not clear (see Gaab, this volume, Sect. 1 "Youth in Gunzenhausen"). But even before his travels to Prague and Italy, he had begun authoring annual calendars; he had, that is, mastered enough mathematical astronomy, to say nothing of the related compendium of astrological knowledge, to enter what was by then a very crowded marketplace of calendar making. With his first publications for the year 1601, Marius had become a calendar maker.

For the next 30 years, Marius would author an annual *Schreibkalender* and a separate astrological prognostication, both calculated for the geographical longitude and latitude of Heilsbronn (after 1609 for Ansbach) and printed in Nuremberg.[17]

---

[15]Eriksen's "astronomical heresies" probably refer to Kepler's work that spring, writing at the behest of Tycho an attack on the astronomical hypothesis, published in 1588, of Nicolaus Raimarus Ursus. See Jardine 1984, pp. 9–28; Christianson 2000, pp. 272–273; Voelkel 2001, pp. 117–120. Tycho's observing instruments had been installed at his house in Prague only in April 1601; the assistants recorded very few observations (solar altitudes, a few positions of Saturn and Jupiter) during the summer of 1601. See Brahe: *Opera*, Vol. XIII., 1926, pp. 253–285; Thoren 1990, p. 446.

[16]Marius to Mästlin, 29 March 1612, transcribed in Zinner 1942, p. 42. See Caspar 1993, p. 119 and Gaab, this volume, Section 5.

[17]Only the 1613 *Prognosticon* would be printed elsewhere, in Ansbach by Paul Böhem. After 1613, all of Marius's publications would be published and printed by his father-in-law, Johann Lauer. Active since 1599 as a "Buchführer" in Nuremberg, Lauer had published (verlegt) all of Marius's

Both genres had, since 1500, become exceedingly popular. Printers across Europe, but especially in the German-speaking lands, issued literally dozens of editions each year in formats that had become very standardized. *Schreibkalender*, small booklets of usually 12–14 leaves, devoted one page to each month, with the verso pages listing for each day the position of the Moon, the planetary aspects (angles between the planet and the Moon, all of which had astrological implications), and a set of marks encoding astrological advice for that day. The recto sides generally were unprinted, allowing space for owners to write in comments as a kind of daybook (hence the name). *Schreibkalender* also marked the dates and times (to the nearest minute) of the new and full moons (syzygies).[18] *Prognostica* (also called *Practica*) might have up to 30+ leaves and described, in prose, elaborate astrological predictions, based on the astronomical times of the syzygies, eclipses, aspects, and the Sun's entry into the four cardinal points for the year. Detailed prognostications for weather, health and disease, politics, religious affairs, agriculture and fertility, mining, and various social groups were offered, derived from interpretations of horoscopes cast for the predicted times of the astronomical events. Printing these periodica provided stability for print shops; authoring them provided long-term employment for calendar makers at courts, towns, and cities across Europe.[19]

Computing the times of syzygy and eclipses was perhaps the most difficult task in mathematical astronomy. To make things easier for the multitude of calendar makers, a number of leading astronomers, starting with Regiomontanus in 1474, had prepared and printed ephemerides, massive books offering daily positions (to minutes) of all the planets, daily aspects, and the times of syzygies and eclipses for long spans of years. Computed from both the Ptolemaic-based Alfonsine Tables and Copernican-based Prutenic Tables, these ephemerides were easily accessible to sixteenth-century calendar makers. They could copy the needed astronomical data from the printed ephemerides and then construct their astrological interpretations on those quantitative foundations. A sixteenth-century calendar maker, thus, did not necessarily need to know much mathematical astronomy.

Marius's Early Calendars

Marius's calendrical editions contain five types of quantitative astronomical data, the computational foundations of which we can examine. The *Schreibkalender* lists dates and times (generally to minutes) of the new and full moons for the year. The

pre-1613 calendars. When in 1613 Lauer opened his own printing shop, the other Nuremberg printers raised legal complaints, which might explain the Ansbach printing of the 1613 Prognosticon. See Diefenbacher, Fischer-Pache 2003, No. 2745–2746, 2758–2759; Matthäus 1969, cols. 1099–1102; Matthäus, this volume; and Zinner 1942, p. 29.

[18]Marius's *Schreibkalender* for 1602 until 1606 include, on the recto pages, the midday planetary longitudes. Since these data are given only to degrees, I cannot identify their sources.

[19]For an introduction to the massive literature on early printed calendars and practica, see Matthäus 1969, cols. 965–1396; Seethaler 1982; Herbst 2012; Green 2012.

early calendars, from 1602 through 1609, also give daily longitudes (to degrees) for the planets and the Moon.[20] The annual prognostica usually repeat the syzygy times but add the eclipses of the year. And they invariably provide times (to minutes) of the Sun's entry into the four cardinal points of the year; for the vernal equinox, Marius always included a woodcut of the horoscope for that time, with house boundaries and planetary positions given to degrees. The prognostica also specify dozens of planetary aspects (to days). Computational astronomy circa 1600 usually worked to a precision of minutes, increasingly even to seconds; hence, Marius's aspects specified to days or longitudes specified to degrees do not necessarily allow us to uncover his sources. But when he offered his data to a precision of minutes, we can generally identify his sources.

Marius's first annual calendar for the year 1601 shows him intent on explicitly evaluating the several sets of astronomical tables or ephemerides available for that year. Although he had prepared a *Schreibkalender* and *Prognosticon auf 1601*, only the latter has survived.[21] Announcing himself on its title page as a "astronomiae studiosus," Marius dated the dedication to Heilsbronn, 29 June 1600; Marius undoubtedly authored his first calendar before traveling to Prague. Interestingly, the 1601 *Prognosticon* shows the young Marius eagerly investigating Tychonic astronomy.

For any sixteenth-century calendar maker, the year's most important event, which would dominate its astrological interpretation, was the entry of the Sun into the first point of Aries, i.e., the spring equinox or in Ptolemy's astrological parlance, the "revolution of the year." For his 1601 *Prognosticon*, Marius presented the horoscope for this event on the verso side of the title page, announcing that the time had been determined "iuxta calculum generosi et magnifici viri Dn. Tychonis Brahe Dani astronomi magni." But rather than simply stating Tycho's time for the 1601 vernal equinox, Marius offered his readers a table comparing the times from three ephemerides against Tycho's, each adjusted to the meridian of Heilsbronn (see Fig. 11.1). Tested were the 1599 ephemerides of David Origanus, professor of mathematics at the university in Frankfurt/Oder, based on the Copernican Prutenic Tables; the 1585 ephemerides of Johann Stadius, professor of mathematics first in Louvain and then in Paris, also Prutenic; and the 1597 ephemerides of Marten Everaert, based on this Bruges physician's own "Belgian Tables."[22] Interestingly, Marius did not include the still widely used ephemerides by Cyprian Leowitz (1557), based on the medieval Alfonsine Tables.

---

[20]After 1609, the *Schreibkalender* apparently were printed in two versions, one with the daily longitudes and the other with either a historical chronicle of fifteenth- and sixteenth-century events in the Margraviate Brandenburg-Ansbach or a table of daily times of sunrise and sunset. Too few exemplars are preserved, however, to confirm this pattern, and I have not specified here which edition of a given *Schreibkalender* I consulted. See Matthäus, this volume, Section 1.

[21]In his *Prog. 1602*, sig. A3^r Marius explicitly mentioned that he had published a *Schreibkalender* and *Prognosticon auf 1601*.

[22]Origanus 1599; Stadius 1585; Everaert 1597. No printed or manuscript copy of a "Belgian Tables" has been found; as far as I know, the astronomical foundations of Everaert's ephemerides have not been analyzed.

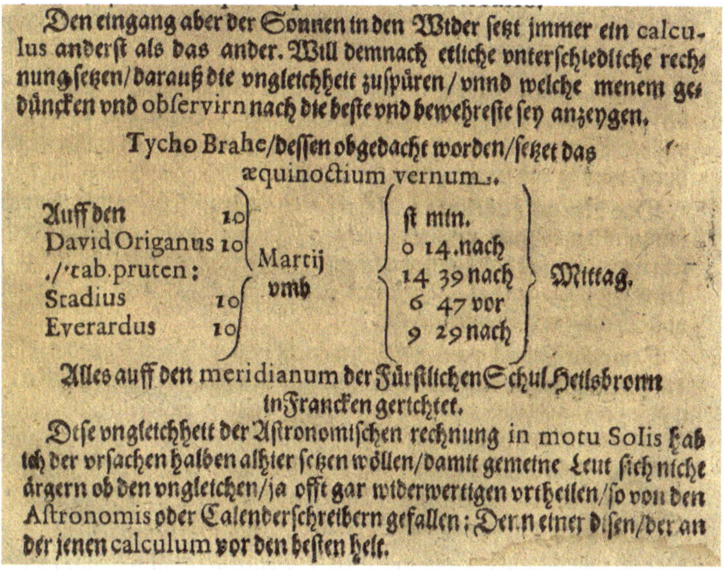

Fig. 11.1 Comparing computations for the spring equinox in 1601. Marius, *Prog. 1601*, sig. A6^{r}. BSB München: Chrlg. 325r, urn:nbn:de:bvb:12-bsb00021158-13

For Origanus and Everaert, Marius correctly converted the times (assuming Heilsbronn is 0;02h west of Nuremberg) according to the tables of local meridians provided in each ephemeris. However Marius seems not to have realized that Stadius's Prutenic-based time should be roughly identical to Origanus's. I convert Stadius' time to 14;40h "nach" (after) [Mittag] (see Fig. 11.1), very close to Origanus's time but far from Marius's value of 6;47h "vor" (before). Either our young calendar maker or his printer erred in this line of the table.

As can be seen from Fig. 11.1, these times vary by more than 20 hours. Horoscopes cast for such times would be completely different and would generate quite divergent astrological predictions. Exhibiting a critical approach to calendar making, Marius explained why he had selected Tycho's time for the revolution of the year (*Prog. 1601*, sig. A6$^{v–r}$)[23]:

[23] "Dise vngleichheit der Astronomischen rechnung in motu Solis hab ich der vrsachen halben alhier setzen wöllen, damit gemeine Leut sich nicht ärgern ob den vngleichen, ja offt gar widerwertigen vrtheilen, so von den Astronomis oder Calendarschreibern gefallen. Denn einer disen, der an der jenen calculum vor den besten helt. Vnter disen vier widerwertigen rechnungen, nach außweissung täglicher observation ist die beste vnd gewisseste Tychonis Brahe, wie solchs weitleufftig zu erweisen wehre, vnd sonderlich mit dem aequinoctio verno voriges 1600 Jars, da ich durch einen gerechten messi[n]gen quadrantem befunden, das die Sonn den ersten Punct deß Widers erreichet hat, den 10 tag Martij zwischen 6. und 7. uhr vor mittag, damit gentzlich vbereinstimmet calculus Tychonis, deme ich dißmals vnd vorthin folgen wil." Using Tycho's solar theory as presented in the *Progymnasmata* (1602), I compute the Sun's entry into Aries on 10 March 1600 at 6:30 a.m. for the meridian of Uraniborg.

> This irregularity in astronomical computation of the Sun's motion I wanted to show so that common people will not be angered by the irregular, yes often even contradictory, [astrological] judgments so liked by astronomers or calendar makers. For one of these considers one of those calculations best. Of these four contradictory computations, the best and most certain is Tycho Brahe's, according to the evidence of daily observation as could be proved in detail, especially with the vernal equinox of the previous year 1600. Using a properly aligned brass quadrant I found the Sun to have entered the first point of Aries on 10 March between 6 and 7 a.m., which completely agrees with prediction of Tycho, that I will here and in the future follow.

When reporting Tycho's time for the winter solstice, Marius further explained why he considered Tycho's solar theory the best. His times were (*Prog. 1601*, sig. A4^{r})[24]:

> [. . .] based on certain, true and infallible computation, as well as on protracted observation, unbelievable effort, diligence and work and on great expense invested on the part of the honorable and widely famous Herr Tycho Brahe through God's help. For hundreds of years many excellent astronomers have demanded such a certain and true calculation for the Sun's motion. Thus we, and all mathematicians as well as all of future posterity, should offer eternal praise and thanks for his outstanding mathematical work.

If we assume that Marius wrote the 1601 *Prognosticon* in the summer of 1600, these comments provide important information about how Marius sought to launch an astronomical career. Indeed, we might consider the 1601 *Prognosticon* as having been crafted explicitly for the purpose of persuading Tycho to accept Marius into his "familia" of assistants in Prague.

Obviously, the young Marius must have been in contact with someone who had access to Tycho's solar theory. As is well known, tables for the final versions of Tycho's new solar and lunar theories would be published only after his death, in the *Progymnasmata* of 1602. By 1590, Tycho had at his press in Hven printed the first part of that book (pp. 1–295) except for an appendix dealing with *De lunae motu restituto* (Tycho to Longomontanus, 21 March 1599, translated in Swerdlow 2009, p. 7).[25] Perhaps Marius already by the summer of 1600 had secured a set of the unbound printed sheets with Tycho's solar theory; or perhaps he managed to extract that theory from a 1599 imprint, authored by a young Tychonic (see below), that included parameters but not tables for the solar theory.

Equally significant, the young Marius in the 1601 *Prognosticon* announced himself as an astronomical observer, claiming that his measurements of solar altitudes, taken with a brass quadrant, had confirmed the Danish astronomer's

---

[24]"Nach einer gewissen, eigentlichen vnd vnfehlbaren rechnung, so durch langwiriges observirn, vngleublich mühe, fleis vnd arbeit, vnnd vber grossen vnkosten dermal eines von dem Edlen vnd weitberümbten Herrn Tychone Brahe durch Gottes hilff ist in das werck gesetzt vnd herfür bracht worden. Nach welchem gewissen vnd eigentlichen calculo in motu solis vil trefflicher Astronomi so vil hundert Jar groß verlangen gehabt haben. Dafür den, wie auch vor andere sein vortreffliche opera Mathematica, mit mir alle Mathematici, wie auch tota futura posteritas, nechst Gott, ewiges lob vnnd danck sagen sollen."

[25]For the complex printing history of the *Progymnasmata*, see Norlind 1970, pp. 144–150; Thoren 1990, 313.

solar calculation. As I have argued previously, many calendar makers by 1600 had complained about divergent predictions of solar motion in the ephemerides; yet very few had tried to measure solar positions to test the computational algorithms, and for good reason.[26] One cannot simply sight the Sun's altitude with a quadrant on 10 March and thereby know the exact time of the equinox to minutes. Rather, Marius would have needed a series of measured midday solar altitudes, taken days before and after the equinox, from which its time could be computed, provided that he knew his geographical latitude, obliquity of the ecliptic, and corrections for atmospheric refraction (data Marius could have extracted from the astronomical literature).[27] Making such solar measurements, usually over years, was required before any systematic observation of the positions of other celestial bodies could be initiated since the Sun's annual path on the ecliptic defines the coordinate framework by which all positions are measured. If Marius in 1600 had been regularly measuring solar altitudes with a brass quadrant, was he thinking of launching a longer-term campaign to measure planetary motions? Or was he simply trying to impress the Tychonics? As far as I know, Marius would never systematically measure midday solar altitudes even if he did occasionally record the times of eclipses. As an observer Marius would not become famous for measuring planetary or solar positions.[28]

A comparison of Marius's times for the Sun's entry into the cardinal points against my computations with Tycho's solar tables[29] of the *Progymnasmata* reveals how the young calendar maker used the Tychonic material (see Table 11.1). Columns 2–5 of Table 11.1 show my computed Tychonic times minus Marius's times, to minutes. Following a tradition begun already in the fifteenth century by Regiomontanus, Marius calculated *tempore apparente*, i.e., corrected the mean astronomical times generated by the tables with Tycho's "equation of days" (maximum 0; 24, 30 h) to give apparent (sundial) time. If Marius had followed exactly my computational procedures, his times (for Heilsbronn) should differ from mine (for Uraniborg) by a constant amount, indicating the difference in longitude Marius assumed between these two places. As can be seen in Table 11.1, Marius in 1601 assumed this difference to be about 5 minutes of time. I do not know how he chose this meridian; tables of geographical latitudes in the ephemerides he consulted

[26]See Kremer 2006.

[27]For an analysis of 30 years of solar observations made by the Nuremberg merchant Bernard Walther a century earlier, the first systematic, long-term set of astronomical observations in medieval Europe, see Kremer 2010. For Tycho's solar observations, see Dreyer 1890, pp. 333–336.

[28]At the end of his latest extant *Prognosticon*, written in 1624 for the year 1629, Marius admitted that his predictions for eclipses 5 years hence might be flawed. "From my solar observations I cannot yet conclude anything with certainty, who knows meanwhile what might happen with the observations" ("Also kann ich auch auß meinen observationibus solaribus noch nichts gewisses schliessen, wer weiß was vnterdessen sich mit den observationibus begeben möchte", *Prog. 1629*, sig. D3^r).

[29]I thank Lars Gislen for sharing with me his initial spreadsheet of Tycho's solar and lunar tables, as presented in the 1602 *Progymnasmata*, that we together debugged. The spreadsheet looks up values in the tables, interpolates, and follows the same procedures that a pencil-and-paper computation would follow.

Table 11.1 Differences between Tycho's and Marius's computed apparent times for the Sun's entry into the cardinal points, 1601–1613 (in minutes)

Year	Capricorn[a]	Aries	Cancer	Libra
1601	4	4	14	6
1602	23	7	23	23
1603	6	7	8	9
1605	6	6	8	9
1606	7	6	8	12
1607	7	6	8	7
1608	18	6	7	6
1609	18	11	6	6
1610	8	8		16
1611	7	8	7	15
1612	7	8	9	21
1613	3	8	6	8

[a]Note that Marius, like most calendar makers, always included the winter solstice for the previous year in his calendars. All the data in col. 2 of this table refer to the previous year. Marius neglected to specify a time for the summer solstice of 1610

provide diverging values, none close to 5 minutes.[30] In 1602, Marius wrote that he had "truly found" (he did not say how) that Heilsbronn is 7 minutes of time west of Uraniborg (*Prog. 1602*, sig. C4[v]). Although three of the four times for 1602 are in serious error, his meridian from 1603 forward is generally about 7–8 minutes from Uraniborg's, even if occasionally he forgot to add the equation of days.[31] In sum, the computational noise in Table 11.1 suggests that the young Marius was a competent calculator with Tycho's solar tables; he may, however, have used an equation of days slightly different from that given in Tycho's *Progymnasmata*.

We should also note that in comparing earlier printed ephemerides against Tycho's new tables and his own solar observations, Marius was following a path previously taken by the Danish astronomer. In the 1602 *Progymnasmata*, Tycho explicitly compared computations with his tables of solar longitudes against those from the Alfonsine and Prutenic Tables, finding by the year 1700 differences reaching more than 26 hours (Brahe: *Opera*, Vol. 2, 1915, p. 91). Such evidence strengthens our speculation that Marius, as he made his first calendar, had access to Tycho's incompletely printed *Progymnasmata*.

The syzygy times in Marius's calendars from 1601 through 1610 are not computed but rather copied from the 1599 edition of Origanus's Prutenic ephemerides. Origanus had set his ephemerides to the meridian of Frankfurt/Oder; his table of

[30]The 1585 Prutentic Tables list Nuremberg as 17 minutes west of Copenhagen (Uraniborg was not named), Stadius's ephemerides as 17 minutes west, Everaert's as 18 minutes west, and Origanus's as 1 minute east of Copenhagen. Origanus's 1609 ephemerides specify Nuremberg as 5 minutes west of Hven.

[31]If I compute Tychonic mean rather than apparent times for 1601 Cancer, 1602 Libra, 1608 Capricorn, 1609 Capricorn, and 1612 Libra, the differences approach Marius's announced meridian for Heilsbronn of 7 minutes west.

places lists Nuremberg at 17 minutes west. Marius consistently shifted Origanus's syzygy times 18 minutes west, as Heilsbronn and Ansbach are slightly west of Nuremberg.[32] Interestingly, the first syzygy time for 1601, specified to minutes, is exactly 18 minutes west of Origanus's time. The final syzygy time for that year, also given to minutes, is computed from Tycho's early lunar theory (see below). The remaining 1601 syzygy times, however, are rounded to the nearest half hour. Did Marius not trust the precision of Origanus's computed times to minutes?[33] Did he think his readers would not desire syzygies to a precision of minutes? After 1601 Marius consistently would list syzygy times to minutes. Perhaps his readers had complained? In any case, Marius would never inform his readers that the syzygy times were Prutenic. Neither did he indicate that Origanus's Prutenic times were mean, not apparent. Indeed, Marius's calendars from 1601 through 1610 all exhibit this inconsistency; their cardinal point times are apparent; the syzygy times are mean.

The dates of planetary aspects listed in the 1601 calendar also closely follow the aspects pages in Origanus's ephemerides. But the two eclipses Marius described differ in magnitudes and times from Origanus's descriptions in ways that suggest that Marius (or his source) independently computed these latter phenomena. As noted above, computing eclipses, especially solar eclipses which require correction for parallax, was quite difficult. Marius in the 1601 *Prognosticon* did not describe his procedures. However in the 1602 *Prognosticon*, he revealed that for the previous year, he had computed the eclipse times (*Prog. 1602*, sig. D2^{v})[34]:

> with the first correction of lunar motion as published by Melchior Jöstel. But because this correction is not yet completed, another was created by Tycho Brahe and his assistants which corresponds in its hypothesis and foundation to the present time. Diligent observers will want to pay attention to see if this hypothesis holds true.

One of Tycho's many assistants, Melchior Jöstel, had worked with the Dane in Wittenberg from 1598 to 1600 and in 1599 had published an early version of Tycho's lunar theory, computing the lunar eclipse of 31 January 1599. Jöstel's imprint, now rare, includes woodcuts showing the geometrical model for the lunar theory and

---

[32]Later Marius explicitly noted that Ansbach is 0;02h west of Nuremberg: "Sequuntur nunc tabulae ipsae, supputatae ad meridianum Onoldinum, quià Noriberga versus occasum distat duobus minutis unius horae" (Marius 1614, sig. F3^{v}; Marius 1614/1916/2019, last lines before the tables).

[33]Origanus's table of places lists Frankfurt/Oder as 38 minutes west of Königsberg, the meridian of the Prutenic Tables. If I drop two outliers (deviations of 49 and 55 minutes), I find for the remaining 23 syzygies of 1601 that Origanus's syzygy times are shifted by an average of 38 minutes from my Prutenic computed times for Königsberg, with a standard deviation in the differences of 2.9 minutes. The differences range from 34 to 46 minutes, which indicates something about the precision Origanus achieved in his Prutenic computations.

[34]"Nach der ersten restitution in motu Lunae, wie sie ist durch D. Melchiorem Jostellum publicirt worden. Weil aber solche restitutio der sachen noch nicht allerdings genug gethan, als ist von Tycho Brahe vnd den seinigen ein andere gemacht worden, nach welcher hypothesi vnnd fundamento diese jetztige Zeit entspringt, fleissige obseruatores wollen achtung darauff haben, wie dise wirde zutreffen." See Jöstel 1599. Copies (VD16 ZV 8685) are known in Dresden, Jena, Hannover, Vienna and Columbia, South Carolina.

listing individual steps of the computation. It does not reduce the models to tables, but the computations would enable a conscientious reader to recover parameters of both the solar and lunar theories. Tycho's solar equation would be simple to compute from the given eccentricity; but the lunar theory, with two epicycles and two small central circles, would be exceedingly difficult to compute, as Noel Swerdlow recently has shown (Swerdlow 2004). If Marius did indeed compute the 1601 eclipses solely from the "first restitution of the lunar motion" offered in Jöstel's small pamphlet, the young Heilsbronn student would have revealed himself to be a skilled geometer and astronomical computer.[35] Or we might guess that Marius, having somehow established contact with Tycho's circle before 1601 (see above), had gained access to manuscript tables that implemented Jöstel's lunar theory.

And we should also note that Marius, in the 1602 *Prognosticon* quoted above, correctly reported that Tycho had abandoned Jöstel's lunar theory. As Swerdlow has indicated, already in March of 1599, Tycho wrote to another assistant that their observations of the 1599 lunar eclipse showed that Jöstel's theory erred by nearly two-fifths of an hour. Marius presumably had learned, in Prague, about Tycho's rejection of Jöstel's model. In 1599–1601, as they struggled to finalize a lunar theory, Tycho and his assistant were keenly interested in observing lunar eclipses; Marius's 1602 exhortation to his readers to test predictions against observed eclipses clearly reflects his engagement with the Tychonic program. And we should recall that Jöstel in his treatise explicitly compared the eclipse predictions of Tycho's theories against those computed from the Alfonsine Tables, Peuerbach's eclipse tables (also Alfonsine), and the Prutenic Tables (Swerdlow 2004, p. 35). Might these comparisons have prompted the young Marius to compare various predictions that we examined in Fig. 11.1? At many points, Marius's earliest *Prognosticon* reflects Tychonic influences.

The dedication to the 1602 *Prognosticon* is dated 21 September 1601 (no location is given), presumably just after Marius had returned from Prague. In this dedication Marius briefly mentioned his journey that summer, undertaken with the "permission" of his patron (Margrave Georg Friedrich) to whom he dedicated the *Prognosticon*. Marius gave his Prague summer a negative rather than positive evaluation. This trip (neither Prague nor Tycho Brahe are mentioned) prevented him from "considering the thing [lunar theory?] more broadly" and disrupted access to "my books known to me" as well as to the "old observations (very important in such matters and nicely noticeable in the annual practica of Caesius)" (*Prog. 1602.*

[35]I have not tried to recompute Marius's computation of the 1601 eclipses; but his times do vary significantly from those in the ephemerides he compared for the spring equinox, and I do not doubt that he (or his source) independently computed the 1601 eclipses. For the 29 November 1601 lunar eclipse, Marius set the mid-eclipse time at 7:03 p.m.; shifted to the Heilsbronn meridian, Stadius's time would be 6:39 p.m., Origanus's 6:41 p.m., and Everaert's 6:08 p.m. (cf. Fig. 11.1). Note also that Jöstel's treatise does not include Tycho's equation of days. If Marius's access to Tycho's solar theory were solely via Jöstel, he would have needed to use some other equation of days to compute the times of the cardinal points, which might explain the pattern of deviations visible in Table 11.1.

sig. A3^r).[36] Marius probably referred here not to astronomical but rather to earlier meteorological and historical observations, which were, he rightly observed, often cited in the weather predictions of prognostica. Georg Caesius (1543–1606), a clergyman in Burgbernheim (near Rothenburg o.T.), had since the 1570s authored annual calendars, also printed in Nuremberg.[37] Marius here acknowledged his competition.

In the 1602 dedication, Marius returned to the problem of astrology's flawed astronomical foundations, a topic that makes his silence about the summer with the Tychonics in Prague even more puzzling (*Prog. 1602*, sig. A3^v)[38]:

> And it certainly happens when someone diligently calculates from today's [astronomical] tables that despite the art being so vast and difficult (which most of the learned astrologers probably never understand) it is nonetheless easy to err, to say nothing of this occurring frequently when today's tables predict a conjunction, opposition or other planetary aspects that differ significantly from the heavens. So it would be highly desirable if corrected ordinary tables, which today some distinguished and famous astronomers plan with high officials, could be completed and if the work would not be disrupted by their death. Then one would have better and more certain tables from which ephemerides could be written and more certain practica could be made, because a few minutes, to say nothing of some degrees, can soon allow the bodies, especially the speedy planets, to reach another place.

The existing tables are difficult to use; they err "pretty far" in predicting planetary positions, occasionally up to degrees. If only, Marius opined, higher powers would patronize the correction of these tables by "eminent and famous astronomers"; if only such work would not be interrupted by death. Was Marius here referring to Nicolaus Reimarus Ursus, who had died in the fall of 1600 in Prague? Or to Tycho Brahe, who would die on 24 October 1601 (a month after the date of Marius's dedication; perhaps Marius altered the text after Tycho's death?).

Later in this *Prognosticon*, Marius illustrated such discrepancies by noting that in December, the Prutenic Tables (he did not mention Origanus) predict no aspects for Mercury, apart from a conjunction with the Sun, although Everaert and Leowitz (the Alfonsine ephemerides he had not used in 1600) predict Mercury in sextile with

[36]"Alten obseruationibus (die viel in solchen sachen thun, vnnd in deß Herrn Caesij järlichen Pratiken wol gespürt wirdt)."

[37]See Caesius, *Prog. 1601*, for literally dozens of references to historical and weather events from the last century, correlated with planetary configurations. For Caesius, see Kempkens in this volume.

[38]"Vnd ist gewiß wenn schon einer auff das allerfleissigst auß den biß daher gebräuchlichen Tabulis calculirt, das gleichwol die Kunst so weitleufftig vnnd schwer (welches obangeregte allzu gelehrte Astrologi wol nit verstehen) das es leichtlich einem fehlen kan, wil geschweigen, das es vielmals geschicht, wann die jetzigen tabulae etwan ein conjunction, opposition oder andere Aspecten der Planeten setzen, solches je zimblich weit im Himmel fehlet. Darumb denn hoch zuwinschen were, daß die correctio tabularum usitatarum, mit welcher auß verlegung hoher Potentaten jetziger Zeit etliche verneme vnnd berühmte Astronomi vmbgegangen, folgend were zu end gebracht, vnnd durch ihren tödlichen abgang das Werck nicht gesperret worden, da hette man als dann bessere vnnd gewissere tabulas, vnnd auß denselbigen Ephemerides schreiben vnd gewissere Practiken, denn jetzt geschehen kan, stellen können, sintemal ein wenig Minuten, wil etlicher Grad geschweigen, bald durchauß ein andern positum sonderlich der geschwinden Planeten, machen können."

Saturn, quartile with Jupiter, and triune with Mars.[39] However, the dates Marius listed for these aspects do not jibe with those listed by Everaert or Leowitz. The young calendar maker was not always careful with his numbers, even as he criticized the existing literature for not agreeing on the planetary positions. In 1602, however, Marius listed the syzygy times to minutes, shifting nearly all of them exactly 18 minutes from Origanus's times, i.e., to the meridian of Heilsbronn.

After leaving Prague, Marius traveled to Italy where he would spend several years studying medicine. His 1603 *Prognosticon* is dated Padua, 28 May 1602. As he had done in 1601 and 1602 (see Table 11.1), he computed from Tycho's solar theory the apparent times for the cardinal points. But rather than copying the syzygy times from Origanus, Marius this year borrowed those times from the ephemerides of Giovanni Magini, editions of which had been printed in Venice in 1582 and 1599. Magini had computed his times from the Prutenic Tables, adding the Prutenic equation of time, information Marius did not disclose for his readers. Marius rounded 17 of Magini's times to the nearest 15 minutes, but did not shift them from the meridian of Venice to that of Heilsbronn (21 minutes), which seems surprising since for 1601 and 1602, he had shifted Origanus's meridian from Frankfurt/Oder to Heilsbronn (Cf. Magini 1599).[40] Marius gave the remaining syzygy times to minutes, in five cases agreeing verbatim with Magini. This pattern confirms Marius's reliance on Magini's ephemerides, which is distinguished by its rather inexact computations from the Prutenic Tables.

Most interesting in the 1603 *Prognosticon* is Marius's reported observational test of an earlier eclipse prediction, a Tychonic move in those years.[41] In his 1601 *Prognosticon*, Marius had predicted a partial solar eclipse for 14 December, to be visible in Heilsbronn at 2:43 p.m., using Jöstel's rendering of Tycho's solar and lunar theories. That eclipse, he wrote, "will occur in Italy in Padua" and reported (his own?) observed times of its beginning and end. Marius converted the observed mid-time of the eclipse to the meridian of Heilsbronn ("about 17 or 18 minutes" west). This observed time, 2:30 p.m., "agrees with the computation in my finished practica," Marius claimed, noting that the "Calculus prutenicus" had erred by an entire hour. "Here we can see that this new restitution of the courses of the Sun and Moon is more certain than anything previously had" (*Prog. 1601*, sig. B6^{v}; cf. *Prog. 1603*, sig. D1^{r}).[42] Marius did not indicate that the "new restitution" was Jöstel's 1599

[39]At one point, (*Prog. 1602*, sig. B3^{r}) Marius mentioned that he had never seen so many planetary aspects grouped together over a two-day period, "and also cannot find such in ephemerides since 1499" ("kan auch in Ephemerides von 1499 an, biß hierher keine finden"), suggesting that he had access to all the printed ephemerides of the sixteenth century. An eighteenth-century catalog of books in the Heilsbronn monastery library lists some, but not all, of these ephemerides. See Hocker (1731, pp. 268–271).

[40]Marius rounded 17 of the 25 values, listing the others to the nearest minute.

[41]Already in his 1596 comet tract, Marius reported his observation of a solar eclipse in May 1593, visible "a good part slower" than the Prutenic prediction by Stadius (Marius 1596, sig A4^{v}).

[42]"Hierauß ist nun zu sehen, wie diese newe restitutio curriculi Solaris & Lunaris also gewiß ist, dergleichen man vor niemals gehabt." By converting the mid-eclipse times announced in the

publication of Tycho's early lunar theory. Did he think that by 1603, readers would know that any "restitution" of astronomical theory was Tychonic?

In the calendars for 1601 to 1603, Marius explicitly announced his allegiance to Tycho's reformed astronomical calculations. He showed himself capable of making quantitative observations, in both position and time, of planetary motions. And if indeed he had access only to Jöstel's 1599 treatise, he silently revealed sophisticated geometrical talents in being able to compute lunar positions directly without access to tables. These calendars also indicate, however, that Marius did not build on a consistent computational foundation. He silently copied Prutenic (Copernican) syzygy times from existing printed ephemerides. He offered the cardinal times in apparent time, never mentioning that the syzygy times were mean. Also he nowhere indicated any interest in the cosmological arrangements of Tycho's or Copernicus's astronomy. In his early calendars, Marius presented himself as an astronomical calculator interested in predictive accuracy, not as a natural philosopher concerned with the framework of the universe.

Learning to Compute Tychonic Syzygies

From 1605 through 1610, Marius did not vary his astronomical sources from the pattern he had developed in the 1601–1603 calendars. He reported on his further observational activities, which remained occasional rather than systematic. Most significant for our purposes, he increasingly did not fully inform his readers of his computational practices; at times, he explicitly misrepresented those practices. Perhaps not surprising in a genre concerned with predicting the future, the rhetoric in Marius's prognostica concerning his astronomical practices, as well as his personal circumstances, can become rather inexact.[43]

For his *Schreibkalender* and prognostica for the years 1605 through 1610, Marius computed the cardinal point times from Tycho's solar theory and an unknown equation of time, exactly as he had done in 1601–1603 (see Table 11.1). He continued to copy the Prutenic syzygy times from Origanus's 1599 ephemerides, shifting meridians 18 minutes west (for 1610, he shifted the meridians 19 minutes).[44] But he began to mix Tychonic syzygy times with the Prutenic times from Origanus; and his syzygy times in the *Schreibkalender* and prognostica are not always identical. During these years, Marius appears to have increasingly distrusted the Prutenic

Prutenic ephemerides of Origanus and Magini to the meridian of Heilsbronn, using meridians listed in these ephemerides, I find predicted times of 1:35 p.m. and 1:22 p.m., respectively. Marius followed his sources correctly.

[43]For a study of the rhetorical structures of sixteenth-century prognosticatory literature, see Bauer 1994.

[44]Either 24 or 25 syzygies occur each year. Marius's syzygy times in his *Schreibkalender* from 1605 to 1609 match Origanus's values, minus 18 minutes, verbatim at least 20 times per year; in 1610, the verbatim match drops to 16 times.

syzygy times. Note, of course, that times of true syzygy cannot be observationally checked except during eclipses. Only occasionally did Marius provide hints at his motives for shifting to Tycho's solar and lunar models for his syzygy predictions.

Although Marius drafted his 1605 calendar in Italy (Padua, 13 June 1604), its syzygy times are not based on Magini's ephemerides as was his 1603 calendar. Twenty of the 1605 times are silently copied from Origanus. Three times, however, deviate significantly from Origanus; in the *Prognosticon* each is indicated as "according to the new and true correction of Tycho Brahe."[45] These times match those I compute from Tycho's solar and lunar theories in the *Progymnasmata*, with orbital rather than ecliptical lunar longitudes, adjusted by Tycho's "equation of time," and a difference between the meridians of Uraniborg and Heilsbronn of 7 minutes.[46] Apparently, Marius by the summer of 1604 had gained access to a copy of the *Progmnasmata* and no longer was dependent on Jöstel's preliminary lunar theory. As he had done previously, Marius explicitly described his cardinal point times as "according to a more true astronomical computation" (Tycho's name is not mentioned), adding (correctly) that the "Prutenic Tables give 12 hours slower" (*Prog. 1605*, sig. B1^{v}). And he computed the 1605 solar eclipse "according to the new restoration of Tycho Brahe," showing many of the intermediate values and a diagram for the parallax (Marius had not provided such computational features in his previous prognostica), "so that one can see the difference between the common tables and this new restoration." Such detail, Marius continued, is also useful (*Prog. 1605*, sig. D3$^{r–v}$)[47]:

> [...] so that common astronomers can practice and become familiar with this calculation, for which my computation can be somewhat helpful. We can easily excuse those who, on account of advanced age and other affairs, cannot be bothered with such subtle details. But for the others, it is a great scandal to simply depend on the [received] ephemerides and to ignore, because of presumed difficulty or laziness, this sincere restoration of the new paths of the Sun and Moon that all mathematicians from the beginning of the world have sought with such great seriousness yet without success.

[45]Times for 10 December 1604, 9 March 1605 (the *prognosticon* value of 3;36 h is Tychonic; the *Schreibkalender* value of 3;04 is copied from Origanus), and 24 March 1605.

[46]Tycho's tables give the lunar longitude in the plane of that body's orbit; an additional computation is required to shift this longitude to the plane of the ecliptic. Marius here did not shift the longitudes. But note that he correctly implemented Tycho's procedures, in the lunar theory, for replacing the "equation of days" with an annual equation, called in the *Progymnasmata* the "equation of time" (maximum 0;9,56 h). For these idiosyncrasies in Tycho's model, see Swerdlow 2009, pp. 24–31.

[47][...] sich die gemeinen Astronomi darinnen vben, vnd denselben calculum ihnen bekandt machen, darzu denn diese meine supputatio etwas behülfflich seyn kan. Allein die wegen hohes Alters vnnd anderer Geschäfft halben solchen subtilen sachen nit können abwarten, seyn leichtlich entschuldiget. Aber den andern ist es ein grosse schande nur also an den Ephemeridibus hangen, vnnd dieser herzlichen restitutionis novae curriculi Solis & Lunae, nach der alle Mathematici von anfang der Welt mit so grossem ernst vnd verlangen gestrebet vnd doch nicht erlanget, wegen vermeinter difficultet oder auß faulheit nicht achten." For Marius's "Supputatio ecclipsis solaris" see sig. D4^{v}–E2^{r}. For a later example of Marius explicitly computing eclipses from Tycho's tables, see *Prog. 1608*, sig. D4^{v}–D5^{v}.

Marius did not tell his readers, however, that 20 of his 1605 syzygy times were copied directly from Origanus's ephemerides! No, it was the other calendar makers whom Marius charged with the "great scandal" of incompetence or laziness.

The first syzygy time in Marius's 1606 *Prognosticon* (for 15 December 1605) is announced as computed "according to the new and true computation of the most distinguished astronomer Tycho Brahe." The time differs from that listed in the 1606 *Schreibkalender* and indeed matches exactly the time I compute using Tycho's solar theory (with no correction to apparent time or shift of meridian). Readers might conclude that Marius had for the year computed Tychonic syzygy times; yet all 24 of the syzygy times in the 1606 *Schreibkalender* are copied verbatim from Origanus. Twenty of these times appear verbatim in the *Prognosticon*; three other times in the latter match Tycho's times, now corrected to apparent time and shifted 6 minutes to the meridian of Heilsbronn.[48] Clearly, Marius after computing the 1605 solar eclipse had mastered Tycho's solar and lunar theories as presented in the *Progymnasmata* (Marius explicitly mentioned this work in the dedication to the 1605 *Prognosticon*), so his "competence" cannot be questioned. Perhaps was he too "lazy" to compute all the syzygy times with those theories?

The same pattern occurs in Marius's 1607 *Prognosticon*, where he announced that the initial syzygy time and all the subsequent quantitative astronomical data were (*Prog. 1607*, sig. B1^{r})[49]:

> [...] according to the correct astronomical computation and the true restoration of Tycho Brahe of the paths of the Sun and Moon (never taken from ephemerides but all eclipses, the Sun's entry into the cardinal points, and the new and full moons in my annual practica are computed by me, with special diligence and effort from Tycho's new tables by means of the doctrine of triangles, tables that other calendar makers ignore either due to difficulty or more likely to ignorance).

Indeed, the first syzygy time, 4 December 1606, plus the times for 16 February, 3 March, and 9 December do appear to have been computed from Tycho's lunar and solar theories as presented in the *Progymnasmata*.[50] Yet Marius's remaining syzygy times for 1607 are copied verbatim from Origanus's Prutenic-based ephemerides. Either Marius did not complete the Tychonic computations or his printer somehow intermingled the two sets of times for 1607. In 1608, Marius again silently copied 20 syzygy times from Origanus; the other five times he computed "from the

48 I should add that Marius here corrected the lunar longitudes to the plane of the ecliptic, an extra computational step that he would not always take (see footnote 46).

49 "[...] nach rechter astronomischer rechnung, vnd nach den wahren restitution Tychonis Brahe in curriculo Solis et Lunae (wie ich denn alle Finsternuß, eingang des Sonnen inn die vier puncta cardinalis mit den vorhergehenden New oder Vollmon, inn meinen järlichen practices, nit auß den Ephimeridibus nehme, sondern mit besonderm fleiß vnd mühe auß den tabulis novis Tychonis vermittelst der doctrina triangulorum rechne, welcher andere Practicanten entweder propter difficultatem, oder viel mehr propter ignorantium calculi nit achten)."

50 If I do not correct the lunar longitudes to the plane of the ecliptic (see footnote 46) and add Tycho's equation of time, my times differ from Marius's (i.e., his meridian) by 9, 14, 8, and 7 minutes.

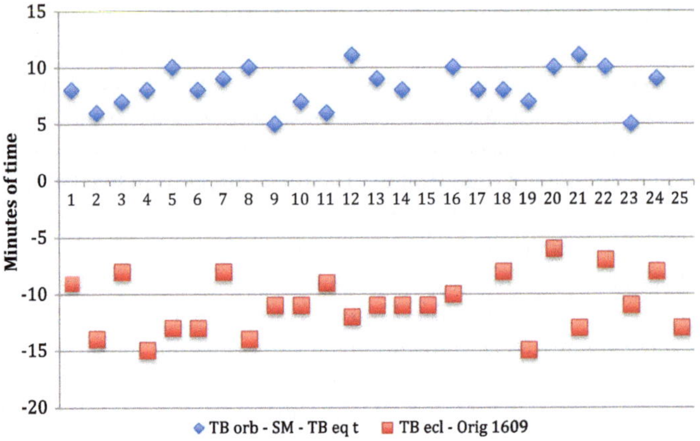

Fig. 11.2 Syzygy times computed from Tycho's tables for 1611 by Marius and Origanus

Tychonic computation" as he stated in 3 of those cases.[51] Despite this continued use of Origanus's ephemerides, Marius increasingly emphasized the divergence in time between Tycho's and Origanus's predictions of the Sun's entry into the winter solstice, noting (correctly) that the former could be 10–12 hours earlier than the latter (*Prog. 1605*, sig. B1^{v}; *Prog. 1606*, sig. B1^{v}).

Not until 1611, 1612, and 1613 would Marius finally compute all the syzygy times from Tycho's tables, using procedures he had occasionally employed in 1606 through 1608, viz., he did not correct the lunar longitudes to the ecliptical plane and added Tycho's "equation of time." As we might expect, Marius did not announce this change to his readers, but its timing can hardly have been accidental. In 1609, Origanus had published a second edition of his ephemerides. As indicated in its title, all the solar and lunar positions are now calculated from both the Prutenic Tables and Tycho's tables in the *Progymnasmata*. Here for the first time readers could compare eclipse and syzygy times computed from these competing models. As he had done in his 1599 edition, Origanus presented all these data in mean, not apparent, time. And Origanus corrected Tycho's lunar longitudes to the ecliptical plane.

A comparison of Origanus's and Marius's 1611 syzygy times reveals a very similar level of computational noise in their work (see Fig. 11.2). I drop two outliers from Marius's and one outlier from Origanus's data. As can be seen, Marius set his meridian about 8 minutes west of Uraniborg, Origanus about 11 minutes east. The standard deviation of Marius's noise is about 2 minutes of time; Origanus's is very similar, about 3 minutes of time. Here for the first time we see Marius computing not just single eclipses but ephemerides-like series. And his computational consistency exceeds that of the era's leading ephemerides maker, David Origanus. In 1612, the

[51]This year Marius again did not correct the lunar longitudes to the ecliptic plane but did not add Tycho's equation of time. My times differ from Marius's by 6, 7, 8, 10, and 8 minutes, for 9 December 1607 and 6 March, 2 June, 31 July, and 14 Sept 1608.

standard deviation in Marius's syzygy times drops to 1 minute of time; in 1613, it is 2 minutes of time. That is, during the years when he made his most important telescopic observations, Marius the calendar maker was showing himself to be an accurate and independent calculator of syzygy and eclipse times, using the solar and lunar tables in Tycho's *Progymnasmata*.

And he never missed the opportunity to criticize other calendar makers who did not base their work on Tychonic computation. In his now notorious 1612 *Prognosticon*, in which he announced his telescopic observations, Marius noted that the Prutenic predictions for the fall equinox differed by 14 hours from (*Prog. 1612*, sig. B1^{v})[52]:

> ... the true restoration of Tycho, which is the best, according to my observations and those of other diligent and famous astronomers, and should justly forbid the handwork of the lazy and coarse calendar makers who remain with the old, error-laden calculation and out of ignorance disregard the new and true correction. For so long they cried and wished for a true restoration but now do not want to think that the lunar path, God be praised, is well corrected. The small remaining defect cannot be compared against the large errors that diligent observers have noticed in other calculations. I demand no power to prescribe but I do complain about the large ignorance and laziness, in general, of some calendar makers who present themselves as important astronomers but in reality do not know how to solve a triangle.

Even before his feud with Galileo, Marius's annual imprints project an aggressive tone that would generate enemies in Nuremberg's crowded printers' market.[53]

---

[52]"[...] die wahre restitution Tychonis, welche nach meiner vnd anderer fleissiger vnd berhümbter Astronomorum observation, die beste ist, vnd solte billich den faulen vnd groben Calender machern, das Handwerck verbotten werden, dieweil sie dennoch bey dem alten vnd irrigen calculo bleiben vnd der neuen vnd eygentlichen Correctur, auß vnwissenheit nit achten, da man doch so lange zeit nach einer rechten Restitution geschrien vnd gewünschet hat, will jetzt nit gedencken deß Monds lauff, der auch nun mehr, Gott lob, so wol corrigirt ist, der gleichen niemals gewesen, denn der sehr geringe defect, so etwan noch vorhanden, nit zu schätzen ist, gegen den grossen Irrthumen, so fleissige Observatores in andern calculis vermercken. Ich begere keiner Herrschaft etwas vorzuschreiben, sondern ich klage über die grosse vnwissenheit vnd faulheit etlicher Calenderschreiber ins gemein, die sich vor stattlich Astronomus außgeben, vnnd aber in warheit nit ein triangulum zu solviren wissen."

[53]I lack space here to discuss the various legal charges raised against Marius's calendars. For the best-known case, a 1610 legal battle between Marius and another local calendar maker, Georg Halbmaier, see Matthäus 1969, cols. 1099–1102; Diefenbacher, Fischer-Pache 2003, No. 2215–2219. The Nuremberg town council impounded all 11,000 copies (!) of Marius's 1610 *Prognosticon*, demanding that its offensive first quire be reprinted. The council's archives do not reveal whether this recall occurred. The copy of the 1610 *Prognosticon* I have examined (WLB Stuttgart HBF 3708) probably contains the uncorrected first quire, for it scurrilously attacks an unnamed author who had translated Latin medical books: "great abuse and damage arises when idiots come behind such German or German-translated trick books; unfortunately it is more than true that many people not only are corrupted but also may even die when common people who can only read ... without understanding and doubt write out the common recipe without correct knowledge of the disease and thus dare to cure people" (... grossen mißbrauchs vnd schadens ... entstehet, wenn Idioten hinter solche Teutsche oder verteutschte Kunstbücher kommen, wie das leyder mehr als wahr ist, daß viel Leut nit allein verderbt werden, sondern auch wol gar vmb das leben kommen, inn deme gemeine Leut die nur lesen können ... vnnd ohn allen verstand vnd

Marius's calendars from 1605 to 1612 reveal other features of his astronomical practices during those years. For example, the dedication to the 1606 *Prognosticon* (returned from Italy, Marius was back in his home town of Gunzenhausen by 12 September 1605) describes, in general terms, some astronomical observations Marius had made in Padua with his student Balthasar Capra, an aristocrat from Milan who in 1607 would provoke Galileo's anger by claiming to have invented a geometrical compass (see Vergara 1992). Using a quadrant and a "mediocre" sextant, financed by Capra and located in his garden,[54] they had observed the 1604 nova, searched (in vain) for comets, and observed the stars and superior planets.

Although he offered no details about the planetary observations, Marius did state their observed location for the nova (to minutes of longitude and latitude) and concluded that it had showed no parallax. Writing now as a natural philosopher, he concluded that the nova must be located above the Moon and be at least 150 times larger than the Earth. More importantly for our purposes, Marius presented himself here as a Tychonic observer. He and Capra had constructed the instruments "after the form and method of the noble and outstanding astronomer Tycho Brahe, as I saw in 1601 when I spent some time with him in Prague." And his treatment of the nova, Marius added, followed the "astronomical mode" of Tycho's examination of the 1577 nova as reported in the *Progymnasmata* (*Prog. 1607*, sig. A3^{r}–A4^{r}).[55] Here for the first time in his calendars, Marius referred explicitly to his summer in Prague, suggesting that he had met Tycho and firmly situating himself among the Tychonics.

By 1612, Marius seems to have decided that Tycho's "restoration" of the solar and lunar calculation was adequate and not in need of further empirical evaluation (never did Marius suggest other criteria against which to test calculations). The 1606 dedication mentions that Marius had observed a 1601 solar eclipse in Padua but gives no results or comparisons of observations against Tycho's (or anyone else's) predictions (recall that in 1602 Marius had urged readers to compare eclipse observations against the predictions). In his 1608 *Prognosticon*, he stated that (unspecified) observations had "already defeated and destroyed" the solar predictions of the Alfonsine, Prutenic, and "Belgian Tables" of Everaert, so that Marius would remain (*Prog. 1608*, sig. B4^{r}):[56]

---

bedencken, gemeine recept, ohn rechte erkandtnuß der Kranckheit, herauß schreiben, vnd also die Leut zu Curirn sich vnterfangen, sig. A2^{v}).

[54] Tycho described both of these instruments in the *Progymnasmata*. See Brahe: *Opera*, Vol. II, 1915, pp. 330–352.

[55] In his *Prog. 1608*, sig. B4^{r}, Marius again wrote of Tycho "whose instruments I not only saw but also used myself."

[56] "Bey der wahren restitution deß Edlen Tychonis Brahe, welche mit den rechten vnd eigentlichen observationibus auff daß geneuest ubereinstimmet, nit allein zu diser vnsrer zeit, sondern auch albereit vor 16 Jaren von dem vortrefflichen Landgräffischen Mathematico, Christophoro Rothmanno vor gewiß vnd eigentlich ist erfunden worden." A decade later, Marius displayed less confidence, writing that "Tycho's doctrine or calculation is, according to him, currently the best available even if it has not reached its perfection, as can be noticed in eclipses." He provided no

with the true restoration of the noble Tycho Brahe that agrees exactly with just and true observations not only in our time but already 16 years earlier as was discovered with certainty and truth by the outstanding princely court mathematician, Christoph Rothmann [of Kassel].

In the same edition, Marius reviewed the adequacy of Tycho's lunar predictions, showing himself to be a scrupulous reader of the *Progymnasmata*. In an appendix to the latter, composed after the solar eclipse of December 1601, Tycho's assistant Longomontanus argued that Tycho's lunar theory would more accurately have predicted that eclipse if the value assumed for the lunar diameter had been slightly larger. Marius in 1608 cited his own observation, made in Ansbach, of the 1607 solar eclipse. Its timing had matched the Tychonic predictions in Marius's 1607 *Prognosticon* (deviations of 2 minutes for the beginning, 10 for the end), but the observed magnitude (slightly more than 2 points) was "a little larger" than the predicted value (1;54 points). Thus, the lunar diameter must be "in truth somewhat larger than specified in [Tycho]'s tables, as the author announced in the appendix" (*Prog. 1607*, sig. A2^{r-v}; 1608, sig. E1^{r}; Brahe: *Opera*, Vol. 2, 1915, p. 147; Vol. 3, 1916, pp. 321–322). This is the first and only example I have found where Marius suggests revising Tychonic theory from observational evidence.[57] Marius called on readers to observe the 1608 solar non-eclipse on 31 July (Origanus and Everaert had predicted an eclipse) and thereby to find "how true is Tycho's restoration and how false and futile are the Belgian Tables praised by the good Everaert in the dedication to his ephemerides" (*Prog. 1608*, sig. D5^{v}). As far as I know, Marius never reported in print any observations of the July 1608 event.

Apparently, however, subsequent eclipse observations raised doubts for Marius about the accuracy of the Tychonic calculations for the luminaries. In 1619, he announced that he would, as usual, use "Tycho's doctrine or computation" for syzygy times "which presently are the best even if they have not achieved perfection, as can be seen during eclipses" (*Prog. 1619*, sig. A4^{r}). No particular eclipse observations are cited, so again we cannot judge Marius's expectations for the accuracy of astronomical predictions.

Marius's prognostica around 1610 also reveal his interest in comparing observed planetary positions against the calculations. Mercury always deviates most from the Prutenic Tables, Marius commented in the 1611 *Prognosticon*, especially when the Sun is in Aries and Mercury is retrograding. To explore this situation, he had observed Mercury in the spring of 1598, 1599, 1606, 1608, and 1609, finding that the Prutenic predicted stationary points could err by 3 to 4 days. Everaert's predictions were worse, making "his correction of Mercury's motion worth not a rotten pear." The largest deviations, Marius concluded, occur when the Sun is "circa

examples, however, so we cannot here learn the standards of accuracy Marius expected from astronomical computation. See *Prog. 1621*, sig. A4^{r}.

[57]In his *Prog. 1621*, sig. C1^{r}, Marius quoted a 1615 letter from Fabricius indicating that the latter's observations showed no solar parallax and noting that Fabricius's 1618 *Prognosticon* (no longer extant) predicted the spring equinox two hours before Tycho's time. But Marius did here did not question the adequacy of Tycho's solar theory.

medias longitudines eccentrici," i.e., when the solar equation is at its maximum (*Prog. 1611*, sig. B4^{v})[58]:

> ... which is derived especially from the solar eccentricity, as I noticed for Mars. Tycho Brahe discussed this in his letters, which I read after my return from Italy and about which I was told in 1601 when in Prague as a student of Tycho. We soon want to hear a more complete report, dear God, by the outstanding imperial mathematician, Johannes Kepler, since his commentary on the motion of Mars is now being printed in Heidelberg.

In perhaps his most technical comment on planetary theory in the calendrical imprints, Marius here argued that Copernicus's solar eccentricity must be flawed. Tycho had identified this problem earlier (Brahe, vol. 6, p. 103); Marius's own observations of Mercury and Mars had pointed toward the same problem. But as far as I know, Marius himself did not try to produce a new calculus for Mercury or Mars based on Tycho's "restitution" of the solar theory. He would leave that task to Fabricius and Kepler; indeed, he referred here to the latter's *Astronomia nova* as being presently printed but gave no indication of knowing how radically that book would revise Tycho's approach to astronomy (see below).

Elsewhere Marius occasionally urged readers to observe the times of planetary conjunctions as a means of testing theories. In 1607, for example, he noted that a conjunction (in both longitude and latitude) of Jupiter and Mars would occur on 4 December 1606 and predicted that the Prutenic Tables "here are not false [...] whoever wishes should pay attention on the evening of this day after sunset to how exactly the Prutenic Tables match [the conjunction]." In 1608, he wrote that the Prutenic Tables predicted a conjunction (in both longitude and latitude) of Saturn and Mercury for 1 February, early in the morning (Origanus's ephemerides sets the conjunction for 3:17 a.m. at the meridian of Frankfurt/Oder). "Whoever desires should carefully observe whether the Prutenic Tables agree or not with the phenomena, for Mercury, especially, has until now eluded all tables as I myself have often observed." Marius cited two other examples to illustrate his doubts about Prutenic predictions, conjunctions of Venus-Regulus and Jupiter-Mars in 1606, but in neither case did he provide any observational details (*Prog. 1607*, sig. B1^{v}; *Prog. 1608*, sig. B2^{v}). Slightly more details appear in the 1610 *Prognosticon*, where Marius reports his "diligentem observationem" of the position of Mars on 10 September 1608, in longitude and latitude (see Fig. 11.3) (*Prog. 1610*, sig. D3^{v}). Not simply a conjunction, this measurement would have required measuring angles between the planet and nearby stars or use of an armillary sphere aligned to the ecliptic. In the *Prognosticon*, Marius listed the data as if he were regularly measuring planetary positions and had no need to inform readers about his procedures. As above, he used the observations to evaluate the competing ephemerides, concluding that Everaert's

[58]"Welches den sonderlich auß der Eccentricitate Solis herführet, wie solches ich auch in Marte vermercket. Dessen gedenckt Tycho Brahae in seinen Epistolis, wie ich solches nach wider kunfft auß Italia darinn gelesen, vnnd dessen auch in Prag Anno 1601 von den damals anwesenten Studiosis Tychonis bin berichtet worden. Vollkommenern bericht wollen wir geliebtes Gott, von dem vortrefflichen Käyserlichen Mathematico Iohanne Kepplero in kurtzem vernemen, sintemal sein commentaria über den motum Martis jetziger zeit zu Heydelberg getruckt werden."

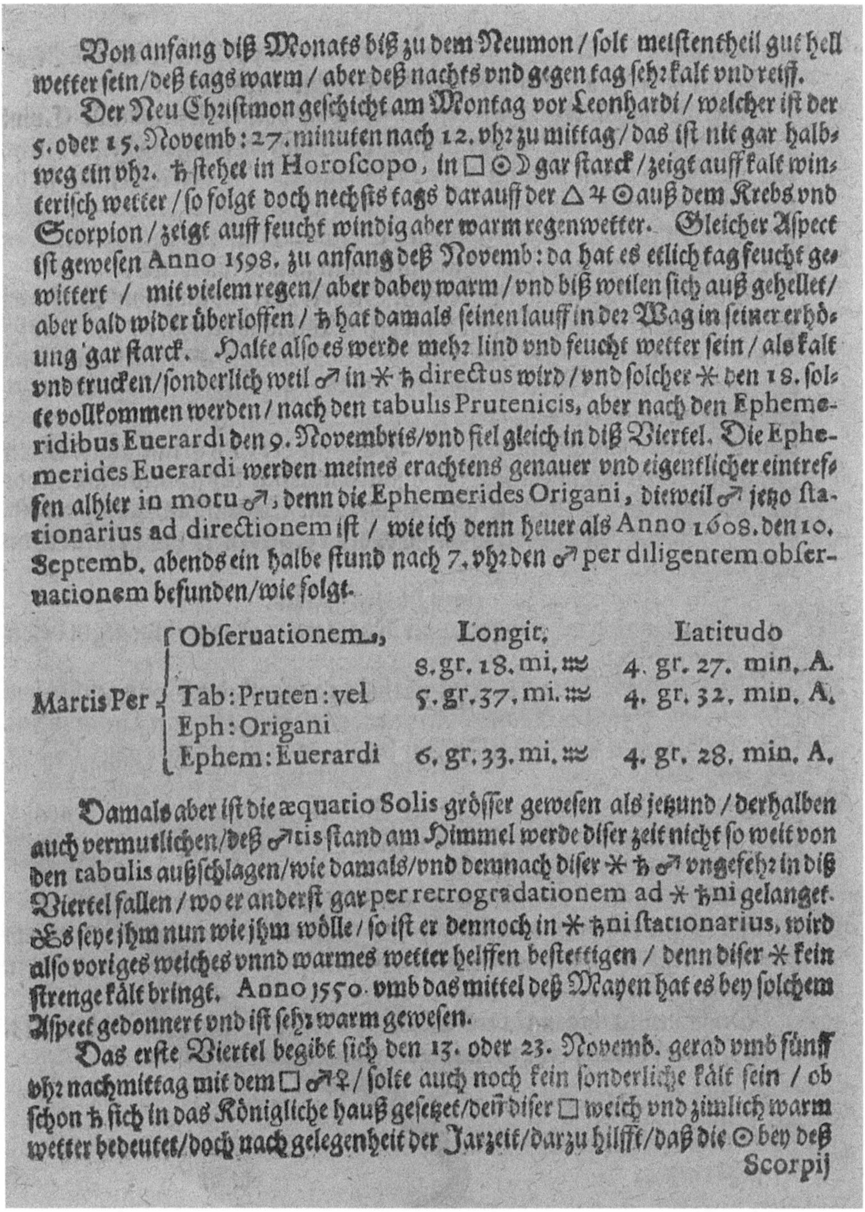

Von anfang diß Monats biß zu dem Neumon / solt meistentheil gut hell wetter sein/deß tags warm / aber deß nachts vnd gegen tag sehr kalt vnd reiff.

Der Neu Christmon geschicht am Montag vor Leonhardi / welcher ist der 5. oder 15. Novemb: 27. minuten nach 12. vhr zu mittag / das ist nit gar halbweg ein vhr. ♄ stehet in Horoscopo, in □ ☉ ☽ gar starck / zeigt auff kalt winterisch wetter / so folgt doch nechsts tags darauff der △ ♃ ☉ auß dem Krebs vnd Scorpion / zeigt auff feucht windig aber warm regenwetter. Gleicher Aspect ist gewesen Anno 1598. zu anfang deß Novemb: da hat es etlich tag feucht gewittert / mit vielem regen/ aber dabey warm / vnd biß weilen sich auß gehellet/ aber bald wider überloffen / ♄ hat damals seinen lauff in der Wag in seiner erhöung gar starck. Halte also es werde mehr lind vnd feucht wetter sein / als kalt vnd trucken/sonderlich weil ♂ in ✱ ♄ directus wird / vnd solcher ✱ den 18. solte vollkommen werden / nach den tabulis Prutenicis, aber nach den Ephemeridibus Euerardi den 9. Novembris/vnd fiel gleich in diß Viertel. Die Ephemerides Euerardi werden meines erachtens genauer vnd eigentlicher eintreffen alhier in motu ♂, denn die Ephemerides Origani, dieweil ♂ jetzo stationarius ad directionem ist / wie ich denn heuer als Anno 1608. den 10. Septemb. abends ein halbe stund nach 7. vhr den ♂ per diligentem obseruationem befunden/wie folgt.

		Longit;	Latitudo
Martis Per	Obseruationem,	8. gr. 18. mi. ♒	4. gr. 27. min. A.
	Tab: Pruten: vel	5. gr. 37. mi. ♒	4. gr. 32. min. A.
	Eph: Origani		
	Ephem: Euerardi	6. gr. 33. mi. ♒	4. gr. 28. min. A.

Damals aber ist die æquatio Solis grösser gewesen als jetzund / derhalben auch vermutlichen/deß ♂tis stand am Himmel werde diser zeit nicht so weit von den tabulis außschlagen/wie damals/vnd demnach diser ✱ ♄ ♂ vngefehr in diß Viertel fallen / wo er anderst gar per retrogradationem ad ✱ ♄ni gelanget. Es seye jhm nun wie jhm wölle / so ist er dennoch in ✱ ♄ni stationarius, wird also voriges weiches vnnd warmes wetter helffen bestettigen / denn diser ✱ kein strenge kält bringt. Anno 1550. vmb das mittel deß Mayen hat es bey solchem Aspect gedonnert vnd ist sehr warm gewesen.

Das erste Viertel begibt sich den 13. oder 23. Novemb. gerad vmb fünff vhr nachmittag mit dem □ ♂ ♀ / solte auch noch kein sonderliche kält sein / ob schon ♄ sich in das Königliche hauß gesetzet/denn diser □ weich vnd zimlich warm wetter bedeutet/doch nach gelegenheit der Jarzeit/darzu hilfft/daß die ☉ bey deß Scorpij

Fig. 11.3 Comparing Marius's observation of Mars against the predictions of the ephemerides. Marius, *Prog. 1610*, sig. D3^{v}. WLB Stuttgart: HBF 3708

was "more exact" than the Prutenics. And he speculated that an incorrect solar theory had caused the large deviations between the sky and predictions for that date since in September the solar equation reaches its maximum. That is, Marius used this single observation to investigate the solar parameters; but he reported no attempt to correct those parameters.

Only once did Marius relate observations to the question of cosmological hypotheses. In the 1607 *Prognosticon*, Marius predicted that in March of that year, Jupiter will appear very large because, when in opposition to the Sun, it is (*Prog. 1607*, sig. C4$^{r–v}$)[59]:

> [. . .] not only in perigee of eccentricity and epicycle according to the old viewpoint, but also according to my, Tycho's and Röslin's hypothesis it is many hundreds of times closer to the Earth than usual when it stands not in opposition to the Sun.

As is well known, in the 1580s, Tycho and Helisaeus Röslin (1548–1616), court physician to Georg Hans I von Pfalz-Veldenz in Alsace, had proposed geo-heliocentric arrangements, a cosmos that Marius here appears to be supporting. In this hypothesis, Jupiter's distance from the central Earth can vary by up to twice the distance between the Earth and Sun, i.e., by hundreds of miles but not by hundreds of times the distance between Jupiter and the Earth. Did Marius misunderstand the hypothesis, or did his printer mistakenly render "meilen" as "mal"? Interestingly, Marius did not mention Nicolaus Reimers Ursus, whose 1588 geo-heliocentric system, like Röslin's, featured a larger Martian orbit that does not intersect with the Sun's and prompted a ferocious priority conflict with Tycho. And Marius did not mention that in Copernicus's heliocentric system, the Jupiter-Earth distance varies exactly as in the geo-heliocentric arrangement (Cf. Gingerich, Westman 1988; Schofield 1989; Granada 1996, 2000, 2002, pp. 137–181, 279–294; Jardine 2008).[60] Did Marius in 1608 not fully understand these systems? Or is he, once again, simply not describing very clearly the foundations of his astronomical views?

In any case, Marius's calendars from 1605 to 1613 show him still copying astronomical data from Origanus's ephemerides but increasingly making his own computations with Tycho's "restoration" of the lunar and solar theories. He reported eclipse observations showing that Tycho's calculations were more accurate than those based on the Prutenic or the Belgian Tables but might still have an incorrect value for the apparent lunar diameter. He also presented scattered planetary observations showing that all Prutenic predictions could be flawed by an incorrect solar eccentricity used by Copernicus. Tycho had died before he could revise the planetary theories; Marius in 1611 suggested that Kepler would complete that task. Marius, however, demonstrated no interest himself in making such revisions. Indeed, he did

[59]"Nit allein in perigno eventrici [sic eccentrici] et epicycli, der alten meinung nach, sondern auch nach meiner Tychonis Brahe, vnd Röselini Hypothesibus den Erden viel hundert mal neher, als sonsten in orten des Himels, wenn er der Sonnen nit entgegen stehet." For Röslin's discussion of how planetary sizes vary in his geo-heliocentric arrangement, see Granada 2012, p. 443.

[60]As noted above, in the 1614 publication of his telescopic observations of Jupiter's moons, Marius claimed to have independently invented the geo-heliocentric hypothesis. Here Marius's assertion is more reserved. For one of the earliest publications of diagrams of the five competing systems (Ptolemy, Copernicus, Ursus, Röslin, Tycho), see Röslin 1597, pp. 53–55.

not even inform his readers when he was computing, or when he was copying from others, the astronomical data for his calendars.

Exploring Tychonic Planetary Calculations in the Calendars from 1614 to 1629

During the second half of his calendar-making career, for the years 1614–1629, Marius did not change the computational foundations of his work.[61] For the solar and lunar motions, he consistently described his calculations as Tychonic. Tycho had not published planetary tables by his death in 1601, and Marius began testing the planetary predictions of the earlier Prutenic ephemerides. Occasionally he used the new tables of David Fabricius, a clergyman in East Frisia who had spent time in Prague with Tycho's *familia*. Marius knew of Kepler's new but still incomplete *Rudolphine Tables* but apparently never employed them (in any case, they were not published until 1627, after Marius's death). He also apparently never used Longomontanus's *Astronomia danica* (1622), a work that provided tables for the planetary motions derived from Tycho's models and parameters. Thus, Marius in his published rhetoric about his mathematical practices would remain Tychonic from his earliest through his latest calendars. And in the public space of calendar making, Marius remained focused on the computational and astrological; only once did he refer to the physical question of a moving Earth, an issue that, he wrote, need not concern the common person and should not enter the calendars.

In 1614, Marius changed the computational practices he had used in the 1610–1613 calendars. He returned to copying his syzygy times verbatim from Origanus's ephemerides, now using the 1609 edition that listed side by side the Tychonic and Prutenic times. Marius copied Origanus's Tychonic times, subtracting 19 minutes to shift the meridian from Frankfurt/Oder to Ansbach and subtracting Tycho's "equation of time."[62] For 1614, Marius's syzygy times are inconsistent. Three of the times do not include the equation of time; another three add rather than subtract that value. Was Marius simply careless here? In 1615, he offered a consistent set of Origanus syzygy times, silently correcting two obvious typographical

[61]All of Marius's calendars privilege the Julian (alte) calendar, placing its dates, starting 1 January, before the Gregorian (newe) dates, starting 11 January. Before he died in 1624, Marius managed to draft calendars for the years 1625–1629. His printer apparently prepared a posthumous, second edition of the 1628 calendar, in octavo, giving only the Gregorian dates and truncating all the times to hours. For the octavo 1628 *Schreibkalender* and *prognostica*, see Stadtarchiv Nürnberg, Av 2584.8; for the quarto 1628 imprints, see WLB Stuttgart, HBF 3726.

[62]For whatever reason, Marius in 1618 shifted the meridian by 18 minutes. A table of Tycho's "Aequationis temporis" was provided by Origanus 1609, Vol. 1, p. 101, the "Aequationis dierum naturalium" on p. 100.

errors in Origanus's times.[63] In 1624, he followed a 10-minute typographical error in Origanus. Marius's 1627 and 1628 calendars (by these dates Marius no longer was alive) contain at least 9 recognizable typographical errors in the syzygy times (i.e., of 10, 20 or 30 minutes). And his final calendar, for 1629, copies two Origanus times verbatim, without subtracting Tycho's equation of time.[64] Recalling Marius's earlier diatribes against calendar makers who copy ephemerides, we must emphasize that the 1614–1629 calendars never inform readers that the Tychonic syzygy times are consistently copied from Origanus!

Origanus's 1609 ephemerides lists both Prutenic and Tychonic times, to seconds, of the Sun's entry into the cardinal points, in both apparent and mean time, yet another example of how conveniently Origanus designed his ephemerides for the astrological or calendrical user. In the post-1614 prognostica, Marius repeatedly indicated that his cardinal point times were "from Tycho's restoration" and I would guess that he copied most of those times verbatim from Origanus's apparent times, shifting the meridian 19 minutes to Ansbach. Nearly half of these times in Marius's 1614–1629 prognostica differ by exactly 19 minutes from Origanus's times (26 cases). However, in 10 cases, the times differ by 20 minutes, in 7 cases by 18 minutes, and in 6 cases by 21 minutes. I cannot therefore exclude the possibility that Marius may have computed at least some of his cardinal point times directly from Tycho's solar tables, perhaps using a slightly different "equation of days." But in any case, he consistently praised the "good Tychonic restoration of lunar motion," adding, more personally, "that I think of it as reasonable, honest, generous on account of the great expenditure of money, effort and work that it entailed" (*Prog. 1616*, sig. B1^r).

Origanus's ephemerides also offer eclipse times from both the Prutenic and Tychonic tables, providing most of the intermediate, computed values. Marius occasionally compared the Prutenic and Tychonic times explicitly; generally, however, he simply copied the Tychonic apparent eclipse times, shifting them 19 minutes to the Ansbach meridian. Our calendar maker, by 1614, had become an ephemerides copier, a practice that he would continue until his death 10 years later.

The other significant feature of Marius's post-1614 calendars is his tentative exploration of Tychonic calculations for the planets and his interactions with the astronomer David Fabricius. As is well known, Tycho died before converting his geo-heliocentric geometry into tables for computing planetary motion. The *Progymnasmata* presents tables only for the Sun and Moon. It would be several Tychonics who would construct new planetary tables based on Tychonic observational data, geo-heliocentric models, and physical assumptions.

For whatever reason, Marius apparently never established contact with Christian Longomontanus (1562–1647), who had served as Tycho's assistant from 1589 to 1597 on Hven and from 1600 to 1601 in Prague and who would hold the chair for

[63]For 3 April 1615, reading 8;44 for Origanus's 8;04; for 28 August 1615, reading 0;01 for Origanus's 6;01.

[64]For 27 April and 25 June 1629

mathematics at the University of Copenhagen from 1607 until his death. Longomontanus completed Tycho's lunar theory and in 1622 published the last major work in Tychonic astronomy, presenting detailed descriptions of the geometrical models, parameters, and tables for all the planetary motions (Longomontanus 1622; cf. Christianson 2000, pp. 313–319; Swerdlow 2009, 2010). I have found no mention of Longomontanus in Marius's calendars.

Marius did, however, occasionally mention and correspond with another Tychonic astronomer, David Fabricius (1564–1617). Born in East Frisia, Fabricius had briefly visited Tycho's *familia* in 1598 in Wandsbeck and in Prague from late May until 3 July 1601, where he met Marius (*Prog. 1610*, sig. C4^{v}).[65] By 1603, Fabricius had assumed a clerical position of Osteel, where he built an observatory equipped with large, Tychonic instruments (an iron quadrant and a semi-sextant). His son, Johannes, brought home a telescope from Leiden in 1610, and together they made observations of sunspots, leading to the first publication on the subject (Fabricius 1611), a publication that was completely ignored by Galileo and Scheiner in their priority dispute over who first observed sunspots with a telescope.[66] Fabricius also authored annual *Schreibkalender* and prognostica. He is best known, however, for his extensive correspondence with Kepler; 49 letters are extant, from 1601 through 1609. James Voelkel has convincingly shown that Fabricius greatly influenced both the tone and structure of Kepler's *Astronomia nova* (1609). Emphatically rejecting Kepler's physical approach and ellipses, Fabricius by 1608 had developed a Tychonic model for Mars. Although composed only of circles, Fabricius's model could represent many features of Kepler's elliptical theory. Unfortunately, Fabricius never published his new model, which is incompletely presented in his letters to Kepler (Wattenberg 1964; Folkerts 2000; Voelkel 2001, Chap. 8; Christianson 2000, pp. 273–276; Apelt 1852, pp. 313–326).[67] Apparently Fabricius had managed to construct similar models for the other planets and to compute a complete set of astronomical tables. Marius used these tables to challenge the Prutenic Tables; but it is not clear whether Marius, by the end of his calendar making, had abandoned the Prutenics for Fabricius's tables.

Marius first mentioned Fabricius in his 1608 *Prognosticon* while discussing various observations of the 1604 nova. Neither Marius in Padua, nor Kepler in Prague or Fabricius in East Frisia had seen any proper motion in the nova, Marius reported, information he could have extracted from published sources such as Kepler's *De stella nova* (1606). In his 1609 *Prognosticon*, Fabricius noted that Prutenic and Tychonic eclipse predictions can differ by half an hour and urged

[65]Referring to a device Fabricius had invented to measure distances to clouds, Marius expressed a desire to use the instrument and report his findings back to Fabricius and thus "to continue the collegiality and friendship and begun in Prague." However, an unpublished Fabricius manuscript refers, in 1599, to receiving a report from "Marius" about a storm in Heilsbronn. If this Marius is our Marius, the two men might have been in correspondence before meeting in Prague. See Bunte 1885, p. 112.

[66]See Reeves and van Helden 2010.

[67]Apelt attempted to reconstruct Fabricius's planetary theory. See Kremer b, forthcoming.

astronomers to calculate eclipses from the latter's "more certain" tables, as had, among others including (Fabricius, *Prog. 1609*, sig. D3^{r})[68]:

> Herr Simon Marius, physician and astronomer in Ansbach, whose diligence and special skill in these arts I myself have seen and experienced as we were together in Prague for some years, spending time together with Tycho Brahe.

Fabricius could have learned about this information from Marius's published calendars.

The first evidence of correspondence between these Tychonic astronomers appears in Marius's 1612 *Prognosticon*, where he referred to a letter from Fabricius concerning the latter's unpublished observations of novae in August of 1596 and February of 1609. In his 1613 *Prognosticon* (dedication dated 30 June 1612), Marius praised Fabricius's *Tabulae motuum planetarum superiorum*, completed in 1610. "God bestow on him a rich and large remuneration for his diligence and work, so that he would, from well-placed expenditures, receive eternal praise and thanks from those coming after" (*Prog. 1613*, sig. B3^{r-v}). Calling for these tables to be printed, Marius had ostensibly used them in manuscript to compute planetary aspects for 1613, occasionally noting where they differed from Prutenic predictions by several days or more. Indeed, Marius now indicated that, back in 1610, he had sent Fabricius his observations of Mars compared against Prutenic predictions (see above, Fig. 11.3). Marius noted that Kepler too was completing the new "Tabulis Rudolphaeis," enthusiastically concluding: "Meanwhile the motion is being derived from various hypotheses by two excellent mathematicians such as have never been available since the beginning of the world" (*Prog. 1612*, sig. A3^{v}; 1613, sig. B3^{r-v}, D1^{r}). Marius did not explain the differing hypotheses of Fabricius and Kepler to his readers.

Fabricius's final three prognostica, for the years 1615–1617, would be "printed and issued" in Nuremberg by Marius's printer (and father-in-law), Johann Lauer. Presumably, Marius had encouraged this arrangement (however Fabricius's tables would never be printed). In a long dedication to his 1615 edition, Fabricius surveyed recent progress in astronomy, noting that "my special master and friend" Marius had newly written him to report telescopic observation of a nova in Cassiopeia. Fabricius gave Galileo priority for discovering the moons of Jupiter but added that Marius had first measured their longitudes and latitudes. "[. . .] such will hopefully enable him, contrary to all expectations, to communicate with posterity and to make a famous name" (*Prog. 1612*, sig. A3^{v}; Fabricius, *Prog. 1615*, sig. A2^{v}–A3^{r}).[69] Encouraging

[68]"Der Herr Simon Marius, Medicus vnd Astronomus zu Ansbach . . . dessen fleiß vnnd besondere geschickligkeit in diesen Künsten ich für der zeit selbst gesehen vnd erfahren, als wir zu Prag für etlichen jahren, eine zeitlang beym seligen Herrn Tychone Brahen, bey einander gewesen." Note that Fabricius in his 1607 *Prognosticon* computed the cardinal point times from Tycho's solar theory in the *Progymnasmata*, without shifting the meridian or converting to apparent times. By 1615, Fabricius gave syzygy times only to days. For such features, Fabricius's calendrical computations were less demanding than were Marius's.

[69]Fabricius's dedication is dated 1 June 1614; apparently he had not yet seen Marius's *Mundus Iovialis*, the dedication to which is dated 18 February 1614 and which was also printed by Lauer.

each other to publish, Marius and Fabricius displayed a level of mutual respect that would never develop between Marius and Kepler (see below).

In the 1615 survey, Fabricius also urged Kepler to publish his new Rudolphine Tables and thereby to keep up with Fabricius's own tables "that I have constructed from my 20 years of diligent observations with great effort of time and labor and would have nearly completed if I could find a liberal patron for this work, which has previously been lacking." Fabricius reported that his Mercury hypothesis differed from the Prutenics by over four degrees; Jupiter agrees most closely; Mars and Venus also differ by "several degrees" (Fabricius, *Prog. 1615*, sig. A3^v–A4^r, B7^v). To demonstrate the need for improved planetary predictions, Fabricius compared data for the Great Conjunction in December of 1603. Kepler in *De stella nova* (1606) reported his measurements of the positions of Saturn and Jupiter in Prague on 18 December, from which he computed the observed time of true conjunction to 7 December at 0:50 a.m. From Tycho's planetary theories (Kepler's versions, unpublished in 1606), Kepler had computed the conjunction for 8 December at 11:50 p.m., i.e., 23 hours later than the observed time. Fabricius's new hypotheses yielded a predicted time of 6 a.m. on 7 December, only 5 hours after the observed conjunction. Fabricius noted that Origanus's Prutenic prediction was for 14 December; Leowitz's Alfonsine prediction was for 19 December; and Johannes Krabbe, the court astrologer in Wolfenbüttel, had constructed his own set of tables (unpublished) that predicted the conjunction for 11 December. "Thus enough to see how far and how noticeable the common tables differ from the truth in both time and place of the great conjunction" (Fabricius, *Prog. 1615*, sig. B8^r. Cf. Kepler, *GW*, Vol. I, 1938, pp.199–201).

As a final example, Fabricius in 1615 referred to an opposition of the Sun and Mars, computed from his new tables for 5 January with the bodies in 26;16 Capricorn and Cancer. Although he did not report the Prutenic prediction (that I compute for 31 December in 29;32 of those signs), Fabricius stressed that (Fabricius, *Prog. 1615*, sig. C4^r)[70]:

> ... in the place of Mars the calculus differs by 2 complete degrees from the true place, thus a large error is concealed in the common tables for the motion of Mars ... just as the highly experienced and diligent Ansbach astronomer, D. Simon Marius, in his 1610 *Prognosticon* recalled such large differences and errors, which he had often found from his own observations of Mars.

As noted above, Marius had reported a deviation of more than three degrees between his observed longitude of Mars and the Prutenic prediction.

By 1616, Marius began comparing predicted times for planetary aspects against Origanus's and Fabricius's tables. In his 1616 *Prognosticon* (dedication dated

[70]"... aberriret der Calculus alhie 2. gantze gr[adus] in loco Mars a veritate, daß also in motu Mars juxta communes tabulas ein grosse vnrichtigkeit stecket [...] wie dann auch der hoch erfahrne vnd fleissige Astronomus zu Onoltzbach, D. Simon Marius in seinem Prognostico über dz 1610 Jar, solcher grossen differenz vnd vnrichtigkeit gedencket, welche er ex propriis observationibus zum offternmal im Mars befunden."

25 January 1615), Fabricius had selected, from among the dozens of aspects he listed, six where his new tables deviated significantly from Origanus's Prutenic predictions.[71] Marius in his 1616 *Prognosticon* (dedication dated 28 May 1615) commented briefly on two of these aspects. He noted that triune Saturn-Mars in January would occur "several days slower" than predicted by Origanus; ". . . from Herr Fabricius's tables or prognostication one can experience the reason for this." Origanus had predicted the triune aspect for 3 February at 2 a.m. and Fabricius for 24 January at 10 p.m. Marius apparently did not realize that Fabricius used the old calendar, Origanus the new. The difference between the two predictions was 20 hours, not "several" days. Marius could have extracted this information from Fabricius's 1616 *Prognosticon.* However, in discussing the quadrature of Jupiter-Mars in March (Fabricius's tables placed the event 30 hours before Origanus's ephemerides), Marius reported that Fabricius, "my especially valuable master and good friend confidentially told me, from his new restoration of the Mars calculation," that the revised midday longitude of Mars for the day in question would deviate by 1;05 from Origanus's prediction, information not listed in Fabricius's *Prognosticon* (*Prog. 1616*, sig. B3^{v}, C2^{r}; Fabricius, *Prog. 1616*, sig. B3^{r}). Clearly by 1615 Fabricius was privately sending Marius computational results from his new tables; or Marius himself was computing positions from a manuscript copy of those tables.

In his calendars for 1618, 1619, 1620, 1621, and 1625, Marius continued to compare the times of aspects of the superior planets predicted by Origanus and by Fabricius's new tables. Each comparison involves Mars, which might suggest that Marius had Fabricius's table only for that planet or that the deviations were the greatest for that planet. The "errors" of the selected aspects reach 1–2 days of time or 1–2 degrees of longitude. Marius simply asserted, without offering any evidence, that Fabricius's predictions were "correct," Origanus's in "error." Without elaborating, Marius also told his readers that Fabricius's "corrections" for Mars agree with those made earlier by Kepler in his *Astronomia nova.* Hence, the conjunction of Jupiter-Mars will occur on 16 March 1620 "according to the corrector Herr David Fabricius, as he communicated to me on a certain date [Fabricius had died in May, 1617] and as Herr Kepler, Imperial Mathematicus and my good friend had earlier published in his Commentary on the motions of Mars. The Prutenic calculation sets this conjunction for 18 March" (*Prog. 1620*, sig. B5^{v}).

From this rhetoric, we might guess that Marius had shifted completely to the new tables of Fabricius for computing his planetary positions. Yet most of his aspects silently remain Prutenic; occasionally he even lists a Martian aspect as "from the Prutenic Tables" without challenging those tables (*Prog. 1627*, sig. D2^{r}). Despite having access to at least a partial set of Fabricius's new tables, Marius still preferred the easier path of copying the Prutenic aspects directly from Origanus's

[71]Interestingly, these aspects feature only the superior planets. Perhaps he had not yet completed tables for the inferior planets? Or perhaps he emphasized the superior planets because Saturn and Mars carry especially ominous astrological meanings?

ephemerides. Like the overwhelming majority of early seventeenth-century calendar makers, Marius copied rather than computed the astronomical data, even as he informed his readers about the new, more accurate, tables of Fabricius and Kepler.

Marius and Kepler

We conclude by considering briefly Marius's notorious relationship with Kepler. Previous scholars have reviewed, in meticulous detail, this relationship as a window into the priority controversy between Marius and Galileo (Klug 1906, pp. 418–424; Wohlwill 1926, Vol. 2, pp. 377–415; Gaab, this volume). My interest here is not with the telescopic observations but with Marius's knowledge of mathematical astronomy and his rhetorical style of self-representation.

Marius first mentioned Kepler, very briefly, in the 1608 *Prognosticon*. The Wolfenbüttel mathematician, Johann Krabbe, had in 1604 published *Newe astronomische Observationes*, in which he proposed an improved solar theory based on his observations. Marius (correctly) noted that Krabbe's solar longitude predictions varied slightly from Tycho's and asserted that he would "remain with the restoration of Tycho." Krabbe could not be trusted as an observer since, Marius charged, he alone had claimed to see proper motion in the new star of 1604, unlike Fabricius, Kepler, and Marius himself in Padua, who had observed no motion in the nova. Kepler had thoroughly discussed various observers and had criticized Krabbe in his *De stella nova et de trigono igneo* (1606). In 1610, Marius suggested that Kepler had opinions on the optical effects that appear during lunar eclipses but added that he had not yet seen Kepler's book on optics (published 1604) and thus could not comment (*Prog. 1608*, sig. B4^{r}; 1610, sig. E2^{v}; Krabbe 1605, sig. D1^{v}; Kepler, *GW*, Vol. I, 1938, pp. 161–162, 471).

As is well known, the Kepler-Marius relationship would become personal and stormy in 1611 when Kepler reprinted in his *Dioptrice*, without Marius's consent, a letter Marius had written in June of that year concerning his plans to publish his telescopic observations.[72] The intermediary in this snafu was a little-known government official in Wolfenbüttel, Nikolaus Vicke, who displayed deep interests in astrology, astronomy, and alchemy.[73] Vicke was corresponding not only with Marius but also with Jöstel, Longomontanus, and Kepler, asking technical questions about how Tycho's observations could be used to improve parameters in Prutenic computations. He was also interested in methods to compute house boundaries for

[72]Kepler's correspondence shows that he had seen Marius's *Tabula directionum novae* (1599) by 1600 and had been informed of Marius's arrival in Prague in 1601. See Gaab, this volume, n. 113, and above.

[73]Apart from his letters to Kepler, very little is known about Vicke. Four large codices document his legal actions in 1603–1610 against an official of the cathedral chapter in nearby Halberstadt, Niedersächsisches Landesarchiv, Wolfenbüttel, 1 Alt 5 No. 102a, b, c, and d, material I have not seen.

horoscopes. In January of 1611, Vicke asked Kepler about the geometrical foundations of the house boundaries that Marius had included in his 1599 *Tabula directionum novae*. Apparently he also wrote Marius, for the latter sent Vicke an autobiographical sketch discussing not only the origin of the tables (see above) but also plans to publish his recently made telescopic observations (Vicke to Kepler, July 1611, Kepler, *GW*, Vol. XVI, 1954, pp. 382–383).[74] Marius's original letter is not extant, but Vicke sent Kepler a chatty summary and lengthy quotation. Presumably it was this letter that first informed Kepler about Marius's telescopic activities.

According to Vicke, Marius worried that Kepler's evaluation of his house tables would be dismissive and offered excuses for their infelicities. In his response to Vicke, Kepler agreed that Marius's tables were difficult to use and then attacked, in increasingly critical and ironic tones, the Ansbacher's report of his telescopic findings. Listing dates, Kepler asserted Galileo's priority for most of the discoveries and accused Marius of immodestly inflating his claims. Reprinting telescopic sections of Marius's letter in the *Dioptrice*, Kepler added even more satirical commentary. After an imperial official intervened on Marius's behalf, Kepler late in 1612 half-heartedly apologized in a lengthy letter to Marius (his only known letter to him) that again defended Galileo's priority. More important for our purposes, Kepler also mentioned his plans to publish an ephemerides and the slow progress being made on the Rudolphine Tables. But he offered no technical details, clearly not wishing to engage Marius, the self-confessed geocentrist, in a discussion of Kepler's new astronomy.[75]

Marius would not respond to Kepler's letter until August of 1613 (likewise, his only known letter to him). He grudgingly accepted Kepler's apologies, complaining that his earlier letter to Vicke had not been intended for publication, and reiterated his defense of a stationary Earth in Tycho's hypothesis. On a more pleasant note, Marius hoped that he might sometime meet Kepler personally (Marius to Kepler, 16 August 1613, Kepler, *GW*, Vol. XVII, 1955, pp. 72–74).

And indeed, relations between the two men would thaw. In his 1612 *Prognosticon* (in which he publicized for the first time his telescope findings), Marius began using new planetary aspects, the quintilis (1/5 of the circle), biquintilis (2/5), and the sesquiquadratus (3/8), proposed by "the outstanding imperial mathematician, Herr Johann Kepler," in his *De stella nova et de trigono igneo* (1606) (*Prog. 1612*, sig. C6^v; Kepler, *GW*, Vol. I, 1938, pp. 189–194, 449. See Bialas 2004, pp. 139–144). Marius did not explain why he had waited until 1612 to introduce the new aspects; neither did he explain Kepler's geometrical reasons for defining the new aspects (Fig. 11.4).

In October of 1613, the two men finally met in Regensburg, where, among other things, they discussed novae, Kepler's unpublished Rudolphine Tables, and

---

[74]Marius would briefly report his telescope observations in his 1612 (dated 1 March 1611) and 1613 (dated 30 June 1612) prognostica before publishing his *Mundus Iovalis* in 1614 (dated 18 February 1614).

[75]Kepler's first *Ephemerides novae ... ex observationibus Tychonis, hypothesibus physicis, et tabulis Rudolphinis* for the years 1617–1620 would be printed irregularly from 1617–1619. His Rudolphine Tables would not be printed until 1627.

Fig. 11.4 (**a, b**) "Meaning of the signs for this almanach," for 1602 and 1615. The later edition includes Kepler's new aspects of quintile, biquintile and sesquiquadratus. Both images are from sig. A2r of the given imprints. Staatsarchiv Nürnberg: Fürstentum Brandenburg-Ansbach, Staats- und Schreibkalender (129), No. 283; WLB Stuttgart: HBF 3713

Kepler's suggested names for Jupiter's moons (names still in use after Marius purposed them in his 1614 *Mundus Iovalis*) (Marius 1614/1916/2019, end of part 1; *Prog. 1615*, sig. C1^{v}). Significantly, Marius in his calendars did not refer to Kepler's new planetary hypotheses until 1619, when (as mentioned above) he noted that Kepler's prediction of Mars' motion agreed with Fabricius's, despite their "differing hypotheses." Yet as usual, Marius did not tell his readers what he knew about their respective theories, adding that "this does not belong here" (*Prog. 1619*, sig. B2^{r}). In 1620, Marius again referred to the close agreement between Fabricius's and Kepler's "correction," now eliding any differences. Here Marius referred to Kepler as "mein guter freund" and mentioned the *Astronomia nova* (1609).[76] Again in the 1625 *Prognosticon*, Marius would predict a Jupiter-Mars opposition "according to the true Mars computation of Kepler or David Fabricius," equating the two men's revised tables (*Prog. 1620*, sig. B5^{v}; 1625, sig. C2^{v}).

In a posthumous 1628 *Prognosticon* (prepared in 1624), Marius referred for the first (and last) time to Kepler's new tables, now not mentioning Fabricius. The triune Saturn-Mercury will probably occur several days later than the Prutenic prediction, Marius wrote (*Prog. 1628*, sig A4^{v})[77]:

> as is known to everyone. My gracious master and good friend Joh. Kepler knows this best from his tables that he perfected with great effort from the observations and foundations of Tycho. It would be greatly desirable were Kepler to publish these tables or to publish ephemerides prepared from them, as he ten years ago promised when I met him in Regensburg. This would earn him, as is only fair, a splendid reward for his effort and work and still not make his tables common.

This aside by Marius is revealing on several counts. First, the Rudolphine Tables (not printed until 1627) were not perfected "ex ... fundamentis Tychonis" but from Kepler's new astronomy of ellipses and physical forces. Already in 1617–1619, Kepler had published new ephemerides for the years 1617–1620, "ex observationibus potissimum Tychonic Brahei, hypothesibus physicis, & tabulis Rudolphinis," as they announce on their title pages. Second, Marius apparently had never learned of the publication of Kepler's ephemerides. Third, Marius did not inform his readers that the tables of his "guter Freund" presume a heliocentric cosmology. And finally, Marius predicted that even if the ephemerides were published, Kepler's tables would not become common knowledge. Did Marius in

[76]Note that Marius also referred to Kepler as "my good friend" in Marius 1619, sig. B2^{v}. See Gaab, this volume, n. 336.

[77]"Wie männiglich bewust ist. Mein günstiger Herr vnd guter Freund Joh. Kepler wird es am besten wissen nach seinen Tabulis so er ex observationibus et fundamentis Tychonis mit grosser mühe perficirt hat, were zu wünschen das solche, oder auß denselben von dem Autore deducirte Ephemerides publicirt würden, wie er vor 10. Jahren zu thun willens gewesen, wie ich von ihm zu Regenspurg verstehen können, so könte er, wie denn billich ist, eine stattliche Ergetzligkeit seiner gehabten Mühe vnd Arbeit haben, vnd dennoch seine Tabuln nicht gemein warden." In *Prog. 1626*, sig. D2^{r} Marius also urged "my good friend" Kepler to publish "his labors on planetary motions," noting how far the observed place of Mercury differed from the Prutenic prediction.

1624 know enough about Kepler's Rudolphine Tables, with their novel physical assumptions (e.g., time represented as areas), logarithms, and approximative techniques, to guess that they would face a rocky reception by contemporary astronomers and astrologers, which indeed happened? (Wilson 1989; Kremer 2012)[78] Or was Marius merely expressing ill will toward his "good friend" (i.e., old enemy)? Alas, Marius did not elaborate that final phrase, abruptly returning, as we have come to expect, to his calendrical predictions: "I return to my purpose."

Marius may have remained ambivalent concerning Kepler's contributions to astronomy. Kepler's opinion of the Ansbach astronomer, however, remained unambivalent. In 1619, a correspondent wrote Kepler to suggest that the material substance of comets derives from sunspots. Kepler responded tersely. A similar belief about sunspots was held by Marius, "who is a detested and audacious seer and more-than-prognosticator, as he himself confesses. May he have his things for himself; if only he were not so burdensome for friends" (Kepler to Johannes Remus Quietanus, 31 August 1619, Kepler, *GW*, Vol. XVII, 1955, p. 376).[79] Kepler apparently had not forgotten his 1611 encounter with Marius!

Conclusions

Most of the content of Marius's calendars is strictly astrological. From the bits and pieces of mathematical astronomy that he included, we have tried to weave a tapestry to illustrate his computational practices. The most prominent feature in that tapestry is Marius's commitment to Tychonic computational astronomy. To the very end, Marius remained a Tychonic calendar maker. Throughout his career, he consistently proclaimed that prognostica and *Schreibkalender* should present quantitative predictions of planetary motions based on whatever "calculations" most closely match the observed heavens. For the motions of the Sun and Moon, Marius judged the Tychonic calculations to best meet that criterion.

For the motions of the planets, the Tychonics would not publish any tables until Longomontanus's *Astronomia Danica* of 1622, a work that apparently appeared too late to catch Marius's attention. Hence, Marius used Prutenic predictions for the planets until 1616 when he occasionally (but not consistently) took dates for the aspects from Fabricius's new tables. But he offered readers no clues about the

---

[78]Cf. Fabricius to Kepler, 27 February 1608, Kepler, *GW*, Vol. XVI, 1954, pp. 127–128, who already then found Kepler's hypothesis "so perplexing and laborious that it can frighten someone off even at first glance" (transl. in Voelkel 2001, p. 208).

[79]"[. . .] caetera vates invisus et audax et plus quam prognostes, ut quidem et fatetur. Habeat sibi res suas seorsim; ne gravis sit amicis." Cf. Wohlwill, 1926, Vol. 2, p. 404; Klug 1906, p. 400. For Marius's various views on the origin of comets, including a speculation that they arise "per adunationem, vel potius conglobationem" from sunspots, see Marius 1619, sig. C2^{v}–C3^{r}. Clearly, Kepler was reading Marius's latest publications (the dedication of his cometary tract is dated 16 April 1619)!

foundations of those tables, and they might well have assumed that Fabricius's tables derived from the Tychonic observational corpus just as they were founded on the geo-heliocentric arrangement favored by Tycho.

Marius also did not inform his readers that he frequently combined Prutenic and Tychonic syzygy times in the same calendrical edition or that he copied such times directly from printed ephemerides. Marius condemned other calendar makers who copied ephemerides as "lazy and incompetent." Yet given the inscrutability of syzygy times, none of Marius's early readers, including Galileo and Kepler, apparently noticed that he too copied ephemerides.

Finally, we have seen that Marius often used imprecise language to describe his observations, his computational procedures, and his personal views of various astronomical or astrological matters. On the one hand, we might expect to find such rhetorical *Spielraum* in astrological prognostica, where authors year after year exposed themselves to the challenge of predicting the future. Marius consistently wrote that non-astrological content did not belong in the prognostica; for example, in one of the later prognostica, he wrote that Copernicus's claim for moving Earth was a matter for "us astronomers ... and not for the common man, for whom this topic is too high and also unnecessary to understand" (*Prog. 1628*, sig. B1^{r}). On the other hand, we should not be surprised that Marius's imprecise language, combined with his combative style, provoked sometimes angry responses from contemporaries such as Galileo, Kepler, and other Nuremberg calendar makers. Indeed, we might go further to suggest that the calendar-making market, especially in printing centers like Nuremberg, drove authors like Marius toward a provocative rhetorical style. Given their profusion of astronomical and astrological factoids, the many prognostica editions printed each year might well have blurred together in the minds of readers. But factoids salted with provocative, even combative asides might have floated above the competition.

For some decades now, historians have attributed at least part of Galileo's combative style to his social role as a courtier (Cf. Biagioli 1993). Perhaps we should do the same for Marius, in his social role as a calendar maker at the court Brandenburg-Ansbach. From 1607 until his death, Marius would sign himself on the title pages of his prognostica as "Fürstlich Brandenburgischer bestellter Mathematicus und Medicinae studiosus."[80] As courtier and calendar maker, Marius surely needed to do more than predict the weather.

Acknowledgments Projects like this can thrive only with collaboration at many levels. With gratitude, I thank the Bayerische Staatsbibliothek, the Württembergische Landesbibliothek, and the Staatsarchiv Nürnberg for permitting publication of their images and staff at those libraries plus the Bibliothek des Germanisches Nationalmuseums, Niedersächsische Landesarchiv-Staatsarchiv Wolfenbüttel, Marienbibliothek in Halle, Universitätsbibliothek Tübingen, and the Universitäts- und Landesbibliothek Darmstadt for their kind assistance. Owen Gingerich, James Voelkel, J.R. Christianson, Lars Gilsen, and Noel Swerdlow provided valuable advice on astronomical questions. Eileen A. Reeves, Jonathan Green, Christoph Mackert, and Michael Shank helped

[80] In the prognostica for 1616–1625, the phrase becomes "... Mathematicus und Medicus."

with the translations. Pietro Daniel Omodeo and Christian Heitzmann offered bibliographical suggestions. The materials assembled by Pierre Leich and colleagues at the *Marius-Portal* (http:// www.simon-marius.net) greatly facilitated access to the sources. For sharing their rich knowledge of Marius and the world of early modern calendars, I am deeply indebted to Klaus Matthäus, Hans Gaab, and Klaus-Dieter Herbst. And I thank the Max Planck Institute for History of Science in Berlin for hosting me as I wrote this paper.

References

Apelt, Ernst Friedrich (1852): *Die Reformation der Sternkunde: Ein Beitrag zur Deutschen Culturgeschichte*. Jena: Friedrich Mauke 1852

Bauer, Barbara (1994): Sprüche in Prognostiken des 16. Jahrhunderts. In: Haug, Walter; Wachinger, Burghart (eds.): *Kleinstformen der Literatur*. Tübingen: Niemeyer 1994, pp. 164–204

Biagioli, Mario (1993): *Galileo, Courtier: The Practice of Science in the Age of Absolutism* Chicago: University of Chicago Press 1993

Bialas, Volker (2004): *Johannes Kepler*. München: Beck 2004

Brahe, Tycho: *Opera Omnia*, ed. J. L. E. Dreyer. 15 vol. Copenhagen: Libraria Gyldendaliana 1913–1929

Bunte, Bernhard (1885): Über David Fabricius. *Jahrbuch der Gesellschaft für bildende Kunst und vaterländische Altertümer zu Emden* 6/2 (1885), pp. 91–128

Büttner, Heinrich-Christoph (1813): *Franconia: Beiträge zur Geschichte, Topographie und Litteratur von Franken*. 2 vol. Ansbach: Gassert 1813

Caesius, Georg (1601): *Prognosticon astrologicon [...] auf das Jahr MDCI*. Nürnberg: Valentin Fuhrmann, 1601

Caspar, Max (1993): *Kepler*. Translated by C. Doris Hellman. New York: Dover 1993

Christianson, John Robert (2000): *On Tycho's Island: Tycho Brahe and His Assistants, 1570–1601*. Cambridge: Cambridge University Press 2000

Diefenbacher, Michael; Fischer-Pache, Wiltrud (eds.) (2003): *Das Nürnberger Buchgewerbe: Buch- und Zeitungsdrucker, Verleger und Druckhändler vom 16. bis zum 18. Jahrhundert*. Nürnberg: Selbstverlag des Stadtarchivs Nürnberg 2003

Dreyer, John Louis Emil (1890): *Tycho Brahe*. Edinburgh: A. & C. Black 1890

Everaert, Marten (1597): *Ephemerides novae et exactae: Ab Anno Incarnationis Domini 1590, ad Annum 1610. Ex novis Tabulis Belgicis Authoris Supputatae. Ad longitudidem 24.0 graduum, latitudinem 51.30 Graduum*. Leiden: J. C. van Dorp 1597

Fabricius, Johannes (1611): *Joh. Fabricii Phrysii De Maculis in Sole observatis, et apparente earum cum Sole conversione, narratio*. Wittenberg: Borner & Rehefeld 1611

Folkerts, Menso (1990): Marius (Mayr), Simon (1573–1624). In *Neue Deutsche Biographie* 16 (1990), p. 217f.

Folkerts, Menso (2000): Der Astronom David Fabricius (1564–1617): Leben und Wirken. *Berichte zur Wissenschaftsgeschichte* 23 (2000), pp. 127–142

Gingerich, Owen; Westman, Robert S. (1988): The Wittich Connection: Conflict and Priority in Late Sixteenth-Century Cosmology. *Transactions of the American Philosophical Society* 78/7 (1988), pp. 1–148

Granada, Miguel A. (1996): *El Debate cosmológico en 1588: Bruno, Brahe, Rothmann, Ursus, Röslin*. Neapel: Instituto Italiano per gli Studi Filosofici 1996

Granada, Miguel A. (ed.) (2000): *Helisaeus Röslin, De Opera Dei creationis seu De Mundo hypotheses, aurifodina philosophica*. Lecce: Conte Editore 2000

Granada, Miguel A. (2002): *Sfere solide e cielo fluido: Momenti del dibattito cosmologico nella seconda metà del cinquecento*. Milan: Edizioni Angelo Guerini e Associati SpA 2002

Granada, Miguel A. (2012): Helisaeus Röslin contre Raymarus Ursus après la publication du De Astronomicis hypothesibus: ses lettres à Herwart von Hohenburg de 1597. In: Noirot, Caroline (ed.): *Omnia in uno: Hommage à Alain-Philippe Segonds*. Paris: Les Belles Lettres 2012, pp. 425–451

Green, Jonathan (2012): Printing the Future: The Origin and Development of the 'Practica'. *Archiv für Geschichte des Buchwesens* 67 (2012), pp. 1–18

Heller, Joachim (ed.) (1548): *Epitome totius astrologiae, Conscripta a Ioanne Hispalensi [John of Seville] Hispani astrologo celeberrimo, ante annos quadringentos, Ac nunc primum in luce edita. Cum Praefatione Ioachimi Helleri Levocopetriaei, Contra astrologiae aduersarios*. Nürnberg: Johann Montanus & Ulrich Neuber 1548

Herbst, Klaus-Dieter (ed.) (2012): *Astronomie – Literatur – Volksaufklärung: Der Schreibkalender der Frühen Neuzeit mit seinen Text- und Bildbeigaben*. Jena: HKD 2012

Hocker, Johann Ludwig (1984): *Bibliotheca Heilsbronnensis*. Nürnberg: Monath 1731

Jardine, Nicholas (1984): *The Birth of History and Philosophy of Science: Kepler's A Defense of Tycho against Ursus with Essays on its Provenance and Significance*. Cambridge: Cambridge University Press 1984

Jardine, Nicholas (2008): *La Guerre des Astronomes: La querelle au sujet de l'origine du système géo-héliocentrique à la fin du XVIe siècle*. 2 vol. Paris: Les Belles Lettres 2008

Jöstel, Melchior (1599): *Lunae deliquium ad hunc Christi annum 1599 [...] Ex generosi et magnifici viri Tychonis Brahe hypothesi et tabulis supputatum*. Wittenberg: Zacharias Lehmann 1599

Kennedy, E. S. (1994): Ibn Mu'ādh on the astrological houses. *Zeitschrift für Geschichte der Arabisch-Islamischen Wissenschaften* 9 (1994), pp. 153–160

Kepler, Johannes: *Gesammelte Werke*, 21 vol. München: C.H. Beck, 1937–2009

Klug, Josef (1906): Simon Marius aus Gunzenhausen und Galileo Galilei. *Abhandlungen der mathematisch-physikalischen Classe der Königlich Bayerischen Akademie der Wissenschaften* 22 (1906), pp. 385–526

Krabbe, Johannes (1605): *Newe astronomische observationes der zweyen obern Planeten Saturni und Iovis, wie auch Martis, Solis, Veneris und Mercurii, sampt eim Bericht von der Grossen Coniunction Saturni und Iovis, Anno 1603*. Erfurt: Beck 1605

Kremer, Richard L. (2006): Copernicus among the Astrologers: A Preliminary Study. In: Folkerts, Menso; Kühne, Andreas (eds.): *Astronomy as a Model for the Sciences in Early Modern Times* (= *Algorismus*; 59). Augsburg: Rauner 2006, pp. 225–252

Kremer, Richard L. (2010): War Bernard Walther, Nürnberger astronomischer Beobachter des 15. Jahrhunderts, auch ein Theoretiker? In: Wolfschmidt, Gudrun (ed.): *Astronomie in Nürnberg*. Hamburg: tredition science 2010, pp. 156–183

Kremer, Richard L. (2012): Mathematical Astronomy and Calendar-Making in Gdańsk from 1540 to 1700. In: Herbst 2012, pp. 477–492

Kremer, Richard L. (n.d.-a) What was new in Simon Marius's *Tabulae directionum novae* of 1599?" forthcoming

Kremer, Richard L. (n.d.-b) Kepler's competition: Longomontanus and Fabricius on the theory of Mars," forthcoming

Lang, Karl Heinrich (1811): *Neuere Geschichte des Fürstenthums Baireuth*, vol. 3. Nürnberg: Monath und Kußler 1811

Longomontanus, Christen Sørensen (1622): *Astronomia Danica*. Amsterdam: Wilhelm Caesius 1622

Magini, Giovanni Antonio (1599): *Ephemerides coelestium motuum ab anno 1598 usque ad annum 1610 secundum Copernici observationes accuratissime supputatae & correctae ad longitudinem inclytae Venetiarum urbis*. Venedig: Damian Zenaro 1599

Marius, Simon (1596): *Kurtze und eigentliche Beschreibung des Cometen oder Wundersterns*. Nürnberg: Paul Kauffmann 1596

Marius, Simon (1599): *Tabulae Directionum Novæ*. Nürnberg: Christoph Lochner 1599

Marius, Simon (1614): *Mundus Iovialis Anno M.DC.IX. Detectus Ope Perspicilli Belgici.* Nürnberg: Johann Lauer 1614

Marius, Simon (1619): *Astronomische vnd astrologische beschreibung deß Cometen [...] 1618.* Nürnberg: Johann Lauer 1619

Marius, Simon (1614/1988): Joachim Schlör, *Mundus Iovialis – Die Welt des Jupiter.* Gunzenhausen: Johann Schrenk 1988

Matthäus, Klaus (1969): Zur Geschichte des Nürnberger Kalenderwesens: Die Entwicklung der in Nürnberg gedruckten Jahreskalender in Buchform. *Archiv für Geschichte des Buchwesens* 9 (1969), cols. 965–1396

Norlind, Wilhelm (1970): *Tycho Brahe: En levnadsteckning med nya bidrag belysande hans liv och verk.* Lund: Gleerup 1970

North, John D. (1986): *Horoscopes and history.* London: Warburg Institute 1986

Origanus, David (1599): *Ephemerides novae annorum XXXVI, incipientes ab anno 1595, quo Ioannis Stadii maxime aberrare incipiunt, et desinentes in annum 1630*, 2 vol. Frankfurt (Oder): Andreas Eichhorn 1599

Origanus, David (1609): *Novae motuum coelestium ephemerides Brandenburgicae, annorum LX, incipientes ab anno 1595, & desinentes in annum 1655, calculo duplici luminarium, Tychonico & Copernicaeo, reliquorum planetarum posteriore elaboratae*, 3 vol. Frankfurt/Oder: Johannes Eichhorn 1609

Pasachoff, Jay M. (2015): Simon Marius's *Mundus Iovialis*: 400th Anniversary in Galileo's Shadow. *Journal for the history of astronomy* 46 (2015), pp. 218–234

Programme de la Société Hollandaise des Sciences a Harlem pour l'année 1898. *Archives néerlandaises des sciences exactes et naturelles*, 2d. ser., 2 (1899), pp. 1–7

Reeves, Eileen and van Helden, Albert (2010): *Galileo Galilei & Christoph Scheiner*: On Sunspots. Chicago: The University of Chicago Press 2010

Regiomontanus (1490): *Tabulae directionum profectionumque.* Augsburg: Ratdolt 1490

Röslin, Helisaeus (1597): *De opera Dei creationis, seu de mundo hypotheses.* Frankfurt a.M.: Claude de Marne und Johann Aubry 1597

Schofield, Christine (1989): The Tychonic and semi-Tychonic world systems. In: Taton, René; Wilson, Curtis (eds.): *Planetary Astronomy from the Renaissance to the Rise of Astrophysics. Part A: Tycho Brahe to Newton* (= *The General History of Astronomy*; 2A). Cambridge: Cambridge University Press 1989, pp. 33–44

Schöner, Johannes (1544): *Scripta clarissimi mathematici M. Ioanis Regiomontani.* Nürnberg: Johann Montanus & Ulrich Neuber 1544

Seethaler, Josef (1982): *Das Wiener Kalenderwesen von seinen Anfängen bis zum Ende des 17. Jahrhunderts: Ein Beitrag zur Geschichte des Buchdrucks.* Dissertation, Universität Wien 1982

Stadius, Johann (1585): *Ephemerides secundùm Antuerpiae longitudinem ex tabulis Prutenicis supputate ab anno 1583 usque ad annum 1606. & ad S D. N. Gregorij XIII. anni reformationem accuratißimè accommodatae.* Lyon: Philip Tinghi 1585

Stöffler, Johann; Pflaum, Jacob (1499): *Almanach nova plurimis annis venturis inservientia.* Ulm: Joannis Reger 1499

Swerdlow, Noel M. (2004): Tycho Brahe's Early Lunar Theory and the Lunar Eclipse of 31 January 1599. *Centaurus* 46 (2004), pp. 1–40

Swerdlow, Noel M. (2009): The Lunar Theories of Tycho Brahe and Christian Longomontanus in the *Progymnasmata* and *Astronomia Danica. Annals of science* 66 (2009), pp. 5–58

Swerdlow, Noel M. (2010): Tycho, Lonomontanus, and Kepler on Ptolemy's Solar Observations and Theory, Precession of the Equinoxes, and Obliquity of the Ecliptic. In: Jones, Alexander (ed.): *Ptolemy in Perspective: Use and Criticism of His Work from Antiquity to the Nineteenth Century* (= *Archimedes*; 23). Berlin: Springer 2010, pp. 151–202

Thoren, Victor E. (1990): *The Lord of Uraniborg: A Biography of Tycho Brahe.* Cambridge: Cambridge University Press 1990

Vergara, Roberto Caffarelli (ed.) (1992): *Il compasso geometrico e militare di Galileo Galilei: Testi, annotazioni e disputa negli scritti di G. Galilei, M. Bernegger e B. Capra*. Pisa: ETS 1992

Voelkel, James R. (2001): *The Composition of Kepler's Astronomia Nova*. Princeton: Princeton University Press 2001

Wattenberg, Diedrich (1964): *David Fabricius. Der Astronom Ostfrieslands (1564–1617)* (= *Archenhold-Sternwarte, Vorträge und Schriften*; 19). Berlin-Treptow: Archenhold-Sternwarte 1964

Weidler, Johann Friedrich (1741): *Historia astronomiae*. Wittenberg: Schwartz 1741

Wilson, Curtis (1989): Predictive Astronomy in the Century after Kepler. In: Taton, René; Wilson, Curtis (eds.): *Planetary Astronomy from the Renaissance to the Rise of Astrophysics. Part A: Tycho Brahe to Newton* (= *The General History of Astronomy*; 2A). Cambridge: Cambridge University Press 1989, pp. 161–206

Wohlwill, Emil (1909/1926): *Galilei und sein Kampf für die copernicanische Lehre*. 2 vol. Leipzig: Leopold Voss 1909/1926

Zinner, Ernst (1942): Zur Ehrenrettung des Simon Marius. *Vierteljahrsschrift der Astronomischen Gesellschaft* 77 (1942), pp. 23–75

Chapter 12
Simon Marius: *Tabulae Directionum Novae*—A First Approach

Thony Christie

In 1599 Simon Marius published his *Tabulae Directionum Novae*. This paper explains what that is and how and why Marius came to write and publish this work.

For a large part of his adult life, Simon Marius was a professional astrologer; he was employed as such at the margravial court in Ansbach from 1606 until his death in 1624. His official title was "Hofmathematicus" (court mathematicus). However, in the early seventeenth century, the terms mathematicus, astronomus, and astrologus were still synonymous. Had Marius been introduced to somebody as court mathematicus, that person would have automatically assumed that Marius was an astrologer and not a mathematician. Marius was first appointed court mathematicus in 1606, after he had begun writing and publishing Schreibkalender (writing calendars) and Prognostica in 1601.[1] However he had already been active as an astrologer for at least 10 years prior to his appointment. In 1596 he had published a comet text, which was dedicated to the margrave of Ansbach—an obvious attempt to win favor with his future employer (Marius 1596).[2] His second astrological publication, his *Tabulae Directionum Novae* (Marius 1599), was once again dedicated to Margrave Georg Friedrich.[3] A comet text was something like a first degree for a budding astrologer, but a *Tabulae Directionum* was, in comparison, his master's exam (Fig. 12.1).

It is important to note that when Marius started writing his *Tabulae Directionum Novae*, probably in 1596, he was only 21 years old and still a student at the

[1]See the chapters by Klaus Matthäus and Richard Kremer in this volume.

[2]See the chapter by Jürgen Hamel in this volume.

[3]ILLUSRISSIMO PRINCIPI AC DOMINO, DOMINO Georgio Friderico, MARCHIONI BRANDENBURGENSI, BORRUSIAE, STETINI, POMERaniæ, Cassubiorum, VVandalorum, Silesiæ Duci in Icgerndorff, &c, Burggrauio Noribergensi, & Principi Rugiæ, &c, Domino suo Clementissimo.

T. Christie (✉)
Ronin Institute, Erlangen, Germany

H. Gaab, P. Leich (eds.), *Simon Marius and His Research*, Historical & Cultural Astronomy, https://doi.org/10.1007/978-3-319-92621-6_12

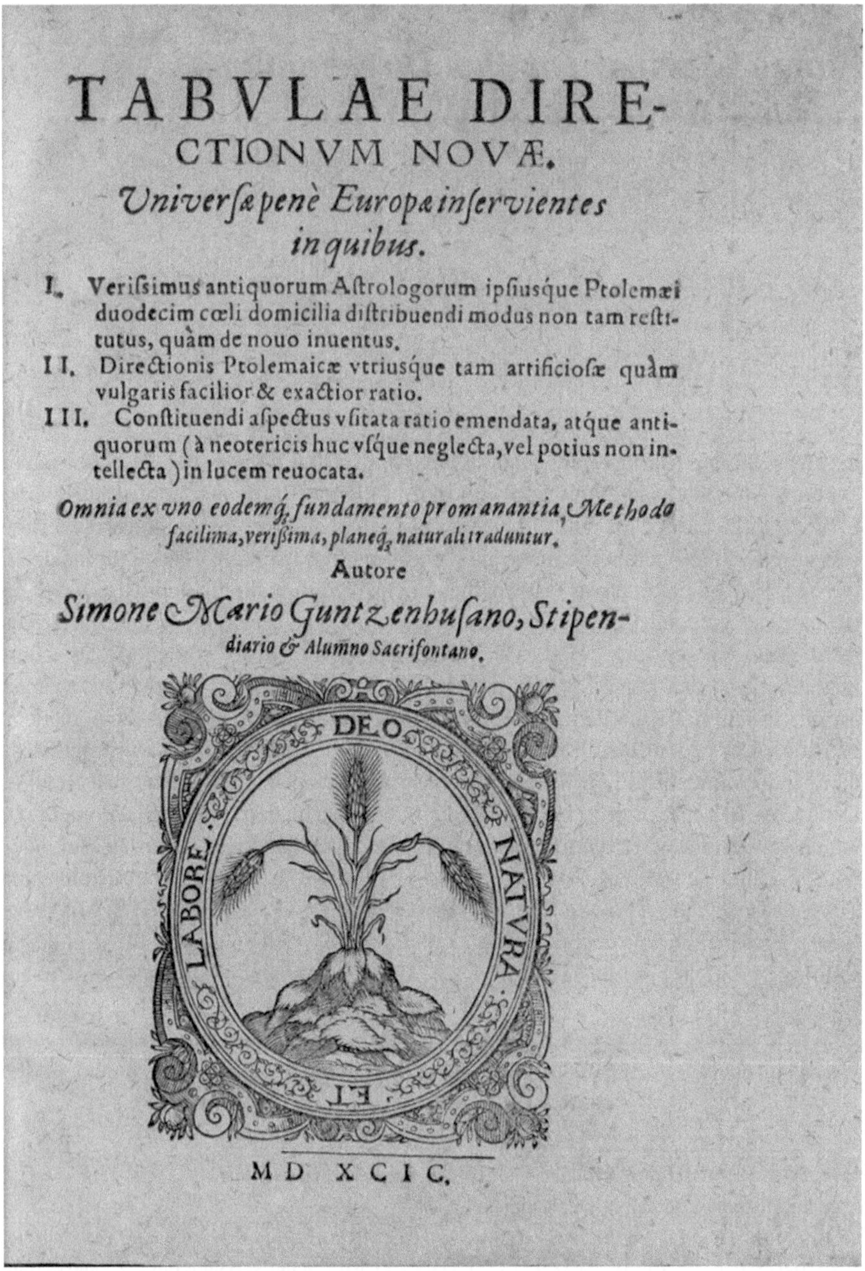

TABVLAE DIRE-
CTIONVM NOVÆ.

Vniversæ penè Europæ inservientes in quibus.

I. Verissimus antiquorum Astrologorum ipsiusque Ptolemæi duodecim cœli domicilia distribuendi modus non tam restitutus, quàm de nouo inuentus.

II. Directionis Ptolemaicæ vtriusque tam artificiosæ quàm vulgaris facilior & exactior ratio.

III. Constituendi aspectus vsitata ratio emendata, atque antiquorum (à neotericis huc vsque neglecta, vel potius non intellecta) in lucem reuocata.

Omnia ex vno eodemq; fundamento promanantia, Methodo facilima, verissima, planeq; naturali traduntur.

Autore

Simone Mario Guntzenhusano, Stipendiario & Alumno Sacrifontano.

MD XCIC.

Fig. 12.1 Title page of the *Tabulae Directionum Novae* 1599; Courtesy of the BSB München: 4. A. gr. B 959 Beibd. 1

Fürstenschule of Heilsbronn. This was a school founded by Margrave Georg Friedrich in 1581 and as stated in the deed of foundation:

> [. . .] that above all the children of the poor and of church and school servants and those of rightful people, who are either poor and without resources or gifted with many children by God, should be admitted to school.[4]

It was supposed to train "religious and moral church, school and civil servants and through their efficacy improve the religious and moral life of the people" (Muck 1880, p. 13f.).[5]

A *Tabulae Directionum* is a handbook written to assist astrologers to determine so-called directions, a very popular method for making prognostications in the Renaissance. In astrology directions is a technique for determining points in time, an important method of prognostication used in the analysis of natal horoscopes. Three different terms are often used in Latin astrology texts *directio*, *prorogatio*, and *progression* for the original Greek *aphesis* (Broecke 2003, p. 227, footnote 1). The calculation of directions serves only one purpose, determining the point of time of the occurrence of events indicated in the natal horoscope.

The earliest description of the method of directions can be found in Claudius Ptolemy's *Apotelematika* (*Tetrabiblos*) Book III.10 and Book IV.10 written in the second century CE. The *Tetrabiblos* was regarded as the most important astrology textbook during the Renaissance, but why the method of directions enjoyed such popularity is not really clear (Broecke 2003, p. 228). There are other methods for determining points of time in a horoscope, and the method of directions is anything other than simple.

In order to ascertain the point of time when a predicted event will occur, the astrologer first has to determine the so-called Promissor (Ptolemy's term is Hyleg, Arabic term is Apheta) on the ecliptic; this can be the time of birth but also diverse other points in time.[6] Having determined the Promissor, the astrologer now has to do the same for the Significator (Ptolemy's term is Anareta); that is the position of a planet or similar on the ecliptic that signifies a given event in the life of the horoscope's subject. Having ascertained the positions of both on the ecliptic, these are now projected onto the celestial equator. The distance in degrees between the two points corresponds to the number of years (one degree equals one calendar year) until the predicted event occurs. The transfer of the two points from the ecliptic to the celestial equator is a moderately complex exercise in spherical trigonometry, and astrologers produced books of mathematical tables in order to simplify the procedure.

[4]"[. . .] daß fürnehmlich armer, sowohl auch der Kirchen- und Schul-Diener und um die Herrschafft wohl verdienter Leut Kinder, die entweder arm und unvermöglich, oder von Gott mit vielen Kindern begabt seyn, in diese Unsere Schule angenommen werden sollen." The foundation letter is printed in Junger (1971), pp. 44–49.

[5]For more detail on the Fürstenschule and Marius's time there, see Chap. 2, section "Marius at the "Fürstenschule" of Heilsbronn".

[6]The whole theory of determining the Promissor is too complex to be dealt with here.

Simon Marius was by no means the only astrologer in the early modern period who wrote and published a *Tabulae Directionum*. The most well-known was produced by another Franconian astrologer, Johannes Müller (1436–1476), better known as Regiomontanus. In order to obtain a picture of the impact and influence of the genre *Tabulae Directionum* during the Renaissance, we will take a brief look at the history of Regiomontanus's volume.

Regiomontanus wrote his *Tabulae Directionum* in 1467, at the request of János Vitéz, the archbishop of Gran (today Esztergom) in Hungary. The work is dedicated to Vitéz. It consists of an explanation in 31 sections and trigonometrical and astronomical tables. The original manuscript written for Vitéz no longer exists; however the large number of other manuscripts in archives and libraries throughout Europe demonstrate the popularity of this work. Regiomontanus wanted to publish his *Tabulae Directionum* in his own printing office in Nuremberg in the 1470s, but he died without being able to realize this wish. The work was first published in print by Erhard Ratdolt in Augsburg in 1490. The book went through 11 editions, the last appearing as late as 1626. The tables were used by Albert von Bredzewo, Nicolaus Copernicus, Johannes Werner, Lucas Gauricus, Georg Tannstetter, and Johannes Kepler among many others. The explanatory section was translated into German, French, and English.[7] This very brief sketch of the history of Regiomontanus's *Tabulae* should serve to illustrate the strong demand for such works.

A central element of Regiomontanus's *Tabulae Directionum* is his new method for the division of the houses,[8] a very important aspect in the casting of horoscopes; the correct method for which was much disputed among practicing astrologers. There are almost as many different methods for dividing the houses in casting horoscopes as there are famous astrologers. Regiomontanus's method found many adepts and remained the dominant method in European astrology until the middle of the seventeenth century. It is exactly here that Marius sees the necessity for a new (novae) *Tabulae Directionum*. He accuses Regiomontanus of having misunderstood Ptolemy and claims that his new tables reject the "repugnant method of Regiomontanus" and not only restore but discover anew the "true foundation of Ptolemy and other ancients" (Marius 1599, sig. A1^{r}–A2^{r}).[9] This doesn't, however, prevent him from taking over three of Regiomontanus's trigonometrical tables, one assumes to save himself the trouble of having to calculate them from scratch. Marius emphasized his critique of Regiomontanus in his *Prognosticon auf 1624* (sig. A2^{r}), saying that he formulated his critique for Regiomontanus's method already in 1599 in his *Tabulae*: "I printed a Latin treatise already 24 years ago, wherein I discarded the method of Regiomontanus concerning his astrological houses, against it I have used the old process with the tempora horaria" ("Nun hab ich aber vor 24. Jahren

[7]This brief description of the publication history of Regiomontanus's *Tabulae* is distilled from Zinner (1990).

[8]On Regiomontanus's *Tabulae Directionum* and his house division, see Broecke (2003) p. 235 FF.

[9]Marius's title page reads: "Verissimus antiquorum astrologorum ipsiusque Ptolemaei duodecim coeli domicilia distribuendi modus non tam restitutus, quam de nouo inuentus."

einen Lateinischen Tractat trucken lassen/darinnen ich den modum Regiomontani mit seinen circulis positionum verworffen/hergegen den alten proceß duch die tempora horaria wider herfür gesucht").

The calculation of mathematical tables in the pre-computer age and even before the invention of logarithm tables was an extremely tedious and time-consuming exercise. Regiomontanus was aided in this work in Gran by the Polish astrologer Marcin Bylica (1433–1493).[10] Bylica took his copy of the *Tabulae Directionum* manuscript together with other works by Regiomontanus back to Krákow University, in those days a major European center for the study of astronomy and astrology, when he left Hungary. Marius tells us himself that he was also aided in the calculation of his tables for the division of the houses by Augustin Lanius, which Marius acknowledges at the end of preface:

> I gave the calculation of the tabulae vero domorum [true house tables] to my special friend and true colleague, the talented young man Aug. Lanius from Ansbach, after I had first shown him the method of calculation. (Marius 1599, sig. C1r)

Lanius was a fellow student at the Fürstenschule, who Marius describes thus in his *Mundus Iovialis*:

> [. . .] an extremely scholarly and well read man, who now lives in Halle in Saxonia as a private man [. . .]. He worked as organist in Heilsbronn in those days and because we had been neighbors and good friends for a long time, he had constant access to my work.[11]

Marius petitioned the margrave in 1598 for help in covering the costs of getting his *Tabulae Directionum* printed, and it appears that there were originally plans at the margravial court to send him to Königsberg to get the book printed. On May 20th 1597, the councilors Nicolaus Stadtmann, Stephan Muhr, Andreas Frobenius, Streuberger, and Johann Gümbelein signed a consideration, in which they recommended sending Marius to Königsberg.[12] Attached was a letter of recommendation to be signed by the margrave. Initially it only says that Marius "submissively asked for the continuation of his studies at other academies, where he could put his newly invented Tabulas directionum in print and then publish."[13] That doesn't sound like Marius expressed the wish to go to Königsberg. The argument was that Marius could correct his opus with better opportunities in our printing house in Königsberg

[10]For more on Bylica and his relationship with Regiomontanus, see Hayton (2007).

[11]"Inter alios autem non postremum locum occupat doctissimus & multae lectionis vir, Dominus Augustinus Lanius, nunc Halae Saxonum privatam vivens, qui eo tempore organedum Heilsbronnae agebat, & propter habitationes vicinas & amicitatem dudum inter nos initiam, fere perpetuus mearum actionum inspector erat" (Marius 1614/1988, p. 100). In *Prognosticon auf 1607* (sig. C4r; cf. Zinner 1942, p. 53). Marius reports about observations of Jupiter, made in 1596 by "me and my good friend Mr Augustinus Lanius, organist of the Monastery at Heilsbronn at that time" ("ich vnd mein guter freundt Herr Augustinus Lanius, damals Organist im Closter Heylsbronn").

[12]The *Bedenken* is printed in Büttner 2 (1813), p. 74f.

[13]"[. . .] unterthenigst angelangt zu Vortstellung seiner Studien uf auf andern academiis, dann seine neuerfundenen Tabulas directionum in Truckh zu geben und hiezu verlag zu thun" (Büttner 2 1813, p. 75).

on his own [...]. However in the end, it was printed by Christoph Lochner in Nuremberg in 1599.[14] Marius's *Tabulae* consists of eight numerical tables for determining the division of a horoscope into houses and for the trigonometry required to determine the direction; these are preceded by 18 rules for their use.

Marius's *Tabulae Directionum* appears not to have had a very good reception among other astrologers and astronomers. Johannes Kepler, who had already seen the *Tabulae* in 1599,[15] was asked several times for his opinion on the tables, and in July 1611, he wrote[16]:

> I have no desire to argue any further about his tables (Marii). What I have already said is sufficient, that they are inconvenient to use, which, as I see, the author admits. [...] I do not intend therefore to agitate against Marius.

In a letter to Bergrat Vicke from Wolfenbüttel, Marius wrote regarding his printed *Tabulae Directionum*: "I practiced astronomy barely for two years without having any teacher for (astronomy-) mathematics" (Kepler XIV, 1954, p. 383). Klug (1906, p. 403) saw a contradiction to the statement above. The *Tabulae* was printed in 1599, but Marius seemingly had worked on them at the latest since 1596.

Despite the fact that his *Tabulae Directionum* was not as successful as he might have hoped, it must still be acknowledged that it was a commendable effort for a young, almost entirely self-taught astrologer. This effort might well have played a role in the margrave's decision to send Marius to study with Tycho in Prague in 1601 and subsequently to Padua to study medicine.

Further research into Marius's *Tabulae Directionum* should be in the form of examination of its originality, i.e., did he borrow more than three tables from Regiomontanus or from other astrologers.[17]

Acknowledgments This brief description of Marius's *Tabulae Directionum Novae* would not have been possible without the extensive research carried out by my colleagues Hans Gaab, Richard Kremer, and Klaus Matthäus for their own contributions to the volume and from which I have profited greatly. I also owe a great debt to Pierre Leich without whom this paper would never have been written at all.

---

[14]For more detail on this story, see Chap. 2, section "Simon Marius in Königsberg?".

[15]Kepler, Vol. 14 (1949), p. 131 (Letter to Herwart von Hohenburg 12th July 1600).

[16]"Sed de ipsius tabulis disceptare ulterius animus mihi non est. Sufficit hoc quod dixi incommodas esse usu, quod video authorem fateri. [...] me publicas adversus ipsum suscepisse inimicitias" (Kepler (1937–2009) Vol. XVI, 1954, p. 388; cf. Klug 1906, p. 403). Johann Georg Herwart von Hohenburg (1553–1622) asked Kepler in a letter from March 18th 1600 for his opinion; he answered on July 12th. David Fabricius (1564–1617) expressed himself on the topic on April 28th 1602 (Kepler (1937–2009) Vol. XIV, 1949, p. 111, 131f., 231). In a letter from February 23rd 1610, Nikolaus Vicke requested from Kepler, to explain him the directions with three examples "und alle drey figuras uia Ptolemaica ex tabulis Simonis Marij zu erigiren." On March 25th, he asked Kepler for the differences in Marius's and Magini's tables (Kepler (1937–2009) Vol. XVI, 1954, p. 290, 376).

[17]Richard Kremer is already undertaking this research; see Kremer (Forthcoming).

References

Broecke, Steven vanden (2003): *The Limits of Influence: Pico, Louvain, and the Crisis of Renaissance Astrology*, Leiden: Brill 2003

Büttner, Heinrich Christoph (1813): Franconia. Beiträge zur Geschichte, Topographie und Litteratur von Franken, Vol. 2. Ansbach: Gassert 1813 [UB Augsburg: 02/IV .15.8.607]

Hayton, Darin (2007): Martin Bylica at the Court of Matthias Corvinus: Astrology and Politics in Renaissance Hungary. *Centarus* 49.3 (2007), pp. 185–198

Junger, Karl (1971): *Die Fürstenschule zu Heilsbronn*. Erlangen: Hespelein 1971

Kepler, Johannes (1937–2009): *Gesammelte Werke*, 21 Vol. München: Beck 1937–2009

Klug, Josef (1906): Simon Marius aus Gunzenhausen und Galileo Galilei: Ein Versuch zur Entscheidung der Frage über den wahren Entdecker der Jupitertrabanten und ihrer Perioden. *Abhandlungen der Mathematisch-Physikalischen Klasse der Königlich Bayerischen Akademie der Wissenschaften* 22 (1906), pp. 385–526

Kremer, Richard (Forthcoming): What was new in Simon Marius's *Tabulae directionum novae* of 1599?

Marius (1614/1988): Schlör, Joachim ed., *Mundus Iovialis – Die Welt des Jupiter. Die Endeckung der Jupitermonde dirch den fränkischen Hofmathematiker und AStronomer im Jahr 1609 – latinisch und deutsch* (= Fränkische Geschichte, Band 4). Gunzenhausen: Johann Schrenk 1988

Marius, Simon (1596): *Kurtze und eigentliche Beschreibung des Cometen oder Wundersterns/So sich in disem jetzt lauffenden Jar Christi unsers Heilands/1596. in dem Monat Julio/bey den Füssen des grossen Beerens/im Mitnächtischen Himmel hat sehen lassen. Gestellet durch Simonem Maierum Guntzenhusanum Alumnum Sacrifontanum*. Nürnberg: Paul Kauffmann 1596, 12 pp. 4°

Marius, Simon (1599): *Tabulae Directionum Novæ: Universæ pene Europæ inservientes in quibus. I. Verissimus antiquorum Astrologorum ispsiusque Ptolemæi duodecim cœli domicilia distribuendi modus non tam restitutus, quam de nouo inuentus. II. Directionis Ptolemaicæ vtriusque tam artificiosæ quam vulgaris facilior & exactior ratio. III. Constituendi aspectus vsitata ratio emendata, atque antiquorum (a neotericis huc vsque neglecta, vel potius non intellecta) in lucem reuocata. Omnia ex uno eodemq; fundamento promanantia, Methodo facilima, verißima, planeq; naturaliter raduntur. Autore Simone Mario Guntzenhusano, Stipendiario & Alumno Sacrifontano*. Nürnberg: Christoph Lochner 1599, 66 Bl. 4°

Muck, Georg (1880): *Geschichte von Kloster Heilsbronn von der Urzeit bis zur Neuzeit*, Vol. 3. Nördlingen: Beck 1880

Zinner, Ernst (1990): *Regiomontanus His Life and Work*, trans. By Ezra Brown, North Holland, Amsterdam: Elsevier 1990; German original edition *Leben und Wirken des Johannes Müller von Königsberg, genannt Regiomontanus*, München: Beck 1938, Osnabrück: Zeller 1968

Zinner Ernst (1942): Zur Ehrenrettung des Simon Marius. Vierteljahresschrift der Astronomischen Gesellschaft 77/1. 1942, pp. 23–75

Chapter 13
Translating the *Mundus Iovialis* into German

Joachim Schlör

Knowledge of the Latin language can open new horizons for the patient student; in our case it can also provide some understanding of the dramatic reorientation of the modern worldview toward the heliocentric system. Thereby, it may convey to young people an exciting impression of a researcher's indefatigable activity striving for highest precision on the threshold of the modern period, to be finally rewarded with the gratification of discovery[1] (Figs. 13.1 and 13.2).

To the modern student, getting a grip on Marius's Latin is not very hard, so that it can be well understood on an intermediate level of study. Scholars [or researchers; scientist is anachronistic] of the early modern period intending to spread their cutting-edge findings, as widely as possible, among the scientific community, made use of this most universally known means of communication. What is more, Marius does not use the Latin of the Middle Ages but follows the example set by the classical authors Caesar and Cicero, as did all humanists and as we do teaching Latin today. So reading the *Mundus* does not pose too great a problem to our students. Of course, it goes without saying that it cannot hold a candle to the refinement of classical literary style (Fig. 13.3; see also Schlör 2012a).

In more recent times, Ernst Zinner, the German astronomer from Bamberg, portrayed Simon Marius's work favorably. At the end of an essay entitled "Retrieving Simon Marius's Honor" (Zur Ehrenrettung des Simon Marius) of 1942, he demands that "in honor of its citizen Simon Marius, an outstanding astronomer, the city of Ansbach should save his main work *Mundus Iovialis* from oblivion, as it did with the more important writings of his contemporaries, and distribute it among the world's great libraries."[2]

[1]For more details, see Bemmer (2015).

[2]Zinner (1942); for more details, see Wolfschmidt (2012).

J. Schlör (✉)
Simon Marius Society, Weißenburg, Germany
e-mail: jschloer@wugnet.de

H. Gaab, P. Leich (eds.), *Simon Marius and His Research*, Historical & Cultural Astronomy, https://doi.org/10.1007/978-3-319-92621-6_13

Fig. 13.1 The Gunzenhausen original of Simon Marius, *Mundus Iovialis* (1614), since 1995 in the city archive; the *Prognosticon auf 1622* gives the same image with an alternative text praising Marius as a conscientious craftsman, a true astrologer, and pious "magician"—redefining the terms of astrologer and magus in favor of Marius in a word play with the famous sorcerer Simon Magus of classical antiquity

Since, apart from Prickard's English translation of 1916 (Marius 1614/1916), there had been no modern version of this work so far, the Simon Marius Gymnasium (SMG) met Zinner's demand in 1987–1988 publishing a Latin and German edition which gives the *Mundus* in a facsimile of the Ansbach original as well as a German translation on the opposing pages of the book. The beautiful original page was included not only for decorative or aesthetic reasons; its choice initials, type fonts, and ornaments emphasize the authentic quality of this print as well; additionally, its design is moderately mirrored in the layout of the German equivalent. The translation was partly a result of students' efforts in a Latin advanced level course at our SMG under my direction. Alois Wilder, the mathematics and physics teacher in charge of the school's observatory for a long time, accompanying our work from the perspective of the natural sciences contributed an epilogue in which he emphasized particularly the precision of Marius's observations.

HOC OPUS, HIC LABOR! Following a raw translation, numerous weekends and holidays had to be devoted to the effort. Questions of scientific content, style, and design had to be answered, correspondence had to be entered into with various archives and libraries, the publisher and the printer had to be talked to, and time and again the manuscript had to be scrutinized for printing errors. Yet, in the end, it was worth the toil, and it had indeed been fun. Marius himself complains in his Prognosticon for 1610: "Dum [. . .] immensum (qu)e animo metimur Olympum, pauperie premimur, patimurque incommoda multa [. . .]" (While we are climbing the

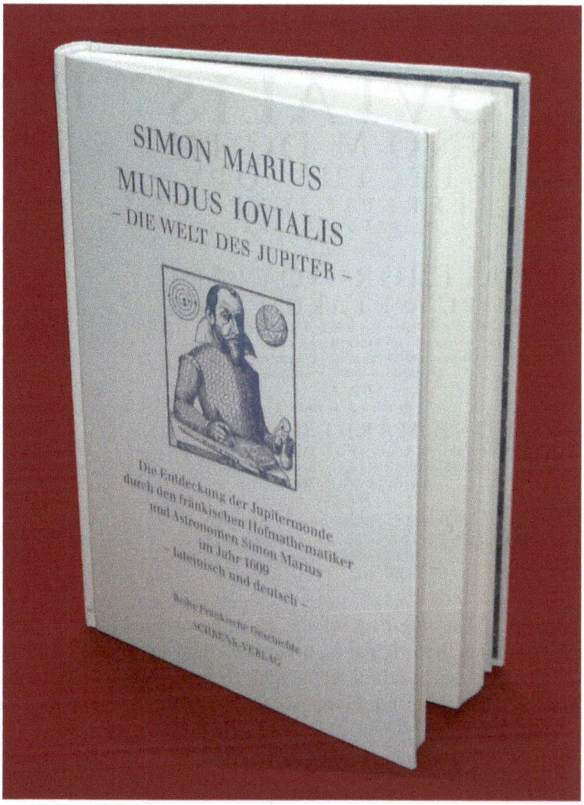

Fig. 13.2 The bilingual edition of Marius's *Mundus Iovialis* (Marius 1614/1988), translated by Joachim Schlör and published by Dr. Schrenk of Gunzenhausen; the title *Mundus Iovialis* here translates as "Die Welt des Jupiter" (Jupiter's World); "Das Jupitersystem" (The Jupiter System) would also be acceptable

giant Olympus [of the sciences], we are hard pressed by poverty and suffer hardship and adversity.) When all was said and done, experts praised our German translation for its neat language and style (Fig. 13.4).

From the outset, a precise and well-readable translation was our aim; a commented edition of the *Mundus* was not intended; in our preface, however, we expressed our hope that the book "will give rise to further research and explanation. (. . .) Perhaps one day this will finally result in a follow-up book of the text volume." The two recently published tomes (Wolfschmidt 2012; Gaab and Leich 2016) now present the interested reader with a wealth of background information, as was hardly imaginable to the initial translators and today fills us with great joy and satisfaction.

Among the relatively numerous copies of the *Mundus*, we found four in our immediate vicinity, namely, in the Nuremberg City Library; the Ansbach castle library, which we relied on in translating; the Wolfenbüttel Herzog August Library, which alone contains an additional section in which Marius defends himself against criticism (Gaab and Leich 2014); and, since 1995, in the city archive of Gunzenhausen. Any specification as to the number of books then printed, however, is not given. Regrettably despite all efforts, I did not succeed at that time in receiving a copy of the Wolfenbüttel Appendix, and therefore I could not include it in our

14 Titelblatt

MUNDUS
IOVIALIS
ANNO M. DC. IX.
DETECTUS OPE
PERSPICILLI
BELGICI,
Hoc eſt,
QUATUOR JOVIALI-
UM PLANETARUM, CUM
THEORIA, TUM TABULÆ, PROPRIIS OB-
SERVATIONIBUS MAXIME FUNDATÆ, EX QUIBUS
ſitus illorum ad Iovem, ad quodvis tempus datum
promptiſſimè & facillimè ſuppu-
tari poteſt.
Inventore & Authore
SIMONE MARIO GUNTZEN-
HUSANO, MARCHIONUM BRANDEN
BURGENSIUM IN FRANCONIÂ MATHE-
matico, puriorisque Medici-
næ Studioſo.
Cum gratia & privil. Sac. Cæſ. Majeſt.
Sumptibus & Typis IOHANNIS LAURI Civis & Bibliopolæ.
Noribergenſis, ANNO
M. DC. XIV.

Titelblatt 15

DIE WELT
DES JUPITER
im Jahre 1609
entdeckt mit Hilfe
des belgischen Fernrohrs,
nämlich
sowohl die Theorie der vier Monde des Jupiter
als auch die Tabellen, welche
durch eigene Beobachtungen sehr gut ab-
gestützt sind und es gestatten, eine
sehr rasche und einfache Berechnung
der Positionen jener Monde zum Jupiter
zu jedem beliebig gegebenen
Zeitpunkt vorzunehmen.
Vom Entdecker & Verfasser
SIMON MARIUS AUS GUNZENHAUSEN,
Mathematiker der Markgrafen von
Brandenburg in Franken
& Anhänger der reineren Medizin.
Mit der Huld & Erlaubnis
der Heiligsten Kaiserlichen Majestät
Druck & Verlag: Johann Lauer,
Bürger und Buchhändler zu Nürnberg,
im Jahre 1614

Fig. 13.3 The title page of the *Mundus*, in our bilingual edition (Marius 1614/1988, p. 14f.), demonstrates the original as a facsimile and the parallel translation on the opposing page. The original boasts a beautiful typographic design in Antiqua type font, partially cursive; it does without images. Publishing his findings only in 1614, Marius lagged behind his competitor Galileo by 4 years

Mundus translation. It was only in 2014 that I was able to study and lend a hand in translating this text, as Hans Gaab and Pierre Leich planned to publish a closer discussion of this vivid example of the angry quarrels and heated controversies among astronomers of that time. Marius, it is true, defends his due diligence as well as his scientific authority without animosity against cunning and furious attacks in a calm and factual way.

The book finally appeared in 1988 as the fourth volume of the “Franconian History” series published by Dr. Schrenk of Gunzenhausen with as many as one thousand copies and, sorry to say, is out of print today (Schlör 2012b). Our German translation was presented to the public in Gunzenhausen in a little celebration on November 10, 1988. In the beautiful Baroque-style hall of the historical hunting lodge, quite a number of students, teachers, and political and educational dignitaries had gathered to get a first-hand impression of Jupiter’s World, as Marius had seen it. It was only shortly after this official event that the Bavarian Broadcasting Company emphasized particularly that with our publication, schoolwork had finally resulted in a practical outcome, which was of public as well as scientific interest. The *Frankfurter Allgemeine Zeitung* regretted the fact that Marius was much slower—not in his findings, but—in presenting them to the public than Galileo and quoted the beautiful text passage starting with “In the year 1608, when the Frankfurt autumn fair was held [. . .]” from our German version (Fig. 13.5).

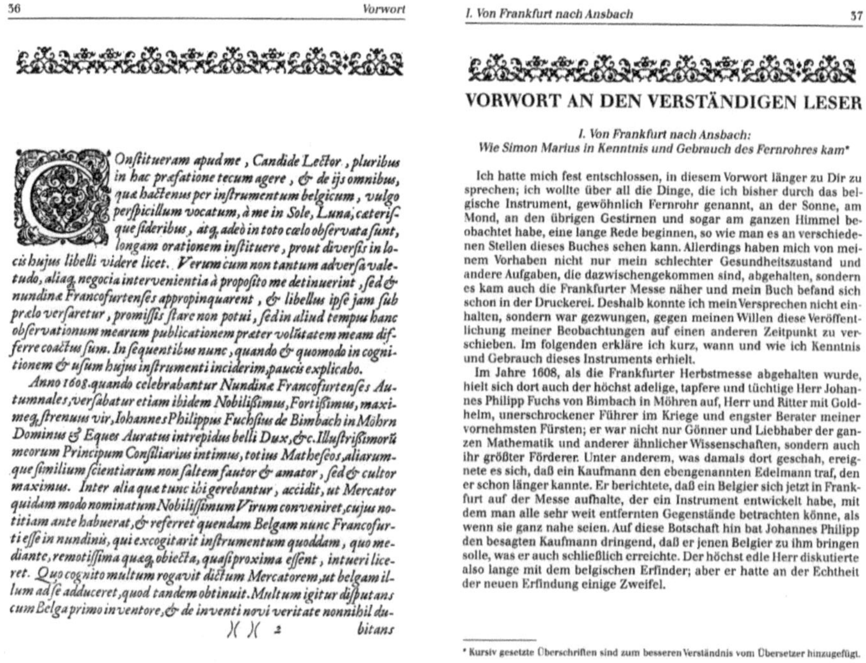

Constitueram apud me, Candide Lector, pluribus in hac præfatione tecum agere, & de ijs omnibus, quæ hactenus per instrumentum belgicum, vulgo perspicillum vocatum, à me in Sole, Luna, cæterisque sideribus, atq. adeò in toto cælo observata sunt, longam orationem instituere, prout diversis in locis hujus libelli videre licet. Verum cum non tantum adversa valetudo, aliaq. negocia intervenientia à proposito me detinuerint, sed & nundinæ Francofurtenses appropinquarent, & libellus ipse jam sub prælo versaretur, promissis stare non potui, sed in aliud tempus hanc observationum mearum publicationem præter volũtatem meam differre coactus sum. In sequentibus nunc, quando & quomodo in cognitionem & usum hujus instrumenti inciderim, paucis explicabo.

Anno 1608. quando celebrabantur Nundinæ Francofurtenses Autumnales, versabatur etiam ibidem Nobilissimus, Fortissimus, maximeq. strenuus vir, Iohannes Philippus Fuchsius de Bimbach in Möhrn Dominus & Eques Auratus intrepidus belli Dux, &c. Illustrissimorũ meorum Principum Consiliarius intimus, totius Matheseos, aliarumque similium scientiarum non saltem fautor & amator, sed & cultor maximus. Inter alia quæ tunc ibi gerebantur, accidit, ut Mercator quidam modo nominatum Nobilissimum Virum conveniret, cujus notitiam ante habuerat, & referret quendam Belgam nunc Francofurti esse in nundinis, qui excogitarit instrumentum quoddam, quo mediante, remotissima quæq. obiecta, quasi proxima essent, intueri liceret. Quo cognito multum rogavit dictum Mercatorem, ut belgam illum ad se adduceret, quod tandem obtinuit. Multum igitur disputans cum Belga primo inventore, & de inventi novi veritate nonnihil du-

)( )(2 bitans

VORWORT AN DEN VERSTÄNDIGEN LESER

I. Von Frankfurt nach Ansbach:
*Wie Simon Marius in Kenntnis und Gebrauch des Fernrohres kam**

Ich hatte mich fest entschlossen, in diesem Vorwort länger zu Dir zu sprechen; ich wollte über all die Dinge, die ich bisher durch das belgische Instrument, gewöhnlich Fernrohr genannt, an der Sonne, am Mond, an den übrigen Gestirnen und sogar am ganzen Himmel beobachtet habe, eine lange Rede beginnen, so wie man es an verschiedenen Stellen dieses Buches sehen kann. Allerdings haben mich von meinem Vorhaben nicht nur mein schlechter Gesundheitszustand und andere Aufgaben, die dazwischengekommen sind, abgehalten, sondern es kam auch die Frankfurter Messe näher und mein Buch befand sich schon in der Druckerei. Deshalb konnte ich mein Versprechen nicht einhalten, sondern war gezwungen, gegen meinen Willen diese Veröffentlichung meiner Beobachtungen auf einen anderen Zeitpunkt zu verschieben. Im folgenden erkläre ich kurz, wann und wie ich Kenntnis und Gebrauch dieses Instruments erhielt.

Im Jahre 1608, als die Frankfurter Herbstmesse abgehalten wurde, hielt sich dort auch der höchst adelige, tapfere und tüchtige Herr Johannes Philipp Fuchs von Bimbach in Möhren auf, Herr und Ritter mit Goldhelm, unerschrockener Führer im Kriege und engster Berater meiner vornehmsten Fürsten; er war nicht nur Gönner und Liebhaber der ganzen Mathematik und anderer ähnlicher Wissenschaften, sondern auch ihr größter Förderer. Unter anderem, was damals dort geschah, ereignete es sich, daß ein Kaufmann den ebengenannten Edelmann traf, den er schon länger kannte. Er berichtete, daß ein Belgier sich jetzt in Frankfurt auf der Messe aufhalte, der ein Instrument entwickelt habe, mit dem man alle sehr weit entfernten Gegenstände betrachten könne, als wenn sie ganz nahe seien. Auf diese Botschaft hin bat Johannes Philipp den besagten Kaufmann dringend, daß er jenen Belgier zu ihm bringen solle, was er auch schließlich erreichte. Der höchst edle Herr diskutierte also lange mit dem belgischen Erfinder; aber er hatte an der Echtheit der neuen Erfindung einige Zweifel.

* Kursiv gesetzte Überschriften sind zum besseren Verständnis vom Übersetzer hinzugefügt.

Fig. 13.4 The preface of the *Mundus Iovialis* (Marius 1614/1988, p. 36f.), in which Marius describes, e.g., how he came to the use of the telescope and made his first observations. The impressive narrative quality of the story of the events at the 1608 autumn fair in Frankfurt lends itself particularly for use in the classroom, e.g., dramatization for the theatre group

How did our author feel about using his German mother tongue, as it was Latin that had always been the language of science? In the humanist age, people were newly excited about the high level of culture of classical antiquity as well as the languages of the Bible, particularly Latin. Quite clearly, Marius also dismisses the idea of publishing a work of science using his native language or translating from Latin, stating that then the common people would lose their respect for the secrets of nature (*Prog. 1610*, preface). Yet, favorably inclined to translating the Bible and other important work, our German rendering of the *Mundus* does not seem fundamentally out of keeping with our astronomer's attitude after all (*Prog. 1611*). What is more, by means of our translation and the inclusion of the original in facsimile, we also want to encourage the study of the Latin original. It is those who strive to acquire not only a smattering but a certain command of Latin today that our book is dedicated to.

In his opening words, Herbert Marius of Vienna, a distant relative of our author, expressed his joy at our students "having turned the main work of our ancestor into readable form and presented our numerous great grandsons and granddaughters of the 13th generation with the opportunity of reading their ancestor's words." The then SMG principal, Werner Pilhofer, was delighted that "the scientific achievements of

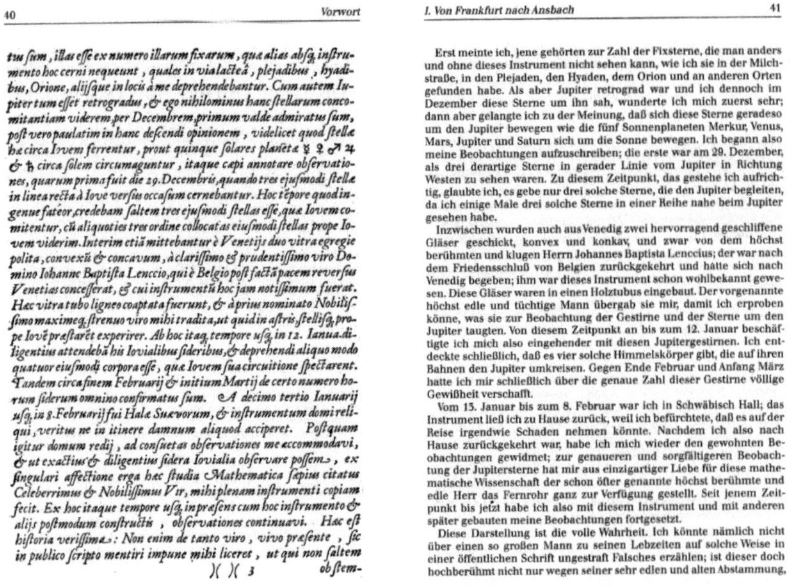

40 Vorwort

tus ſum, illas eſſe ex numero illarum fixarum, quæ alias abſq; inſtrumento hoc cerni nequeunt, quales in via lacteâ, plejadibus, hyadibus, Orione, aliiſque in locis à me deprehendebantur. Cum autem Iupiter tum eſſet retrogradus, & ego nihilominus hanc ſtellarum concomitantiam viderem per Decembrem, primum valde admiratus ſum, poſt vero paulatim in hanc deſcendi opinionem, videlicet quod ſtellæ hæ circa Iovem ferrentur, prout quinque ſolares planetæ ☿ ♀ ♂ ♃ & ♄ circa ſolem circumaguntur, itaque cæpi annotare obſervationes, quarum prima fuit die 29. Decembris, quando tres ejuſmodi ſtellæ in linea recta à Iove verſus occaſum cernebantur. Hoc tēpore quod ingenue fateor, credebam ſaltem tres ejuſmodi ſtellas eſſe, quæ Iovem comitentur, cū aliquoties tres ordine collocatas eiuſmodi ſtellas prope Iovem viderim. Interim etiā mittebantur è Venetijs duo vitra egregie polita, convexū & concavum, à clariſſimo & prudentiſſimo viro Domino Iohanne Baptiſta Lenccio, qui è Belgio poſt factā pacem reverſus Venetias conceſſerat, & cui inſtrumentū hoc jam notiſſimum fuerat. Hæc vitra tubo ligneo coaptata fuerunt, & à prius nominato Nobiliſſimo maximeq; ſtrenuo viro mihi tradita, ut quid in aſtris, ſtelliſq; prope Iovē præſtarēt experirer. Ab hoc itaq; tempore uſq; in 12. Ianua. diligentius attendebā his Iovialibus ſideribus, & deprehendi aliquo modo quatuor eiuſmodi corpora eſſe, quæ Iovem ſua circuitione ſpectarent. Tandem circa finem Februarij & initium Martij de certo numero horum ſiderum omnino confirmatus ſum. A decimo tertio Ianuarij uſq; in 8. Februarij fui Halæ Suævorum, & inſtrumentum domi reliqui, veritus ne in itinere damnum aliquod acciperet. Poſtquam igitur domum redij, ad conſuetas obſervationes me accommodavi, & ut exactius & diligentius ſidera Iovialia obſervare poſſem, ex ſingulari affectione erga hæc ſtudia Mathematica ſæpius citatus Celeberrimus & Nobiliſſimus Vir, mihi plenam inſtrumenti copiam fecit. Ex hoc itaque tempore uſq; in præſens cum hoc inſtrumento & alijs poſtmodum conſtructis, obſervationes continuavi. Hæc eſt hiſtoria veriſſima: Non enim de tanto viro, vivo præſente, ſic in publico ſcripto mentiri impune mihi liceret, ut qui non ſaltem

)( )(3 obſtem-

I. Von Frankfurt nach Ansbach 41

Erst meinte ich, jene gehörten zur Zahl der Fixsterne, die man anders und ohne dieses Instrument nicht sehen kann, wie ich sie in der Milchstraße, in den Plejaden, den Hyaden, dem Orion und an anderen Orten gefunden habe. Als aber Jupiter retrograd war und ich dennoch im Dezember diese Sterne um ihn sah, wunderte ich mich zuerst sehr; dann aber gelangte ich zu der Meinung, daß sich diese Sterne geradeso um den Jupiter bewegen wie die fünf Sonnenplaneten Merkur, Venus, Mars, Jupiter und Saturn sich um die Sonne bewegen. Ich begann also meine Beobachtungen aufzuschreiben; die erste war am 29. Dezember, als drei derartige Sterne in gerader Linie vom Jupiter in Richtung Westen zu sehen waren. Zu diesem Zeitpunkt, das gestehe ich aufrichtig, glaubte ich, es gebe nur drei solche Sterne, die den Jupiter begleiten, da ich einige Male drei solche Sterne in einer Reihe nahe beim Jupiter gesehen habe.

Inzwischen wurden auch aus Venedig zwei hervorragend geschliffene Gläser geschickt, konvex und konkav, und zwar von dem höchst berühmten und klugen Herrn Johannes Baptista Lenccius; der war nach dem Friedensschluß von Belgien zurückgekehrt und hatte sich nach Venedig begeben; ihm war dieses Instrument schon wohlbekannt gewesen. Diese Gläser waren in einen Holztubus eingebaut. Der vorgenannte höchst edle und tüchtige Mann übergab sie mir, damit ich erproben könne, was sie zur Beobachtung der Gestirne und der Sterne um den Jupiter taugten. Von diesem Zeitpunkt an bis zum 12. Januar beschäftigte ich mich also eingehender mit diesen Jupitergestirnen. Ich entdeckte schließlich, daß es vier solche Himmelskörper gibt, die auf ihren Bahnen den Jupiter umkreisen. Gegen Ende Februar und Anfang März hatte ich mir schließlich über die genaue Zahl dieser Gestirne völlige Gewißheit verschafft.

Vom 13. Januar bis zum 8. Februar war ich in Schwäbisch Hall; das Instrument ließ ich zu Hause zurück, weil ich befürchtete, daß es auf der Reise irgendwie Schaden nehmen könnte. Nachdem ich also nach Hause zurückgekehrt war, habe ich mich wieder den gewohnten Beobachtungen gewidmet; zur genaueren und sorgfältigeren Beobachtung der Jupitersterne hat mir aus einzigartiger Liebe für diese mathematische Wissenschaft der schon öfter genannte höchst berühmte und edle Herr das Fernrohr ganz zur Verfügung gestellt. Seit jenem Zeitpunkt bis jetzt habe ich also mit diesem Instrument und mit anderen später gebauten meine Beobachtungen fortgesetzt.

Diese Darstellung ist die volle Wahrheit. Ich könnte nämlich nicht über einen so großen Mann zu seinen Lebzeiten auf solche Weise in einer öffentlichen Schrift ungestraft Falsches erzählen; ist dieser doch hochberühmt nicht nur wegen seiner sehr edlen und alten Abstammung,

Fig. 13.5 Marius's first observation of Jupiter's moons (Marius 1614/1988, p. 40f.), which was made by means of lenses from Venice; here our author depicts the gradual process in which his discoveries develop

an outstanding citizen of the town of Gunzenhausen, the patron of our school, finally received their due attention."

Obviously the scientific community had in the meantime become aware of the book, as well, and discussed it in various publications. Diester Stöffler from the Münster Institute of Planetology drew his colleagues' attention to it. Half a year later, Alto Brachner of the Munich German Museum wrote that he had now set out to study the *Mundus*, which he greatly enjoyed. He considered it a shame that it took the SMG to translate this "from a historical point of view not uninteresting work"—"in our gigantic, noisy, and 'athletic' hustle and bustle of culture and science." Soon our book became known beyond the German borders, as Albert van Helden of the Rice University, Houston, Texas, wrote in the *Journal for the History of Astronomy* (Volume 21, 1990, p. 371f.): "The German text of this volume is a faithful and competent rendering, and it will be of considerable use to scholars of the subject. [...] Joachim Schlör is to be commended on undertaking this project and providing us with a good translation of this controversial book. What a wonderful way to make Latin relevant to one's students!"

Only 6 months after publication, more than half of the books printed had found a buyer. Who would have hoped for such a success during the preparations? Obviously the toil had been worthwhile, a worth, however, of the purely idealistic kind. It was Marius himself that realized how little was sometimes earned with spiritual labor (*Prog. 1610*) (Fig. 13.6).

Fig. 13.6 The original of the *Mundus Iovialis* acquired by Gunzenhausen in 1995, obviously in a good condition, stamped "C. Jacobi," supposedly an earlier owner (Carl Gustav Jacobi, German mathematician, b. 1804, d. 1851, studied philosophy, Latin, and mathematics at the University of Berlin). The inside covers show commentaries in tight handwriting

In 1994, 6 years after the appearance of our bilingual edition, through a stroke of luck—not merely for the city of Gunzenhausen and the SMG—Hermann Neumann, then head of the Latin department of the SMG and known bibliophile, thumbing through a catalogue of a Berlin antiquarian bookshop, discovered a well-preserved copy of the *Mundus Iovialis* among the tomes offered. Straightaway the flabbergasted connoisseur contacted the then mayor of Gunzenhausen, who in turn informed the local Sparkasse manager. They without any hesitation purchased the precious tome for a respectable sum of 24,000 DM in April 1995, sharing the cost equally (Schlör 1995). So one of the rare original copies of the main work by the now famous son of the city and the name giver of the SMG came into the possession of Gunzenhausen.

An impression of our bilingual edition is given here by means of some selected pages, which are presented with the permission of the publisher Dr. Schrenk. On the left, the original Latin page is opposed to the respective page giving the German translation. German subtitles have been added in italics.

After the translation work had been published, I developed three concepts for Latin school lessons based on the *Mundus*, i.e., a text collection, as well as two study group projects: "Simon Marius, the Franconian Galileo—A Latin text collection for the intermediate and advanced level" of the German Gymnasium and the two classroom tuition projects: "Simon Marius, the Franconian Galileo: 'Tunc aspexi stellas'—A study group project on Simon Marius, *Mundus Iovialis*, Praefatio; revision of the noun forms" and "Simon Marius, the Franconian Galileo: 'Mundus detectus ope perspicilli'—A study group project on Simon Marius, *Mundus Iovialis*,

Simon Marius
Der fränkische Galilei

Eine lateinische Textsammlung
die Mittel- und Oberstufe

Herausgegeben von Joachim Schlör

Fig. 13.7 Frontispiece of "Simon Marius, the Franconian Galileo—A Latin text collection for the intermediate and advanced level" (Schlör 2008), which is meant to pave the way into the classroom. Here the title of "Der fränkische Galilei" (The Franconian Galilei) is used for the first time; Wolfschmidt (2012) followed this example

Prima et Secunda Pars; general grammar revision." These materials can easily be obtained from the *Marius-Portal* (www.simon-marius.net) (Fig. 13.7).[3]

This modern approach to learning is intended to be enjoyable for students and to provide them with opportunities for working on their own. The projects are based on main *Mundus Iovialis* passages taken from the Latin text collection, which are provided with sublinear comments and language hints, parallel texts, partly in facsimile, illustrations, and tasks for the study of language, content, and historical context. Comments for teachers, a glimpse of the Bavarian curriculum, as well as detailed work tasks and solution sheets for the students are meant to provide useful assistance for lessons (Figs. 13.8 and 13.9).

[3]For more details, see Gaab and Leich (2016).

Texte zum Wandel des antiken Weltbilds in der frühen Neuzeit: Simon Marius 11

„Simon Marius, der fränkische Galilei - Eine lateinische Textsammlung"

Praefatio ad candidum lectorem: Die Vorrede

AUF DER FRANKFURTER MESSE

1 Constitueram apud me, candide lector, pluribus in hac praefatione tecum agere, et de iis omnibus, quae hactenus per instrumentum Belgicum, vulgo „perspicillum" vocatum, a me in sole, luna, ceterisque sideribus atque adeo in toto caelo observata sunt, longam orationem instituere, prout diversis in locis huius libelli videre licet.

2 Verum cum non tantum adversa valetudo aliaque negotia intervenientia a proposito me detinuerint, sed Nundinae Francofurtenses appropinquarent, et libellus ipse iam sub praelo versaretur, promissis stare non potui, sed in aliud tempus hanc observationum mearum publicationem praeter voluntatem meam differre coactus sum. In sequentibus nunc, quando et quomodo in cognitionem et usum huius instrumenti inciderim, paucîs explicabo.

3 Anno 1608, quando celebrabantur Nundinae Francofurtenses Autumnales, versabatur etiam ibidem nobilissimus, fortissimus maximeque strenuus vir, Ioannes Philippus Fuchsius de Bimbach in Möhrn, Dominus et eques auratus, intrepidus belli dux, et ceterum illustrissimorum meorum principum consiliarius intimus, totius matheseos aliarumque similium scientiarum saltem fautor et amator, sed et cultor maximus.

----

1 candidus: glänzend weiß, lauter, redlich; geneigt, freundlich – **pluribus** <*verbis*>: wortreich, intensiv – **agere**: abhandeln, sprechen – **hactenus**: bisher – **vulgo**: in der Umgangssprache – **atque adeo**: und sogar – **instituere**: *incipere* – **pro-ut**: so wie – **in locis**: klass. Latein?

2 adversa valetudo: schlechter („widriger") Gesundheitszustand; **Nundinae Francofurtenses** <*Autumnales*>: die Frankfurter <Herbst->Messe (urspr. an Christi Himmelfahrt, 15. August) – **sub praelo versari**: sich unter dem Druck-stock befinden – **promissis stare**: zu einem Versprechen stehen, ein V. halten – **praeter** + Akk: vorbei an; gegen – **in sequentibus**: im Folgenden – **in usum alicius rei incidere**: „in den Gebrauch einer Sache geraten" – **paucîs**: s. oben zu *pluribus*

3 Philipp Fuchs von Bimbach in Möhrn: mächtiger Beamter am Hof von Markgraf Joachim Ernst v. Brandenburg-Ansbach (1603–25) – **auratus**^{0}: <= *aurum*; „Ritter vom goldenen Sporn" / mit vergoldeter Rüstung / mit Goldkette; „Ritterschlag" für besondere Verdienste, bes. im 16. Jahrhundert – **intrepidus**: unerschrocken (*trepidare, tremere*: zittern) – **consiliarius**^{0}: <= *consilium* – **totius matheseos**: Gen. zu *tota mathesis* – **fautor**: <= *favêre*; Gönner – **cultor**: <= *colere*: Förderer (Die hochgestellte Null „^{0}" markiert Wörter, die die Schüler nicht lernen brauchen.)

Fig. 13.8 Example page from the text collection (Schlör 2008); a sublinear commentary helps with vocabulary the modern student is not usually familiar with and gives some background information

3 Mundus Iovialis: Das Titelblatt

MUNDUS
IOVIALIS
ANNO M. DC. IX.
DETECTUS OPE
PERSPICILLI
BELGICI,
Hoc eſt,
QUATUOR JOVIALI-
UM PLANETARUM, CUM
THEORIA, TUM TABULÆ, PROPRIIS OB-
SERVATIONIBUS MAXIME FUNDATÆ, EX QUIBUS
ſitus illorum ad Iovem, ad quodvis tempus datum
promptiſſimè & facilimè ſuppu-
tari poteſt.
Inventore & Authore
SIMONE MARIO GUNTZEN-
HUSANO, MARCHIONUM BRANDEN
BURGENSIUM IN FRANCONIÂ MATHE-
matico, puriorisque Medici-
næ Studioſo.
Cum gratia & privil. Sac. Cæſ. Majeſt.
Sumptibus & Typis IOHANNIS LAURI Civis & Bibliopolæ.
Noribergenſis, ANNO
M. DC. XIV.

Aufgaben zum Titelblatt:
- *Transkribiere den Text!*
- *Vergleiche die Seite in Form und Inhalt mit Galileis* Sternenboten.

Mundus Iovialis: Titelblatt - Sidereus Nuncius (1610): Titelblatt

SIDEREVS
NVNCIVS
MAGNA, LONGEQVE ADMIRABILIA
Spectacula pandens, ſuſpiciendaque proponens
vnicuique, præſertim verò
PHILOSOPHIS, atq; ASTRONOMIS, quæ à
GALILEO GALILEO
PATRITIO FLORENTINO
Patauini Gymnaſij Publico Mathematico
PERSPICILLI
*Nuper à ſe reperti beneficio ſunt obſeruata in LVNÆ FACIE, FIXIS IN-
NVMERIS, LACTEO CIRCVLO, STELLIS NEBVLOSIS,
Apprime verò in*
QVATVOR PLANETIS
Circa IOVIS Stellam diſparibus interuallis, atque periodis, celeri-
tate mirabili circumuolutis; quos, nemini in hanc vſque
diem cognitos, nouiſſimè Author depræ-
hendit primus; atque
MEDICEA SIDERA
NVNCVPANDOS DECREVIT.

VENETIIS, Apud Thomam Baglionum. M DC X.
Superiorum Permiſſu, & Priuilegio.

Aufgaben zum Titelblatt:
- *Transkribiere den Text und übersetze ihn.*
- *Vergleiche das Titelblatt mit dem des Mundus Iovialis.*

Fig. 13.9 Example page from the text collection (Schlör 2008); the student has first to characterize the two pages and then compare the respective corresponding elements; specific interest is given to the aesthetic design and the role of the telescope

Here I present a few examples of worksheets to be used in class from the fourth year of Latin studies at a German high school. If you are interested in further and more detailed information, please turn to Wolfschmidt's and Gaab/Leich's books on Simon Marius (Wolfschmidt 2012; Gaab and Leich 2016) (Figs. 13.10 and 13.11).

Simon Marius – Der fränkische Galilei: „Deprehendi stellulas exiguas“
***Lernzirkel zu S. Marius, Mundus Iovialis, Praefatio*; Wh. der Nomina**

Arbeitsplan

smg

Schüler – 1

Vorname: ____________ Familienname: ________________

(Vorlauf Titel; Porträt, Praefatio 1–3)

Stunde:	Aufgaben:	erle-digt:
1	Übersetze *Mundus Iovialis*, Praefatio 4–6 (Inter *alia, quae ... discessum est*) ● Studiere die Aufgaben und Fragen zu diesem Text auf der Rückseite dieses Blattes; Notizen im Lateinheft! ● Trage die Vokabeln ein und lerne sie. ● GR: Studiere die Bildung der Substantive der 3. Deklination (z.B. *Grammadux*); bilde jeweils alle Kasus im Plural, lateinisch und deutsch: *similia vitra – ähnliche Linsen.*	
3	Übersetze *Mundus Iovialis*, Praefatio 11–13 (*Hac ratione ... admiratus sum*) ● Studiere die Aufgaben und Fragen zu diesem Text auf der Rückseite dieses Blattes; Notizen im Lateinheft! ● Trage die Vokabeln ein und lerne sie. ● GR: Studiere die („unregelmäßige“) Steigerung der Adjektive *bonus, malus, magnus, parvus, multi* (z.B. *Grammadux*); bilde jeweils Grundstufe, Komparativ und Superlativ im Nominativ Singular und Plural, deutsch und lateinisch: *a) Singular: das gute Fernrohr: bonum perspicillum => das*	

Fig. 13.10 (left): Example page from the classroom tuition project "Simon Marius, the Franconian Galileo: 'Tunc aspexi stellas'—A study group project on Simon Marius, *Mundus Iovialis*, Praefatio; revision of the noun forms" (Schlör 2010a); the student translates the text passage of the *praefatio* and does some language work

	bessere F.: melius perspicillum => das beste Fernrohr: optimum perspicillum; Plural: ... *b) der schlechte Handwerker – c) der kleine Stern (sidus)*	
4	Übersetze *Mundus Iovialis*, Praefatio 14–17 (*Post vero ... experirer*)	
	● Studiere die Aufgaben und Fragen zu diesem Text auf der Rückseite dieses Blattes; Notizen im Lateinheft!	
	● Trage die Vokabeln ein und lerne sie.	
	● GR: Studiere die Formen von *hic, haec, hoc* (z.B. *Grammadux*); bilde jeweils alle Kasus im Singular, lateinisch und deutsch: *haec opinio – diese Meinung*.	
5	Übersetze *Mundus Iovialis*, Praefatio 18–20 (*Ab hoc itaque ... discontinuavi*)	
	● Studiere die Aufgaben und Fragen zu diesem Text auf der Rückseite dieses Blattes; Notizen im Lateinheft!	
	● Trage die Vokabeln ein und lerne sie.	
	● GR: Studiere die Formen des Partizips Präsens (z.B. *Grammadux*); bilde jeweils alle Kasus im Singular, lateinisch und deutsch: Marius domum rediens – Marius, der nach Hause zurückgeht (zurückging).	
6	Besprechungsstunde; Wissenstest	

Fig. 13.10 (continued)

Simon Marius – Der fränkische Galilei: „Mundus detectus ope perspicilli"
***Lernzirkel* zu S. Marius, *Mundus Iovialis, Prima et Secunda Pars;* GR-Wh.**

Arbeitsplan

Schüler – 1 smg

	Vorname: ____________ Familienname: ________________		
Teil:	**Aufgaben:**	**erle-digt:**	
1	**Übersetze *Mundus Iovialis*, Prima Pars, 1–7 *(Prima pars ... diametri terrestris)***		
	● Studiere die Aufgaben und Fragen zu diesem Text auf der Rückseite dieses Blattes; Notizen im Lateinheft!		
	● Trage die Vokabeln ein und lerne sie.		
	● GR: Studiere die Formenbildung der Adjektive der 3. Deklination (z.B. *Grammadux* § 70); welche Besonderheit haben die Formen des Partizips Präsens Aktiv (z. B. *Grammadux* § 138)? Bilde alle Kasus in Singular und Plural (nur lateinisch) von: *universalis consideratio et theoria conveniens*		
2	**Übersetze *Mundus Iovialis*, De nominibus 13–14 (*Forsitan autem ... imprimis Ovidius*)**		
	● Studiere die Aufgaben und Fragen zu diesem Text auf der Rückseite dieses Blattes; Notizen im Lateinheft!		
	● Trage die Vokabeln ein und lerne sie.		
	● GR: Studiere die Bildung des Passivs (z. B. *Grammadux* § 92–98). Setze ins Aktiv und übersetze: a) tractabitur – b) determinabitur – c) explicabuntur – d) subiungetur – e) satisfieri – f) arguitur – g) captus est – h) comprehenduntur – i) ficta sunt – j) moventur		
3	**Übersetze *Mundus Iovialis*, De nominibus 15–17 (*Itaque non ... nunc secunda*)**		
	● Studiere die Aufgaben und Fragen zu diesem Text auf der Rückseite dieses Blattes; Notizen im Lateinheft!		
	● Trage die Vokabeln ein und lerne sie.		
	● GR: Studiere die Formen der Deponentien (z.B. *Grammadux* § 119). Setze ins Perfekt und übersetze: a) videor – b) versantur – c) arbitror – d) sequantur – e)		

Fig. 13.11 (right): Example page from the classroom tuition project "Simon Marius, the Franconian Galileo: 'Mundus detectus ope perspicilli'—A study group project on Simon Marius, *Mundus Iovialis*, Prima et Secunda Pars; general grammar revision" (Schlör 2010b)

4	Übersetze *Mundus Iovialis*, De septem phenomenis, bis IV (*Secunda pars ... remotioris tardiorem*)		
	● Studiere die Aufgaben und Fragen zu diesem Text auf der Rückseite dieses Blattes; Notizen im Lateinheft!		
	● Trage die Vokabeln ein und lerne sie.		
	● GR: Studiere die Steigerung der Adjektive (z.B. *Grammadux* § 75–76). Bilde den Positiv, Komparativ und Superlativ im Nominativ Singular Maskulinum, deutsch lateinisch, von diesen Adjektiven: *a) der weit entfernte: remotus – b) der schnelle: celer – c) der langsame: tardus – d) der gute*		
5	*Übersetze Mundus Iovialis, De septem phenomenis, V bis VII (Post plurimas ... sunt deprehensa)*		
	● Studiere die Aufgaben und Fragen zu diesem Text auf der Rückseite dieses Blattes; Notizen im Lateinheft!		
	● Trage die Vokabeln ein und lerne sie.		
	● GR: Markiere und übersetze nach den Regeln: a) *mundus^Iovialis (omnibus^mortalibus incognitus)* *b) assumptâ aquilae figurâ* *c) Iupiter quarum furtivo amore captus* *d) per amicitiam inter nos tunc initam* *e) planetae modo orientales modo occidentales a Iove exsistentes*		
6	Besprechungsstunde; Wissenstest		

Fig. 13.11 (continued)

References

Bemmer, Ariane (2015): Schule ist keine Lehrlingsfabrik", *Der Tagesspiegel*, 03.03.2015

Gaab, Hans; Leich, Pierre (2014): Marius' Replik auf Scheiner – Der Anhang zum Mundus Iovialis von Simon Marius, *Globulus – Beiträge der Natur- und kulturwissenschaftlichen Gesellschaft e. V.*, 18 (2014), Eichstätt 2014, pp. 11–14

Gaab, Hans; Leich, Pierre (eds.) (2016): *Simon Marius und seine Forschung*, Leipzig: Akademische Verlagsanstalt 2016

Marius, Simon (1614/1916): A.O. Prickard: The 'Mundus Jovialis' of Simon Marius, in *The Observatory* 39 (1916), pp. 367–381, 403–412, 443–452, 498–504

Marius, Simon (1614/1988): *Simon Marius – Mundus Iovialis, Die Welt des Jupiter, Die Entdeckung der Jupitermonde durch den fränkischen Hofmathematiker und Astronomen Simon Marius im Jahr 1609, lateinisch und deutsch.* Hrsg. und bearb. von Joachim Schlör, naturwiss. begleitet und mit einem Nachw. vers. von Alois Wilder, Gunzenhausen: Schrenk-Verlag 1988

Schlör, Joachim (1989): Neu erschienen und viel beachtet: SIMON MARIUS, MUNDUS IOVIALIS, All denen gewidmet, die heute Latein lernen, in *Jahresbericht des Simon-Marius-Gymnasiums Gunzenhausen 1988/89*, Gunzenhausen 1989

Schlör, Joachim (1995): MUNDUS IOVIALIS des Simon Marius von Sparkasse angekauft, in *Jahresbericht des Simon-Marius-Gymnasiums Gunzenhausen 1994/95*, Gunzenhausen 1995

Schlör, Joachim (2008): Simon Marius, der fränkische Galilei – Eine lateinische Textsammlung für die Mittel- und Oberstufe, Gunzenhausen 2008 (to be obtained through the webpage www.simon-marius.net, Secondary Literature)

Schlör, Joachim (2010a): Simon Marius – Der fränkische Galilei: 'Tunc aspexi stellas', Lernzirkel zu Simon Marius, Mundus Iovialis, Praefatio; Wiederholung der Formen der Nomina, Gunzenhausen 2010; to be obtained through the Marius-Portal (www.simon-marius.net)

Schlör, Joachim (2010b): Simon Marius – Der fränkische Galilei: 'Mundus detectus ope perspicilli', Lernzirkel zu Simon Marius, Mundus Iovialis, Prima et Secunda Pars; allgemeine Grammatik-Wiederholung, Gunzenhausen 2010; to be obtained through the Marius-Portal (www.simon-marius.net)

Schlör, Joachim (2012a): Pro 'In Europa' lege 'Io, Europa' – Simon Marius' Mundus Iovialis: Errata, Fehler, Missverständnisse, in *SMG-Jahresbericht 2011/12*, pp. 145–147, Gunzenhausen 2012

Schlör, Joachim (2012b): Mundus Iovialis – Die Welt des Jupiter, in: Wolfschmidt (2012), pp. 50–71

Wolfschmidt, Gudrun (ed.) (2012): *Simon Marius, der fränkische Galilei, und die Entwicklung des astronomischen Weltbildes*, Hamburg: tredition 2012

Zinner, Ernst (1942): Zur Ehrenrettung des Simon Marius, in *Vierteljahrszeitschrift der Astronomischen Gesellschaft* 77/1 (1942), pp. 23–75

Chapter 14
Prickard's English Translation of *Mundus Iovialis*, Completed

Albert Van Helden

Until 1903, Simon Marius had a poor reputation among the English. In *The History of Physical Astronomy*, Robert Grant had gone so far as to label him an "impudent pretender" to the discovery of the moons of Jupiter. This changed after the publication, in 1903, of Jean Abraham Chrétien Oudemands and Johannes Bosscha's "Galilée et Marius" in the *Archives Néerlandaises des Sciences Exactes et Naturelles*, when W.T. Lynn, a prolific author of popular books and articles on astronomy and biblical chronology, took up his cause. In a letter published in *The Observatory* of June 1903, Lynn reviewed the paper, repeating the conclusions of Oudemans and Bosscha, that Marius's observations were more accurate than those of Galileo, that Marius never claimed to have been the first to observe Jupiter's moons, and that this misunderstanding was based on the fact that Marius used the Julian, not the Gregorian, calendar. Lynn, who should have known better, also repeated the Oudemans and Bosscha's claim that anyone who had a telescope in 1609 "could hardly fail to notice the little stars near Jupiter, which continued observation would show were moving around the planet" (p. 255).

To Lynn's satisfaction, Oudemans (and Bosscha) had "fully established that there is no reason to doubt the genuineness of the observations of the satellites of Jupiter or the independence of his discovery," and he defended his position the subsequent issues of *The Observatory* and then turned his attention to Antonio Favaro, who had weighed in in favor of Galileo in the pages of *The Observatory* on one of Lynn's statements. Where Oudemans and Bosscha had hinted that perhaps a letter from Kepler to Galileo, with a supposed positive opinion of Marius's *Mundus Iovialis*, had been destroyed by Galileo or perhaps suppressed by editors off his works, Lynn

A. Van Helden (✉)
Rice University, Houston, TX, USA

University of Utrecht, Utrecht, The Netherlands
e-mail: a.vanhelden@uu.nl

H. Gaab, P. Leich (eds.), *Simon Marius and His Research*, Historical & Cultural Astronomy, https://doi.org/10.1007/978-3-319-92621-6_14

had written: “Professor Oudemans points out that reference *was* made to Kepler, but that the answer seems to have been suppressed.”

Antonio Favaro, the editor of the magisterial “Edizione Nazionale” of *Le opere di Galileo Galilei*, always jealous to protect Galileo’s reputation, could hardly afford to let Lynn’s insinuation pass unanswered (27, 1904, pp. 199–200). The editors of *The Observatory* attached Lynn’s reply to Favaro’s letter, in which Lynn stood fast. If Galileo had suggested writing to Kepler about Marius, in 1614 (although he was advised against this by his colleagues), he must have done so, and perhaps, therefore, Kepler preferred not to reply. At any rate, “Prof. Oudemans has clearly proved that Marius had for some time been in possession of a telescope quite equal to showing these satellites; and that being so, why should we doubt that one who is known in many ways to have been a careful and accurate observer did actually see them?” (27, 1904, p. 201).

Lynn then took J. J. Fahie to task for stating in his (then recently published) *Galileo, his life and works* (1903) that Marius had “arrogat[ed] to himself the merit of two of Galileo’s astronomical discoveries” (p. 46), and Galileo was not to get any justice in the amateur (and professional?) circles of astronomy in England. And for the time being, in England, Marius got equal billing with Galileo on the discovery of Jupiter’s satellites. As an illustration, I cite *The Planet Jupiter* (1958), Bertrand Peeke wrote:

> [U]ntil recently it was generally accepted that Galileo had the undisputed right to be named their discoverer. Historical research, however, has made out a strong case for the claim that Simon Marius should be regarded as an independent discoverer, in that he had probably been observing these objects at the same time as Galileo or even a month or two earlier and certainly, he stated, before the news had reached him that they had already been discovered. (p. 255)

This, then, was the dominant opinion (albeit always with a qualification) among British astronomers and historians of astronomy, and the English translation of *Mundus Iovialis* occupied a central place. It appeared in the pages of *The Observer* in 1916, done by Arthur Octavius Prickard. Now Prickard was no amateur astronomer, or professional historian: he was an Oxford don known for his research on the plays of Aeschylus and his well-known *Longinus on the Sublime*, of which at least 25 editions have appeared since its publication in 1906. I have not been able to discover why Prickard undertook this translation and can only assume that it was a casual project he undertook upon his retirement.

But Prickard did not translate all of *Mundus Iovialis*. He omitted everything that did not bear on the discovery of Jupiter’s moons and paid no attention to the tables at the end of the work. Nor did he translate the text Marius added after the publication of Christopher Scheiner’s *Disquisitiones Mathematicae* (1614) in which Scheiner called him a Calvinists and a pretender. I have completed Prickard’s translation, with the added sections printed in bold letters, to make sure that Prickard will not be blamed for any mistakes I may have made.

Chapter 15
Priority, Reception, and Rehabilitation of Simon Marius: From the Accusation of Plagiarism to the Marius-Portal as His Virtual Collected Works

Pierre Leich

Abstract Although he was the margravial court astronomer in Ansbach and was responsible for at least three notable achievements in the history of astronomy, the history of science has paid little attention to Simon Marius. Outside of the Netherlands, Marius may have been the first-ever professional astronomer to hear about the telescope, which was only presented to the world in September 1608. He claimed to have independently devised the Tychonic world system, and he observed the moons of Jupiter at roughly the same time as Galileo. Of course, there are reasons for this lack of attention. Marius could begin to exploit his early knowledge of the telescope only when his patron was able to purchase one. Tycho Brahe had already published his geo-heliocentric system years earlier, and whereas Galileo Galilei published his observations of the Jupiter moons in 1610, Marius first published in 1614, bringing down a charge of plagiarism on his head. The accusation of plagiarism was largely accepted by contemporary astronomers, and Marius had to wait until the early twentieth century before the quality of his telescopic observations and their independence were finally proved. His *opus magnum* was therefore translated comparatively late and only into very few languages. Many of his other writings and calendars remained difficult to access until the *Marius-Portal* was launched in 2014.

At least three facts about the margravial court astronomer Marius are remarkable: firstly, through his patron, Hans Philipp Fuchs von Bimbach, he was probably the first professional astronomer outside of the Netherlands to become aware of the telescope and its optical structure, and he also was one of the firsts to make highly significant astronomical observations with this new instrument. The interpretation of Marius's observations led to important arguments regarding the description of the

P. Leich (✉)
Simon Marius Society, Nuremberg, Germany
e-mail: pierre.leich@simag-ev.de

H. Gaab, P. Leich (eds.), *Simon Marius and His Research*, Historical & Cultural Astronomy, https://doi.org/10.1007/978-3-319-92621-6_15

world system, even though Marius, for empirical reasons, could never force himself to adopt Copernicanism.

Secondly, already in 1596, according to his own account, Marius had argued against the Ptolemaic system and supported a system, which then became known to him 1 year later as the Tychonic system. Based on the state of knowledge at the time, which was not sufficient to prove heliocentrism, the kinematically equivalent system was quite progressive. It could explain the apparent planetary loops, as well as such telescopically observable phenomena as the moons of Jupiter, the phases of Venus, and the varying planetary disks, without simultaneously contradicting certain assumptions of Aristotelian-Ptolemaic natural philosophy, as well as common sense.

And thirdly, Marius presumably started to observe the Jupiter system in the end of 1609. That October, Jupiter was a worthwhile object for every observational astronomer for several reasons: the beginning of its retrograde motion and the most northern position on October 9th, the closest approach between the Earth and Jupiter on the 6th, and the opposition on December 8th, as well as the long visibility around the winter solstice. As a calendar-maker, Marius knew about these circumstances, which makes the simultaneous discovery of the four Jupiter moons together with Galileo in January 1610 quite possible and plausible.

Unfortunately, Marius failed to announce his scientific priority for these three fields in time. We have to rely on his later reports, which were discredited by Galileo's weighty word, leading to Marius's being known as plagiarizer for three centuries until the accusations were shown to be untenable. The following article reconstructs the reception of his contributions from his first observations to the development of the *Marius-Portal* as his virtual "collected works."

In his main work, the *Mundus Iovialis* from 1614, Marius reports how his patron, Colonel (later General) Hans Philipp Fuchs von Bimbach (ca.1567–1626), made contact with a "Belgian" at the Autumn Fair in Frankfurt, "who had invented an instrument by means of which the most distant objects might be seen as though quite near" (Marius 1614; Marius 1614/1916/2019, preface). Since one of the lenses was cracked, he couldn't return home with a functioning instrument but instead brought a description of it back to Ansbach. This led to Marius's knowledge of this invention, only shortly after the spectacle-maker Hans Lipperhey from Middelburg approached Maurits van Nassau, the stadtholder and also commander-in-chief of the land and naval forces of the United Provinces, to show him a telescope.[1]

[1]According to Ernst Wohlwill's research in the Frankfurt archives, the Frankfurt Autumn Fair started on the 12th of September in 1608.Wohlwill II, 1926/1969, appendix III, p. 347. Hermann Grotefeld (1891) says it closed on 8th of September: "unser frauwen tag so der alden messe uslutet (Frankf. Stadtarch.), da am 8. Sept. die alte Messe ausgeläutet wird, die am 15. Aug. eingeläutet war," p. 69. Opposed to this we find in Johann Philipp Orths *Ausfürliche Abhandlung von den berümten zwoen Reichsmessen so in der Reichsstadt Frankfurt am Main järlich gehalten werden* (Orth 1765) the assumption that the Autumn Fair began first on 8th of September: "Aus welchem iezoangefürten gar deutlich zu erkennen, daß dieser vor alters übliche gebrauch, besonders bei der herbstmesse, daß sie auf Marienhimmelfart ein- und Mariengeburt ausgeläutet worden, ongeachtet diese messe, nach obangezogenen ser warscheinlichen gründen, von gar langen zeiten her, nach leztem festtage, erst ihren anfang genommen und noch iezo nimt, meistens beibehalten worden sei,

At the end of September, The Hague was the scene of an important peace conference, where the desire of the Republic of the Seven United Netherlands to gain sovereignty from Spain, free trade with the East and West Indies, as well as religious autonomy, led to a 12-year-long truce the following spring. We know from an unidentified spectacle-maker's letter of introduction from September 25, 1608, that the "Gecommitteerde Raden" (councilmen) of Zeeland requested their delegates to set up an audience with the Dutch chief negotiator. Soon afterward, a presentation took place, which Ambrogio Spinola (Marqué de los Balbases), commander of all Spanish troops in the Netherlands, also attended. On October 2, 1608, Lipperhey presented a patent application for his invention to the States General.

Marius's account corresponds with the fact that a patent was denied, because in October, Jacob Adriaan von Alkmaar (also Jakob Adriaanszoon Metius)—the brother of Adriaan Adriaansz, called Metius—and another unidentified spectacle-maker also raised patent claims. Obviously, they were already familiar with the knowledge about the potential of those lens systems,[2] which lends further credibility to the report about the "Belgian" at the Autumn Fair.

Lipperhey's achievement was less the invention of the telescopic principle, but more the application of improved manufacturing technique at the turn of the century and the introduction of a diaphragm, reducing the effects of the spherical and chromatic aberration (cf. Willach 2007) (Fig. 15.1).

The presentation became public knowledge via a French pamphlet with the title "Ambassades du Roy de Siam envoyé à l'Excellence du Prince Maurice, arrivé à La Haye le 10 Septemb.1608," reporting the arrival of Lipperhey as well as the appraisal by the attendant nobles. It was above all quickly obvious that because of the simple construction principle, secrecy in the matter was futile. Although this pamphlet has no publication date, the state of knowledge in the article narrows its release down to after October 5 but before the middle of October (cf. Zuidervaart/Zoomers 2008, p. 18). Paolo Sarpi (1552–1623), theologian from Venice, a friend of Galileo with a deep interest in optics, read the report,[3] according to his own statement, in November 1608.[4] With an underlying skepticism, he corresponded about this with, among others, Francesco Castrino and Jacques Badovere in Paris (Giacomo Badoer, around 1575–ca. 1620) who in turn informed Galileo, as noted by him in the *Sidereus*

gleichwie solcher noch, bis auf den heutigen tag, mithin über 250. jare, unverrückt vortwäret," p. 546. For further information of the dating, see the chapter contributed by Dick (Chap. 3) in this volume.

[2]Girolamo Fracastoro (ca. 1478–1553) described already in Fracastoro (1538) *Homocentrica* (sig.18^{v}) that two lenses in series let the image appear larger and closer. Other conceptual designs utilize a mirror and a lens.

[3]Or the reprint, which appeared by Jean Gazeau in Lyon in November 1608.

[4]Sarpi to Castrino, 9th of December 1609, in: Busnelli (1928, p. 1069); reprinted in Sarpi (1931, p. 15).

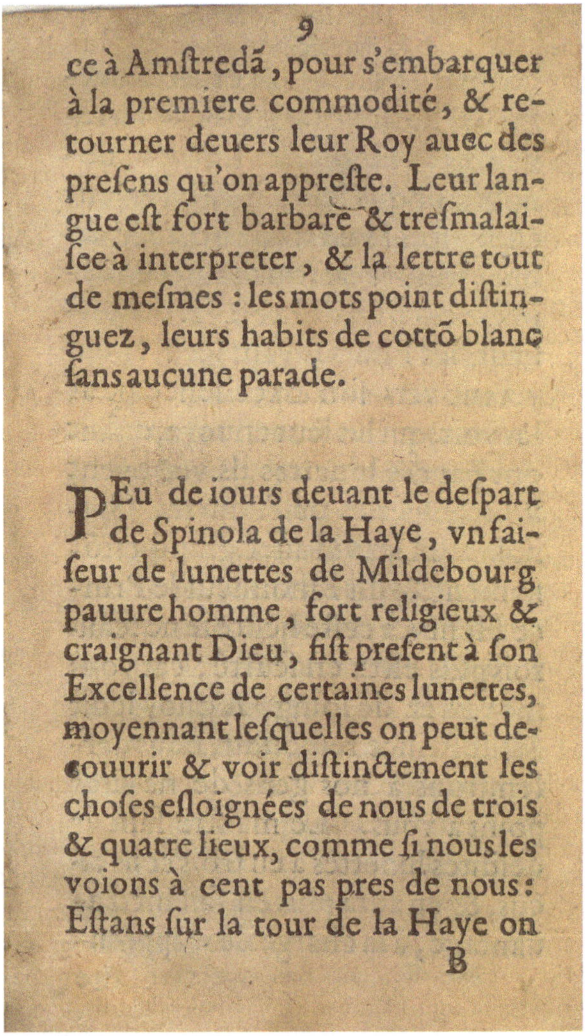
9
ce à Amſtredã, pour s'embarquer
à la premiere commodité, & re-
tourner deuers leur Roy auec des
preſens qu'on appreſte. Leur lan-
gue eſt fort barbare & treſmalai-
ſee à interpreter, & la lettre tout
de meſmes : les mots point diſtin-
guez, leurs habits de cottõ blanc
ſans aucune parade.

PEu de iours deuant le deſpart
de Spinola de la Haye, vn fai-
ſeur de lunettes de Mildebourg
pauure homme, fort religieux &
craignant Dieu, fiſt preſent à ſon
Excellence de certaines lunettes,
moyennant leſquelles on peut de-
couurir & voir diſtinctement les
choſes eſloignées de nous de trois
& quatre lieux, comme ſi nous les
voions à cent pas pres de nous:
Eſtans ſur la tour de la Haye on
B

Fig. 15.1 The pamphlet "Ambassades du Roy de Siam envoyé à l'Excellence du Prince Maurice, arrivé à La Haye le 10 Septemb. 1608" reports on pp. 9–11 (sig. $B1^r$–$B2^r$) about the presentation of an instrument with certain lenses, with which even the stars, usually invisible to us because of their small size and our weak visual acuity, can be seen. I thank the owner of a copy of this pamphlet for making it publicly available: Louwman Collection of Historic Telescopes, Den Haag

Nuncius.[5] Between May and July 1609, Galileo began to take these rumors about the new glasses seriously.

---

[5]Galilei (1610/1989, p. 37). Galilei reported this to Benedetto Landucci dated 29th of August 1609 also, and regarding the sequence of events, in slightly altered form in *Il Saggiatore* (1623).

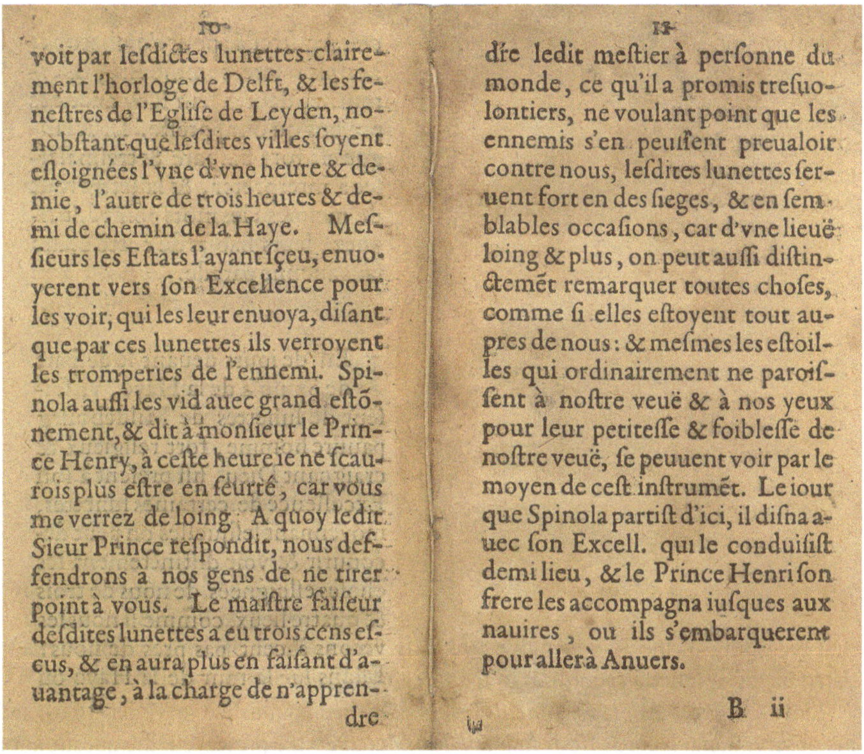
10

voit par leſdictes lunettes claire-
ment l'horloge de Delft, & les fe-
neſtres de l'Egliſe de Leyden, no-
nobſtant que leſdites villes ſoyent
eſloignées l'vne d'vne heure & de-
mie, l'autre de trois heures & de-
mi de chemin de la Haye. Meſ-
ſieurs les Eſtats l'ayant ſçeu, enuo-
yerent vers ſon Excellence pour
les voir, qui les leur enuoya, diſant
que par ces lunettes ils verroyent
les tromperies de l'ennemi. Spi-
nola auſſi les vid auec grand eſtõ-
nement, & dit à monſieur le Prin-
ce Henry, à ceſte heure ie ne ſçau-
rois plus eſtre en ſeurté, car vous
me verrez de loing. A quoy ledit
Sieur Prince reſpondit, nous def-
fendrons à nos gens de ne tirer
point à vous. Le maiſtre faiſeur
deſdites lunettes a eu trois cens eſ-
cus, & en aura plus en faiſant d'a-
uantage, à la charge de n'appren-
dre

11

dre ledit meſtier à perſonne du
monde, ce qu'il a promis treſuo-
lontiers, ne voulant point que les
ennemis s'en peuſſent preualoir
contre nous, leſdites lunettes ſer-
uent fort en des ſieges, & en ſem-
blables occaſions, car d'vne lieuë
loing & plus, on peut auſſi diſtin-
ctemẽt remarquer toutes choſes,
comme ſi elles eſtoyent tout au-
pres de nous: & meſmes les eſtoil-
les qui ordinairement ne paroiſ-
ſent à noſtre veuë & à nos yeux
pour leur petiteſſe & foibleſſe de
noſtre veuë, ſe peuuent voir par le
moyen de ceſt inſtrumẽt. Le iour
que Spinola partiſt d'ici, il diſna a-
uec ſon Excell. qui le conduiſiſt
demi lieu, & le Prince Henri ſon
frere les accompagna iuſques aux
nauires, ou ils s'embarquerent
pour aller à Anuers.

B ii

Fig. 15.1 (continued)

Telescopic Observations

Simon Marius, however, couldn't utilize his half-year edge, because a reconstruction with the help of lens makers from Nuremberg failed despite the use of plaster casts as templates. He only received a "Belgian" telescope (from the Netherlands and probably from Delft) in summer 1609 and began observing the heavens from then on. When he was allowed to take the instrument home at the end of November, he observed Jupiter for the first time. It stood in opposition to the Sun, and Marius soon discovered tiny stars behind and in front of Jupiter, and also in a straight line with it (Marius 1614, sig. 2^{v}; Marius 1614/1988, p. 38).

In print, Marius mentions the Jupiter moons for the first time in the 1611 published *Prognosticon auf 1612* and dates his observations "from the end of December 1609 to 1610" (*Prog. 1612*, sig. B1^{v}) and also "from the end of December 1609 to the mid of April 1610" (*Prog. 1612*, sig. C3^{r}). In the *Prognosticon auf 1613*, he gives distances and orbital periods of the Jupiter moons, and in 1614, his comprehensive representation of Jupiter's world was published in the *Mundus Iovialis*. In this, he identifies December 29, 1609 (Marius 1614, sig. 3^{r}, B4^{r}; Marius 1614/1988, p. 40, 86), Julian date (= January 8th 1610, Gregorian), as the day of his

Fig. 15.2 A calendar compilation with the *Prognosticon Astrologicum auf das Jahr 1612* by Simon Marius from the State Archives Nuremberg; Picture taken by: Norman Anja Schmidt

first recording of the Jupiter moons, exactly 1 day after Galileo, who had already dated his descriptions according to the Gregorian calendar.

Since Galileo's *Sidereus Nuncius* had already been published in March 1610, and his observational records from January 7 to March 29 are preserved, the question for the priority regarding the Jupiter moons is clearly decided in favor of Galileo, and there are also no known notifications in Marius's correspondence that are dated before March 1610 (Fig. 15.2).

Also in the question of priority of sunspot observations, Galileo sees himself as the victor, when he lets Salviati declare in the *Dialogo*: "The first discoverer and observer of the sunspots, and all other new celestial phenomena, was our friend from the Accademia dei Lincei. He discovered them in the year 1610 [...]."[6] In 1632, he dates his first observations in Padua and Florence as having been made in July or August 1610. His records of dark spots on the Sun are handed down at least from February to April 1612, and in August 1612, he mentions them briefly in the preface to his discourse about the movement of swimming bodies *Discorso intorno alle cose, che stanno in sù I'acqua*; a more detailed analysis was issued by the Accademia Lincei in March 1613.

Already in May and later in October 1612, Galileo had reacted to *Three letters about sunspots,* which Markus Welser had received from Christoph Scheiner in January 1612, printed under the pseudonym "Apelles." We can assume that Galileo's backdating was intended to secure his priority.

[6]*Dialogo sopra i due massimi sistemi del mondo, Tolemaico e Copernicano*, Galilei (1632, p. 337): "Fù il primo scopritore, & osseruatore delle Macchie solari, si come di tutte l'altre nouità celesti, il nostro academico Linceo; e queste scopers' egli l'anno 1610"

Simon Marius had observed sunspots since August 1611. In the *Prognosticon Astrologicum auf 1613*, he reports that Ahasvero Schmidner[7] had shown them to him and that in October, without being more precise, he "had thought of another way to look at the Sun with the named instrument without hurting his face during the daylight" (*Prog. 1613*, sig. A4^{v}). Marius never claimed priority for the sunspots and left this dispute to Galileo and Scheiner.

Simon Marius in the *Mundus Iovialis* (1614) and Johannes Kepler in the *Ephemerides Novae* (1618) both give credit for the discovery to Johann Fabricius, who was the first to publish about sunspots at the Autumn Fair 1611 with his *De maculis in Sole observatis*. Although he doesn't indicate, when he saw the sunspots for the first time, his father, David Fabricius, states in his *Prognosticon auf das Jahr 1615* that the observation took place on February 27, 1611.

But all of them were probably preempted by Thomas Harriot (1560–1621), who saw sunspots for the first time on December 8/18, 1610, although he never published the fact (cf. Zinner 1943, p. 266; Pilz 1977, p. 266). Kepler even observed a sunspot on May 17/27, 1607, but believed it to have been the planet Mercury.[8] A rotating Sun confirmed Kepler's idea of a flow of forces, with which the Sun entrained the planets. In any case, the spots contradicted the dogma of an immaculate Sun.

Marius's discovery of the Andromeda nebula on December 15/25, 1612, with a telescope was acknowledged. He was the first European to describe the pale gloss that couldn't be resolved into single stars, and he compared the nebula's gloss to a burning candle's light shining through a translucent horn (Marius 1614, sig. 4^{r}; Marius 1614/1988, p. 44). Of course, the true nature of the galaxy was still hidden from him, and he discussed a distant comet as a possible explanation. What was special about this description was not so much the "discovery" itself—a skilled observer may recognize M 31 or NGC 224 with the bare eye—but the fact that the astronomers of the day now viewed the nebulae as regular objects of their field. The earliest description can be found in a parchment manuscript on the fixed stars by the Persian astronomer Abd ar-Rahmann as-Sufi (Al Sufi), from around 964, which wasn't known in Europe yet (Strohmaier 1984, p. 50).

Sighting the Venus phases was also a head-to-head-race, whereby Galileo observed them more than a quarter of a year earlier than Marius. Galileo deposited his knowledge about the variation in luminosity in the form of an anagram with the ambassador of Tuscany in Prague, Giuliano de Medici (1574–1636), in a letter from December 11, 1610. After Galileo had become convinced of Venus's moonlike phases, he wrote to Christoph Clavius in Rome and to Bendetto Castelli (Galileo X, 1965, pp. 499–502, document 446 and pp. 502–504, document 447) about his discovery on December 30 and sent another letter to Giuliano de Medici with a decoding of the anagram on January 1, 1611. While Venus phases are also predicted in Ptolemaic astronomy, the visible progression, observed with a telescope for the

[7]About Schmidner, see the chapter by Gaab (Chap. 2) in this book.

[8]Cf. Kepler (1609); there have been pretelescopic observations in Europe as well as in China for two millennia.

first time, was only explainable with Venus orbiting the Sun. Of course, there was no information on how the Sun-Venus system moved, but a first "stage win" had been achieved.

The phases of Venus are mentioned by Marius in a letter to Nikolaus von Vicke in summer 1611. In print he first talks about them in the *Prognosticon auf 1612* (Dedication 3-1-1611) and delivers an appropriate illustration (*Prog. 1612*, sig. A3^r), so he at least receives the fame for the first published depiction. Galileo goes even further in the *Il Saggiatore* from 1623. In the dedication from June 1612 for his *Prognosticon auf 1613* Marius notes, that Mercury is lit by the Sun in the same manner as Venus and the Moon.[9] However, this may have been more of a conclusion than a clear observation. Early observations of the phases of Venus were also made by Thomas Harriot, the Jesuit astronomers in Rome and Nicolas-Claude Fabri de Peiresc.

Arguments for the Tychonic Astronomical System

The findings made possible by the telescope encouraged Marius in his belief of correctness of the Tychonic system, which, according to his own statement in the *Mundus Iovialis*, he had already been aware of in 1596, without knowledge of Tycho's work, and which he had disclosed to the consistory of Ansbach (Marius 1614, sig. C3^{r-v}; Marius 1614/1988, p. 98f.).

Unfortunately, no manuscript confirming this claim has survived. It should be noted, however, that from 1613 at the latest, Marius had better arguments than Galileo that Mercury, Venus, Mars, Jupiter, and Saturn orbit the Sun.

This assertion was empirically proven by the phases of Venus, although strictly speaking the relation to the Sun can only be deduced for Venus itself. But because of the similarity in motion, a transfer to Mercury may be quite legitimate. For a convinced Copernican like Galileo, there was no doubt that this was true for the outer planets. Nonetheless, a convincing argument would have been desirable. Marius found such an argument for Jupiter. He noticed that, assuming a uniform motion of the newly discovered moons around Jupiter, that this uniformity is maintained, respective not of the Earth but of the Sun. In the *Mundus Iovialis* he states that his observations "are proof, that Jupiter doesn't have the Earth as its center, but the Sun."[10]

Assuming what holds true for one outer planet also holds true for the other planets, then this necessarily results in a Tychonic or a Copernican system. Since both Marius and Galileo ignored Kepler's laws, and Marius found the star sizes

[9] *Prog. 1613*, sig. A3^r: "Das erste ist nun / dass ich auch vermerket / daß Mercurius gleicher weise von der Sonnen erleuchtet werde / wie die Venus vnnd der Monn".

[10] Marius (1614), sig. E2v; Marius (1614/1988, p. 132): "Verum observationes meæ ... Iovem non terram sed Solem pro centro habere"; similar Marius (1614), sig. B3v; Marius (1614/1988, p. 84).

resulting from the assumption of a Copernican system unacceptable,[11] the choice for Marius was clear.

Of course, this leads to his opposing the supporters of Copernicanism, and thus Marius finds himself historically on the losing side after the formulation of Newton's law of universal gravitation and the subsequent acceptance of heliocentrism. Since, to this day, the history of science tends toward a glorification of heroes, the historians of astronomy never really cared to take a closer look at Marius's research. This would have been more than justified, however, since already during Marius's lifetime there were several prominent criticisms of his work.

The first "review" of Marius's efforts can be found in Kepler's *Dioptrice* from 1611, in which he delivers a correct optic along with a theory for lens systems, thus creating the necessary conditions for the telescope's foundation as a legitimate research instrument. When the printing of Kepler's work was delayed, he used the time to publish three letters from Galileo with enthusiastic explanations in the foreword. Without Marius's knowledge, he published another letter from Marius to Nikolaus von Vicke, which Kepler had received as a copy from von Vicke in July.[12] It states:

> Third, I will prove, that Venus [just like the Moon] is illuminated by the Sun and that it becomes horned and half, which I have thoroughly observed and seen with the use of the Belgian Perspicill from the end of last year to April of this year.

This message is introduced by Kepler with the following words (Kepler IV, 1941, p. 353f.):

> Since in science the nations never have a lack of competitiveness or denigration and a lot of people here in Germany will demand proof from Germans, I hereby notify them about these subjects with the letter of a German, which also indicates that it wasn't malicious of Galileo to care for his own interest, and at least communicate his invention to us in Prague in the form of an anagram.

This statement understandably angered Marius, because now he also appeared as plagiarist in the case of the Venus phases, even though he only mentioned his research in a private letter and neither knew about Galileo's Venus observations, nor claimed priority. Marius almost never mentioned the phases of Venus again. Only in the *Prognosticon auf 1614* does he state that "The Venus [...] will remain morning star until January 1615, will now become crescent again, and turn its spikes to a decline" (sig. C3^{v}; similar sig. C2^{r}), and in the *Prognosticon auf 1627* he points out very briefly: "[...] whoever has a good Perspicill will find her shortly before dawn" (*Prog. 1627*, sig. D1^{r}).

Given that Kepler should have been aware of Marius's visit to Brahe in Prague in 1601, why did he make such a fuss about Galileo's priority in the question of the

---

[11] See the chapters by Graney and Leich (Chaps. 8 and 9) in this book.

[12] Kepler XVI (1954) (Letters 1607–1611), p. 383: "Tertio demonstrabo, Venerem non secus illuminari à Sole ac Lunam eamque corniculatam διχότομον etc. reddi, prout à fine anni superioris usque in Aprilem praesentis, à me ope perspicilli belgici multoties et diligentissimè obseruata et uisa est [...]".

phases of Venus. It must have been very disappointing for Kepler that someone, who made the same convincing observations as Galileo had, ultimately opts for the Tychonic system. In the letter to Vicke, Marius explains programmatically (Kepler XVI, 1954 (Briefe 1607–1611), p. 382f.):

> First, I assert the immobility of earth, whereby personal issues are put aside, in fact only the arguments against Copernicus's justifications are examined, which in our times, are endorsed and seriously acknowledged as correct by Kepler and the Paduan mathematician Galileo.

Marius's citation of the Holy Bible as an argument as well as the size of the celestial bodies, the phases of Venus, and the "Jovial planets" must have been a no-go for Kepler, turning Marius into an even worse enemy than those who only knew about the new telescopic observations by hearsay Marius must have been particularly annoyed to discover that Kepler, in his *Dissertatio cum Nuncio Sidereo nuper ad mortales misso à Galilaeo Galilaei Mathematico Patavino* (Kepler 1610), without having made any own observations of the moons so far, committed to the authenticity of Galileo's observations and the credibility of his conclusions. This failure of Kepler to personally verify the observations was quite unfortunate for Marius, who was, after all, the only astronomer with comparable observational records to Galileo's nonetheless opposed Copernicanism.

Marius also didn't have any more luck with Christoph Scheiner (1573–1650), although he was a supporter of the Tychonic system and someone who had suffered attacks by Galileo. In 1614, Scheiner and his student Stefan Locher published the work *Disquisitiones mathematicae* (Mathematical Investigations) in Ingolstadt. In it they discussed the arguments for and against the Ptolemaic, Tychonic, and Copernican systems. Although Scheiner doesn't doubt the truth of the observations, published in the *Sidereus Nuncius*, and even mentions that he has observed the phases of Venus and sunspots together with his student Johann Baptist Cysat since March 1611, he doesn't follow Galileo's interpretation. He only accepts Copernicanism as a hypothesis and favors the Tychonic system. In Chapter 39, "De Jove" Scheiner speaks about the discovery of the moons of Jupiter and notes on page 78[13]:

> The admirable company of Jupiter, which Galileo, the outstanding Italian mathematician, discovered with the greatest skill a few years ago (in vain and in contrary a Calvinist tries for the first time in this year to unashamedly convince) had rightly fascinated the whole host of astronomers.

Although Scheiner relies on Marius a few pages further (p. 83) when he talks about the latitudinal deviations of the Jupiter moons, he doesn't give him any credit and still denounces Marius as plagiarist when it comes to the moons of Jupiter (Gaab and Leich 2014).

[13]"Jovis Comitatus admirabilis ab hinc paucis annis D. Galilaeo Mathematico Italo praestante solertissime primum detectus, (frustra enim feroque nimis contrarium Calvinianus quidem hoc primum anno & importune satis persuadere conatur) in sui admirationem totum Astronomorum gymnasium merito rapuit."

Scheiner sent a copy of the *Disquisitiones* to Galileo and called his attention to the *Mundus Iovialis*. He also added that Galileo would probably be amused by Marius's arrogance and certainly would correct his mistakes. Galileo didn't reward this service but instead mocked Scheiner in his *Dialogo* of 1632, where he disputes Scheiner's arguments against the Earth's rotation on the second day.

Galileo only entered the fray in his *Il Saggiatore* in 1623 but then all the more fiercely[14]:

> Four years after the publication of my Starry Message, the same man was unabashed enough, after further years of research, to make himself the author of things that I myself had already described and published years ago. Under the title 'The Moons of Jupiter' he blatantly claimed that he discovered how the Medici planets orbit Jupiter before me.

Most certainly Galileo was uncomfortable with Marius circulating the competing term "Brandenburger Stars" for those objects, which he had dedicated to the house of Medici. Even if Marius was partially to blame, two of Galileo's accusations are at least exaggerated. Galileo assumed that Marius's claim of December 29, 1609, as the date of his first record was evidence of malicious intent. First, it should have been obvious for every scholar during the period of sectarian disputes that a publication in a Protestant domain makes use not of Gregorian but of Julian dates. In the *Mundus Iovialis*, Marius mentions in the same sentence where he noted the date December 29, 1609, that he uses the Julian calendar,[15] which is further clarified by an example (Marius 1614, sig. F3^{v}; Marius 1614/1988, p. 152). In his methodical notations, he explains his use of Prutenic tables,[16] and even a double entry can be found (Marius 1614, sig. D4^{r}; Marius 1614/1988, p. 118), with his calendars having columns both for was usual for that time.

Galileo also rewarmed a proven and already punished plagiarism by Baldessare Capra. He had been a student of Marius during his time in Padua. In 1607, he published *Usus et Fabrica Circini cuiusdam Proportionis*, which turned out to be a Latin translation of Galileo's manual for a military compass from 1606.[17] Although Galileo didn't invent the proportional compass but only improved it, Capra never credits Galileo and instead claimed to be the inventor. However, Galileo is wrong, when he implicitly accuses Marius of the authorship of this plagiarism and added, "he instantly departed to his homeland, maybe to avoid punishment, and left his

---

[14]"Questo istesso, quattro anni dopo la publicazione del mio Nunzio Sidereo, avvezzo a volersi ornar dell'altrui fatiche, non si è arrossito nel farsi autore delle cose da me ritrovate ed in quell'opera publicate; e stampando sotto titolo di Mundus Iovialis etc., ha temerariamente affermato, sé aver avanti di me osservati i pianeti Medicei, che si girano intorno a Giove." *Il Saggiatore*, Rom 1623, sig. A2^{r}/p. 3.

[15]Marius (1614), sig. C2^{v}; Marius (1614/1988, p. 96): "juxta Calendarium Iulianum".

[16]Marius (1614), sig. E4^{r}; Marius (1614/1988, p. 138): "Annum & diem ordior à media nocte pręcedente Calendas Ianuarij, more Romano, quem etiam Reniholdus in suis tabulis retinuit".

[17]"Le operazioni del compasso geometrico, et militare"; in Galilei II (1965, pp. 363–424). It follows Capras's text (pp. 425–511) and "Difesa di Galileo Galilei ... Contro alle Calumnie & imposture di Baldessar Capra." Venestra: Tomaso Baglioni 1607 (pp. 513–601).

student behind,"[18] because Marius had already left Italy in summer 1605, and there is no proof that Marius had been the author.

The events are described more neutrally by David Fabricius. He considered it to be[19]

> memorable, that Galileo Galilei, an Italian, with the help of this telescope discovered four little planets around and next to Jupiter, which no astronomer before knew about or mentioned. What Herr Simon Marius so far has also observed of these new little planets' orbits alongside longitude and latitude, he will hopefully publish and make for himself a laudable name.

For almost three centuries, it was left to personal taste of the reader, to view Marius as a plagiarist or the discoverer of the moons of Jupiter—both for equally inadequate reasons. Most authors, including Giovanni Riccioli, Pierre Gassendi, Giovanni Domenico Cassini, Christiaan Huygens, François Arago, and Antonio Favaro side with Galileo. When the Protestant estates of the Holy Roman Empire adapted the Gregorian calendar in 1700, knowledge of the dating problem obviously faded, and, especially in the German-language area, voices were raised more frequently in Marius's favor. Johann Gabriel Doppelmayr noted in 1730 that Marius "set eyes [on the moons of Jupiter] somewhat earlier than Galileo [...]" (Doppelmayr 1730/1972, p. 90, footnote y), and in 1850 Alexander von Humboldt mentions the court mathematician, "who saw the Jupiter satellites nine days earlier than Galileo" (Humboldt 1850, p. 315).

Rehabilitation

In this assessment, two questions need to be handled separately. The first is to whom the priority should be attributed; the second questions the independence of the respective researches. While the first point, as elucidated above, is resolved with customary and reasonable standards in the sciences with the publication of the *Sidereus Nuncius*, the second point is addressed by the Société Hollandaise des Sciences. On January 1, 1900, they issued a prize question: to what extent Galileo's accusation of plagiarism against Marius was justified.

Only one answer was submitted. Josef Klug, a high school teacher from Nuremberg, sent his extensive answer in favor of Galileo on 235 pages with a lot of interesting details, though his assertions were unfortunately highly tendentious and even almost deluded at times. His contribution was therefore rejected, and the jury of the Netherlands Academy of Sciences decided to answer the question themselves. After an in-depth examination, the decision of Jean Abraham Chrétien Oudemans

[18] "[...] e subito, forse per fuggir il castigo, se n'andò alla patria sua, lasciando il suo scolare, come si dice, nelle peste; contro il quale mi fu forza, in assenza di Simon Mario [...]." *Il Saggiatore*, sig. A2^r/p. 3.

[19] Fabricius, David: *Prognostikon auf 1615*.

Fig. 15.3 Jean Abraham Chrétien Oudemans (in the white suit), in the year 1874 [the ToV was in 1874, not 1875], as a participant in the Netherlands transit of Venus expedition on Réunion; next to him Friedrich-Ernst van de Sande Bakhuyzen (1848–1918, younger brother of H. G. van de Sande Bakhuyzen, the director of the observatory Leiden, who prepared the expedition, but didn't participate himself); on the right side of the man with the helmet: the astrophotographer Pieter Jan Kaiser (1838–1916, son of the former director of the observatory Leiden, Frederik Kaiser) and Hilaire Gabriel Bridet (1818–1896), the military meteorologist and head of the local observatory. Picture: Bibliothèque départementale de La Réunion via Wikimedia Commons

(1827–1906), Hendricus Gerardus van de Sande Bakhuyzen (1838–1923), and Jacobus C. Kapteyn (1851–1922) was unambiguous: Galileo's accusation does not have any basis in fact.[20]

Since Antonio Favaro,[21] the editor of Galileo's collected works, showed surprise that anybody could doubt the authority of Galileo's outrage toward Marius, Oudemans and Johannes Bosscha (1831–1911) decided in 1903 to publish the arguments that had led to such an unexpected judgment. This work, entitled "Galilee et Marius," may be viewed as Marius's rehabilitation (Fig. 15.3).

The authors showed Galileo's accusations to be invalid. Marius had produced independent work that should be taken seriously and that in some respects even

[20]Oudemans and Bosscha (1903, p. 115): "les accusations de Galilée n'ont aucun fondemet sérieux".

[21]*Le Opere di Galileo Galilei*, Edizione Nazionale sotto gli auspicii di Sua Maestà il Re d'Italia, I–XX, Firenze (G. Barbèra Editrice) 1890–1909.

Fig. 15.4 Johannes Bosscha in the year 1911; Picture: nvt (Prins der Geillustreerde Bladen) via Wikimedia Commons

surpassed that of Galileo. Their arguments are based on Marius having come closer to the modern recalculated data than Galileo and his having interpreted some phenomena correctly in contrast to Galileo.

For example, Galileo traced the differences of the moons' brightness to an atmosphere on Jupiter that weakens the moons' reflected light. Marius instead believed "that their different size is due to their different position to Jupiter and the Sun in relation to the Earth" (Marius 1614, sig. D2^r; Marius 1614/1988, p. 110) and is therefore an effect of the illumination by the Sun and the reflection from Jupiter. Marius also calls attention to the fact that if Galileo's assumption of an atmosphere around Jupiter is right, then "the fourth moon could never be seen close to Jupiter from Earth's distance" (Marius 1614, sig. D1^r; Marius 1614/1988, p. 106), although it is weakest at the largest elongation (Fig. 15.4).

"Furthermore the German astronomer gave an explanation for the latitudinal deviation of the moons, for which Galileo couldn't present plausible arguments."[22] "Galileo tried for the first time to give an explanation of this phenomenon only in his offensive in 1623 against Marius; this is, however, inaccurate. The moons' orbits are by no means parallel to the ecliptic. Granted, the angle of Jupiter's orbit contributes to this impression, but only insignificantly. On the contrary, Marius attributed this phenomenon to the angle of the moons' orbits respective Jupiter's orbit, but made the mistake of viewing this angle as always oriented in the same direction towards

[22]Oudemans and Bosscha (1903, p. 139): "De plus, l'astronome allemand avait donné une explication du mouvement en latitude, dont Galilée n'avait donné aucune raison plausible ..."; cf. p. 147f. and 162.

the Sun" (Oudemans and Bosscha 1903, p. 148f.). A few pages further, the authors explain that "For this reason we think, that the observations by MARIUS regarding the latitudinal deviation of the moons were not only entirely new for that time, but they also were as accurate as they could be with the available limited resources" (Oudemans and Bosscha 1903, p. 154). As far as the orbital periods of the moons are concerned, Oudemans/Bosscha determined: "The values given by MARIUS are sufficient enough to prove them to be the results of his own observations."[23] They postulate a development of the values given in the *Prognosticon auf 1613* in comparison to those in the *Mundus Iovialis*: "Predominantly in MARIUS's values for the first and the fourth moon remarkable advances in estimating the periods can be found" (Oudemans and Bosscha 1903, p. 156).

Also, during his observations, Galileo, as already mentioned above, did not take into consideration the reference to the Jupiter system's movement respective to the Sun. Regarding this matter, Oudemans/Bosscha state the following about Marius: "By contrast he indeed took the parallax into account, meaning the difference between the directions Jupiter—Sun and Jupiter—Earth. This is never mentioned in GALILEO's writings and proves the autonomy of MARIUS's work."[24]

Since the *Mundus Iovialis* was hard to come by at that time and only accessible in Latin, Oudemans/Bosscha introduce Marius's statements in detail and conclude "that MARIUS not only preempted GALILEO with the publication of numeric values, but also his comparable measurements were more accurate than those available to GALILEO at the time of publication of MARIUS's work" (Oudemans and Bosscha 1903, p. 161).

Oudemans/Bosscha didn't forget to discuss the apparent planar appearance of the stars, induced by scintillation, diffraction, and the optical properties of the early telescopes. They regret that Marius interpreted what today we call Airy-discs as real discs of stars, which explains his rejection of the Copernican assumption of the huge distance of the stars and instead favored the Tychonic system. Although the authors at first doubted the visibility of the "false" discs with Marius's instruments, they used a replica to convince themselves that the discs were clearly visible to him and therefore accepted Marius as the discoverer of these "false discs" (Oudemans and Bosscha 1903, p. 164f.).

In an appendix Oudemans reviews the accuracy of the tables and draws the conclusion: "When omitting the year 1609, for which no observations were made, then the preceding review shows that the biggest mistake was made with the third moon, whose deviation is roughly equal to 84" in 1610. All the other mistakes are well under $1'$ of arc. Indeed, one cannot demand more precise tables, made by sketches without the use of micrometers and tables constructed from lunar eclipses"

[23]Oudemans and Bosscha (1903, p. 155): "Mais les valeurs mêmes publiées par Marius suffisent pour démontrer qu'elles sont bien le résultat de ses propres observations."

[24]Oudemans and Bosscha (1903, p. 157): Par contre, il eut égard à la 'parallaxe', c'est à dire à la différence entre les directions Jupiter-Soleil et Jupiter-Terre, dont aucune mention n'avait été faite dans les écrits de Galilée, ce qui à lui seul offre une preuve de l'orginalité du travail de Marius."

(Oudemans and Bosscha 1903, pp. 168–172, here p. 172). Of course Marius—as well as Galileo—didn't take elliptical orbits, Kepler's second law, the gravitational attraction of Saturn, or the speed of light into consideration, but the discrepancies thus caused are minor. Bosscha, in a second appendix, which he himself edited, tried to check Galileo's method by calculation, as formulated in his *Sidereus Nuncius* and with the use of replicas of Marius's and Galileo's instruments. He summarized: "One may not be surprised when Marius remarks that he wasn't able to apply Galileo's method for measuring the angular distances of the stars. One rather wonders if Galileo ever tried to apply his recommended method himself" (Oudemans and Bosscha 1903, pp. 173–179, here p. 177f.).

In appendix V, about the alleged observation of an eclipse of one of Jupiter's moons by Galileo on January 12, 1609, Oudemans/Bosscha remark: "The comparison of the moons' positions given by Galileo with the calculated ones shows that Marius indeed can state Galileo's first observations to be very inaccurate. Moon II moves further away than the third, and moon I couldn't be seen at all" (Oudemans and Bosscha 1903, pp. 187–189, here p. 189).

The examination by Oudemans and Bosscha doesn't make Marius the first discoverer of the moons of Jupiter, but, 280 years after Galileo's accusations, it secures justice regarding the recognition of Marius's precise observations as well as his theoretical knowledge.

In reaction to this analysis, Josef Klug felt obliged to publish his rejected thesis, "Simon Marius from Gunzenhausen and Galileo Galilei. An attempt to determine the true discoverer of the moons of Jupiter and their periods."[25] After the death of Oudemans, in the absence of additional facts, Bosscha supplemented their first work with a second article (Bosscha 1907), "Simon Marius: Réhabilitation d'un astronome calomnié," illustrating the inadmissibility of Klug's argument.

The German chemist Emil Wohlwill (1835–1912) took no notice of either French article in the Dutch journal. His long-standing engagement with the history of science, especially with Galileo, drew him, however, to Klug's comments. This impelled him in 1910 to write the text "The deception of Simon Marius from Gunzenhausen," published as appendix III of the second edition of *Galilei und sein Kampf für die copernicanische Lehre*.[26] In the preface of the revised publication of 1969, Hans-Werner Schütt writes, considering Wohlwill's account of Galileo's confrontation with the Church, that today it seems "as if the picture's contours are

[25] "Simon Marius aus Gunzenhausen und Galileo Galilei. Ein Versuch zur Entscheidung der Frage über den wahren Entdecker der Jupitertrabanten und ihrer Perioden" was ready in 1904 and published in *Abhandlungen der mathematisch-physikalischen Klasse der Königlich Bayerischen Akademie der Wissenschaften*, 22 (1906), II. Abt., München 1906, pp. 385–526.

[26] Wohlwill's first volume has the title "Bis zur Verurteilung der copernicanischen Lehre durch die römischen Kongregationen" (Till the condemnation of the Copernican doctrine by the Roman Congregation), Hamburg/Leipzig 1909, and appeared posthumously in Vol. 2 "Nach der Verurteilung der copernicanischen Lehre durch das Dekret von 1616" (Following the condemnation of the Copernican Doctrine through the Decree of 1616), Leipzig 1926, with appendix III, pp. 343–426.

too hard" and that "the problems appear today more differentiated, than they were presented by Wohlwill."[27] It is hardly surprising that Wohlwill took the side of Galileo and draws upon the correspondence by Marius, Kepler's writings, and the plagiarism of Capra.

Already in 1901, Antonio Favaro[28] called attention in the Italian language area to the prize question by the Société Hollandaise des Sciences, followed by J. C. Rodolphe Radau (1835–1939) in 1904 in the *Bulletin Astronomique*. And in the English language area, William Thynne Lynn (1835–1911) publicized the analysis by Oudemanns/Bosscha through four articles (cf. Lynn 1903a, b, 1904, 1909) and strengthened the interest in the *Mundus Iovialis*.

With the translation of Marius's main work into English by Arthur Octavius Prickard (1843–1939), published 1916 in *The Observatory*, a more objective and international analysis was possible (cf. Marius 1614/1916, translated is to sig. F3^{v}). Prickard, knowing Oudemans, Bosscha, and Klug, and having heard of the second edition and its epilogue, written in 1615,[29] also introduced biographical clues the following year in *Note on "Simon Marius" and the "Mundus Jovialis"* (cf. Prickard 1917). Regarding the question of whose observations were made earlier, Pickard concluded: "It is probably hopeless, without fuller evidence, to attempt to come to a conclusion upon the main question which should be fair to Marius" (Prickard 1917, p. 121).

New facts came to light through Ernst Zinner (1886–1970) only in 1942. The former director of the Dr. Karl Remeis-observatory in Bamberg (1926–1953) wrote a rehabilitation (cf. Zinner 1942), including not only a survey of letters and printed works but also, after extensive research in libraries, comprehensive passages from letters and rare prognostica, in which Marius reports again and again about his telescopic observations. Zinner closes his article with a call for a German translation of the *Mundus Iovialis* (Fig. 15.5).

Joachim Schlör (*1946; cf. Marius 1614/1988), a Latin teacher at the Simon-Marius-High School in Gunzenhausen, undertook this task with his Abitur course over the period 1985–1987, and in 1988 his physics teacher colleague, Alois Wilder, joined the effort, providing scientific support. The bilingual edition was also a facsimile publication of the main work, based on the copy of the State Library of Ansbach. As a result, these efforts significantly bolstered the active interest in Marius and his work in Germany. (The author of this article himself derived great

---

[27]Wohlwill I, 1926/1969, unnumbered, penultimate page of the foreword: "Es scheint uns heute, als sei dieses Bild in etwas zu harten Konturen gezeichnet. Die Kirche etwa und ihre speziellen-auch internen-Probleme wirken heute differenzierter, als Wohlwill sie dargestellt hat".

[28]Cf. Favaro (1901), English translation: Favaro (1904); cf. also Favaro (1917–1918).

[29]The copy in the Herzog August Library Wolfenbüttel is a second edition with a three-page appendix, which follows directly on from the errata. In this Marius defends himself against Scheiner's attacks. Cf. Gaab and Leich (2014). In September 2017, the author was able to inspect the copy in the Library of the Evangelisches Predigerseminar in Wittenberg, which also contains this appendix.

Fig. 15.5 The translator of the *Mundus Iovialis*, Joachim Schlör (second from the left), at a festive event in honor of Marius in Ansbach with Alois Wilder, Pierre Leich, Günther Löffladt, and Thony Christie on February 18, 2014; Picture taken by: Rudolf Laux

benefit from this translation into his mother tongue, and he is indebted to the translators for their efforts, contributing to his study of Marius.)

In 2009, the International Year of Astronomy, Simon Marius was naturally already included alongside other Franconian astronomers, including Regiomontanus and Clavius, in the traveling exhibition "Astronomy in the Nuremberg Metropolitan Region—History, Research and Public Observatories," which toured through 22 cities in Northern Bavaria. The scientific stage play "SCIENCE-Fiction: The Kepler conference"[30] by the Nuremberg author Chriska Wagner—a cultural highlight of the German contributions—gave Marius his stage debut, and the passage about the moons of Jupiter was taken from the *Mundus Iovialis*.

For the Simon-Marius-High school in Gunzenhausen the Cauchy-Forum-Nuremberg conceived the conference "Simon Marius at the Turning Point of Astronomy." The 2012 anthology *Simon Marius, der fränkische Galilei, und die Entwicklung des astronomischen Weltbildes* (Wolfschmidt 2012), which also contained a bibliography of the works and secondary literature (Leich and Wolfschmidt 2012), was based on these lectures. For the astronomy portal *Astronomie in Nürnberg*,[31] operated by

[30]On the stage Anne Devries played the Cosmic Spirit "Canis Marsi"; Duke Meyer danced projections of the conference as "Prof. Din. Acreaur," an expert from the planet Jupiter; Chriska Wagner played "Interstellar Geist" from Kepler's early science fiction novel "Somnium–der Traum"; and Sigi Wekerle played Johannes Kepler's favorite snowflake "Nix Nicis." Staging was by Ingo Schweiger.

[31]www.astronomie-nuernberg.de, Menu "Geschichte".

the Nuremberg Astronomical Society (NAG), Hans Gaab compiled a comprehensive "History of astronomy in Nuremberg," among those 500 and more astronomers, mathematicians, globe makers, etc. There is also an entry about Marius that contained at the end of 2012 a complete overview of his publications, including the calendars. Meanwhile, this overview has been expanded and moved to the *Marius-Portal*.

The *Marius-Portal* as a Virtual *Collected Works*

Nevertheless, the situation remained unsatisfactory, insofar as only a few of the 6 works, 58 calendars,[32] and 6 smaller texts were traceable, let alone easily accessible. Rectifying this situation became the goal of the *Marius-Portal* through a compilation and verification of the complete works, secondary literature, media coverage, lectures, and websites, all electronic sources about Marius, together in a central multilingual website accessible to international sciences. Crucial to this was to win over the owners of the original works for a digitization and the willingness to place those digital copies at the *Marius-Portal*'s disposal or provide corresponding links.

In order to create a framework and to generate the necessary publicity, a PR strategy was planned, using the 400th anniversary of Simon Marius's main work and inviting numerous cooperation partners to collaborate. Eventually 66 cooperation partners contributed to the anniversary carrying out 60 events—mainly lectures, presentations, tours, some exhibitions,[33] and a conference. The resonance from seven press releases—four of those in English—was over 250 reports at home and

[32]Five calendars have disappeared and one is only preserved as fragments.

[33]From January 13 to February 28 in the Gotische Halle des Stadhauses in Ansbach, the exhibition "Sonne Mond und Marius" (Sun, Moon, and Marius) took place. The State Library Ansbach (Schlossbibliothek) presented in cooperation with the town of Ansbach and the Ansbach Art Society, with items loaned from the town archives and the Margravial Museum "Die 4 Monde des Jupiter—die Entdeckung des Simon Marius in Anbach 1614–2014" (The 4 Moons of Jupiter – The Discovery of Simon Marius 1614–2014) from February 3 to March 4. From 17th to 30th of September lured the exhibition visitors "Fränkische Astronomen der Frühen Neuzeit" (Franconian Astronomers of the Early Modern Period) into the University Library Erlangen-Nuremberg. In the 20th of September, during the Marius Conference in the Planetarium Nuremberg, the roll-up displays from the traveling exhibition "Astronomie in der Metropolregion Nürnberg – Geschichte, Forschung und Volkssternwarten" (Astronomy in Nuremberg Metropolitan Region – History, Research and Public Observatories) for the International Year of Astronomy 2009 were on view, and Rudolph Pausenberger demonstrated the observational possibilities of the early seventeenth century with an exhibit on the Moons of Jupiter. For the 21st of February, the students of the Simon Marius Gymnasium had prepared an exhibition on the life and work of Marius. The student of the Beruflichen Oberschule Ansbach (BOS) (Occupational High School) developed the exhibition "Zum Jupiter aufblicken" (Looking Up to Jupiter), which could be viewed from March to July. Almost all of the year, "Sonne, Mond und Marius – Austellungen von Kinderkunstwerk" (Sun, Moon, and Marius – Exhibition of Children's Art) from the Russian Youth Art School Obraz in

abroad. A special highlight was the naming of an asteroid by the International Astronomical Union (IAU), for which Olivier Hainaut, Kurt Hopf, Hans-Ulrich Keller, and Gero Rupprecht campaigned.

It was the author's pleasure to be active as initiator, and, after preliminary consideration, he together with Hans Gaab and Klaus Matthäus visited the former director of the State Archives of Nuremberg, Gerhard Rechter (who sadly died and far too early). Together we inquired about the willingness of the State Archives to be a partner for the planned Marius conference in a follow-up to the International Year of Astronomy, and the feasibility of the digitization of the Marius holdings. Although the finances couldn't be clarified at this first meeting, the State Archives supported the development of the *Marius-Portal* from the beginning; this support was continued seamlessly by the new director, Peter Fleischmann.

In the middle of 2011, a small task force developed a rough concept, which was then formulated in 2012. The domain simon-marius.net was registered in March 2012, and by the middle of the year, the portal concept was complete. Norman Anja Schmidt took over the administration, and the Nuremberg agency Kaller & Kaller developed the visual design.

In March 2012 it was decided that the Nuremberg Astronomical Society (NAG) should take over the management of this project. As central coordinator for all IYA activities in the Nuremberg Metropolitan Region, which received nationwide recognition, they had gained experience in regional cooperation (Fig. 15.6).

Having determined the owners of the still extant Marius works and having compiled the majority of the secondary literature, in summer 2013, the NAG made contact with the German archives, libraries, and publishers. In most cases, the institutions were very forthcoming, whereby, because of the large quantities of material that they provided, the State Archives Nuremberg, the Municipal Library Nuremberg, and the Bavarian State Archives must receive a special mention.[34] Many calendars of which only a single copy is known worldwide were digitized. In the course of our work, we became aware of the existence of further copies of the *Mundus Iovialis*, raising the number of known copies to 34.

The Simon-Marius-Anniversary 2014 started with opening events in Nuremberg and Ansbach, the Marius-Day in Gunzenhausen, and activation of the *Marius-Portal*

Protvino near Moscow was on display. A large portrait of Marius remains in the Cosmonaut museum in Moscow through the mediation of the intercultural magazine *Resonanz*.

[34]The *Marius-Portal* is much obliged to Biblioteca Nazionale Centrale di Firenze, Bibliothek des Evangelischen Predigerseminar, Wittenberg, Deutsches Museum, Emmy Riedel Buchdruckerei und Verlag, ETH-Bibliothek Zürich, Franckh-Kosmos Verlag, Herzog August Bibliothek Wolfenbüttel, Herzogin Anna Amalia Bibliothek Weimar, Landesbibliothek Coburg, NABI Verlag, Private collection Wolfgang Marius, Österreichische Nationalbibliothek, Ratsschulbibliothek Zwickau, Staatliche Bibliothek Ansbach, Stadt- und Schulbücherei Gunzenhausen, Stadtarchiv Ansbach, Stadtarchiv Gunzenhausen, Stadtarchiv Nürnberg, Jay and Naomi Pasachoff Collection, Stadtarchiv Rothenburg ob der Tauber, Stadtbücherei Ansbach, Universitäts- und Landesbibliothek Sachsen-Anhalt, Universitätsbibliothek Augsburg, Universitätsbibliothek Erlangen-Nürnberg, Universitätsbibliothek Wien, Verlag Bayerische Staatszeitung, Verlag Harri Deutsch, Verlag Nürnberger Presse, and Verlag tradition.

Fig. 15.6 Opening event of the Simon Marius Anniversary on 10th of February in the Nicolaus-Copernicus-Planetarium: Wolfgang Eckart (Director of the Education Campus Nuremberg), Julia Lehner (Head of the Nuremberg Department of Culture), Pierre Leich, Hans-Ulrich Keller (Long-time Director of the Stuttgarter Planetarium), Dieter Hölzl (President NAG), Thomas A. H. Schöck (Chancellor of the University of Erlangen-Nürnberg) and Dieter Schoch (Chairman of the STAEDTLER Trust; Photo: Ramon Görke

in the Kaisersaal of the State Archives Nuremberg on February 18—exactly 400 years after the date in the dedication of the *Mundus Iovialis*. The *Marius-Portal* was launched by Pierre Leich, Peter Fleischmann, Stephan Kellner, and Ralph Puchta together pushing a big red button. The starting switch was of course merely symbolic but in the background, webmaster Norman Anja Schmidt simultaneously switched the web access from the test website to the final portal design. Another photo opportunity for the press was created by the representatives of the Municipal Archives of Ansbach and Gunzenhausen, the University Library Erlangen-Nuremberg, and the Municipal Library Nuremberg presenting valuable works (Fig. 15.7).

The digitization of Marius's main works is now complete, as each of his works is available at least once electronically. Of the calendars from 1601 to 1629, only a few are missing. Meanwhile there are 210 secondary literature articles, 400 media reports, and 139 lectures and exhibits recorded.

At the end of the anniversary year, the Nuremberg Astronomical Society transferred the operation of the *Marius-Portal* (whose menu is now available in 33 languages) to the newly founded Simon Marius Society. The society endeavors to make the research of the margravial mathematician, physician, and astronomer available to

Fig. 15.7 Pierre Leich, Peter Fleischmann (director State Archives Nuremberg), Stephan Kellner (Bavarian State Archives), and Ralph Puchta (vice president of Nuremberg Astronomical Society) at the portal's activation; picture: Mark Kaller

Fig. 15.8 The members of the Simon Marius Society by the Inaugural Meeting in December 2014 at the Regiomontanus-Observatory Nuremberg: Joachim Schlör, Reinhard Neumann, Günter Löffladt, Rudolf Laux, Caroline Merkel, Ulrich Kiesmüller, Chriska Wagner, Ralph Puchta, Pierre Leich, Rudolf Pausenberger, Hans Gaab, Torsten Sommer, Jürgen Krüger, Klaus Matthäus, Michael Pragal, Norman Anja Schmidt, and Thony Christie

academic scholars, educators, and the general public and to further develop the portal for a regional and international audience of interested laypersons and scholars (Fig. 15.8).

References

Bosscha, Johannes (1907): Simon Marius. Réhabilitation d'un astronome calomnié, *Archives Néerlandaises des Sciences Exactes et Naturelles,* vol. 12 (1907), serie 2a, La Haye, pp. 258–307, 490–528

Busnelli, Manlio Duilio (1928): Un Carteggio Inedito di Frau Paolo Sarpi con L'Ugonotto Francesco Castriono, *Atti del Reale Istituto Veneto di Scienze, Lettere ed Arti* 87, parte seconda (1927–28)

Doppelmayr, Johann Gabriel (1730): *Historische Nachricht von den Nürnbergischen Mathematicis und Künstlern.* Nürnberg: Peter Conrad Monath 1730. Reprint Hildesheim: Olms 1972

Favaro, Antonio (1901): Galileo Galilei e Simone Mayr, *Bibliotheca Mathematica,* 3. Folge, 2 (1901), Leipzig, pp. 220–223

Favaro, Antonio (1904): Galileo and Marius, *The Observatory. A review of astronomy* 27 (1904), pp. 199–200

Favaro, Antonio (1918): A proposito di Simone Mayr, *Atti e Memorie della R. Accademia di scienze, lettere ed arti in Padova,* n.s. 34 (1917–18), pp. 17–19

Fracastoro, Girolamo (1538): *Homocentrica: Eiusdem de causis criticorum dierum per ea quae in nobis sunt.* Venedig 1538

Gaab, Hans; Leich, Pierre (2014): Marius' Replik auf Scheiner. Der Anhang zum Mundus Iovialis von Simon Marius, *Globulus – Beiträge der Natur- und kulturwissenschaftlichen Gesellschaft e. V.,* 18 (2014), pp. 11–14

Galilei, Galileo (1623): *Il saggiatore.* Rom: Mascardi 1623

Galilei, Galileo (1632): *Dialogo sopra i due massimi sistemi del mondo, Tolemaico e Copernicano.* Florenz: Landini 1632

Galilei, Galileo (1965): *Le Opere di Galileo Galilei,* Vol. II, ed. by Antonio Segni. Florenz: Barbéra 1965

Galilei, Galileo (1965): *Le Opere di Galileo Galilei,* Vol. X, ed. by Antonio Segni. Florenz: Barbéra 1965

Galilei, Galileo (1610/1989): *Sidereus Nuncius, or The Sidereal Messenger,* translated with introduction, conclusion, and notes by Albert van Helden. Chicago/London: Chicago University Press 1989

Grotefeld, Hermann (1891): *Zeitrechnung des Deutschen Mittelalters und der Neuzeit,* 1. Band: Glossar und Tafeln, Hannover 1891

Humboldt, Alexander von (1850): *Kosmos. Entwurf einer physischen Weltbeschreibung,* 3. Band, Kapitel 7, Stuttgart/Tübingen: Cotta 1850

Kepler, Johannes (1609): *Phaenomenon singulare seu Mercurius in Sole.* Leipzig: Schurerus 1609

Kepler, Johannes (1610): *Dissertatio cum Nuncio Sidereo nuper ad mortales misso a Galilaeo Galilaeo Mathematico Patavino.* Prag: Sedesanus 1610

Kepler, Johannes (1938–2009): *Gesammelte Werke,* 21 vols. München: Beck 1938–2009

Klug, Josef (1906): Simon Marius aus Gunzenhausen und Galileo Galilei: ein Versuch zur Entscheidung der Frage über den wahren Entdecker der Jupitertrabanten und ihrer Perioden, *Abhandlungen der Mathematisch-Physikalischen Classe der Königlich Bayerischen Akademie der Wissenschaften* 22 (1906), pp. 385–526

Leich, Pierre; Wolfschmidt, Gudrun (2012): Simon Marius – Werke und Literatur. In: Wolfschmidt 2012, pp. 365–377

Lynn, William Thynne (1903a): Simon Marius and the satellites of Jupiter, *The Observatory. A review of astronomy* 26 (1903), London, pp. 254–256

Lynn, William Thynne (1903b): Galilée et Marius, *The Observatory. A review of astronomy* 26 (1903), London, pp. 389–390

Lynn, William Thynne (1904): Galileo and Marius, *The Observatory. A review of astronomy* 27 (1904), London, pp. 63–64

Lynn, William Thynne (1909): Simon Mayr, *The Observatory. A review of astronomy* 32 (1909), London, pp. 355–356.

Marius, Simon (1614): *Mundus Iovialis Anno M.DC.IX. Detectus Ope Perspicilli Belgici.* Nürnberg: Johann Lauer 1614

Marius, Simon (1614/1916): Arthur Octavius Prickard: The 'Mundus Jovialis' of Simon Marius, *The Observatory. A review of astronomy* 39 (1916), part 1: no. 504 (September), pp. 367–381, part 2: no. 505 (October 1916), pp. 403–412, part 3: no. 506 (November 1916), pp. 443–452, part 4: no. 507 (December 1916), pp. 498–503

Marius, Simon (1614/1988): Joachim Schlör, *Mundus Iovialis – Die Welt des Jupiter. Gunzenhausen*: Johann Schrenk 1988

Marius, Simon (1614/1916/2019): The World of Jupiter, translated by A.O. Prickard and Albert van Helden, in Simon Marius and his Research, Hans Gaab and Pierre Leich (eds.), Cham: Springer 201

Oudemans, Jean Abraham Chrétien; Bosscha, Johannes (1903): Galilee et Marius, *Archives Néerlandaises des Sciences Exactes et Naturelles*, Ser. II, T. VIII, La Haye 1903, pp. 115–189

[Orth, Johann Philipp:] *Ausfürliche Abhandlung von den berümten zwoen Reichsmessen so in der Reichsstadt Frankfurt am Main järlich gehalten werden* [...]. [Frankfurt am Main] 1765. Reprint: [Frankfurt am Main:] Brönner [ca. 1994]

Pilz, Kurt (1977): *600 Jahre Astronomie in Nürnberg*. Nürnberg: Hans Carl 1977

Prickard, Arthur Octavius (1917): Note on "Simon Marius" and the 'Mundus Jovialis', *The Observatory. A review of astronomy* 40 (1917), pp. 119–122

Radau, J. C. Rodolphe (1904): Revue des publications astronomiques. Oudemanns et Bosscha. – Galilée et Marius, *Bulletin Astronomique*, Serie I, vol. 21 (1904), Paris, pp. 119–120

Sarpi, Fra Paolo (1931): *Lettere ai Protestanti*, ed. by Manlio Duilio Busnelli, Bari 1931, Volume secondo

Scheiner, Christoph; Locher, Johann Georg (Respondent) (1614): *Disquisitiones Mathematicae, De Controversiis et Novitatibus Astronomicis*. Ingolstadt: Angermaria 1614

Strohmaier, Gotthard (1984): *Die Sterne des Ab dar-Rahman as-Sufi.* Hanau/Main: Müller & Kiepenheuer 1984

Willach, Rolf (2007): Der lange Weg zur Erfindung des Fernrohrs. In: Hamel, Jürgen; Keil, Inge: *Der Meister und die Fernrohre: Das Wechselspiel zwischen Astronomie und Optik in der Geschichte* (= *Acta Historica Astronomiae*; Vol. 33), Frankfurt a.M.: Harri Deutsch 2007, pp. 34–126

Wohlwill, Emil (1926): *Galilei und sein Kampf für die copernicanische Lehre*, vol. 2. Leipzig: Leopold Voss 1909–1926. Reprint Wiesbaden: Sändig 1969

Wolfschmidt, Gudrun (Ed.) (2012): *Simon Marius, der fränkische Galilei, und die Entwicklung des astronomischen Weltbildes* (= *Nuncius Hamburgensis – Beiträge zur Geschichte der Naturwissenschaften*; Band 16). Hamburg: tredition 2012

Zinner, Ernst (1942): Zur Ehrenrettung des Simon Marius, *Vierteljahrsschrift der Astronomischen Gesellschaft* 77 (1942), pp. 23–75

Zinner, Ernst (1943): *Entstehung und Ausbreitung der Copernicanischen Lehre*, 2nd ed., München: Beck 1988 (first published Erlangen 1943)

Zuidervaart, Huib (Ed.) (2008): Zoomers, Henk (transcribed, translated and introduced), *Embassies of the King of Siam sent to His Excellency Prince Maurits, arrived in The Hague on 10 September 1608. An early 17th century newsletter, reporting both the visit of the first Siamese diplomatic mission to Europe and the first documented demonstration of a telescope worldwide.* Wassenaar: Louwman Collection of Historic Telescopes 2008

Chapter 16
A Word of Caution About the "Rehabilitation" of Simon Marius

Albert Van Helden and Huib Zuidervaart

Following the title of Johannes Bosscha's essay (1907): "Simon Marius. Réhabilitation d'un astronome calomnié," the word "rehabilitation" is often used in studies about the astronomer Simon Marius (1573–1624). Having studied the original essays by Bosscha and his colleague Jean Oudemans and having examined the file with correspondence and calculations surrounding the original prize question that became the occasion for their essays,[1] we want to make some general comments about this aspect of Marius studies.

Johannes Bosscha Jr. (1831–1911) was an important force in the beginning of the so-called second golden age of Dutch science.[2] He was intricately involved in the reformation of secondary and higher scientific education in the Netherlands and personally supported such rising stars as Heike Kamerling Onnes (1853–1926) and Johannes Diderik van der Waals (1837–1923). As a member of the Koninklijke Nederlandse Akademie van Wetenschappen (the Royal Netherlands Academy of Arts and Sciences), secretary of the Hollandsche Maatschappij der Wetenschappen (Dutch Society of Sciences), and editor of its periodical, the *Archives Néerlandaises des Sciences Exactes et Naturelles*, he was at the very center of Dutch science policy

---

This note is a comment on the section "Rehabilitation" in Pierre Leich's paper, "Priorität, Rezeption und Rehabilitation von Simon Marius," *Simon Marius und seine Forschung* ed. H. Gaab and P. Leich (Leipzig: Akademische Verlagsanstalt, 2016), pp. 407–439, at 421–428 = "Priority, Reception and Rehabilitation of Simon Marius—From the Accusation of Plagiarism to the Marius Portal as his Virtual Collected Works" in this edition.

[1] Noord-Hollands Archief, Haarlem: archive no. 444 (Hollandsche Maatschappij der Wetenschappen), inv. no. 455; idem, archive no. 720 (personal papers of Professor Johannes Bosscha, 1831–1911), inv. nos. 33, 35, 48–53, 64–64, 68–71. See also Bruijn 1977.

[2] The phrase "second golden age" of Dutch science is introduced in Willink 1991. See also Willink 1998.

A. Van Helden (✉) · H. Zuidervaart
University of Utrecht, Utrecht, Netherlands
e-mail: a.vanhelden@uu.nl

H. Gaab, P. Leich (eds.), *Simon Marius and His Research*, Historical & Cultural Astronomy, https://doi.org/10.1007/978-3-319-92621-6_16

and politics. One of his close friends was Jean Abraham Chrétien Oudemans (1827–1906), professor of astronomy at the University of Utrecht and director of its observatory, whom he frequently called on for aid in refereeing papers submitted to the *Archives*.

Not only was Bosscha a pivotal character in the rise of Dutch science to prominence, he was also passionately interested in its history. Although as a scientist he realized well that only results, not the identity of the scientist, was what mattered, as a proud Dutchman, he wanted recognition for the contributions Dutch scholars had made to the progress of science. One of these, Simon Stevin (a Fleming by birth), drew Bosscha's special attention. He found that among his manifold interests, Stevin had occupied himself with the study of motion. He had dropped weights from a height in order to prove the Aristotelian ideas of motion wrong—*and he had done so before Galileo Galilei had done so*. Galileo received all the credit for progress in the study of motion during at the turn of the seventeenth century. And for Bosscha, Galileo therefore became a figure who had to be taken down a few pegs. He pursued this goal with passion. His promotion of Simon Marius must be seen in this context.

The prize question posed in the 1898 volume of the *Archives* called for a "scientific" study of the historical sources to determine whether Galileo's accusation of plagiarism against Marius was justified. The parallel with the Stevin-Galileo question is obvious. Could it be—and Bosscha apparently believed so—that Marius had indeed discovered Jupiter's four moons *before* Galileo but had been denied credit for it? Was Marius perhaps vilified by Galileo and his followers, thus casting him into the purgatory of astronomy? The wish is the father to the thought: the prize question was a loaded one.

If Bosscha had expected a number of submissions, he was disappointed. Only one entry was received, a 235-page treatise in a difficult German script, by Josef Klug, "Gymnasialprofessor" in Nürnberg, arguing the case for Galileo, not Marius. Although Bosscha could have refused the entry because of the script, he decided to accept it and send it to referees. He chose Oudemans, Ernst Frederik van de Sande Bakhuyzen (1848–1918), the assistant director of the Leiden Observatory, and Jacobus Kapteyn (1851–1922), professor of Astronomy in Groningen—a well-qualified trio of referees.

The procedure was that the entry (without the cover page with the name of the author) was sent to the first referee, together with a cover letter from Bosscha, the first referee would then send the file, *including his report* and Bosscha's letter to the second referee, and so on. In his cover letter, Bosscha made it perfectly clear what his judgment was. From a modern perspective, this was anything but blind refereeing, and although all three referees filed lengthy reports, the outcome was predictable. Klug's essay was not deemed worthy for the prize. Klug did not receive a letter informing him of the decision; he had to read for himself in the *Archives*.

In the announcement of the verdict, Bosscha stated that the referees had done their own research into the question in order to arrive at their decisions. This statement opened the door for Antonio Favaro, the editor of the Edizione Nazionale of *Le Opere di Galileo Galilei*, to challenge Bosscha to publish the results. After all, this new research should not be hidden from the scientific world. Rising to the challenge,

Bosscha and Oudemans now produced the article "Galilée et Marius" published in the *Archives* in 1903. Meanwhile, Klug corresponded with Favaro, who encouraged him to publish his refused essay elsewhere. It appeared in print in 1905. Bosscha responded with the essay cited at the beginning of this article, written after the death of Oudemans. The question is do the two essays in the *Archives*, the first by Oudemans and Bosscha in 1903 and the second by Oudemans in 1907, add up to a "rehabilitation" of Marius (or *Ehrenrettung* as Ernst Zinner put it in 1942). What exactly is at stake?

Galileo charged Marius with several transgressions. First, Marius did not observe Jupiter's moons before Galileo did, and he used the Julian calendar to confuse the readers. Indeed, he claimed, Marius did not observe the moons in 1610 and probably never at all. Further, according to the tables in *Mundus Iovialis*, the latitudes were correct only for the years 1612–1613, but incorrect for 1610. Ignoring Galileo's comment about the Julian calendar, the use of which an attentive reader should have noticed, the question remains: did Marius observe the moons of Jupiter starting in January 1610 or even December 1609 (Gregorian)?

Oudemans and Bosscha state that once the telescope had been invented and spread over Europe, anyone who directed it to the heavens could hardly miss the bright little stars around Jupiter. Current knowledge about the first few years of the telescope argues against this. The efforts of others to make telescopes good enough to detect the satellites after the publication of *Sidereus Nuncius* show that it was not until the autumn of 1610 that a few others managed with telescopes made by Galileo and Antonio Santini to verify Galileo's discovery of the moons. There is no record of anyone in the Dutch provinces observing the satellites until 1614.[3] Marius's first documented observation of Jupiter's satellites is from late December 1610.

Second, there is no adequate explanation for the error in Marius's tables about the latitudes of the satellites. Although it is true that Galileo's own explanation for the latitude deviations—he supposed that the orbital planes were parallel to the ecliptic—was incorrect, this does not let Marius off the hook. Oudemans and Bosscha were unable to come up with an explanation of this error. All they could do was to cite Marius's own statement in *Mundus Iovialis* that he had come to considering the latitudes only lately.

Bosscha's aim of taking Galileo down a few pegs led him into a proxy war with the Galilean establishment. In view of this prejudice, and the compromised refereeing process abundant clear from the Bosscha archive in Haarlem, we believe that care must be taken in accepting the conclusions of Bosscha and Oudemans. Clearly, then, further research on Marius's observations is needed before we can begin to speak of a "rehabilitation."

[3]"Oock werden door denselven perspective ofte verre ghesichten [by Jacob Adriaensz Metius] ghesien eenighe dwalende sterren ofte planeten, die haer ganck ontrent Jupiter hebben" (Metius 1614, pp. 3–4). English translation Helden 1977, p. 48.

References

Bosscha, Johannes (1907): *Simon Marius. Réhabilitation d'un astronome calomnié*, *Archives Néerlandaises des Sciences Exactes et Naturelles*, Série II, tome 12 (1907), pp. 258–307, 490–528

Bruijn, J.G. de (1977): *Inventaris van de prijsvragen uitgeschreven door de Hollandsche Maatschappij der Wetenschappen 1753–1917.* Haarlem/Groningen 1977, no. 1137

van Helden, Albert (1977): *The Invention of the Telescope*, *Transactions of the American Philosophical Society* 67:4 (1977), pp. 1–67

Klug, Josef (1906): *Simon Marius aus Gunzenhausen und Galileo Galilei. Ein Versuch zur Entscheidung der Frage über den wahren Entdecker der Jupitertrabanten und ihrer Perioden*, *Abhandlungen der mathematisch-physikalischen Klasse der Königlich Bayerischen Akademie der Wissenschaften* 22:2 (1906), pp. 385–526

Metius, Adriaen (1614): *Institutiones astronomicae et geographicae / Fondamentale ende grondelijcke onderwysinge van de sterrekonst ende beschryvinghe der aerden door het ghebruyck van de hemelsche ende aerdtsche globen.* Franeker 1614

Willink, Bastiaan (1991): *Origins of the Second Golden Age of Dutch Science after 1860: Intended and Unintended Consequences of Educational Reform*, *Social Studies of Science* 21:3 (1991), pp. 503–526

Willink, Bastiaan (1998): *De Tweede Gouden Eeuw. Nederland en de Nobelprijzen voor natuurwetenschappen 1870–1940.* Amsterdam 1998

Zinner, Ernst (1942): *Zur Ehrenrettung des Simon Marius*, *Vierteljahrsschrift der Astronomischen Gesellschaft* 77 (1942), pp. 23–75

Chapter 17
The Marius Portal: Simon Marius as Digital Human in the Twenty-First Century

Norman Anja Schmidt and Pierre Leich

Abstract The Simon Marius Anniversary 2014 was aimed at giving the Franconian astronomer proper credit and promoting him and his work within the region and to the worldwide history of astronomy community. An integral project was the creation of a web portal as a twenty-first-century representation of Simon Marius. All his original works, secondary literature, international reporting, and websites about him are collected on www.simon-marius.net, which enjoys international regard. Work on the portal is in progress under the authority of the Simon Marius Society, newly founded in 2014.

Motivation and Aim

Simon Marius was undoubtedly an important astronomer and scientist of the early seventeenth century. He combined personal observations made with one of the first telescopes, theoretical reflections, and the knowledge of his time, publishing this in several works and a series of yearly calendars and prognostica. Additionally, he is connected to the Franconian region by birth and the principle stations of his life.

The publication of their scientific results plays a crucial role for the reputation and recognition of a scientist to the present day. However, in the early twenty-first century, 400 years after Simon Marius's death, it is no longer sufficient to exist as dusty old books in the specialties cabinets of libraries. In the age of the Internet and of online technologies, a presence in the new media is crucial to being appreciated. Initiated by the editors of the present book, a small group of people interested in the history of astronomy in Franconia got together in 2012 with the aim of creating this media presence for Simon Marius.

N. A. Schmidt (✉) · P. Leich
Simon Marius Society, Nuremberg, Germany
e-mail: norman.anja.schmidt@simag-ev.de

H. Gaab, P. Leich (eds.), *Simon Marius and His Research*, Historical & Cultural Astronomy, https://doi.org/10.1007/978-3-319-92621-6_17

Pivotal was the creation of an Internet portal, which should represent Simon Marius in the twenty-first century. His personality and life should be highlighted in a short biography, and his works, as far as possible in complete digital copy, should build the core of the website. In addition to this, everything should be collected by, to, and about Marius, which has been created during the last 400 years. Simon Marius should be carried over into the present as a "Digital Human," to create the opportunity for many people to read up on or do research about Simon Marius without too much effort and without the necessity of going to a special library. From the very beginning, multilingual menu navigation was planned together with a translation of the important biographical contents, in order to facilitate an international dispersion in as many languages as possible.

Development

As early as 2009, in the aftermath of the International Year of Astronomy (IYA), the idea was born to investigate the life and work of the regional astronomer Simon Marius. In 2011, this began to take form when, on the initiative of one of the editors, the possibilities of holding a conference and creating a web portal were discussed in a small circle. The contemplated occasion was the quartercentennial of the publication (more exactly, the date of the dedication) of Simon Marius's main work, the *Mundus Iovialis*, on February 18, 2014 (Fig. 17.1).

Fig. 17.1 Klaus Matthäus, Pierre Leich, Ralph Puchta, Dieter Hölzl, and Hans Gaab reaching agreement on the financial accountability for the Simon Marius Anniversary 2014

In March 2012, under the direction of the Nürnberger Astronomische Gesellschaft (Nuremberg Astronomical Society, NAG), the development of a basic concept for the planned presentations, exhibitions, and publications, with an Internet presentation leading the way, was set in motion (see Fig. 17.1). The NAG had gained experience of regional collaboration as coordinator of all IYA activities throughout the Nuremberg Metropolitan Region, receiving nationwide acknowledgment. At the same time, the main domain name for the *Marius-Portal*, "simon-marius.net," was registered. At this time, the project nucleus consisted of Hans Gaab, Pierre Leich, Klaus Matthäus, Ralph Puchta, and Norman Anja Schmidt.

The second half of that year was used to cultivate and expand institutional and strategic contacts. For this, a 30-page exposé[1] and a presentation were created. A number of prominent advocates gave testimonials and greetings, which are also viewable in the portal.[2]

Additionally, a structural concept for the planned web portal was created, and a first prototype was programmed by Norman Anja Schmidt, comprising a few submenus and a small number of articles with the focus on functionality. An intended student project for the creation of an appealing graphic layout unfortunately did not materialize, and so the design question was deferred in autumn 2012. However, it was certain that the website, at least the navigation, should be available in at least eight different languages, including, of course, German and English.

The first presentation on the "Simon-Marius-Jubiläum 2014" (Simon Marius Anniversary 2014) was given in November 2012 for the "Forum Wissenschaft der Europäischen Metropolregion Nürnberg" (Science Forum of the Nuremberg Metropolitan Region). This was followed by a presentation to the board of trustees of the NAG at the beginning of December. The turn of the year was dedicated to the preparation of a "Letter of Intent" of key partners, applications to the public purse, and a concept for sponsorship. Through the funding by 14 institutions, notably the Department of Cultural Affairs of the City of Nuremberg, the Hermann Gutmann Foundation, and the Staedtler Foundation, a budget of about $20,000 became available.

On February 6, 2013, a working group "Simon Marius" within the NAG was formed. In addition to the participants mentioned above, the inner circle consisted of Thony Christie, Günther Görz, Klaus Herzig, Günter Löffladt, Rudolf Pausenberger, and Torsten H. Sommer. During the later stages, Alexander Biernoth, Eva-Maria Broermann, Christian J. Büttner, Ulrich Heber, Ulrich Kiesmüller, Rudolf Laux, Caroline Merkel, Joachim Schlör, Eckehard Schmidt, Georg Seiderer, Olga Sinzev, Sotirios Xognos, Franz Zitzelsberger, and others also contributed.

Three discussion meetings in the cities of Ansbach (see Fig. 17.2a), Gunzenhausen (see Fig. 17.2b), and Nuremberg (see Fig. 17.2c) with a total of about 60 people involved in science, regional and historical studies, or education followed this kickoff. In February, Mayor Carda Seidel invited most fittingly to the

[1]http://www.simon-marius.net/index.php?lang=en&menu=12#2

[2]http://www.simon-marius.net/index.php?lang=en&menu=9#2

Fig. 17.2 (**a**)–(**c**) Meetings with cooperation partners of the Simon Marius Anniversary 2014 in Ansbach, Gunzenhausen, and Nuremberg

Fig. 17.2 (continued)

Simon Marius Hall in the Onoldia Convention Center Ansbach, followed by Mayor Joachim Federschmidt and District Administrator Gerhard Wägemann inviting to Gunzenhausen, and Head of the Department of Culture Julia Lehner inviting to Nuremberg in June. The circle of active individuals, societies, and institutions was extended considerably by these meetings.

By mid-2013, the website was functional and already filled with most of the significant content, and the first translations had also already been integrated. At the aforementioned meeting in Nuremberg in June 2013, a first public demonstration was given, and it became clear that the design question now had to be addressed.

In late summer 2013, sufficient funds were available to endow all sub-projects for the Simon Marius Anniversary, diverse advertising material could be prepared, and press and public relations work could be started.

Additionally, the web design for the *Marius-Portal* could be developed together with Patrizia Strnad and Mark Kaller of the Kaller & Kaller design studio, Nuremberg, and implemented by Norman Anja Schmidt (see Fig. 17.3). Up to the final release of the website, more and more entries were collected and integrated or, where appropriate, linked to external sources. Additionally, the code of the website engine was refined and amended. Altogether, more than 200 working hours went into implementing the web portal and installing all the previously collected data.

On February 18, 2014, exactly 400 years after the date given in the dedication of the *Mundus Iovialis*, the launch of the website www.simon-marius.net took place in the Kaisersaal of the State Archives Nuremberg. By a live server reconfiguration and reload, the preliminary, worksite-gray countdown page was switched over to the entirely functional *Marius-Portal*.

Fig. 17.3 Patricia Strnad and Mark Kaller (right) present the web design to Norman Anja Schmidt and Pierre Leich

On this occasion, representatives of the Municipal Archives of Ansbach and Gunzenhausen, the University Library Erlangen-Nürnberg, the Municipal Library Nuremberg, the State Library Ansbach, and the State Archives Nuremberg presented precious original works by Simon Marius and other contemporaries (see Fig. 17.4). Afterward, the conservers of the originals socialized with the initiators and creators of the web portal as representatives of the twenty-first-century Simon Marius and with some invited guests for lively discussions (see Figs. 17.5 and 17.6).

It had been planned for the menu navigation to be available in 8 languages for the launch, but 16 languages were already available due to various contributions, and this has been extended to 32 since. Especially helpful with connections to translators were Pit Hauge (Esperanto group Nuremberg), Ulrich Heber (Dr. Karl Remeis-Observatory Bamberg), Gunter Lorenz (Sprachenzentrum der Universität Erlangen-Nürnberg—Language Center of the University of Erlangen-Nuremberg), Thomas A. H. Schöck (former university chancellor), Daniel Werner (the Bavarian-Indian Centre for Business and University Cooperation) and Yan Xu-Lackner (Konfuzius-Institut Nürnberg-Erlangen—Confucius Institute Nuremberg-Erlangen) as well as the Nürnberg-Loge, Karl Benz (Rüsselsheimer Sternfreunde—Star Friends Rüsselsheim), and the Nürnberger Astronomische Arbeitsgemeinschaft (Nuremberg Astronomical Consortium, NAA). The many

Fig. 17.4 Werner Mühlhäußer (Municipal Archive Gunzenhausen) with an original of the *Mundus Iovialis*, Christine Sauer (Municipal Library Nuremberg) with the *SchreibCalender auf 1612*, Wolfgang Reddig (Municipal Archive Ansbach) with an engraving by Merian, Ute Kissling (State Library of Ansbach) with a calendar, Herbert Schott (State Archives Nuremberg) with the *Prognosticon auf 1612*, and Christina Hofmann-Randall (University Library Erlangen-Nürnberg) with *De vita et fatis Simonis Marii mathematici quondam Brandenburgici* of 1775

Fig. 17.5 Audience at the launch of the Simon Marius portal in the Kaisersaal of the State Archives Nuremberg on February18, 2014; Picture taken by: Mark Kaller

Fig. 17.6 Pierre Leich, Thony Christie, Norman Anja Schmidt, and Klaus Matthäus together with some IT equipment at the portal launch; Picture taken by: Rudolf Laux

translators[3] made their contributions free of charge also patiently providing occasional supplements.

Since the project had taken up a decent momentum through public outreach and relevant contacts, a number of digitized original works by Simon Marius could be integrated directly into the portal right from the start or in the months thereafter.

As well as the opening event of the Simon Marius Anniversary, the portal launch, a festive evening, and a celebration on the occasion of the naming of an asteroid, a conference was held at the Nicolaus-Copernicus-Planetarium Nuremberg in autumn

[3]Gholamreza Azarhoushang (Farsi/Persian), Khongorzul Batmunkh (Mongolian), Monica Biasiolo (Italian), Maria Butan (Romanian), Thony Christie (English), Mustafa M. Danpullo (Hausa), Milan Dimitrijevic (Serbian), Michael Ecke (Japanese), Emre Eren (Turkish), Joachim Fux (Norwegian), Rainer Gröbel (French), Sylvia Atalla Hanna (Arabic), Heini Hänninen-Garzia (Finnish), Pit und Sabine Hauge (Esperanto), Friedrich Hofmann (Swedish), Nicolaas J. R. van Eikema Hommes (Dutch), István Jankovics (Hungarian), Sneha Kabburi (Hindi), Shubhangi N. Katkar (Hindi), Kon Kim (Korean), Joanna Kwiatkowski (Polish), Michael Lackner (Chinese), Diana Lagier de Milani (Spanish), Miroslav Malovec (Czech), Jan und Lene Niemann (Danish), Oksana Okulova (Ukrainian), Kavyo Jigar Panchal (Hindi), Edith Pilska (Polish), James Rezende Piton (Portuguese), Rezarta Reimann (Albanian), Joachim Schlör (Latin), Helmut Sebastian (Esperanto), Mey Seifan (Arabic), Olga Sinzev (Russian), Đurđica Skok (Croatian), Augustin Skopal (Slovak), Edit Slezákné Tar (Hungarian), Torsten H. Sommer (English, Japanese), Heizo Takamatsu (Japanese), Katya Tsvetkova and Milcho Tsvetkov (Bulgarian), José Juan Ventura Usó (Spanish), Namitha Venkatesh (Hindi), Jarosław Włodarczyk (Polish), Yi Wu (Chinese), Sotirios Xognos (Greek), Yan Xu-Lackner (Chinese) and Jiayue Xun (Chinese).

Fig. 17.7 Founding board of the Simon Marius Society: Hans Gaab, Klaus Matthäus, Joachim Schlör, Rudolf Laux, Pierre Leich, Norman Anja Schmidt, and Ulrich Kiesmüller in December 2014

2014. The members of the NAG and its partners, altogether 165 individuals, accomplished almost 3000 voluntary working hours to create the *Marius-Portal* and to make the events of the Marius year possible.

The working group "Simon Marius" of the NAG, established in February 2013, finally became the germ for the Simon Marius Society, which was established on December 12, 2014 (see Fig. 17.7)—so to speak as a finale of the Marius year—and its home in the Internet can also be found on the *Marius-Portal*,[4] as it should be.

Technology

At an early stage of development, a rough structure of the portal was designed, and based on this, some basic technical decisions were taken.

Multiauthor capabilities or a general account and privilege system were deemed unnecessary, since the content was to be edited by a small group of people. On the other hand, a simple and feasible way to easily present parts of the portal in many different languages, let alone the varying progress of many translations, seemed paramount. As a precaution, the administration of different language versions of all entries and a dynamical response with respect to the selected language were included. The structure of the website could not be determined in detail but had to be continually restructured following the developing experience and thematic progress of the people involved. The data structure and scope of the different subpages

[4]http://www.simon-marius.net/index.php?lang=en&menu=14

varied widely in some cases, so a simple template would not have been flexible enough and could not reflect some expected and all unexpected special cases.

Therefore, early on the decision was made not to adapt any available content management system but to self-build the portal based on a system of PHP scripts and an SQL database. Basic functions for the HTML page frame and the processing of the tables for the individual submenus were compiled in a PHP library. The PHP scripts then dynamically generate the HTML code, based on a few parameters like language or sort criterion. Some, but not all, of the capabilities of HTML 5 were exploited. DIV elements were used as little as possible, and lists were used almost throughout instead of tables to improve the accessibility.

Right from the beginning, function, content, and design were kept as far as possible separate. In consequence, the design, developed much later, could be implemented with relative ease by introducing the according tags in the HTML source code of the related menu functions and the creation of a CSS 2.1 style sheet (three in fact for non-Latin and right-to-left scriptures). A device-dependent presentation (responsive design) has not been realized due to limited human resources, but as the logs show (for more details see below), more than 80% of the used viewports are at least XGA (1024 × 768) anyway, although with smaller screens on the rise.

The creation and maintenance of the content are done, in the first place, by one of the editors, supplemented by input from a few other members of the Simon Marius Society. All tabular data is kept in an offline database, and their extracts are fed into the online SQL database as necessary. Direct editing is feasible through a "phpMyAdmin"[5] interface. This database usually comprises one table for the content of each submenu and additional tables for keeping the versions of terms and text in every available language in Unicode.

For data protection reasons, the complete website, including all fonts and media files, are kept on a single virtual web server. Cookies and JavaScript were not used except for the email function and for the acquisition of some anonymized technical information for the optimization of the web portal.

Structure and Contents of the Webpage

The landing page[6] (see Fig. 17.8) at www.simon-marius.net and a few other registered domain names presents a short text about Simon Marius, which is available in all 32 languages, as is the complete menu navigation.

Throughout all subpages, three menus serve for the navigation within the website. The vertical main menu on the left page margin leads to the important submenus, and a small horizontal menu on the top right margin leads to minor menus or such dedicated to special target groups like press or members.

---

[5]https://www.phpmyadmin.net, GPL license.

[6]http://www.simon-marius.net/index.php?lang=en&menu=1

Fig. 17.8 Landing page of the Simon Marius portal (English version)

Below that, utilizing the full width of the viewport, the menu for the language selection is situated. With a click on the individual language button, which reads the language name in the according font together with an explaining bubble in English at "mouse over," the whole menu navigation together with the biographical and teaser texts can be switched to the selected language. This is followed by a bar showing the only known picture of Simon Marius and stating his name, professions, and lifespan.

The main content area is arranged in a narrow picture column to the left and a wider, horizontally scalable text column. The data is usually presented in the form of a short article or key points together with a preview picture. In some cases, the data is subdivided into categories, which can be directly accessed through a special menu above the content area or can be sorted in different ways accessed through that menu.

The web portal consists of 19 subpages besides the landing page. Altogether, as of September 2017, they consist of over 1000 articles with almost as many preview pictures, mostly as an image link to internal media files or external web pages.

In "Simon Marius—Life and Research,"[7] a longer biographical text gives an overview of the stations of Marius's life and the main topics of his research, together with a compilation of the honors he received during his life and posthumously. The text has been translated into almost all the provided portal languages.

"Complete Works and Occasional Writings"[8] compiles all known works by or with the participation of Simon Marius (see Fig. 17.9). All original copies of these, which have become known to us, are listed together with location and shelf number. No less than 37 copies of the main work *Mundus Iovialis*, distributed throughout the whole world, have been registered to date. The calendars and prognostica for almost 30 years can be found mainly in German libraries and archives (though unfortunately not completely).

A range of scans were kindly provided by some of these institutions, led by the State Archives Nuremberg, the Municipal Library Nuremberg and a private collector, for publication within the Simon Marius portal. 32 original works are available exclusively on this subpage for information and research on Simon Marius.

During the locating of the original works and comparing the various copies, some initial interesting results were obtained. Calendars and prognostica were produced in different versions for Catholic and Protestant areas. Most of the works presented in the *Marius-Portal* are titled *Alter und Neuer SchreibCalender* (Old and New Writing Calendar), i.e., for Protestant areas, but for the year 1628, only two copies of *Neuer und Alter SchreibCalender* (New and Old Writing Calendar), i.e., for Catholic areas, are available.[9] For the year 1609, however, both versions are available.[10,11] Also, some discrepancies in print years were found, e.g., the *Prognosticon Astrologicum auf das Jahr 1613*[12] was printed in the same year (last page), whereas most of the

[7]http://www.simon-marius.net/index.php?lang=en&menu=2

[8]http://www.simon-marius.net/index.php?lang=en&menu=3

[9]Marius (1627), two available copies, see portal.

[10]Marius (1608), see online.

[11]Marius (1608a), see online.

[12]Marius (1613), p. 20^r (two available copies, see portal and online).

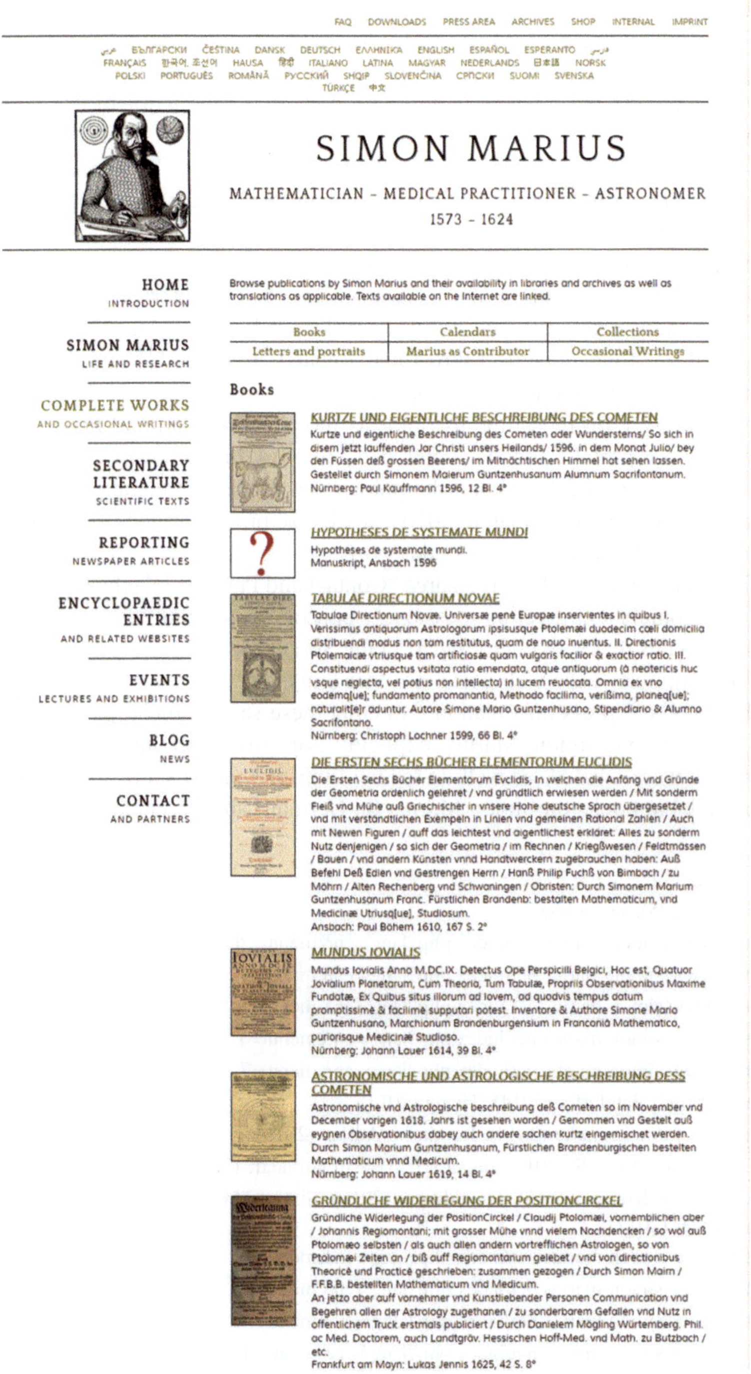

Fig. 17.9 Subsite of the web portal "Complete Works and Occasional Writings" of Simon Marius

other prognostica were printed in the year before their validity, as were the calendars. Also calendars in different formats such as wall calendars have been discovered.[13] These new insights will successively be transferred to the *Marius-Portal*, and a rearrangement of the calendar part will be done after completion of the work on this volume.

Also in preparation is the subpage "Letters and Portraits,"[14] where mainly letters from, to, and about Simon Marius will be compiled and transcribed from handwriting into searchable text.

In addition to Marius's works, the Simon Marius Society is compiling all material and events that have Simon Marius as a subject. As of September 2017, 205 entries in "Secondary Literature—Scientific texts,"[15] 385 "Reporting—Newspaper Articles,"[16] 110 "Encyclopedic Entries and Related Websites,"[17] and 131 "Events—Lectures and Exhibitions"[18] have been registered.

The subpage on reporting also reflects the presswork around the launch of the Simon Marius portal[19] and the conference[20] in 2014, the summary of the Marius year at the beginning of 2015,[21] and the publication of the conference proceedings and Marius anthology in German in 2016[22] in diverse printed articles in regional as well as national newspapers and journals.

Together with this, there is an area "Contact and Partners"[23] which gives room for the testimonials and greeting messages, sponsors, contributors, partners, as well as all individuals that actively took part in the Simon Marius Anniversary or still do so through support of the web portal or other things.

For FAQs,[24] press information[25] (among these seven official press releases) and for material on Simon Marius explicitly for download and usage,[26] separate subpages were created, as well as for the imprint[27] and a shop,[28] where currently, two available books on Simon Marius (in German) are presented.

---

[13]Marius (1625), see online.

[14]http://www.simon-marius.net/index.php?lang=en&menu=20

[15]http://www.simon-marius.net/index.php?lang=en&menu=4

[16]http://www.simon-marius.net/index.php?lang=en&menu=5

[17]http://www.simon-marius.net/index.php?lang=en&menu=6

[18]http://www.simon-marius.net/index.php?lang=en&menu=7

[19]Auer (2014), Helldörfer (2014), Tjiang (2014).

[20]nn (2014), Frankenradar (2014), Kaminski, Adrian (2014).

[21]Main-Post (2014), BR (2014), pm/af (2014), astronomie.de (2015), mn (2015).

[22]hma (2016), Kratzer (2016), Orgeldinger (2016), Simon Marius und seine Forschung (2016), Paul, Günter (2017).

[23]http://www.simon-marius.net/index.php?lang=en&menu=9

[24]http://www.simon-marius.net/index.php?lang=en&menu=11

[25]http://www.simon-marius.net/index.php?lang=en&menu=13

[26]http://www.simon-marius.net/index.php?lang=en&menu=12

[27]http://www.simon-marius.net/index.php?lang=en&menu=10

[28]http://www.simon-marius.net/index.php?lang=en&menu=16

Finally, a blog[29] based on "WordPress"[30] was integrated into the *Marius-Portal* allowing the introduction and highlighting of special proceedings within the portal; also, the comment section of the blog offers an opportunity for interaction.

Altogether, 718 media files are accessible through the *Marius-Portal*, comprising scans of original works, but mainly secondary literature and press coverage, as well as a few videos.

Home of the Simon Marius Society

The whole subpage "Internal"[31] is dedicated to the Simon Marius Society, which was founded at the very end of 2014 and which then assumed responsibility for the Simon Marius portal. Among other items, the charter and the membership roster can be found here (although only in German).

The activities of the Simon Marius Society and its predecessors during the Marius Anniversary 2014 are documented on the "Archives"[32] subpage. Here, information and accounts of the official kickoff on February 10, 2014, in the Nicolaus-Copernicus-Planetarium Nuremberg; the launch of the *Marius-Portal* on February 18 in the State Archives Nuremberg (see Fig. 17.10); the festive evening in Ansbach; the domain of Simon Marius, on the evening of the same day; the Marius Day at the Simon-Marius-Gymnasium, Gunzenhausen; Marius's place of birth, on February 21; a celebration on the naming of an asteroid after Simon Marius at the Regiomontanus-Observatory Nuremberg on April 23; the conference "Simon Marius und seine Zeit" (Simon Marius and his Times) on September 29, 2014, again in the Planetarium (including videos of all talks); and finally the presentation of the anthology *Simon Marius und seine Forschung* (Simon Marius and his Research, as proceedings of the conference) on October 13, 2016, in the Campe bookstore, Nuremberg, are aggregated. Altogether, about 240 photos and 10 videos can be found here.

Visitors and Reception

Based on the log data of the web server, daily, monthly, and yearly access statistics are automatically created with "awstats"[33] and presented in the internal section of the website (see Fig. 17.11). It should be noted, however, that due to German/European data protection laws, for anonymization, only the first two octets of the IPv4

[29]http://www.simon-marius.net/index.php?lang=en&menu=8

[30]https://wordpress.org, GPL.

[31]http://www.simon-marius.net/index.php?lang=en&menu=14

[32]http://www.simon-marius.net/index.php?lang=en&menu=15

[33]http://www.awstats.org, GPL license.

Fig. 17.10 Photo collection of the portal launch in the archive section of the *Marius-Portal*

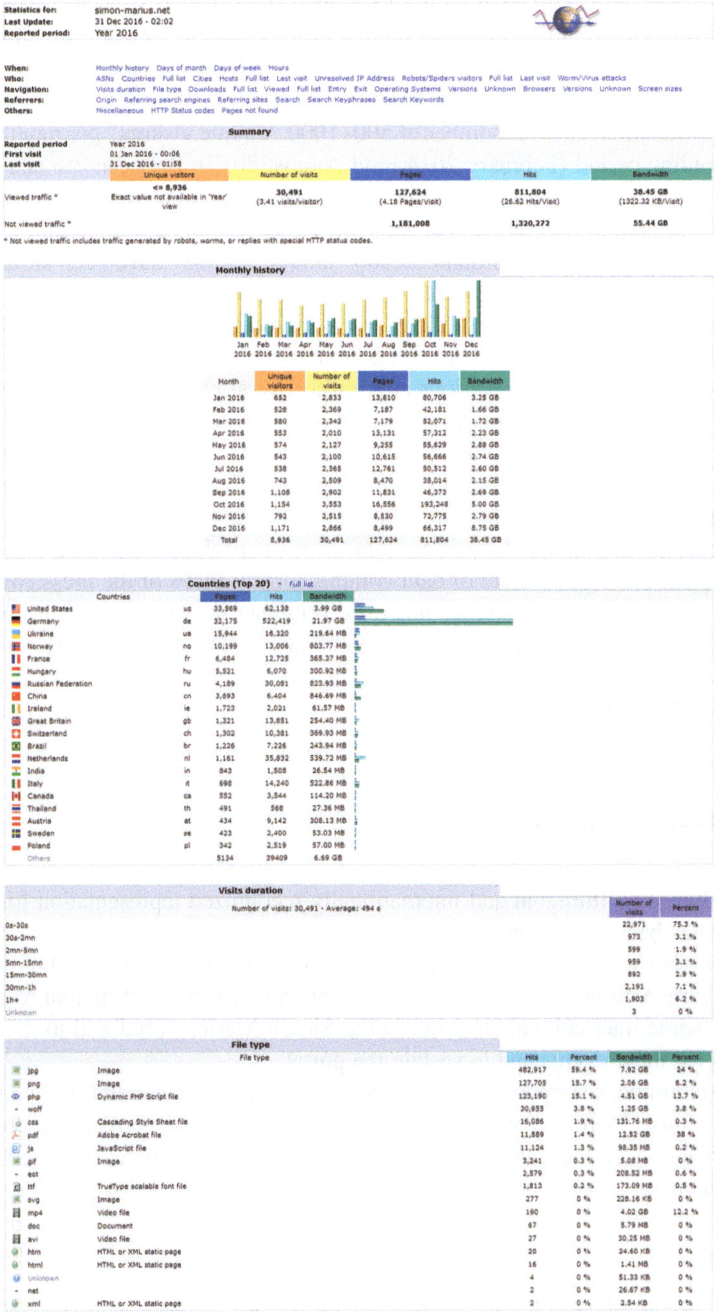

Statistics for:	simon-marius.net
Last Update:	31 Dec 2016 - 02:02
Reported period:	Year 2016

When: Monthly history Days of month Days of week Hours
Who: ASNs Countries Full list Cities Hosts Full list Last visit Unresolved IP Address Robots/Spiders visitors Full list Last visit Worm/Virus attacks
Navigation: Visits duration File type Downloads Full list Viewed Full list Entry Exit Operating Systems Versions Unknown Browsers Versions Unknown Screen sizes
Referrers: Origin Referring search engines Referring sites Search Search Keyphrases Search Keywords
Others: Miscellaneous HTTP Status codes Pages not found

Summary

Reported period	Year 2016
First visit	01 Jan 2016 - 00:06
Last visit	31 Dec 2016 - 01:58

	Unique visitors	Number of visits	Pages	Hits	Bandwidth
Viewed traffic *	<= 8,936 Exact value not available in 'Year' view	30,491 (3.41 visits/visitor)	127,624 (4.18 Pages/Visit)	811,804 (26.62 Hits/Visit)	38.45 GB (1322.32 KB/Visit)
Not viewed traffic *			1,181,008	1,320,272	55.44 GB

* Not viewed traffic includes traffic generated by robots, worms, or replies with special HTTP status codes.

Monthly history

Jan 2016 Feb 2016 Mar 2016 Apr 2016 May 2016 Jun 2016 Jul 2016 Aug 2016 Sep 2016 Oct 2016 Nov 2016 Dec 2016

Month	Unique visitors	Number of visits	Pages	Hits	Bandwidth
Jan 2016	652	2,833	13,610	80,706	3.25 GB
Feb 2016	528	2,369	7,187	42,181	1.66 GB
Mar 2016	580	2,342	7,179	52,071	1.72 GB
Apr 2016	553	2,010	13,131	57,312	2.23 GB
May 2016	574	2,127	9,255	55,629	2.88 GB
Jun 2016	543	2,100	10,615	56,666	2.74 GB
Jul 2016	538	2,365	12,761	50,512	2.60 GB
Aug 2016	743	2,509	8,470	38,014	2.15 GB
Sep 2016	1,108	2,902	11,831	46,373	2.69 GB
Oct 2016	1,154	3,553	16,556	193,248	5.00 GB
Nov 2016	792	2,515	8,530	72,775	2.79 GB
Dec 2016	1,171	2,866	8,499	66,317	8.75 GB
Total	8,936	30,491	127,624	811,804	38.45 GB

Countries (Top 20) - Full list

Countries		Pages	Hits	Bandwidth
United States	us	33,569	62,138	3.99 GB
Germany	de	32,175	522,419	21.97 GB
Ukraine	ua	15,944	16,320	219.64 MB
Norway	no	10,199	13,006	803.77 MB
France	fr	6,484	12,725	365.37 MB
Hungary	hu	5,521	6,070	300.92 MB
Russian Federation	ru	4,189	30,081	823.93 MB
China	cn	3,893	6,404	846.69 MB
Ireland	ie	1,723	2,021	61.57 MB
Great Britain	gb	1,321	13,851	254.40 MB
Switzerland	ch	1,302	10,381	369.93 MB
Brazil	br	1,226	7,226	243.94 MB
Netherlands	nl	1,161	35,832	539.72 MB
India	in	843	1,508	26.54 MB
Italy	it	698	14,240	522.86 MB
Canada	ca	552	3,544	114.20 MB
Thailand	th	491	568	27.36 MB
Austria	at	434	9,142	308.13 MB
Sweden	se	423	2,400	53.03 MB
Poland	pl	342	2,519	57.00 MB
Others		5134	39409	6.69 GB

Visits duration

Number of visits: 30,491 - Average: 494 s	Number of visits	Percent
0s-30s	22,971	75.3 %
30s-2mn	973	3.1 %
2mn-5mn	599	1.9 %
5mn-15mn	959	3.1 %
15mn-30mn	892	2.9 %
30mn-1h	2,191	7.1 %
1h+	1,903	6.2 %
Unknown	3	0 %

File type

File type		Hits	Percent	Bandwidth	Percent
jpg	Image	482,917	59.4 %	7.92 GB	24 %
png	Image	127,705	15.7 %	2.06 GB	6.2 %
php	Dynamic PHP Script file	123,190	15.1 %	4.51 GB	13.7 %
woff		30,955	3.8 %	1.25 GB	3.8 %
css	Cascading Style Sheet file	16,086	1.9 %	131.76 MB	0.3 %
pdf	Adobe Acrobat file	11,589	1.4 %	12.52 GB	38 %
js	JavaScript file	11,124	1.3 %	98.35 MB	0.2 %
gif	Image	3,241	0.3 %	5.08 MB	0 %
eot		2,579	0.3 %	208.52 MB	0.6 %
ttf	TrueType scalable font file	1,813	0.2 %	173.09 MB	0.5 %
svg	Image	277	0 %	228.16 KB	0 %
mp4	Video file	190	0 %	4.02 GB	12.2 %
doc	Document	67	0 %	5.79 MB	0 %
avi	Video file	27	0 %	30.25 MB	0 %
htm	HTML or XML static page	20	0 %	24.60 KB	0 %
html	HTML or XML static page	16	0 %	1.41 MB	0 %
Unknown		4	0 %	51.33 KB	0 %
net		2	0 %	26.67 KB	0 %
xml	HTML or XML static page	2	0 %	2.54 KB	0 %

Fig. 17.11 Excerpts of the access statistics of the *Marius-Portal* for the year 2016 with total numbers and statistics per month, visit duration, country, and file type

addresses are logged and can be used for evaluation, which leads to an underestimation of the important number of "unique visitors" and to an overestimation of the stay-time. In addition, a notable amount of the hits has to be attributed to cracking attempts targeted at the embedded blog.

The evaluation shows a number of 500–1000 "unique visitors" per month for the whole online period (February 2014 until August 2017). Between 7% and 20% of these visitors stay for at least 15 min, i.e., request page contents consecutively for that time, amounting to about four truly interested visitors per day, with a slightly positive slope. Altogether, around 35,000 "unique visitors" have been attracted.

The peaks in the page requests correspond in some cases with activities of the Simon Marius Society and the related press coverage. In April 2014 and October 2016, Simon Marius received some press coverage in the wake of two DPA (German press agency) dispatches in various articles in several nationwide and even international newspapers or web portals,[34] which also pointed to the *Marius-Portal* and obviously increased the number of page visits by a factor of two for a few months.

The page requested mostly is the landing page in German, followed by about half as many requests in English. The subpages are requested mainly in German. The larger part of the audience is apparently of German-speaking origin. German visitors also request the biggest share of data volume, mostly PDFs of the press coverage.

The majority of page requests, however, originate from the USA, most probably due to web crawlers of US-owned search engines, followed by Germany and many other countries, including Russia.[35] The circle of users of the Simon Marius portal undoubtedly extends beyond Germany.

Conclusion and Future

Altogether, a multilingual and internationally recognized representation for Simon Marius has been created in the form of the *Marius-Portal*.

A number of events organized by the Simon Marius Society (SiMaG) and the Nuremberg Astronomical Society (NAG) were able to draw substantial regional as well as some international attention to the Simon Marius portal and to raise funds needed for these events and to sustain the portal.

The continuous support and development of the portal is guaranteed by members of the Simon Marius Society. An extension by historical correspondence and a reorganization of the calendar section are in preparation, which will give some new insights.

[34] epd (2014), The Hindu (2014), Corum (2016), bild.de (2016), Focus Online (2016), Kratzer (2016), Redd (2016), Fischer (2016), Paul (2017).

[35] Country evaluation uses GeoLite data created by MaxMind, available from http://www.maxmind.com

On the technical end, some optimization work has to be done for a better view on portable devices, and the migration to an IPv6-capable server (together with some software updates) is in preparation.

Altogether, the original idea, i.e., to bring Simon Marius from the seventeenth into the twenty-first century, could be implemented in form of the *Marius-Portal.* Even more, this could be realized in many languages. Gratitude goes to the initiators, sponsors, and many committed volunteers.

References

astronomie.de (2015): Marius-Jubiläumsjahr zieht erfolgreiches Fazit, 2 January, *Astronomie.de.* Available from: https://www.astronomie.de/neuigkeiten/marius-jubilaeumsjahr-zieht-erfolgreiches-fazit. [20 November 2017]

Auer, Horst M. (2014): Marius, der fränkische Galilei, *Nürnberger Nachrichten*, 19 February, p. 17

bild.de [dpa/lby] (2016): Neuer Sammelband über fränkischen Astronomen Simon Marius, 5 October, *bild.de.* Available from: http://www.bild.de/regional/aktuelles/bayern/neuer-sammelband-ueber-fraenkischen-astronomen-48133662.bild.html. [20 November 2017]

BR (2014): Simon Marius und die Jupitermonde, *BR.de* [website], 29 December. No longer available.

Corum, Jonathan (2016): Jupiter and Its Moons, *The New York Times* [website], 1 July, updated 5 July 2016. Available from: http://www.nytimes.com/interactive/2016/07/01/science/space/jupiter-and-its-moons.html?_r=0. [20 November 2017]

epd [Tjiang, Thomas] (2014): epd, Ehrenrettung für mittelfränkischen Astronomen, *Nürnberger Nachrichten*, 2 January, p. 19 and teaser p. 1

Fischer, Daniel (2016): Ein Monument für den „fränkischen Galilei", *Skyweek Zwei Punkt Null*, 27 December. Available from: https://skyweek.wordpress.com/2016/12/27/ein-monument-fur-den-frankischen-galilei. [20 November 2017]

Focus Online [dpa] (2016): Neuer Sammelband über fränkischen Astronomen Simon Marius, *Focus Online*, 5 October. Available from: http://www.focus.de/regional/nuernberg/wissenschaft-neuer-sammelband-ueber-fraenkischen-astronomen-simon-marius_id_6029441.html. [20 November 2017]

Frankenradar (2014): Simon Marius und die Astronomie in Franken: Zwei Tagungen zur Astronomiegeschichte, *Frankenradar*, 4 September. Available from: http://www.frankenradar.de/nachrichten/09-2014/simon-marius-und-die-astronomie-in-franken-zwei-tagungen-zur. [20 November 2017]

Helldörfer, Clemens (2014): Pionier statt Plagiator. Neues Internetportal würdigt Simon Marius, *Nürnberger Zeitung*, 19 February, p. 11

hma [Auer, Horst M.] (2016): Im Schatten von Galileo Galilei, *Nürnberger Nachrichten*, 12 October, p. 13

Kaminski, Adrian (2014): Tagungsduett schließt Simon-Marius-Jubiläum ab, *Spektrum.de* [website], 9 September. Available from: http://www.spektrum.de/news/tagungsduett-schliesst-simon-marius-jubilaeum-ab/1307485. [20 November 2017]

Kratzer, Hans (2016): Pünktchen beim Jupiter, *Süddeutsche Zeitung*, 21 October, p. 47

Marius, Simon ([no year, ca.] 1608): *Alter und Newer SchreibCalender [...] auff daß Jar [...] 1609*, Johann Lauer, Nürnberg, copy of Municipal Library Nuremberg, sig. Will. VIII. 167d(1)

Marius, Simon ([no year, ca.] 1608a): *Newer und Alter SchreibCalender [...] auff das Jahr [...] 1609*, Johann Lauer, Nürnberg, copy of National Library of Austria Wien, sig. 306.979-B

Marius, Simon (1613): *Prognosticon Astrologicum [...] auff das Jahr [...] 1613*, Johann Lauer, Ansbach, copy of University Library Tübingen, sig. 80.4-OR, p. 20^r

Marius, Simon ([no year, ca.] 1625): *Alt und New Almanach [...] auff das Jahr [...] 1626*, single sheet print, Johann Lauer [not verified], Nürnberg [not verified], copy of the Germanisches Nationalmuseum Nuremberg, grapical collection, inventory HB 8842, capsule 1241

Marius, Simon ([no year, ca.] 1627): *Newer und Alter Schreibkalender [...] auff das Jahr [...] 1628*, Johann Lauer, Nürnberg, copy of Municipal Archive Nuremberg, sig. Av 2584.8 and copy of Municipal Library Nuremberg, sig. Hist. 8. 6884

Main-Post [reda] (2014): Simon Marius ins Bild gerückt, *Main-Post* [website], 28 December. Available from: http://www.mainpost.de/regional/franken/Simon-Marius-ins-Bild-gerueckt;art1727,8501004. [20 November 2017]

mn (2015): Hof-Astronom auf Topniveau, *Nürnberger Nachrichten*, 2 January, p. 30

nn (2014): Simon Marius und seine Zeit, *Nürnberger Nachrichten*, 4 September, p. 14

Orgeldinger, Mathias (2016): Fränkischer Himmelsforscher, *Nürnberger Zeitung*, 22 October, p. 11

Paul, Günter (2017): Simon Marius war präziser als Galilei, *Frankfurter Allgemeine Zeitung*, 17 January, no. 14, p. 12

pm/af [Frank, Andrea] (2014): Simon-Marius-Jubiläumsjahr geht zu Ende, *Fränkische Landeszeitung*, 31 December, without page

Redd, Nola Taylor (2016): Who really discovered Jupiter's four large moons?, *Astronomy magazine* [website], 15 December. Available from: http://www.astronomy.com/news/2016/12/jupiters-four-large-moons. [20 November 2017]

Simon Marius und seine Forschung [Leich, Pierre] (2016): „Simon Marius und seine Forschung" erschienen, *Sterne und Weltraum*, December 2016, pp. 92–93

The Hindu (2014): Galileo's German rival gets glory at last, *The Hindu* [website], 26 February, updated 18 May 2016. Available from: http://www.thehindu.com/todays-paper/tp-in-school/galileos-german-rival-gets-glory-at-last/article5726815.ece. [20 November 2017]

Tjiang, Thomas [epd] (2014): Endlich rehabilitiert, *Altmühl-Bote*, March 8, p. 37

Chapter 18
The Franconian Asteroid 7984 Marius

Thomas Müller

7984 Marius is a minor planet located in the asteroid main-belt region between Mars and Jupiter, which formed about 4.5 billion years ago when the planets emerged from the dense gas and dust disk around the Sun. The asteroid has an orbit with a semi-major axis of 2.63 AU, an eccentricity of 0.20, and an inclination of 9.1°, with an orbital period of 4.27 years. Light curves indicate a rotation period of 3.54 h and a slightly elongated shape. A radiometric analysis of infrared observations shows that 7984 Marius has a low albedo of about 6% and a diameter of approximately 10.5 km. The low albedo points toward a possible carbonaceous surface composition. Its absolute magnitude is 13.6 mag, and the object reaches a maximum brightness of about 16.4 mag during favorable oppositions.

Discovery, Numbering, and Naming

The small body "1980 SM" was discovered by the Czech astronomer Zdenka Vávrová at the Klet' Observatory[1] in south Bohemia on September 29, 1980. The original sky images covering 20 square degrees were taken with the 0.63-m Maksutov telescope (see Figs. 18.1 and 18.2). The photographic plates combine two 20-min exposures of the same region of the sky, with a 3-arcminutes displacement between the exposures. Each star produces in this way a pair of parallel short lines; the moving small bodies produce nonparallel lines. The inspection of the plates is typically done with a microscope, which allows the discovery of asteroids; the coordinates are extracted via optoelectronic tools. "1980 SM" was detected on

[1]Official observatory code: 046 (see also MPC 5615).

T. Müller (✉)
Max-Planck-Institute for Extraterrestrial Physics, Garching, Germany
e-mail: tmueller@mpe.mpg.de

H. Gaab, P. Leich (eds.), *Simon Marius and His Research*, Historical & Cultural Astronomy, https://doi.org/10.1007/978-3-319-92621-6_18

a b

Fig. 18.1 (**a**) The 0.63-m Maksutov telescope at the Kleť Observatory (1977–1996) (Credit: Jana Tichá), (**b**) Zdenka Vávrová (*1945) worked for more than 20 years as an astronomer at the Kleť Observatory in Southern Bohemia, Czech Republic. She discovered more than 100 asteroids and was officially recognized as co-discoverer of the short-period comet 134P/Kowal-Vávrová (Source: http://www.skaw.sk/comet-discoverer-zdenka-vavrova.htm)

Fig. 18.2 The Kleť Observatory is located at České Budějovice in South Bohemia, Czech Republic, at approximately 1070 m above sea level. The official observatory code is "046 Kleť Observatory, České Budějovice." Source: https://www.klet.org

September 29, on October 1, and again on October 3, 1980. On the same photographic plate, there are another seven new asteroids visible (1980 SJ, 1980 SK, 1980 SL, 1980 SN, 1980 SO, 1980 SP, 1980 SQ). In addition, it was possible to determine the position of "2052 Tamriko." The collection of multiple object detections from that campaign was then reported to the Minor Planet Center[2] (MPC) at the Harvard-Smithsonian Center for Astrophysics in Cambridge, MA, USA (see snapshot of the M.P.C. 5615 from December 1, 1980). The brightness of "1980 SM" at the moment of discovery was estimated to be 17.0 mag.

Extracted from *M. P. C. 5615 1980 DEC. 1* (the * indicates the discovery epoch):

Object	Date	UT	R. A. (1950)	Decl.	Mag.	Obs.
1980 SM*	1980 09	29.82643	23 57 05.27	+07 24 04.0	17.0	046
1980 SM	1980 09	29.83928	23 57 04.50	+07 24 02.8		046
1980 SM	1980 10	01.81006	23 55 12.13	+07 23 08.0		046
1980 SM	1980 10	01.82487	23 55 11.23	+07 23 07.5		046
1980 SM	1980 10	03.82083	23 53 19.73	+07 21 56.5		046
1980 SM	1980 10	03.83501	23 53 18.92	+07 21 54.8		046

In 1991 more observations of "1980 SM" were conducted from the Palomar Observatory and at the Steward Observatory in the USA and then 1995 at the Landessternwarte Thüringen in Tautenburg, Germany. Two years later the orbit of "1980 SM" was improved significantly with observations taken again at the Klet' observatory, which led to the MPC assignment of the fixed number "7984" in 1997. Finally, in 2014, the Committee for Small-Body Nomenclature of the IAU honored the work of Simon Marius (1573–1624), court astronomer in Ansbach, by giving the name[3] "Marius" to the small body 7984 (1980 SM). The official citation was published in the Minor Planet Circular 87545 from March 16, 2014 and is also documented in the Addentum to the *Dictionary of Minor Planet Names* (Schmadel 2015). The name "Marius" is equivalent to the fixed asteroid number "7984," and both can be used when looking for archive entries or for orbit calculations.

Citation published in the MPC 87545:

7984 Marius

1980 SM. Discovered Sept. 29, 1980, by Zdeňka Vávrová at Kleť Observatory (Hvězdárna Kleť) in South Bohemia, Czech Republic (Minor Planet Circular 87545, Mar. 16, 2014)

Simon Marius (1573–1625) was court astronomer in Ansbach. He discovered the Jovian satellites, but published his observations after Galileo, who accused him of plagiarism and ruined his reputation. Marius suggested the names still used today for the satellites and also measured the diameter of the Andromeda Galaxy.

Meanwhile, there are about 1500 individual position measurements available in the MPC archive.[4] Among the entries is also a detection from September 8, 1980 (Crimea-Nauchnij Observatory, observatory code 095), about 20 days before the

[2]http://www.minorplanetcenter.net/iau/mpc.html

[3]"SM" in the provisional designation of 7984 (1980 SM) was probably the reason for linking this object to Simon Marius. The details for the selection process and the origin of the citation text are not known.

[4]http://www.minorplanetcenter.net/db_search/

official discovery (see entry in MPC 5615 marked by *). But the measurement was only reported in August 1982, well after Vávrová's discovery. The number of MPC entries for 7984 is steadily growing, mainly because of regular detections by ongoing survey programs or by projects searching for new objects. As of October 2017, there are about 750.000 asteroids known, more than 500.000 have a fixed number which means that their orbits are well established, and about 21.100 asteroids have a name approved by the International Astronomical Union.

The Orbit of 7984 Marius

7984 Marius is located on an elliptical orbit in the asteroid main belt between Mars and Jupiter (Fig. 18.3). The small body needs 4.27 years for a complete revolution around the Sun. The orbit is inclined by 9.1° with respect to the ecliptic plane; it has an eccentricity of 0.20 and a semi-major axis of 2.63 AU, resulting in a minimum

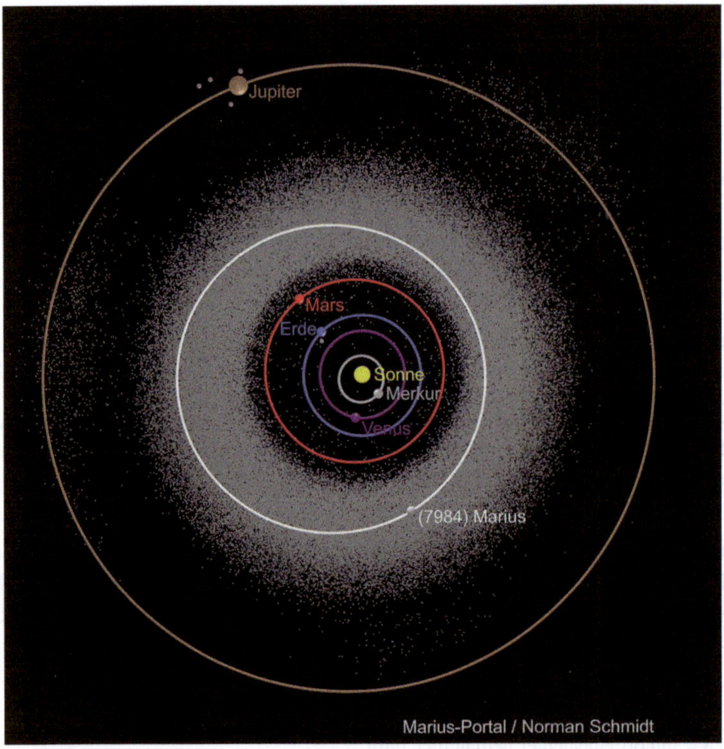

Fig. 18.3 The orbit of the asteroid 7984 Marius as seen from above the ecliptic (Sun-Earth-plane). The orbits of the planets Mercury, Venus, Earth, Mars, and Jupiter are shown, as well as the asteroid belt with the 4.27-year orbit of 7984 Marius. The gray dots represent the asteroid belt between Mars and Jupiter, and the two groups of Trojan objects in the Lagrangian points along Jupiter's orbit. Source: Marius-Portal/Norman Schmidt NOSCC

Earth-Marius distance of 1.11 AU on rare occasions. The orbit calculation of the JPL HORIZONS ephemerides services[5] is currently (October 2017) based on 1444 individual observed astrometric positions, covering about 37 years or almost 9 full orbits of Marius around the Sun (see also table with orbital elements). The object reaches regularly a brightness of 17–18 mag in opposition, in some cases also brighter (see figure with Marius's brightness over the next years). Up to now, no minor planet is named after Galileo Galilei, so close encounters of the rivals in space are therefore excluded. However, in December 2016, the asteroid 7984 had a very close apparent encounter with the Galilean satellites within 5° on the sky, but the true distance was well above 2 AU, and there was no risk of a revival of the past dispute on the discovery of the Jupiter satellites.☺

Orbital elements

ε	0.199		Eccentricity
A	2.635	AU	Semi-major axis
q	2.112	AU	Perihelion distance
i	9.044	°	Inclination
Ω	340.307	°	Longitude of the ascending node
ω	28.429	°	Argument of periapsis
M	228.319	°	Mean anomaly
t_p	2458572.024	JD	Epoch
T	4.28	Years	Orbit duration
H	13.6	Mag	Absolute brightness

The orbital elements of 7984 Marius. Source: http://ssd.jpl.nasa.gov/ (Oct. 07, 2017)

Composition and Properties

The location of the orbit of a small body is indicative for its most likely composition. In the inner asteroid belt (orbits with semi-major axis smaller than 2.5 AU), we find silicate-rich materials mixed with metals, mainly iron and magnesium. These objects appear bright, with albedos[6] of 10–22 %, and they are classified as S-type objects. In the outer belt region, we find predominantly C-type objects with a dark, carbon-rich surface (albedo between 3 and 10%). In the middle region at around 2.7 AU, we encounter often M-type objects with metallic surfaces with a high iron-nickel content, reflecting about 10–20% of the incoming sunlight. Based on the orbit of Marius, it is therefore not possible to assign a unique taxonomic type. For the

[5]http://ssd.jpl.nasa.gov

[6]The albedo determines how much sunlight is reflected and how much is absorbed by the surface of a given object. An albedo of 10% means that 10% of the sunlight is reflected.

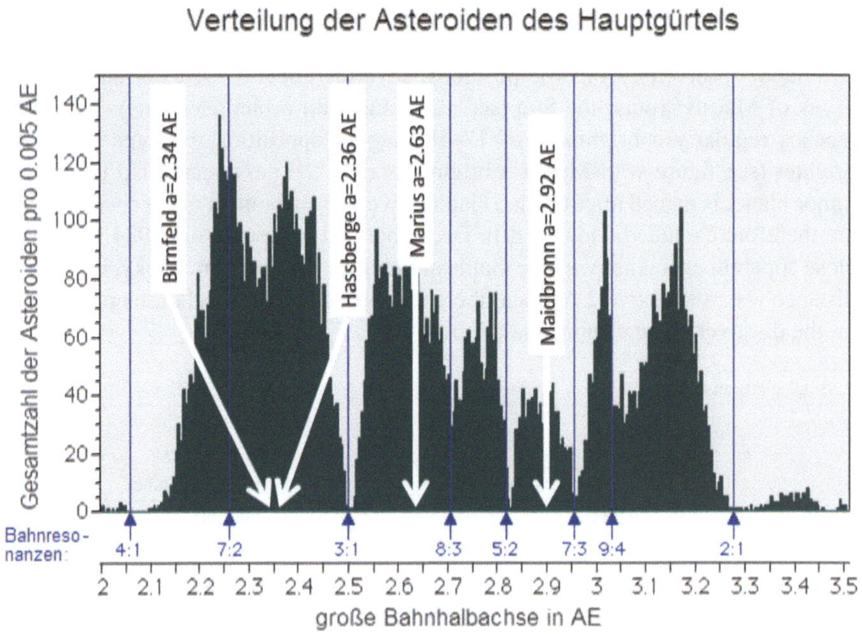

Fig. 18.4 The distribution of small bodies in the asteroid main belt. The recently named "Franconian" objects Birnfeld, Hassberge (both in the inner belt region), and Maidbronn (outer belt) are shown together with Marius (middle part of the asteroid belt)

recently baptized "Franconian asteroids," 365130 Birnfeld and 365131 Hassberge, the situation is more favorable; both lie within 2.3 AU from the Sun and have very likely silicate-rich surfaces. Also the youngest member of the "Franconian" family—410928 Maidbronn—is easier to classify; with a semi-major axis of 2.92 AU, it is located in the outer belt regions where the C-types dominate (see also distribution of the asteroids in the main belt in Fig. 18.4). A study of the orbital elements (and the long-term evolution) allows us to group many objects into collisional families. There are strong indications that these families formed from a much larger parent body during a disruptive collision, often billion years ago. However, Marius does not belong to any of the currently known collisional families.

7984 Marius appears as a point-like source on the sky, similar to most other asteroids. Even with the biggest telescopes, it is not possible to resolve the size and shape of these bodies. However, it is possible to estimate an asteroid's size from its brightness at a given distance from the Sun and the observer. One has to consider that a large and very dark object can appear to the observer as bright as a much smaller body with a high-albedo surface. From the analysis of many asteroids, we know that the albedo lies typically between 3% (very dark object) and 50% (half of the incoming sunlight is reflected), but the majority are close to 10%, while C-types have typically 7% albedo and S-types about 20%. In practical terms, astronomers simply use the object's absolute magnitude H (describing the theoretical brightness of an object at 1 au distance from the Sun and the Earth) and calculate the possible size range for low- and high-albedo assumptions. The asteroid Marius is listed with $H = 13.6$, mag which leads to a possible size between 3.8 km (high albedo of 50%) and 15.3 km (low albedo of 3%). The absolute brightness of 324 Bamberga ($H =$ 6.82 mag)—the biggest Franconian asteroid—would result in a size range between 81 and 332 km, with the recently named objects Birnfeld ($H = 17.7$ mag) and Hassberge ($H = 17.8$ mag) in the range 0.5–2.2 km and Maidbronn ($H = 16.6$ mag) about 0.9–3.9 km.

More accurate size estimates are possible by applying the radiometric method. The energy from the Sun arriving on the surface of a small body is partially reflected (depending on its albedo); the rest is absorbed and reemitted as thermal radiation. The reflected light can be described by the H-magnitude; the thermal emission can be measured at infrared wavelengths (thermal-infrared cameras). The combination of both quantities allows us to derive the object's size and albedo. Several infrared satellites (IRAS, AKARI, WISE) surveyed the sky over the last decades and detected many thousand asteroids. These measurements (available in specific archives) were combined with the object's H-magnitude, and the derived size and albedo properties were published. 7984 Marius was measured multiple times by AKARI and WISE (see Fig. 18.5, Usui et al. 2011; Mainzer et al. 2011; Masiero et al. 2011, 2012, 2014). The combined dataset gave a size determination of 10.5 ± 0.5 km and an albedo of 6%. It is therefore very likely that the surface is covered with dark carbonaceous materials, mixed with organic components, possibly also water ice or minerals modified by water, similar to the CI and CM meteorites, the so-called carbonaceous chondrites. It is interesting to note that 324 Bamberga has a similar albedo and was classified as a C-type asteroid, which is also the most likely classification of Marius. However, Bamberga is known to be much bigger, with a radiometric size of about 230 km, and owing to its high brightness at opposition, it was already discovered in the late nineteenth century.

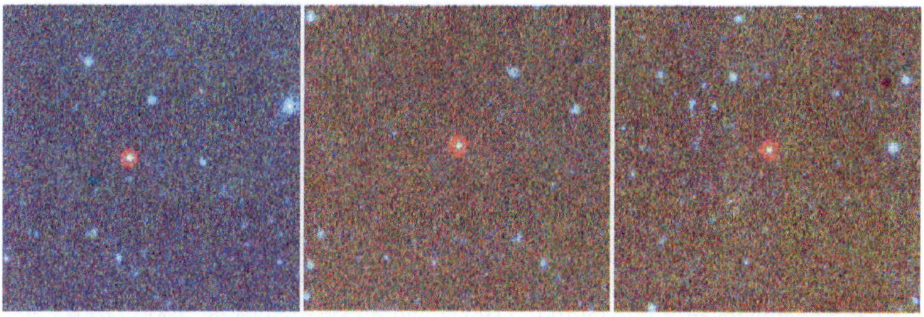

Fig. 18.5 Infrared observations of Marius with the WISE satellite. These images are produced by combining four different filters. The radiometric technique allows us to derive the object's size and albedo from such infrared measurements. The three different epochs have identifiers (from left to right): 06205b149, 06209b149, 06221b149

Fig. 18.6 Time-delayed measurements of Marius in three filters at visual wavelengths. The asteroid moves from left to right during the filter sequence. The measurements were taken by the NEAT system on July 25, 2001 between 06:20 and 07:00 UT (IDs: 20010725062117c, 20010725063718c, 20010725065134c). Source: "Near Earth Asteroid Tracking" archive at http://skyview.gsfc.nasa.gov/cgi-bin/skymorph/mobs.pl

The almost 1500 entries in the MPC archive are mainly for astrometric purposes to improve the object's orbit, but some of the measurements are taken in standard *V* or *R* filter systems (Fig. 18.6). These calibrated filter measurements show that the asteroid 7984 changes its brightness by 0.2–0.3 magnitudes within a few hours. This points to a slightly elongated shape, and the brightness variations are caused by the object's rotation. Good-quality light curves from 2011 and 2012 indicate a rotation period of 3.5341 $\pm$ 0.0006 hours, with an amplitude of about 0.15 magnitudes (Waszczak et al. 2015). This data confirms the slightly elongated shape. For a more detailed study of the rotation axis and shape, one would need many more high-quality light curve measurements. This goal could be achieved with the help of amateur observers. The next opportunity for observations will be in July 2018; Marius will reach 17 magnitudes, and it will be available for almost the entire night for observers in central Europe (Leich 2015).

For more spectacular images of 7984 Marius and a more complete characterization, one would have to launch an interplanetary mission. Up to now, more than ten asteroids have either been seen in a close encounter or have been visited by a dedicated mission. Some prominent examples are as follows: DAWN visit at Ceres (2015–2017) and Vesta (2011–2012), Rosetta flyby at 2867 Šteins (2008) and 21 Lutetia (2010), Galileo mission to 243 Ida (1993) and 951 Gaspra (1991), NEAR Shoemaker at 253 Mathilde (1997) and 433 Eros (1998, 2000, 2001), and Hayabusa rendezvous and sample return at 25143 Itokawa (2005). Two additional sample-return missions will happen in the near future. In summer 2018 the Hayabusa-2 mission (launch was in late 2014) will arrive at the asteroid 162173 Ryugu and the OSIRIS-REx mission (launch was in September 2016) at 101955

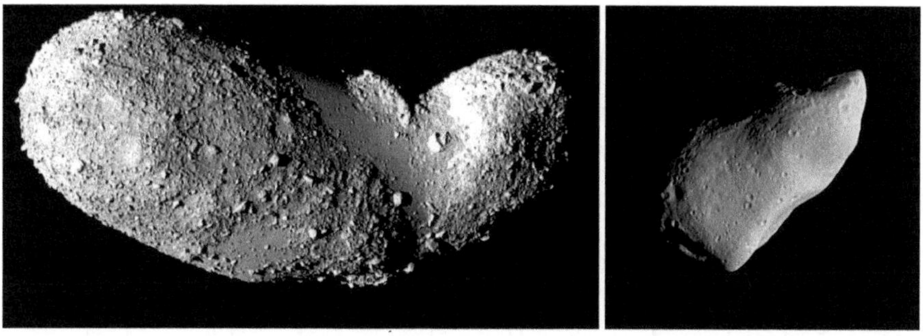

Fig. 18.7 Left, asteroid 951 Gaspra (18. 2 km × 10.5 km × 8.9 km), S-type; Galileo mission 1991 [NASA/USGS]; right, asteroid 25143 Itokawa (535 m × 294 m × 209 m), S-type; Hayabusa mission 2005 [Japan Aerospace Exploration Agency JAXA]

Bennu. The Hayabusa-2 mission is going to be very exciting because it will visit a C-type asteroid for the first time. Marius is about 10 times larger than Ryugu, but has a very similar albedo and is very likely of the same spectral type. The surface composition of both objects could therefore be very similar. The size of Marius compares very well with the Galileo mission target 951 Gaspra, which has a size of 18.2 km × 10.5 km × 8.9 km (see Fig. 18.7). However, Gaspra is an S-type with a different kind of surface and an albedo of 22%.

Marius has a very well-determined orbit and is located in the middle of the asteroid belt between Mars and Jupiter. Its orbit is controlled by gravitational forces (Sun, planets, close encounters with large asteroids), which makes the orbit predictable and very stable over long timescales. Non-gravitational forces, like the Yarkovsky and YORP effects related to the anisotropic emission of the thermal radiation, play a minor role. These secondary forces are mainly relevant for very small objects below 10 km in size. It is therefore expected that 7984 Marius will orbit the Sun for the next millions or even billions of years, maybe with small perturbations following close encounters with other large asteroids. Occasionally Marius will "meet" other Franconian asteroids at least in a projected view from the Earth. Many asteroids are known to have small satellites, and it might turn out that Marius also has a small moon. What should we name it, "Ansbach" or "Gunzenhausen"? The satellite name has to be related to the name of the main body according to the IAU rules.

Observability

Ongoing sky surveys detect 7984 Marius frequently and provide the measured coordinates to the MPC for improving the object's orbit. Apart from these automatic observations, there are no dedicated scientific projects existing in the context of Marius. However, the asteroid reaches frequently 18 or even 17 magnitudes at

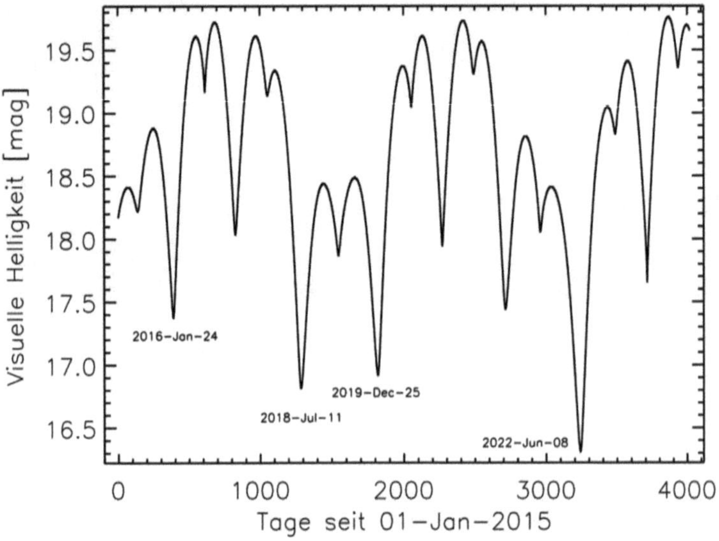

Fig. 18.8 The apparent brightness of 7984 Marius between January 2015 and December 2025. During favorable oppositions the object reaches 17.5 mag, in June 2022 up to 16.4 mag. Smaller numbers for the brightness (in magnitudes) indicate a larger brightness in the sky. Visibility from central Europe will be very limited during the opposition in July 2018

oppositions (see Fig. 18.8; Leich 2015). This makes Marius an attractive target for experienced amateur astronomers with powerful telescopes. Marius will be nicely observable in July 2018, in December 2019 (Fig. 18.9), and especially in June 2022. In these periods it would be possible to measure the object in standard filter systems, required for the determination of the (very likely) surface composition. Monitoring the object's light curve over several nights will help to confirm the estimated rotation period and allow us to derive an approximate spin-axis orientation and the object's shape. It would also be interesting to search for possible stellar occultation events. If the predicted orbit of Marius directly hits the coordinates of a given star (as seen from a given observer on Earth), then the star would disappear for a few seconds and reappear, a stellar eclipse produced by Marius! These events are regularly measured by networks of amateur astronomers to map the projected shadow of a given asteroid. The measurements lead directly to a size and shape determination of the occulting asteroid. Also satellites and rings have already been found using this technique, which would be too faint to be observed otherwise.

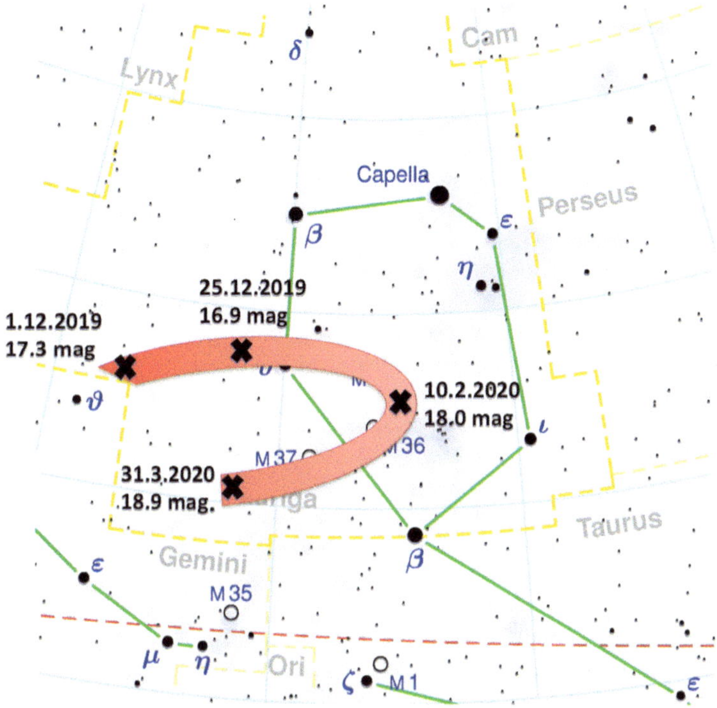

Fig. 18.9 The apparent orbit of 7984 Marius between December 2019 and March 2020 as seen from central Europe. Marius will reach 16.9 mag and will be observable by amateurs for almost the entire night in that phase

References

Leich, Pierre (2015): Zur Sichtbarkeit von (7984) Marius. Ein Aufruf an die Beobachter, *Regiomontanusbote* 28 (2015), issue 4, pp. 10–11

Mainzer, Amy et al. (2011): NEOWISE Studies of Spectrophotometrically Classified Asteroids: Preliminary Results, *The Astrophysical Journal* 741 (2011), issue 2, article id. 90, 25 pp.

Masiero, Joseph R. et al. (2011): Main Belt Asteroids with WISE/NEOWISE. I. Preliminary Albedos and Diameters, *The Astrophysical Journal* 741 (2011), issue 2, article id. 68, 20 pp.

Masiero, Joseph R. et al. (2012): Preliminary Analysis of WISE/NEOWISE 3-Band Cryogenic and Post-Cryogenic Observations of Main Belt Asteroids, *The Astrophysical Journal* 759, issue 1, article id. L8, 5 pp.

Masiero, Joseph R. et al. (2014): Main-belt Asteroids with WISE/NEOWISE: Near-infrared Albedos, *The Astrophysical Journal* 791, issue 2, article id. 121, 11 pp.

Schmadel, Lutz D.: *Dictionary of Minor Planet Names*, Berlin: Springer Verlag, Addendum to 6th Edition: 2012–2014, ISBN 978-3-319-17677-2, 2015

Usui, Fumihiko et al. (2011): The Asteroid catalog using AKARI, *Publications of the Astronomical Society of Japan* 63 (2011), pp. 1117–1138

Waszczak, A.; Chang, C.-K.; Ofek, E. O. et al. (2015): Asteroid Light Curves from the Palomar Transient Factory Survey: Rotation Periods and Phase Functions from Sparse Photometry, *The Astronomical Journal 150*, issue 3, article id. 75, 35 pp.

Chapter 19
Planet Model 1: 50 Billion

Rudolf Pausenberger

Abstract Galileo's and Marius's telescopic observations are simulated by a model at a scale of 1:50 billion. It is one of several hands-on exhibits that were built for a mobile exhibit about the history of astronomy in a school project.

Jupiter and his moons are observed through a telescope from the position of the earth, i.e., from a distance of 13 m. The celestial bodies, especially the small moons (0.06 mm in diameter), are represented by holes in construction paper, engraved by a laser cutter. They are illuminated from behind. The exhibits show the constellations during some historic nights following the 7th of January 1610 that led to the conclusion of orbiting satellites.

Mobile Exhibit as a School Project

Between 2010 and 2013, pupils of the Gymnasium in Lauf a.d. Pegnitz designed and built two travelling exhibitions on the history of astronomy and scientific discoveries made in Nuremberg from the turn of the Middle Ages up to the modern era (Pausenberger 2014). The aim was to give visitors an opportunity to experience the ideas of great discoverers by means of their own hands-on experiments and thus to actually be able to grasp them. The exhibits that illustrate the discovery of Jupiter's moons were also built during this project (Fig. 19.1).

Simon Marius and the Discovery of the Jupiter's Moons

The planet Jupiter has been well known since ancient times and can be seen clearly in the night sky with the naked eye. Its moons, however, are too faint for this and require telescopic observation. As soon as this was invented, Galileo Galilei and Simon Marius, independently from each other, discovered Jupiter's four greatest

R. Pausenberger (✉)
Christoph-Jacob-Treu-Gymnasium, Lauf a.d. Pegnitz, Germany

H. Gaab, P. Leich (eds.), *Simon Marius and His Research*, Historical & Cultural Astronomy, https://doi.org/10.1007/978-3-319-92621-6_19

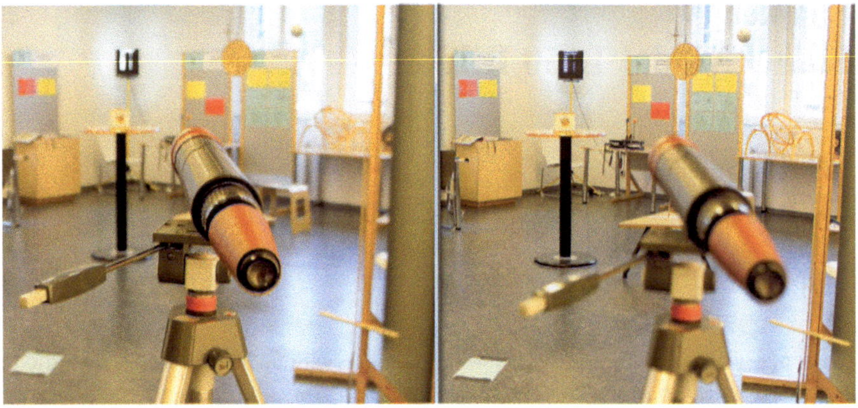

Fig. 19.1 A stereo image of the travelling exhibition (Pausenberger/physik), built in the Phenomena in Flensburg: in the foreground, one can see the telescope directed toward the light box in the back. Squint cross-eyed to observe the 3D effect

moons in January 1610. Our aim was to recreate their visual impression in an experimental setup. For exhibition rooms, a distance from the earth, i.e., the position of the observer, to the model of Jupiter of 13 m is suitable, which equals a scale of 1:50 billion. Then the radius of Callisto's orbit is 38 mm, Io's 8 mm. If all the astronomical dimensions are scaled by the same factor, the angles and thus the optical impressions remain the same. Visitors can therefore realistically compare the naked eye view with their observation through the telescope. In the first case, they will perceive only the vague outlines of the great planet as a disk; in the second one, they will be able to identify its moons clearly.

The exhibit shows the constellation of the celestial bodies in several phases during the historic week beginning January 7, 1610, revealing the crucial changes that finally suggest the conclusion of orbiting moons (Fig. 19.2).

Technical Implementation

On the chosen scale, the radii of the four moons lie between 0.03 and 0.05 millimeters. We tested ways to illuminate holes of this size from behind. Perforating an aluminum foil lying on a hard surface with a thin pin, pupils quickly reached the limits of manual reproducibility, especially with regard to a circular edge. Alternatively, in a simple photographic reduction on negatives, the contrast was not strong enough for a brightly luminous planet in front of the black sky. On the other hand, the use of a laser cutter was not a problem for the pupils; a vector graphic only needs to be drawn in the correct dimensions. This machine cuts in black paper simply with the usual driver command "print"—in really perfect quality. A laser cutter is

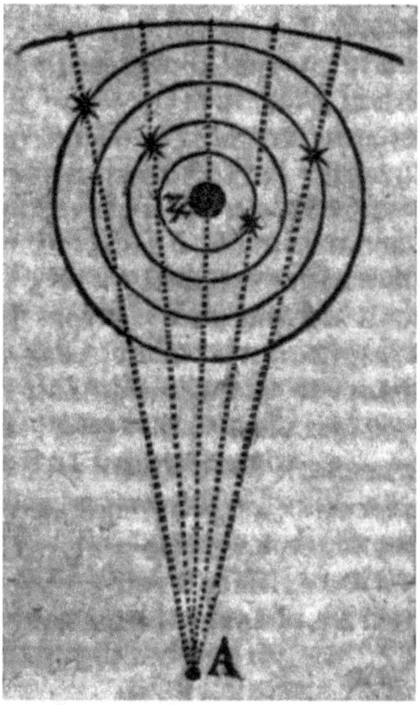

Fig. 19.2 Simon Marius concludes from his observations in *Mundus Iovialis* that the four moons revolve around Jupiter on circles (Marius 1614). Point A is the position of the astronomer's eye. From *Prognosticon Astrologicum auf das Jahr 1612*, Staatsarchiv Nürnberg (Prog. 1612) (Stains touched up)

accessible to the public in a *Fablab*, a high-tech workshop with computer-controlled machines, usually organized as a registered club (Fig. 19.3).

The exhibition visitors can decide for themselves how deeply they wish to investigate the historical background. A multilevel explanation offers information on the use, history, and background of the experiment (Fig. 19.4).

However, the pinhole, which is drawn to scale, would be somewhat too large after cutting, since the laser cannot be focused infinitely narrowly and the cutting width depends on the laser power. On the other hand if the beam of a normal laser pointer is diffracted on a straight test section, the (half) width of the cutting line can be determined with the aid of the interference figure and then the hole radius can be correspondingly reduced in the vector graphics (Fig. 19.5).

Model of the Solar System

In this way, it is not only possible to illustrate the world of Jupiter but the entire solar system: The earth here has a diameter of 0.25 mm and is 3.0 m away from the sun; 8 mm next to her is our moon.

What would be the German government debt (Wiki), piled up in 2.3-mm-thick 1-Euro coins? 2 trillion euros would then amount to over 10 cm. What the model shows is also true; it is a mountain, 13 times the distance to the moon! Even the 340 billion euro financial support for Greece yield more than twice the distance to the lunar orbit.

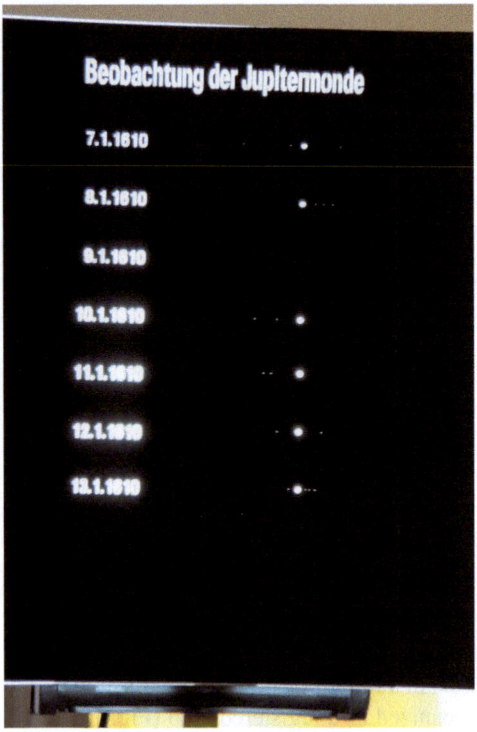

Fig. 19.3 The stencil made of construction paper in front of a spotlight. This stencil is illuminated from the back by a matt acrylic plate as a diffusing disk with a spotlight and viewed through a telescope

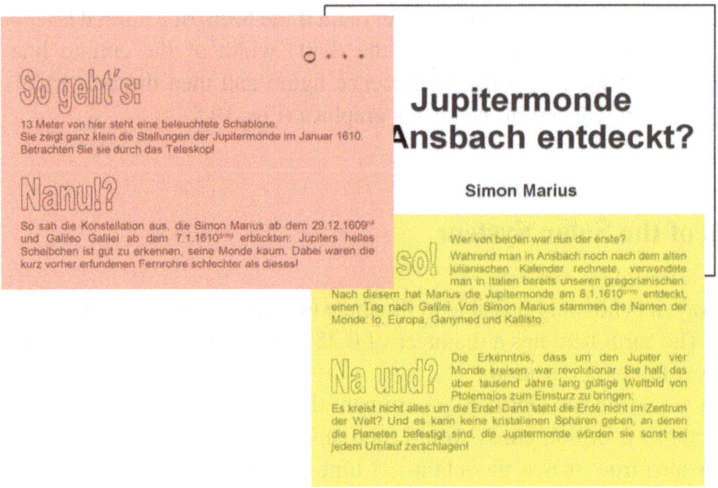

Fig. 19.4 Four steps of explanation: "So geht's," How to; "Ach so," I see . . .; "Nanu!?," Hello, what's this?; "Na und?," So what?

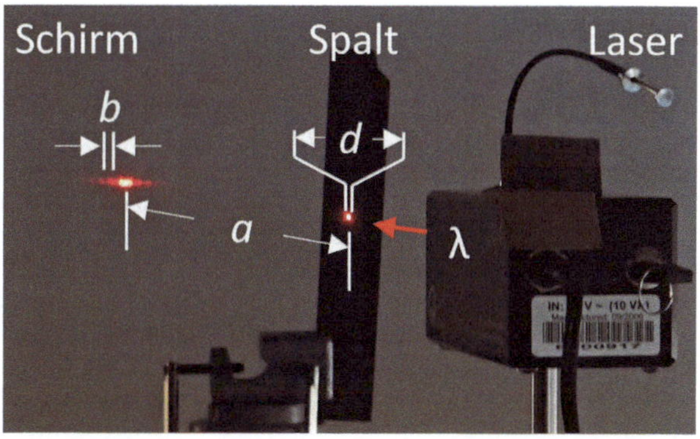

Fig. 19.5 The gap width ("Spalt") of $d = \lambda a/b = 2 \times 0.029$ mm results from the wavelength $\lambda = 633$ nm and the dimensions $a = 100$ cm and $b = 11$ mm

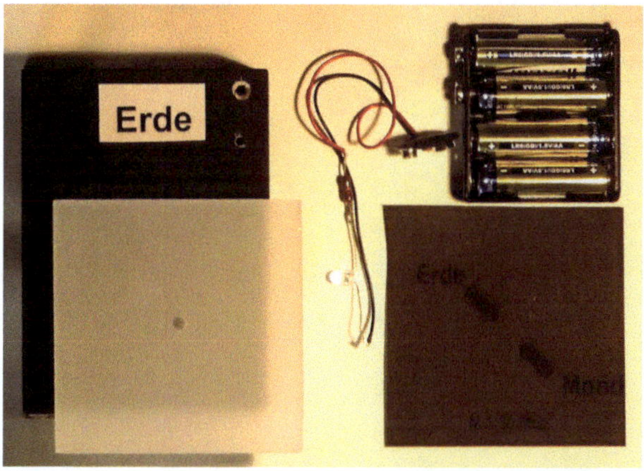

Fig. 19.6 The LED is inserted from behind into the hole in the dull acrylic disc; the black stencil with the motif is positioned on top of that. The holes for earth ("Erde") and moon would be smaller than a pixel and are therefore not recognizable

Back from the astronomical debt to the astronomical model. The small holes for the planets are illuminated from behind with a white LED. The brightness with which the celestial bodies appear to us in reality is, however, not suitable, for the exhibition space will hardly be as dark as at the night sky after the astronomical twilight. Even then the dark adaptation of the visitors' eyes would take an unreasonably long time. If, however, a dimly lit room is chosen for the presentation, a relatively high brightness for the exhibit itself is uncritical. Therefore, the spotlight behind the template of Jupiter's moons can serve as a reference value for all other objects. Saturn, for example, appears to be darker because it is further away from us. This is evident in the scale model as well. On the other hand, it is weaker because of its distance from the sun. This can be achieved by a suitable series resistor for the light-emitting diode or by a suitably gray-printed foil as a filter. A reddish hue is appropriate for the corresponding Mars foil (Fig. 19.6).

This and other exhibits of the travelling exhibition can be borrowed for such presentations or for educational purposes (Pausenberger/physik).

References

Fablab: Website of the Fablab Nürnberger Land http://nueland.de

Marius, Simon (1614): *Mundus Iovialis Anno M.DC.IX. Detectus Ope Perspicilli Belgici.* Nürnberg: Johann Lauer 1614

Pausenberger, Rudolf (2014): Schüler bauen interaktive Wanderausstellungen: "Astronomie beGreifen" und "Renaissance trifft Physik" in: Manfred Schukowski, *Mittelalterliche Astronomische Großuhren*, Leipzig 2014

Pausenberger/physik: A more complete presentation of this project can be found at www.physik.de.rs

Wiki: Government dept https://de.wikipedia.org/wiki/Staatsverschuldung_Deutschlands

Chapter 20
"Sun, Moon, and Marius"

An Exhibition with Images from the Youth Art School "Obraz" in Protvino (Russia)

Olga Sinzev

"Bright beautiful Andromeda," "A beautiful galaxy," "The fiery Andromeda Galaxy," "Wonderful phenomenon," "Fascination Cosmos," "Tycho Brahe Observatory," "Astronomical Research in Ansbach" ... and of course "The greatest moons of Jupiter" and "The astronomer Simon Marius." The artists of these images and of 25 other original works that are dedicated to the findings of the Franconian astronomer Simon Marius and astronomical researches in general are students of the Youth Art School "Obraz" from Protvino near Moscow. Thirty participants of the project, aged between 4 and 21, worked as a cohesive team on the pictures sent to Nuremberg in November 2013. Their images, in which different techniques of painting were used, formed the thematic focus of a mobile exhibition that was displayed in the Nuremberg planetarium from mid-September 2014 to the end of March 2015.

When margravial court astronomer Simon Marius from Gunzenhausen was given the chance to use a "Belgian" telescope in the summer of 1609, he observed from Ansbach comets, planets, stars, and nebulas. He is supposed to have discovered the largest moons of Jupiter—Io, Ganymede, Europa, and Callisto—in a December night from the castle tower.

The moons of Jupiter were not the only discoveries Marius made with the recently developed telescope. He was the first European, who described the Andromeda Galaxy. The naming of Jupiter's moons after characters from antique mythologies originates from Marius and was taken up again in the twentieth century (Fig. 20.1).

Tamara Kuleshova, art teacher at the Youth Art School "Obraz" in Protvino near Moscow, regarded this as a fascinating subject for her students. Young artists in Protvino worked for about 2 months on images that are dedicated to the findings of the Franconian astronomer from the beginning of the seventeenth century and astronomical researches in general. They collected historical material and became familiar

O. Sinzev (✉)
Resonanz Magazine, Nuremberg, Germany
e-mail: redaktion@resonanz-medien.de

H. Gaab, P. Leich (eds.), *Simon Marius and His Research*, Historical & Cultural Astronomy, https://doi.org/10.1007/978-3-319-92621-6_20

Fig. 20.1 The principal of the Obraz Art School in Protvino, Tamara Kuleshova, with her students

Fig. 20.2 Obraz team visiting Radio Podmoskovye in November 2013

with the architecture and typical cultural traditions. They learned the constellations precisely and dealt with the construction of the telescope. Even the astronomical observatories and observational instruments of Tycho Brahe were discussed. All the participants in the project—aged between 4 and 21—worked as a united team. "In the course of the work the children dived deeper and deeper into the subject astronomy and demonstrated amazing fantasy," Kuleshova explained (Fig. 20.2).

In November 2013 34 images were sent to the project's media partner—the intercultural magazine *Resonanz* in Nuremberg. The pictures, created using various painting techniques, were displayed in the mobile exhibition, "Sun, Moon, and Marius." It was shown at the end of 2013 at the Museum for Russian language in Nuremberg and at the beginning of 2014 in the Gothic Chamber of Ansbach Town Hall. From mid-September 2014 to the end of March 2015, the well-attended exhibition was displayed in the Nuremberg planetarium.

The modern Marius portrait, painted by Maria Dementeva for the *Resonanz* cover (issue November 2013), was given by the editors, after publication, into the care of the Memorial Museum of Cosmonautics in Moscow and was handed over to the director, former Russian cosmonaut Alexander Lasutkin, in December 2013. Now the then 15-year-old is studying painting and theater at the Moscow State Academic Art Institute named after VI Surikov and portrayed Simon Marius, as he is shown on his only known portrait; however, in front of him lies not his telescope and his book but the orbital plan of the four largest Jupiter moons. In his hand he does not hold compass and alembic that identify him as a mathematician and physician but his article *Kurtze und eigentliche Beschreibung des Cometen* (Short and proper Description of the Comet) of 1596. An armillary sphere and a sextant with telescope are lying next to him. Several constellations can be seen in the background.

The project gathered a lot of attention and appreciation in the Russian media. In August 2015 the intercultural magazine *Resonanz* from the Nuremberg Metropolitan Region was awarded the international media prize BeBiLin.RU-2015 in the category "The best presentation of bilingualism in mass media"—inter alia for the bilingual reporting about the Marius anniversary year. The award was presented by the Kazan Federal University in cooperation with numerous international partners.

The magazine *Resonanz* had already accompanied the International Year of Astronomy 2009 that was proclaimed by the United Nations. It reported about the Nicolaus-Copernicus-Planetarium Nürnberg and the Uluhg Beg Observatory in Samarkand, Uzbekistan. Repeatedly articles appeared about the Long Night of

a b c

Fig. 20.3 **(a–c)** Darja-Gorodnaja dealt with Tycho Brahe's astronomical observatory Uraniborg on Hven, a Danish island in the Øresund; Maria Dementeva supplied the cover picture for the German edition of this anthology; Ksenia Rischova in front of her sky picture

Fig. 20.4 (**a–c**) Slava Mihaltsov painted the Great Bear constellation and the Jovian system; Vika Eshova depicted Tycho Brahe and Simon Marius; Polina Harlamova used Johannes and Elisabeth Hevelius 1673 in Gdansk with a sextant, as a role model

Fig. 20.5 (**a–c**) Russian cosmonaut Alexander Lasutkin in December 2013 with Simon Marius on the cover of the magazine Resonanz; media prize BeBiLin.Ru-2015 was awarded to the magazine Resonanz in the category "The best presentation of bilingualism in mass media"; view of the Simon Marius Exhibition at the Nuremberg planetarium

a
b

Fig. 20.6 (**a**, **b**) The founder of the Obraz Art School in Protvino, Tamara Kuleshova, with her pupil Maria Dementeva, Valeri Borisov—Lord Mayor of the City of Science Protvino—(1st from the left), and the book author Nikolai Popravko. On March 20, 2016, the school looked back on 30 successful years with a great celebration. In January 2017 a copy of the Marius book arrived in Protvino and some of the pupils found themselves

Sciences Nuremberg-Fürth-Erlangen and the Day of Science of the Nuremberg Metropolitan Region. Since September 2013, 12 articles reported on various aspects of the Simon Marius Anniversary. Furthermore the publisher NABI provided the translation of the Marius Portal menu into Russian, and it was fully operable at the initial presentation of the portal on February 18, 2014, at the Nuremberg States Archives. Finally, Maria Dementeva's portrait of Marius adorns the cover of this book (Figs. 20.3, 20.4, 20.5 and 20.6).

Simon Marius: Works[1]

Marius, Simon (1596)
Kurtze und eigentliche Beschreibung des Cometen oder Wundersterns / So sich in disem jetzt lauffenden Jar Christi unsers Heilands / 1596. in dem Monat Julio / bey den Füssen deß grossen Beerens / im Mitnächtischen Himmel hat sehen lassen. Gestellet durch Simonem Maierum Guntzenhusamum, Alumnum Sacrifontanum. Nürnberg: Paul Kauffmann 1596
http://www.simon-marius.net/index.php?lang=en&menu=3&id=2

Marius, Simon (1599)
Tabulae Directionum Novæ. Universæ penè Europæ inservientes in quibus I. Verissimus antiquorum Astrologorum ipsisusque Ptolemæi duodecim cœli domicilia distribuendi modus non tam restitutus, quam de nouo inuentus. II. Directionis Ptolemaicæ vtriusque tam artificiosæ quam vulgaris facilior & exactior ratio. III. Constituendi aspectus vsitata ratio emendata, atque antiquorum (à neotericis huc vsque neglecta, vel potius non intellecta) in lucem reuocata. Omnia ex vno eodemq [ue]; fundamento promanantia, Methodo facilima, verißima, planeq[ue]; naturalitr aduntur. Autore Simone Mario Guntzenhusano, Stipendiario & Alumno Sacrifontano. Nürnberg: Christoph Lochner 1599
http://www.simon-marius.net/index.php?lang=en&menu=3&id=6

Marius, Simon (1610)
Die Ersten Sechs Bücher Elementorum Evclidis, In welchen die Anfäng vnd Gründe der Geometria ordenlich gelehret / vnd gründtlich erwiesen werden / Mit sonderm Fleiß vnd Mühe auß Griechischer in vnsere Hohe deutsche Sprach übergesetzet / vnd mit verständtlichen Exempeln in Linien vnd gemeinen Rational Zahlen / Auch mit Newen Figuren / auff das leichtest vnd aigentlichest erkläret: Alles zu sonderm

[1]Marius, Simon (Prog. 1601–1629): Information on all calendars of Simon Marius can be found at https://www.simon-marius.net/calendars

H. Gaab, P. Leich (eds.), *Simon Marius and His Research*, Historical & Cultural Astronomy, https://doi.org/10.1007/978-3-319-92621-6

Nutz denjenigen / so sich der Geometria / im Rechnen / Kriegßwesen / Feldtmässen / Bauen / vnd andern Künsten vnnd Handtwerckern zugebrauchen haben: Auß Befehl Deß Edlen vnd Gestrengen Herrn / Hanß Philip Fuchß von Bimbach / zu Möhrn / Alten Rechenberg vnd Schwaningen / Obristen: Durch Simonem Marium Guntzenhusanum Franc. Fürstlichen Brandenb: bestalten Mathematicum, vnd Medicinæ Utriusq[ue], Studiosum. Ansbach: Paul Böhem 1610
http://www.simon-marius.net/index.php?lang=en&menu=3&id=4

Marius, Simon (1614)
Mundus Iovialis Anno M.DC.IX. Detectus Ope Perspicilli Belgici, Hoc est, Quatuor Jovialium Planetarum, Cum Theoria, Tum Tabulæ, Propriis Observationibus Maxime Fundatæ, Ex Quibus situs illorum ad Iovem, ad quodvis tempus datum promptissimè & facilimè supputari potest. Inventore & Authore Simone Mario Guntzenhusano, Marchionum Brandenburgensium in Franconiâ Mathematico, puriorisque Medicinæ Studioso. Nürnberg: Johann Lauer 1614
http://www.simon-marius.net/index.php?lang=en&menu=3&id=1

Marius, Simon (1614/1916)
Prickard, A.O., The 'Mundus Jovialis' of Simon Marius, *The Observatory. A review of astronomy* 39 (1916), pp. 367–381, 403–412, 443–452, 498–503
http://www.simon-marius.net/index.php?lang=de&menu=4#Prickard_1916

Marius, Simon (1614/1988)
Schlör, Joachim (ed.): *Mundus Iovialis – Die Welt des Jupiter. Die Entdeckung der Jupitermonde durch den fränkischen Hofmathematiker und Astronomen Simon Marius im Jahr 1609 – lateinisch und deutsch* (= *Fränkische Geschichte*, Vol. 4). Gunzenhausen: Johann Schrenk 1988
http://www.simon-marius-gymnasium.de/s-marius/mundus-iovialis-die-welt-des-jupiter-die-zweisprachige-ausgabe

Marius, Simon (1614/1916/2019)
Marius Simon, The World of Jupiter, translated by A.O. Prickard and Albert van Helden, in *Simon Marius and his Research*, Hans Gaab and Pierre Leich (ed.), Cham: Springer 2019

Marius, Simon (1619)
Astronomische vnd Astrologische beschreibung deß Cometen so im November vnd December vorigen 1618. Jahrs ist gesehen worden / Genommen vnd Gestelt auß eygnen Observationibus dabey auch andere sachen kurtz eingemischet werden. Durch Simon Marium Guntzenhusanum, Fürstlichen Brandenburgischen bestelten Mathematicum vnnd Medicum. Nürnberg: Johann Lauer 1619
http://www.simon-marius.net/index.php?lang=en&menu=3&id=5

Marius, Simon (1625)
Gründliche Widerlegung der PositionCirckel / Claudij Ptolomæi, vornemblichen aber / Johannis Regiomontani; mit grosser Mühe vnnd vielem Nachdencken / so wol auß Ptolomæo selbsten / als auch allen andern vortrefflichen Astrologen, so von Ptolomæi Zeiten an / biß auff Regiomontanum gelebet / vnd von directionibus

Theoricè und Practicè geschrieben: zusammen gezogen / Durch Simon Mairn / F.F. B.B. bestellten Mathematicum vnd Medicum. An jetzo aber auff vornehmer vnd Kunstliebender Personen Communication vnd Begehren allen der Astrology zugethanen / zu sonderbarem Gefallen vnd Nutz in offentlichem Truck erstmals publiciert / Durch Danielem Mögling Würtemberg. Phil. ac Med. Doctorem, auch Landtgräv. Hessischen Hoff-Med. vnd Math. zu Butzbach / etc. Franckfurt am Mayn: Lukas Iennis 1625

http://www.simon-marius.net/index.php?lang=en&menu=3&id=7

Prog. yyyy

Calendars and Prognostica for 1601–1629 see *Marius-Portal*, Menue Complete Works | Calendars.

www.simon-marius.net/calendars

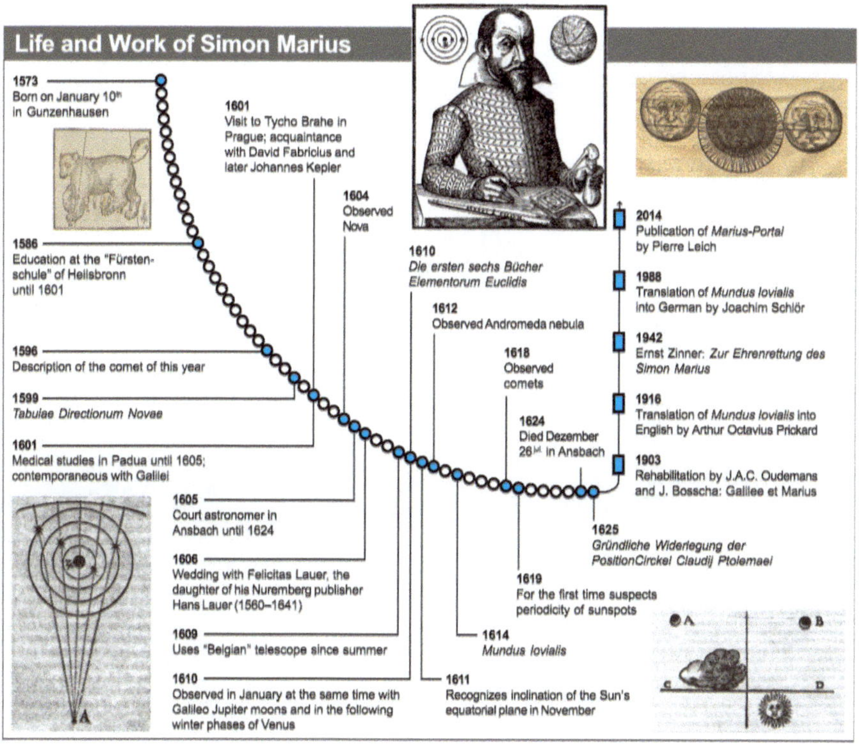

About the Authors

Thony Christie is a, predominantly Internet, freelance historian of science, whose main interest is the history of the mathematical sciences in the Early Modern Period, including, however, astrology, medicine, and alchemy. He has blogged for many years as the *Renaissance Mathematicus*, was for several years the managing editor of the monthly history of science blog carnival *On Giant's Shoulders*, and is the creator, compiler, and editor of the weekly, histories of science, technology, and medicine, Internet links list *Whewell's Gazette*, as well as being English language editor of the *Marius-Portal*. He specializes in debunking myths in the history of science and presenting historically accurate popular accounts of that history. His most recent publications were in *AEON* the digital Ideas magazine and in *Viewpoint* the digital magazine of the British Society for the History of Science. He resides in Spardorf in Middle Franconia, Germany.

H. Gaab, P. Leich (eds.), *Simon Marius and His Research*, Historical & Cultural Astronomy, https://doi.org/10.1007/978-3-319-92621-6

Wolfgang R. Dick, born in 1955 in Greiz, studied astronomy in Kharkov (Ukraine). He worked at the Institute for Astrophysics in Potsdam (1982–1991), at the Bonn University Observatory (1991–1992), and at the Federal Agency for Cartography and Geodesy (BKG), Potsdam Branch Division (1992–2000). Since mid-2000, he has been a member of the scientific staff at the Central Bureau (maintained by BKG in Frankfurt am Main) of the International Earth Rotation and Reference Systems Service. From 1992 to 2014 he was Secretary of the Working Group for the History of Astronomy of the German Astronomical Society. He is co-founder and co-editor of *Acta Historica Astronomiae.* Recently, he published the expanded second edition of the *Biographical Index of Astronomy* together with Wilhelm Brüggenthies.

Hans Gaab, born in 1956 in Ansbach, is a teacher for mathematics and physics at the Labenwolf-Gymnasium in Nuremberg and lives with his wife in Fürth. During 1986–1989, he worked as a math teacher for the Deutscher Entwicklungsdienst DED (German Development Service) in the capital of Tanzania, Dodoma. He has been studying local history of astronomy for many years and has published numerous articles. In 2006, he was awarded the Silver Medal "Bene Merenti de Astronomia Norimbergensi" of the Nuremberg Astronomical Society. In 2010, he gained his doctorate for a thesis about Abdias Trew (1597–1669). It was published as volume 42 of the *Acta Historica Astronomiae*. In 2015, he published an extensive work about Albrecht Dürer's star charts. From 2015 to 2017 he was Vice President of the Simon Marius Society.

Christopher M. Graney is professor of physics and astronomy at Jefferson Community and Technical College in Louisville, Kentucky (USA) and the author of the recent book *Setting Aside All Authority: Giovanni Battista Riccioli and the Science Against Copernicus in the Age of Galileo* (University of Notre Dame Press, 2015). His recently completed translation of the 1614 *Disquisitiones Mathematicae* of Christoph Scheiner (respondent: Johann Georg Locher) will be published at the end of 1617 by the University of Notre Dame Press.

Ph. Jürgen Hamel, born in 1951 in Stralsund, studied philosophy and history in Leipzig and gained his doctorate for a thesis on the early history of astrophysics. During 1978–1991, he worked at the Archenhold-Observatory in Berlin where he inter alia worked on the editions of the Collected Works of Copernicus and of Kepler. During 1998–2000, he worked at the Museum of Astronomy and Technological History in Kassel. He held lectures at the University of Landau on the history of astronomy and introduction to astronomy for many years. Hamel's special fields are the history of astronomy in the late Middle Ages, Early Modern Era and eighteenth/nineteenth century, astronomical bibliography, astronomy and cultural history as well as the history of astronomical monitoring tools. He has published numerous works about scientific and cultural history in Mecklenburg-West Pomerania, especially the area Stralsund-Barth. Hamel is co-founder and co-editor of the journal series *Acta Historica Astronomiae* and also editor-in-chief of the journal *Astronomie+Raumfahrt im Unterricht* (Astronomy & Aerospace at school). Hamel participated in the organization of numerous exhibitions, partly as curator, in Stuttgart, Vienna, Berlin, Rostock, and Stralsund.

Dieter Kempkens, born in 1952, studied history and German philology in Bochum. He works in adult education for an international company. His main research topic is the history of the early modern era in Europe, especially the relevance of periodical media for the explanation of historic processes and the establishment of a critical public in the emerging knowledge-based society. He has published several works about this, most recently *Der "Aviso" von 1609 als Quelle für den Beginn des Jülich-klevischen Erbfolgestreits* (The "Aviso" of 1609 as a source for the beginning of the Juelich-Kleve war of succession, 2010) and *Der Erfolg der Prognostica auf dem Buchmarkt in der frühen Neuzeit* (The success of Prognostica in the book trade of the early modern era, 2014).

Richard L. Kremer, born in 1952 in Nebraska (USA) studied physics and religion at Goshen College, Indiana (USA) and earned his Ph.D. in History of Science from Harvard University with a thesis about "The thermodynamics of life and experimental physiology, 1770–1880." Since 1985, he has been teaching courses in the history of science, medicine, and technology at the Dartmouth College, New Hampshire (USA). He did several research stays in Munich, Berlin, Leipzig, Wolfenbüttel, Paris, and London. Focuses of his researches are the history of mathematical astronomy from Ptolemaeus to Kepler with particular attention on astronomical boards, instruments, and calculation methods as well as the history of printing of astronomical second rank literature (calendars, prognostica). Since 2007, he has been Reviews Editor of the *Journal for the History of Astronomy*.

Pierre Leich, born in 1960, is a freelancer in science communication and the history of astronomy. For 10 years, he was chairman of the art fair ART Nuremberg, edited an art magazine for 6 years and for 4 years he was managing director of Theatresports World Championship to FIFA WM 2006™. Since 2003, Leich has been project manager of the Long Night of Sciences in Nürnberg/Fürth/Erlangen and for the Science Day of the Nuremberg Metropolitan Region. Leich is honorary vice chairman of the Cauchy-Forum-Nürnberg, curator of the Nuremberg Astronomical Society, member of the advisory board of the Nicolaus-Copernicus-Planetarium Nuremberg and member of the Science Forum of the Nuremberg Metropolitan Region. In the "International Year of Astronomy 2009," he was branch manager for Northern Bavaria. Since 2014, he has been editor of the *Marius-Portal* and President of the Simon Marius Society.

Klaus Matthäus, born in 1940, completed a bookseller apprenticeship and studied history and book studies, Munich and Erlangen. After that, he worked for almost 40 years as a bookseller in Erlangen. He was a partner in the bookstores Theodor Krische, Mencke-Blaesing, Palm & Enke and from 1992 onwards of the university bookshop Jena. His thesis "About the history of the Nuremberg calendars. The development of the Nuremberg annual calendars printed in book form" was published in *Archiv für Geschichte des Buchwesens*, Vol. 9 (1969). In 2001, Hans Gaab and Klaus-Dieter Herbst induced him to pick up the calendar-theme once again. Additional publications followed, inter alia to Johann Christoph Sturm, Johann Jakob Christoffel von Grimmelshausen and to the history of book trade in Erlangen. Since 2014, he has been committee member for calendars of the Simon Marius Society.

Thomas Müller, born in 1966, is a scientist at the Max-Planck-Institute for Extraterrestrial Physics in Garching. He works in the field of solar system research and he is specialized in Infrared Astronomy. He studied physics and astronomy at the Universities of Würzburg and Albuquerque (New Mexico). In 1997, he gained his doctorate at the Max-Planck-Institute for Astronomy in Heidelberg on the subject of thermophysical modelling of asteroids. Since 2016, he has been leading a European research project on the characterization of small bodies (minor planets, comets, moons, and trans-Neptunian objects) in our solar system. The minor planet 8793 carries the name Thomasmueller in recognition of his contributions to the study of minor planets in the thermal infrared.

Dagmar L. Neuhäuser M.A. was born in 1966 in Frankfurt/Main. Since 1994, she has written and produced numerous radio features, especially for the Bayerischer Rundfunk (Bavarian Broadcasting) at the crossroads of arts and sciences. Since 2013, she has been active in studying the practice and relevance of historical observations of celestial phenomena as well as Terra-Astronomy with various scientific publications and presentations (www.astro.uni-jena.de/index.php/terra-astronomy.html), inter alia Neuhäuser, D.L. & Neuhäuser, R., A red cross appeared in the sky, and other celestial signs: Presumable European aurorae in the mid AD 770s were halo displays, *Astronomical Notes* Vol. 336, pp. 913–929 (2015). She is married to Ralph Neuhäuser.

Ralph Neuhäuser, born in 1966 in Soest/Westphalia, is ordinary full professor of astrophysics and institute director at the University of Jena. He is the author of numerous peer-reviewed publications in scientific journals on star and planet formation as well as neutron stars, since 2013 mainly on Terra-Astronomy, namely using historical observations for reconstruction of solar activity and Galactic supernovae (www.astro.uni-jena.de/index.php/terra-astronomy.html). He maintains a network for inter-disciplinary collaboration with scholars of, e.g., history, philology, Semitic studies, East-Asian language. He is married to Dagmar L. Neuhäuser.

Jay M. Pasachoff, born in 1943 in New York City, is Field Memorial Professor of Astronomy at Williams College, Williamstown, Massachusetts, USA, and Director of its Hopkins Observatory, the oldest extant astronomical observatory in the United States. He is a former Chair of the Historical Astronomy Division of the International Astronomical Union and on the Organizing Committee of the IAU's Commission on the History of Astronomy; he is current Chair of the Working Group on Solar Eclipses of the International Astronomical Union. He was a President of the Commission on Education and Development of the International Astronomical Union. He is coauthor of *The Cosmos: Astronomy in the New Millennium* (4th edition, 2014, Cambridge University Press), *Nearest Star: The Surprising Science of Our Sun* (2nd edition, 2014, Cambridge University Press), and *The Sun* (Reaktion Books and University of Chicago Press, 2017). In 2013, he received the Janssen Prize of the Société Astronomique de France and in 2003, he received the Education Prize of the American Astronomical Society.

Rudolf Pausenberger, born in 1963, is a physicist and physics teacher at the Gymnasium Lauf a.d. Pegnitz and co-founder of the hands-on-museum "Turm der Sinne" in Nuremberg. He wrote his diploma thesis about speckle interferometry at the Max-Planck-Institute for Radioastronomy in Bonn. Afterwards, he was active at the Institute of Physical Chemistry at the University of Erlangen-Nuremberg. Since 2005, he has been district manager for physics of the Bavarian Federation of Philologists and was a founding member of the Simon Marius Society. Rudolf Pausenberger often carries out school projects, e.g., about the history of astronomy (www.physik.de.rs). His projects are widely regarded. In 2013, he received the award for Bavaria's best Project-Seminar; in 2016, he was acknowledged as the most actively engaged teacher in his district.

Joachim Schlör, Study Director ret., born in 1946 in Aschaffenburg, studied Latin and English philology in Würzburg and was a Gymnasium teacher for Latin and English at the Simon-Marius-Gymnasium. He was mentor for English and Latin. Within his advanced course in Latin during 1985–1987 Marius's main work *Mundus Iovialis* was translated into German. In 1988, he published the bilingual edition in book form and subsequently developed various teaching projects together with a collection of reading and learning circles to work with Simon Marius as an author in schools. Currently, he is translating letters to and from Simon Marius from Latin into German as vice-president of the Simon Marius Society.

Norman Anja Schmidt, born in 1975 in Nürnberg, studied chemistry and astronomy in Erlangen. At the time of publication (2019), the name of "Norman Schmidt" has been changed to "Norman Anja Schmidt". Currently working on a Ph.D. thesis about X-ray absorption spectroscopy of organic molecules and as an IT specialist at the Chair of Physical Chemistry II in Erlangen. Since 2003, Norman has also been working as a freelance IT expert. A long-term interest in practical astronomy manifests in, inter alia, being a member of the guide team at the Regiomontanus-Observatory in Nuremberg. Norman has been in charge of the technical side of the *Marius-Portal* since 2013 and has been serving as executive secretary of the Simon Marius Society since 2014.

Olga Sinzev (russ.: Sizintseva), born in 1968 in Tashkent (Uzbekistan) has lived in Nuremberg for 15 years as a journalist, translator, and author. She studied journalism and media science at the National University of Uzbekistan (Mirzo Ulug'bek nomidagi O'zbekiston Milliy Universiteti) in Tashkent and at the University of Applied Sciences Mittweida (Saxony). She is founder and chief editor of the intercultural magazine *RESONANZ*, which has been published in Nuremberg since 2005. The magazine received several awards, inter alia the international Media Prize BeBiLin.RU-2015 for "The best image of bilinguals in mass media."

Albert Van Helden, professor emeritus, Rice University and the University of Utrecht. Born in 1940 in The Hague, Netherlands, he studied at Stevens Institute of Technology, the University of Michigan, and Imperial College (University of London). He taught the history of science at Rice University from 1970 to 2001, and then at the University of Utrecht until 2005. From 1998 to 1999 he served as president of the History of Science Society. His researches in the history of astronomy include *The Invention of the Telescope* (1977, 2008), *Measuring the University: Cosmic Dimensions from Aristarchus to Halley* (1985), *The Sidereal Messenger of Galileo Galilei* (1989), and numerous articles. He is currently working with colleagues on translations of important Galileo texts into Dutch. He lives in Leiden.

Huib J. Zuidervaart, born in 1951 is a (partly retired) senior researcher at the Huygens Institute for the History of the Netherlands in Amsterdam, an institution of the Royal Netherlands Academy of Arts and Sciences. From 2007 to 2016 he was Editor-in-chief of *Studium. Tijdschrift voor Wetenschaps- en Universiteitsgeschiedenis*, the Belgian-Dutch journal for the history of science, medicine, and universities. Zuidervaart's main field of research is the history of physics and astronomy in Early Modern Europe, with a focus on the history of scientific instruments and collections. His researches in the history of astronomy include the books *Van 'Konstgenoten' en Hemelse Fenomenen. Nederlandse Sterrenkunde in de Achttiende Eeuw* [Dutch astronomy in the eighteenth century] (Rotterdam 1999) and [together with Rob H. van Gent], *Between Rhetoric and Reality: instrumental practices at the Astronomical and Meteorological Observatory of the Amsterdam Society 'Felix Meritis', 1789–1889* (Hilversum 2013). Together with Oscar Matsuura he is currently working on the book *The New World's Tycho. The astronomical work of Georg Marggrafe (1610–c.1644) in Dutch Colonial Brazil.* For his bibliography, see: https://www.huygens.knaw.nl/zuidervaart/.

Author Index

A
Adriaan, J., 391
Albert, 15
Aldrighetti, A., 96
Alpini, P., 96
Apian, P., 216
Aristarchus, 13
Augenio, O., 96

B
Balthasar Bernhold, J., 68
Bimbioli, A., 97
Böheim, P., 117
Bosscha, J., 198, 256, 387, 401, 413
Boyn, P., 215
Brahe, T., 7, 83, 210, 240, 258, 325
Brenz, J.H., 76
Brunn, J., 113
Brunnius, L., 7
Buddeus, J.F., 61
Bürgi, J., 215
Bylica, M., 367

C
Caesius, G., 104, 179, 276, 335
Campolongo, E., 96
Capra, B., 99, 342, 399
Casserio, J., 95
Castelli, B., 265
Christian IV, 150
Clavio, A., 112
Cnod, M., 74
Codomann, J., 67
Copernicus, N., 3, 257
Cosimo II, 197
Cysat, J.B., 253, 398

D
Digges, T., 242

E
Eckebrecht, P., 119
Eisen von Haymen, S., 73
Eriksen, J., 84, 325
Ernest, J., 2
Ernst of Mansfeld, 219
Ernst, J., 97, 139, 311
Esper, J.F., 63
Euclid, 70, 165
Everatus, M., 83

F
Fabricius ab Aquapendente, H., 94
Fabricius, D., 8, 87, 228, 253, 303, 349, 395
Fabricius, J., 228, 251, 395
Falk, N., 75
Favaro, A., 387, 401, 414
Francisci, A., 69
Frederic, G., 15
Friedrich, G., 60, 145, 179, 261, 278, 322, 325, 365
Friedrich, J., 97
Frobenius, A., 74
Frölich, D., 305
Fuchs von Bimbach, A.V., 164
Fuchs von Bimbach, H.P., 93, 139, 250, 390
Fuchs, J.P., 4

H. Gaab, P. Leich (eds.), *Simon Marius and His Research*, Historical & Cultural Astronomy, https://doi.org/10.1007/978-3-319-92621-6

G

Galileo, G., 6, 102, 192, 242, 252, 303, 341, 394
Guilandino, M., 96
Gümbelein, J., 79
Gunkfelder, W., 23
Gurkfelder, W., 68

H

Harmansz, E., 251
Harriot, T., 233, 253, 395
Hartmann, J., 64
Hertel, J., 67
Hirchbauer, G., 23
Hirschbauer, M.G., 68
Hocker, J.L., 62
Hoffmann, C., 90
Hoffmann, P., 182
Hohenstein, J., 73
Holtzmann, W., 112
Horky, M., 305
Huygens, C., 242

I

Ingoli, F., 246

J

Jakob, 57
Jakob Kurz of Senftenau, 84
Janssen, S., 160
Jöcher, C.G., 62
Jöstel, M., 81, 333
Jungius, J., 232

K

Kauffmann, P., 76
Kepler, B., 85
Kepler, J., 8, 81, 120, 254, 353, 368, 395
Köbel, J., 111
Krabbe, J., 302, 353
Krebs, J.F., 63
Kurz, S., 112

L

Laelius, L., 75
Lancius, A., 324
Lanius, A., 23, 77, 367
Lansbergen, P., 242
Lauer, F., 106
Lauer, J., 81, 277
Lenck, J.B., 114
Leonhard, 58
Leovitius, C., 77
Leowitz, C., 182
Leuccius, J.B., 5
Limmer, C., 69
Lipperhey, H., 160, 250, 390
Locher, G.J., 243
Locher, S., 398
Lochner, C., 81
Longomontanus, C., 348
Lorenzini, A., 101

M

Maestlin, M., 228
Magini, G., 336
Maria von Eyb, 82, 277
Mästlin, M., 81, 210
Meelführer, J., 75
Melanchthon, P., 182
Mencke, J.B., 61
Menius, M., 80
Metius, J.A., 160, 250, 391
Michael, 57
Milich, J., 186
Minadoi, G.T., 97
Mögling, D., 108
Moritz of Nassau, 249
Muhr, S., 79
Mummius, S., 73

N

Negro, A., 96
Neser, J., 23, 69

O

Oertel, G.C., 62
Origanus, D., 81, 328
Oudemans, J.A.C., 198, 387, 400, 414

P

Pape, J., 120
Pellegrini, G.P., 97
Peucer, C., 181
Praetorius, J., 80
Prickard, A.O., 195, 372, 388, 405
Priester, J.H., 76

R
Raffael, F., 71
Raffael, M.F., 75
Ramponi, L., 265
Raphael, F., 23
Regiomontanus, 119, 324, 366
Reimers Ursus, N., 346
Reinhold, E., 2, 182
Rhodius, A., 86
Röslin, H., 346
Rothmann, C., 217, 242
Rudio, E., 96

S
Sarpi, P., 391
Sassonia, E., 96
Saxonius, P., 233
Scheiner, C., 41, 230, 244, 253, 303, 388, 394
Scheubel, J., 109
Schmid, W., 109
Schmidner, A., 116, 253, 395
Schmidnerus, A., 228
Schulin, J., 104, 276
Seyfridt, G., 275
Snel, R., 251
Stadius, J., 83, 323
Stadtmann, N., 73
Stathmion, C., 182
Stevin, S., 414
Stieber, M., 75
Stöffler, J., 323

V
Vávrová, Z., 437
Vigenza, A., 97
Vocke, J.A., 108
Vogtherr, G., 58
von Glogau, J., 71
von Rechenberg, B., 182
von Vicke, N., 254, 353, 368, 396

W
Wacker von Wackenfels, J.M., 8
Walther, B., 323
Weidler, J.F., 62
Welchendorff, B., 70
Wickner, A., 75, 180
Wilhelm IV, 210
Wohlwill, E., 404

Z
Zedler, H., 62
Zinner, E., 371
Zupi, G.B., 256

Subject Index

A
Andromeda, 7, 243, 256, 395

B
Bimbach, 4

C
Camera Helioscopica, 230
Christian, 2
Comets, 77, 185, 205, 256

F
Frankfurt autumn fair, 4, 374
Frankfurt fair, 155, 250, 390
Fürstenschule, 61, 279, 322, 363

H
Heilsbronn, 23, 276, 325, 365

J
Jupiter, 5, 191, 252, 393, 449

K
Königsberg, 78, 367

M
Maunder Grand Minimum, 224
Mercury, 209, 256, 323
Mizar, 245, 265

N
New star, 99, 314, 353

P
Padua, 88–105, 144, 192, 250, 278, 326, 368, 394
Parallax, 245, 263

S
Saturn, 182, 240
Solar eclipse, 207
Stadius, 209
Streuberger, 79
Sunspots, 224, 259, 394

V
Venus, 209, 259, 395

W
Wittenberg, 228

H. Gaab, P. Leich (eds.), *Simon Marius and His Research*, Historical & Cultural Astronomy, https://doi.org/10.1007/978-3-319-92621-6

MIX
Papier aus verantwortungsvollen Quellen
Paper from responsible sources
FSC® C105338

If you have any concerns about our products,
you can contact us on
ProductSafety@springernature.com

In case Publisher is established outside the EU,
the EU authorized representative is:
Springer Nature Customer Service Center GmbH
Europaplatz 3, 69115 Heidelberg, Germany

Printed by Libri Plureos GmbH
in Hamburg, Germany